Water Productivity in Agriculture

Limits and Opportunities for Improvement

1.

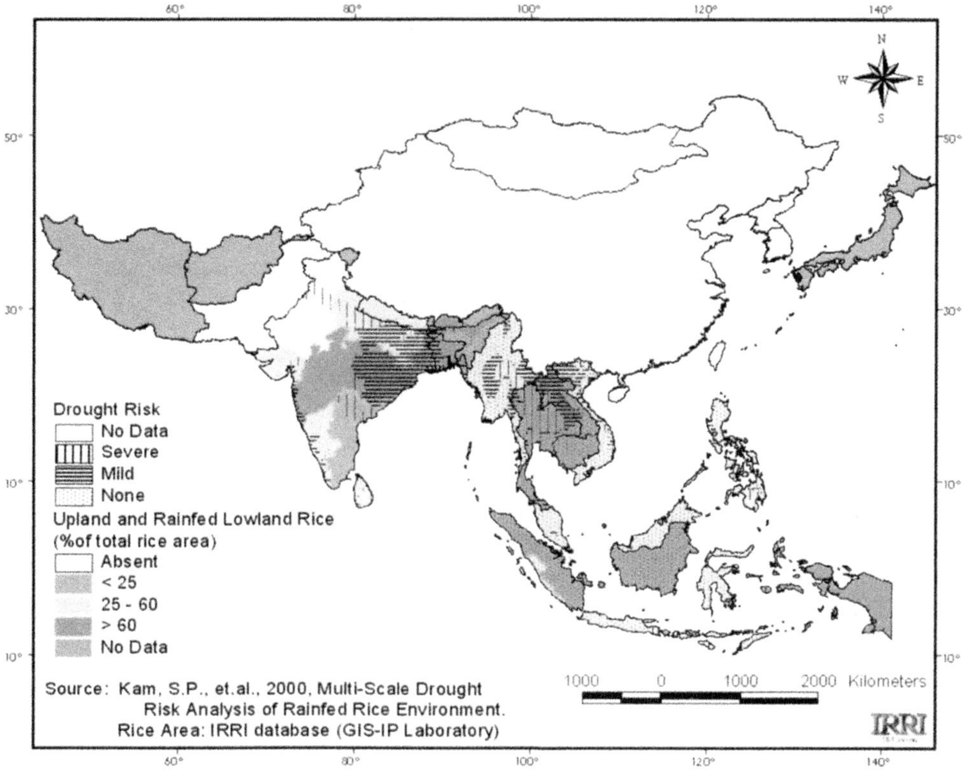

2A.

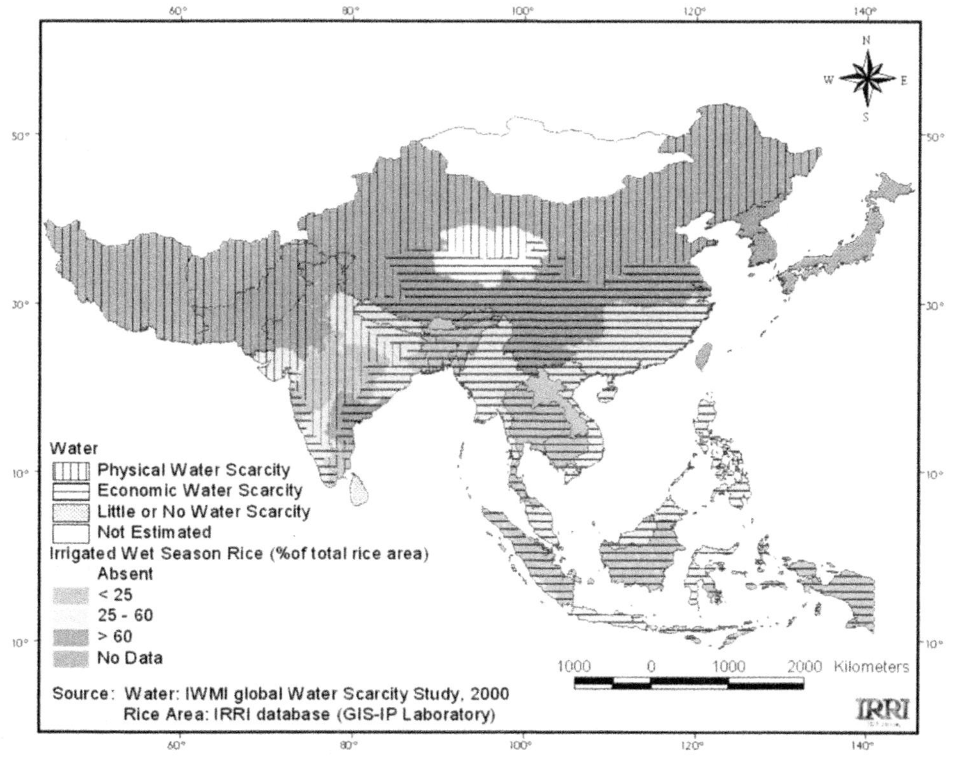

Plate 1. Drought risk in rain-fed rice areas in Asia.

Plate 2A. Projected water scarcity in 2025 in wet-season (summer) irrigated rice areas in Asia.

2B.

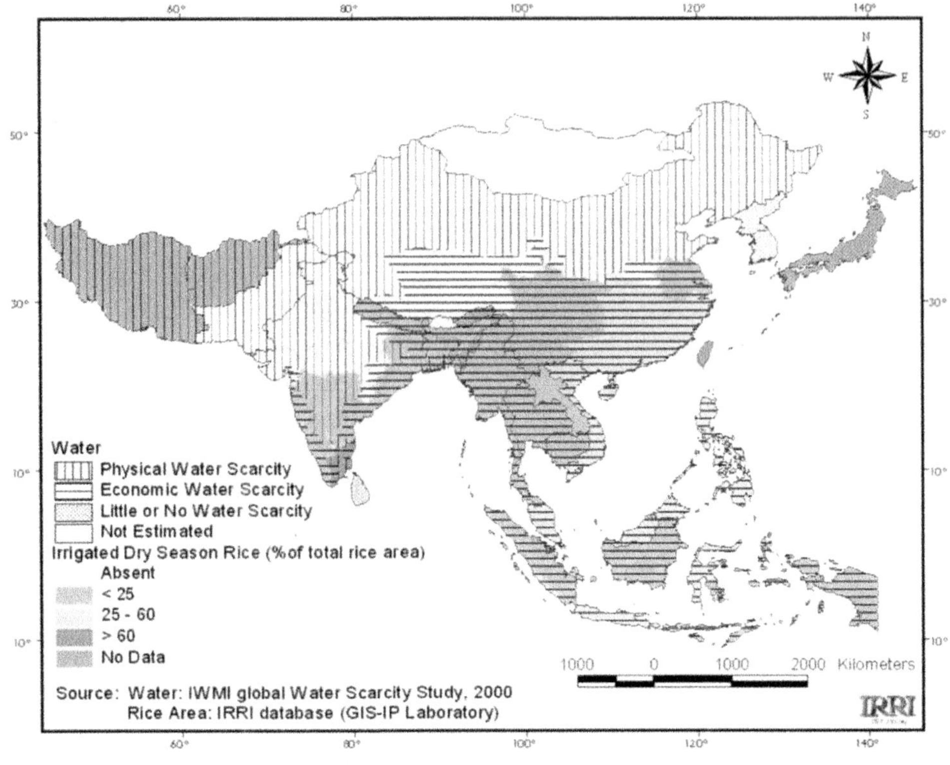

3.

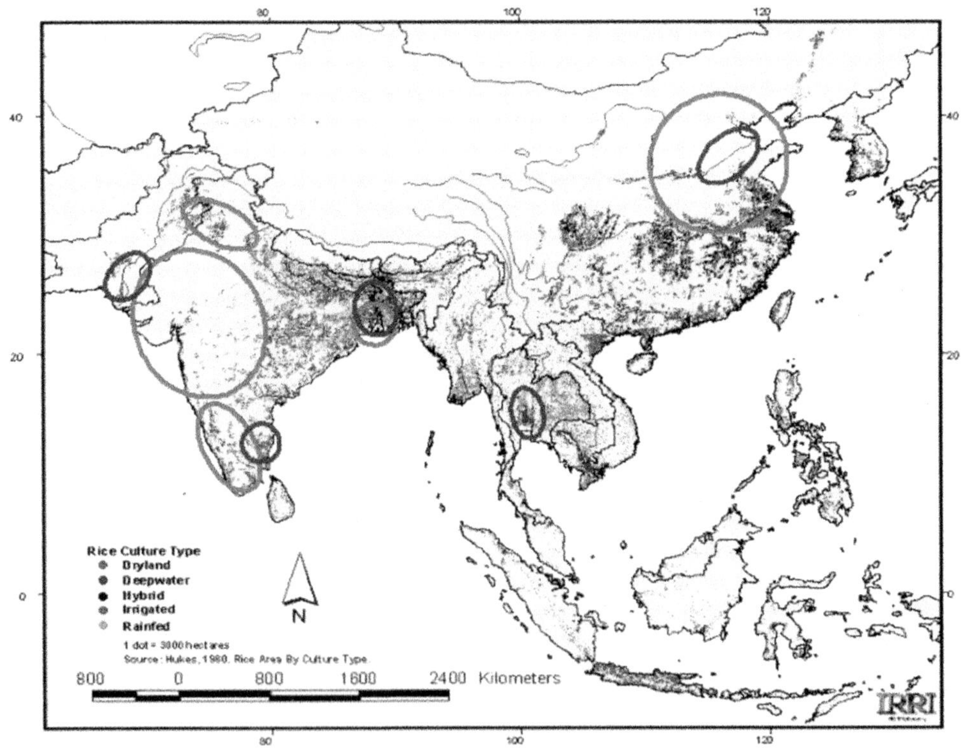

Plate 2B. Projected water scarcity in 2025 in dry-season irrigated rice areas in Asia.

Plate 3. Major hot spots of groundwater depletion (red circles) and reduced river-water flows (blue circles) in Asia.

Water Productivity in Agriculture

Limits and Opportunities for Improvement

Edited by

Jacob W. Kijne, Randolph Barker and David Molden

*International Water Management Institute,
Colombo, Sri Lanka*

CABI Publishing

In association with the
International Water Management Institute

CABI is a trading name of CAB International

CABI Head Office
Nosworthy Way
Wallingford
Oxfordshire OX10 8DE
UK

Tel: +44 (0)1491 832111
Fax: +44 (0)1491 833508
Email: cabi@cabi.org
Web site: www.cabi.org

CABI North American Office
875 Massachusetts Avenue
7th Floor
Cambridge, MA 02139
USA

Tel: +1 617 395 4056
Fax: +1 617 354 6875
Email: cabi-nao@cabi.org

Published in association with:

International Water Management Institute
PO Box 2075
Colombo
Sri Lanka

A catalogue record for this book is available from the British Library, London, UK.

Library of Congress Cataloging-in-Publication Data
Water productivity in agriculture : limits and opportunities for improvement / edited by Jacob W. Kijne, Randolph Barker and David Molden.
p. cm. – (Comprehensive assessment of water management in agriculture series ; 1)
Includes bibliographical references (p.).
ISBN 0 985199 669 8 (alk. paper)
1. Water in agriculture. 2. Irrigation farming. 3. Crops and water. I. Kijne, J. W. II. Barker, Randolph. III. Molden, D. J. IV. Series.
S494.5.W3 W365 32003
631.7—dc21

2002153454

ISBN-13: 978-0-85199-669-1
ISBN-10: 0-85199-669-8

First published 2003
Transferred to print on demand 2011

Printed and bound by CPI Group (UK) Ltd, Croydon, CR0 4YY

Contents

Contributors vii

Series Foreword: Comprehensive Assessment of Water Management in Agriculture Series ix

Improving Water Productivity in Agriculture: Editors' Overview xi
Jacob W. Kijne, Randolph Barker and David Molden

1. **A Water-productivity Framework for Understanding and Action** 1
David Molden, Hammond Murray-Rust, R. Sakthivadivel and Ian Makin

2. **Economics of Water Productivity in Managing Water for Agriculture** 19
Randolph Barker, David Dawe and Arlene Inocencio

3. **The Concept of Efficiency in Water-resources Management and Policy** 37
David Seckler, David Molden and R. Sakthivadivel

4. **Rice Production in Water-scarce Environments** 53
T.P. Tuong and B.A.M. Bouman

5. **Managing Saline and Alkaline Water for Higher Productivity** 69
N.K. Tyagi

6. **Water Productivity under Saline Conditions** 89
Jacob W. Kijne

7. **Opportunities for Increasing Water Productivity of CGIAR Crops through Plant Breeding and Molecular Biology** 103
John Bennett

8. **Management of Drought in ICRISAT Cereal and Legume Mandate Crops** 127
R. Serraj, F.R. Bidinger, Y.S. Chauhan, N. Seetharama, S.N. Nigam and N.P. Saxena

9. **Water Productivity in Rain-fed Agriculture: Challenges and Opportunities for Smallholder Farmers in Drought-prone Tropical Agroecosystems** 145
John Rockström, Jennie Barron and Patrick Fox

10. **World Water Productivity: Current Situation and Future Options** 163
Ximing Cai and Mark W. Rosegrant

11. Improving Water Productivity in the Dry Areas of West Asia and North Africa 179
Theib Y. Oweis and Ahmed Y. Hachum

12. Efficient Management of Rainwater for Increased Crop Productivity and Groundwater Recharge in Asia 199
S.P. Wani, P. Pathak, T.K. Sreedevi, H.P. Singh and P. Singh

13. Water Productivity in Forestry and Agroforestry 217
C.K. Ong and B.M. Swallow

14. Water Productivity and Potato Cultivation 229
Walter T. Bowen

15. Rice–Wheat Cropping Systems in the Indo-Gangetic Plains: Issues of Water Productivity in Relation to New Resource-conserving Technologies 239
Peter R. Hobbs and Raj K. Gupta

16. Land and Water Productivity of Wheat in the Western Indo-Gangetic Plains of India and Pakistan: a Comparative Analysis 255
Intizar Hussain, R. Sakthivadivel and Upali Amarasinghe

17. Reform of the Thai Irrigation Sector: is there Scope for Increasing Water Productivity? 273
François Molle

18. Upscaling Water Productivity in Irrigated Agriculture Using Remote-sensing and GIS Technologies 289
Wim Bastiaanssen, Mobin-ud-Din Ahmad and Zubair Tahir

19. Improving Water Productivity through Deficit Irrigation: Examples from Syria, the North China Plain and Oregon, USA 301
Heping Zhang

Appendix A. A Note on Transpiration 311
David Seckler

Appendix B. Note on Agronomic Practices for Increasing Crop Water Productivity 319
Jacob W. Kijne

Index 323

Contributors

M.-ud-D. Ahmad, *Water Resources Engineer (GIS/RS), International Water Management Institute, 12KM Multan Road, Chowk, Thokar Niaz Baig, Lahore 53700, Pakistan.*

U. Amarasinghe, *International Water Management Institute, PO Box 2075, Colombo, Sri Lanka; Street: 127, Sunil Mawatha, Pelawatte, Battaramulla, Sri Lanka.*

R. Barker, *International Water Management Institute, Mailing Adress: PO Box 2075, Colombo, Sri Lanka; Street: 127, Sunil Mawatha, Pelawatte, Battaramulla, Sri Lanka.*

J. Barron, *Department of Systems Ecology, Stockholm University, S-10691 Stockholm, Sweden.*

W. Bastiaanssen, *International Institute for Aerospace Survey and Earth Sciences, PO Box 6, Hengelosestraat 99, 7514 AE Enschede, The Netherlands.*

J. Bennett, *International Rice Research Institute, DAPO Box 7777, Metro Manila, Philippines.*

F.R. Bidinger, *International Crops Research Institute for the Semi-Arid Tropics, Patancheru 502 324, Andhra Pradesh, India.*

B.A.M. Bouman, *International Rice Research Institute, DAPO Box 7777, Metro Manila, Philippines.*

W.T. Bowen, *International Fertilizer Development Center, Resource Development Division, PO Box 2040, Muscle Shoals, AL 35662, USA.*

X. Cai, *International Food Policy Research Institute, 2033 K Street, NW Washington, DC 20006, USA.*

Y.S. Chauhan, *International Crops Research Institute for the Semi-Arid Tropics, Patancheru 502 324, Andhra Pradesh, India.*

D. Dawe, *International Rice Research Institute, DAPO Box 7777, Metro Manila, Philippines.*

P. Fox, *Department of Systems Ecology, Stockholm University, S-10691 Stockholm, Sweden.*

R.K. Gupta, *Rice–Wheat Consortium for the Indo-Gangetic Plains, CG Centre Block, National Agricultural Science Center (NASC) Complex, DP Shastri, New Delhi, India.*

A.Y. Hachum, *Professor of Water Management and Irrigation Engineering, University of Mosul, Iraq; Currently: Visiting Scientist, International Center for Agricultural Research in the Dry Areas (ICARDA), PO Box 5466, Aleppo, Syria.*

P.R. Hobbs, *Natural Resources Group, Centro Internacional de Mejoramiento de Maize y Trigo, 611 Bradfield Hall, Cornell University, Ithaca, NY 14853, USA.*

I. Hussain, *International Water Management Institute, Mailing Address: PO Box 2075, Colombo, Sri Lanka; Street Address: 127, Sunil Mawatha, Pelawatte, Battaramulla, Sri Lanka.*

A. Inocencio, *International Water Management Institute, Mailing Address: Private Bag X813, Silverton 0127, South Africa; Street Address: 141 Cresswell Street, 0184 Silverton, Pretoria, South Africa.*

J.W. Kijne, *International Water Management Institute, Rose Cottage, 3 Cherry Tree Farm Lane, Hemel Hempstead, Hertfordshire HP2 7HS, UK.*

I. Makin, *International Water Management Institute, Mailing Address: PO Box 2075, Colombo, Sri Lanka; Street Address: 127, Sunil Mawatha, Pelawatte, Battaramulla, Sri Lanka.*

D. Molden, *Comprehensive Assessment of Water Management in Agriculture, International Water Management Institute, Mailing Address: PO Box 2075, Colombo, Sri Lanka; Street Address: 127, Sunil Mawatha, Pelawatte, Battaramulla, Sri Lanka .*

F. Molle, *International Water Management Institute, Mailing Address: PO Box 2075, Colombo, Sri Lanka; Street Address: 127, Sunil Mawatha, Pelawatte, Battaramulla, Sri Lanka.*

H. Murray-Rust, *Integrated Water Management for Agriculture, International Water Management Institute, c/o ICRISAT, Patancheru, India.*

S.N. Nigam, *International Crops Research Institute for the Semi-Arid Tropics, Patancheru 502 324, Andhra Pradesh, India.*

C.K. Ong, *World Agroforestry Centre and Sida's Regional Land Management Unit, International Centre for Research in Agroforestry, PO Box 30677, Nairobi, Kenya.*

T.Y. Oweis, *Natural Resources Management Program, International Center for Agricultural Research in the Dry Areas, PO Box 5466, Aleppo, Syria.*

P. Pathak, *Soil and Water Management, International Crops Research Institute for the Semi-Arid Tropics, Patancheru 502 324, Andhra Pradesh, India.*

J. Rockström, *UNESCO–IHE Institute for Water Education, Delft, The Netherlands; Currently: Coordinator of WaterNet, University of Zimbabwe, PO Box MP 600, Mount Pleasant, Harare, Zimbabwe.*

M.W. Rosegrant, *International Food Policy Research Institute, 2033 K Street, NW Washington, DC 20006, USA.*

R. Sakthivadivel, *International Water Management Institute, 33, 4/14 I East Street, Kamaranga, Thiruvanmiyur, Chennai I 600041, Tamil Nadu, India.*

N.P. Saxena, *International Crop Research Institute for the Semi-Arid Tropics (ICRISAT), Patancheru 502 324, Andhra Pradesh, India.*

D. Seckler, *International Water Management Institute, 3234 West Fairview Place, No. 1, Denver, CO 80211, USA.*

N. Seetharama, *NRCS (ICAR), Hyderabad, and Senior Scientist II, Natural Resources Conservation Service, Indian Council for Agricultural Research, International Crops Research Institute for the Semi-Arid Tropics, Patancheru 502 324, Andhra Pradesh, India.*

R. Serraj, *International Crops Research Institute for the Semi-Arid Tropics, Patancheru 502 324, Andhra Pradesh, India.*

H.P. Singh, *Central Research Institute for Dryland Agriculture, Santoshnagar, Hyderabad 500 059, Andhra Pradesh, India.*

P. Singh, *International Crops Research Institute for the Semi-Arid Tropics, Patancheru 502 324, Andhra Pradesh, India.*

T.K. Sreedevi, *Andhra Pradesh Rural Livelihoods Project, Government of Andhra Pradesh, Arctic Paleo-River Discharge (APARD), Rajendranagar, Hyderabad 500 059, Andhra Pradesh, India.*

B.M. Swallow, *International Centre for Research in Agroforestry, PO Box 30677, Nairobi, Kenya.*

Z. Tahir, *International Water Management Institute, 12KM Multan Road, Chowk Thokar Niaz Baig, Lahore 53700, Pakistan.*

T.P. Tuong, *International Rice Research Institute, DAPO Box 7777, Metro Manila, Philippines.*

N.K. Tyagi, *Central Soil Salinity Research Institute, Zarifa Farm, Karnal-132001, Haryana, India.*

S.P. Wani, *Water and Agro-biodiversity Management, International Crops Research Institute for the Semi-Arid Tropics (ICRISAT), Patancheru 502 324, Andhra Pradesh, India.*

H. Zhang, *CSIRO Plant Industry, Commonwealth Scientific and Industrial Research Organization, Underwood Avenue, Floreat Park, Private Bag 5, Wembley, WA 6913, Australia.*

Series Foreword: Comprehensive Assessment of Water Management in Agriculture Series

To find solutions to the water problems already facing many developing countries, we need a better understanding of how we have used water to grow food and to improve rural livelihoods. We need to know which investments in water for rainfed and irrigated agriculture have reduced poverty and increased food security – and which have not. We need to better understand not only the benefits of irrigation, but also the costs in terms of environmental degradation and pollution.

The Comprehensive Assessment of Water Management in Agriculture, an international research, capacity-building and knowledge-sharing programme, takes stock of the past 50 years of water development for agriculture, the water management challenges that communities are facing today and solutions that people have developed. The results of this research will enable farming communities, governments and donors to make better-quality investment decisions to meet food and environmental security targets in the near future and over the next 25 years.

The Assessment is done by a coalition of partners which includes 11 Future Harvest agricultural research centres supported by the Consultative Group on International Agricultural Research, the Food and Agriculture Organization of the United Nations (FAO), and partners from some 40 research and development institutes globally. Currently, the Governments of The Netherlands, Switzerland, Australia and Taiwan, and the Rockefeller Foundation have supported this work.

The primary research findings will be presented in a series of books that will form the Comprehensive Assessment of Water Management in Agriculture. The books will cover a range of vital topics in the area of water, agriculture, food security and ecosystems – the entire spectrum of developing and managing water in agriculture, from fully irrigated to fully rainfed lands. They are about people and society, why they decide to adopt certain practices and not others, and, in particular, how water management can help poor people. They are about ecosystems – how agriculture affects ecosystems, the goods and services ecosystems provide for food security, and how water can be managed to meet both food and environmental security objectives. This is the first book in the series.

Effectively managing water to meet food and environmental objectives will require the concerted action of individuals from across several professions and disciplines – farmers, water managers, economists, hydrologists, irrigation specialists, agronomists and social scientists. The material presented in this book represents the first effort that brings this diverse group of people together to present a truly cross disciplinary perspective on water productivity. The complete set of books should be invaluable for resource managers, researchers and field implementers. They will provide source material from which policy statements, practical manuals and educational and training material can be prepared.

David Molden
Series Editor
International Water Management Institute
Sri Lanka

Improving Water Productivity in Agriculture: Editors' Overview

Jacob W. Kijne, Randolph Barker and David Molden
International Water Management Institute, Colombo, Sri Lanka

Introduction

One of the critical challenges of the early 21st century will be the resolution of the water crisis. This crisis is defined by scarcity of water, water-driven ecosystem degradation and malnutrition. In spite of massive water-development efforts for food security, the poor are affected the most, because they do not have the resources to obtain or maintain access to reliable and safe water. In the quest for improved access to water and food security, tremendous resources have been invested in developing water for agricultural uses. Yet we know that, with the growing demand for water for industry and municipalities, combined with environmental problems, there will be less water for agriculture in the future.

We hold that the solution to the water crisis is to be found in how water is developed and managed. Increasing the productivity of water means, in its broadest sense, getting more value or benefit from each drop of water used for crops, fish, forests and livestock while maintaining or improving ecosystems and the services they provide. Within agriculture, this means obtaining more production or value from every drop. We must increase the productivity of existing water resources and produce more food with less water. Increases in water productivity provide a means both to ease water scarcity and to leave more water for other human and ecosystem uses.

This book provides state-of-the-art knowledge on how to increase the productivity of water in agriculture. It provides concepts, methodologies, constraints and examples drawn from a wealth of experience from developing and developed countries. The book demonstrates that increasing water productivity will provide a focal point for practitioners and researchers from a variety of social science and physical science backgrounds.

Water-use Efficiency and Water Productivity

The first task in understanding how to increase water productivity is to understand what it means. As presented by Molden, Murray-Rust, Sakthivadivel and Makin in Chapter 1, the definition is scale-dependent. For a farmer, it means getting more crop per drop of irrigation water. But, for society as a whole, concerned with a basin or country's water resource, this means getting more value per unit of water resource used. Increasing water productivity is then the business of several actors working in harmony at plant, field, irrigation-system and river-basin levels.

Crop water productivity means raising crop yields per unit of water consumed. Over the past three decades, this has been achieved largely through higher crop yields per hectare. But, with the declining crop-yield growth, attention has turned to the potential offered by improved management of water resources. Although there is considerable scope for increasing water productivity through this avenue, it is not as large as is commonly thought. As argued by Seckler, Molden and Sakthivadivel in Chapter 3, the amount of reuse (or recycling) of water is often underestimated. When reuse is taken into account, the options for further increases in water productivity are much smaller than were expected at first.

Seckler *et al.* side with those who find the traditional definition of irrigation efficiency misleading. They distinguish between what they refer to as the 'classical' and the 'neo-classical' concept of irrigation efficiency. Classical irrigation efficiency is defined as the crop water requirement (actual evapotranspiration minus effective precipitation) divided by the water withdrawn or diverted from a specific surface-water or groundwater source. 'Losses' in this approach include transpiration and evaporation (evapotranspiration), but also seepage, percolation and runoff, processes in which the water is not consumed. These latter so-called 'losses' may be captured or recycled for use elsewhere in the basin. Thus, classical measures of efficiency tend to underestimate the true efficiency and ignore the important role of surface irrigation systems in recharging groundwater and providing downstream sources of water for agriculture and other ecosystem services.

Seckler *et al.* agree with others that the word 'efficiency' has outlived its usefulness in the field of water-resource policy and management. Willardson *et al.* (1994) introduced the concept of consumed fractions. Others, e.g. Perry (1996), Burt *et al.* (1997) and Molden (1997) have referred to beneficial and non-beneficial depleted or consumed fractions. These are important distinctions that need to be kept firmly in mind throughout these discussions on limits and opportunities for improvements in crop water use.

Throughout this book the reader should be aware of the distinction between crop water productivity and water productivity at the basin level. Crop water productivity is defined in either physical or monetary terms as the ratio of the product (usually measured in kg) over the amount of water depleted (usually limited to crop evapotranspiration, measured in m^3). Occasionally – for example, in the context of supplemental irrigation – there is a felt need to express the productivity of the applied irrigation water. In that case, the denominator refers to irrigation water only, not to rainfall. Obviously, values of irrigation-water productivity cannot be compared with water productivity with depleted water in the denominator.

Basin water productivity takes into consideration beneficial depletion for multiple uses of water, including not only crop production but also uses by the non-agricultural sector, including the environment. Here, the problem lies in allocating the water among its multiple uses and users. Priority in use involves the value judgement of either the allocating agency or society at large and may be legally determined by water rights.

Productivity

The classical concept of irrigation efficiency as used by engineers omits economic values. To determine optimum-level irrigation efficiency, the economist would want to know the value of irrigation water and the cost of increased control or management that would permit a reduction in diversion. As water becomes scarce, increasing crop water productivity or reducing diversions would make sense if the water 'saved' could be put to higher-valued uses. But higher water productivity does not necessarily lead to greater economic efficiency. Moreover, water productivity or yield per unit of water, like yield per unit of land, is a partial productivity of just one factor, whereas the most encompassing measure of productivity used by economists is total factor productivity. The following definitions may help in understanding the differences between various productivity parameters.

Pure physical productivity is defined as the quantity of the product divided by the quantity of the input – for example, yield per hectare or yield per cubic metre of water either diverted or depleted. Combined physical and economic productivity is defined in terms of either the gross or the net present value of the crop divided by the amount of water diverted or depleted.

Economic productivity is the gross or net present value of the product divided by the value of the water diverted or depleted, which can be defined in terms of its opportunity cost in the highest alternative use.

Barker, Dawe and Inocencio address these issues in Chapter 2. The authors give examples of basins where unexpected off-site effects and externalities confound possible changes in water management intended to reduce water diversions. They do so by analysing the relationship between water productivity and economic efficiency and by investigating the possible role of water policies, such as water pricing, and institutions. Just as increased water savings do not necessarily result in increased water productivity, so also increased water productivity does not necessarily result in higher net returns at the farm or basin level. As the examples illustrate, it needs to be determined whether proposed water-management practices or technologies designed to increase water productivity and economic efficiency at the farm level translate into water-productivity and economic-efficiency gains at the system or basin level. Especially when basins become closed (basins are closed when all available water is depleted, i.e. rendered unavailable for further use),[1] setting the priority in the allocation of water among competing uses may reflect either political power at the basin level or a value judgement on the part of society. While the farmer may measure the benefits of increased water productivity in economic terms, valuing beneficial depletion in terms of reallocation of limited water supplies among competing uses and users at the basin level is an important but far more complex undertaking.

Scale Considerations

Water use and management in agriculture encompass many different scales: plants, fields, farms, delivery systems, basins, nations and continents. The focus of attention shifts according to the scale we are considering, from photosynthesis and transpiration, through water distribution and delivery, to allocation between various uses and between nations sharing the same basin.

In the classical irrigation efficiency concept, scale-dependent efficiency is commonly used: application efficiency (the ratio of the water delivered to the root zone over the water delivered to the field); conveyance efficiency (the ratio of the water delivered to the field over the supply of water delivered into the canal from the source); and project efficiency (the overall efficiency of the irrigation system).

The last term usually refers to the ratio of the total water consumption over the amount of water diverted to the system, regardless of how many times the water may have been reused within the system. It is recognized that production per unit of water is an important parameter for irrigation managers, but it is not comparable across scales or readily comparable across locations. However, it can be a useful indicator of performance over time. An increase in production per unit of water diverted at one scale does not necessarily lead to an increase in productivity of water diverted at a larger scale.

In Chapter 1, Molden, Murray-Rust, Sakthivadivel and Makin address these scale issues and discuss water accounting as a means of generalizing about water use across scales and of better understanding the terms in both the numerator and the denominator of the water-productivity ratio. They illustrate this with several water-accounting diagrams applicable to different scales and provide a helpful glossary of terms. Farmer-based strategies to increase water productivity at plant, field, farm, system and basin

[1] This is also the case when water flows to so-called sinks, i.e. into a sea, saline groundwater or another location where it is not readily or economically recoverable for reuse.

level are summarized in a table. Several of these are discussed in more detail in subsequent chapters.

Water Productivity in Rice Cultivation

Cultivation of rice in flooded fields (paddies) is very water-demanding. Declining water availability is seen as a threat to the sustainability of irrigated rice-based production systems. In Chapter 4, Tuong and Bouman explore ways of producing rice with less water. Finding such alternatives, they assert, is essential for food security and sustaining environmental health in Asia. Irrigation methods that require less water, such as saturated-soil culture and alternate wetting and drying can reduce unproductive outflows and raise water productivity at the field level without a reduction in crop yield per hectare.

Other approaches that may increase water productivity include the incorporation of the C_4 photosynthetic pathway into rice, the use of molecular biotechnology to enhance drought-stress tolerance and the development of 'aerobic rice', which refers to rice varieties that yield well under non-flooded conditions. (The potentials for plant breeding and molecular biology are discussed in more detail by Bennett in Chapter 7.) The authors contend that a shift towards aerobic rice will affect water conservation, soil organic-matter turnover, nutrient dynamics, carbon sequestration, weed ecology and greenhouse-gas emissions. Some of these changes lead to greater crop water productivity and are seen as positive; others, such as the release of nitrous oxide from the soil, are seen as having a negative impact.

Water Productivity Under Saline and Alkaline Conditions

The use of saline or alkaline water in crop production enlarges the available water resource but at the cost of lower yields and possible long-term effects on soil structure and soil productivity. Growing plants in saline soil or with saline or alkaline irrigation water presents another example of a trade-off for which the benefits and costs are likely to vary among locations.

Tyagi, in Chapter 5, discusses field-level measures that can be combined with the use of saline/alkaline irrigation water to enhance its productivity and mitigate its adverse effects. Such measures include the choice of the best cropping sequence, conjunctive use with good-quality canal water, water-table management, rainwater conservation in precisely levelled basins and chemical amelioration of alkaline water. The illustrations are taken mainly from the rice–wheat cropping system in the monsoonal climate with moderate rainfall (400–600 mm), as occurs in north-west India. Water transfer, water markets and the disposal of saline water with its basin-level implications are also discussed. Practical examples illustrate the importance of presowing irrigation and the advantage of growing crops during the winter season when soil salinity is less and the evaporative demand is lower than during the pre-monsoonal summer season.

Knowledge of the leaching requirement, i.e. the amount of water that needs to pass through the root zone to maintain an acceptable salt level without unnecessary percolation losses, would help to determine whether increases in crop water productivity are feasible. Kijne, in Chapter 6, describes the difficulties both in determining the leaching requirement and, once known, in accurately applying the desired amount of water. Applying more water than needed causes the groundwater table to rise, which could lead to waterlogging. Evapotranspiration and leaching, which together constitute the beneficial depletion of the water resource under saline growing conditions, are linked through the yield–water–salinity production function. This relationship between yield and amount and quality of the applied water is not well known under field conditions, where crops are subjected to periodic and simultaneous water and salt stress and to non-uniform water application. Moreover, the feedback mechanism that lowers evapotranspiration when plants become more affected by soil salinity adds a further degree of complexity to the relationship between yield and salinity. Accordingly, knowing how much water to apply is important in terms of the sustainability of irrigated agriculture on saline soils.

Plant Breeding for Enhanced Water Productivity

Plant breeding over the last century has indirectly increased the productivity of water (in combination with other production factors) because yields have increased with no additional water consumption. Improved varieties have come from conventional breeding programmes where selection has been for yield per unit of land. Most of the increases have been due to improvements in the harvest index (the ratio of marketable product to total biomass or the so-called grain-to-straw ratio), which may now be approaching its theoretical limit in many of our major crops (Richards *et al.*, 1993). The development of an appropriate phenology by genetic modification, so that the durations of the vegetative and reproductive periods are matched as well as possible with the expected water supply or with the absence of crop hazards, is usually responsible for the most significant improvements in yield stability. Planting, flowering and maturation dates are important in matching the period of maximum crop growth with the time when saturation vapour-pressure deficit is low, and these characteristics may be genetically modified. One way of genetically increasing water productivity is to modify canopy development in order to reduce evaporation from the soil surface. Hence, much work has been done on the selection for large leaf area during the vegetative period to increase early vigour.

Biotechnology is considered to have great potential for the development of drought- or salt-tolerant crops, but this potential has not been fully realized yet. In addressing these topics in Chapter 7, Bennett observes that the slow progress in breeding for drought tolerance is often attributed to the genetic complexity of the trait and its interaction with the environment. Complementary approaches taken to address this issue include improving the environmental simulations used for germplasm screening and analysis, defining how the impact of water deficit on growth and yield components changes during the growth stages and discovering the regulatory genes underlying the plant's responses to water deficit. One promising approach to discovering the genes responsible for drought effects on yield components is quantitative trait loci (QTL) analysis. Such studies tend to focus on indirect effects, such as the inhibition of panicle development by hormonal signals from stressed leaves and roots and the inhibition of carbon flow from leaves to the developing grain. For example, it may be possible to prevent early drought-induced shedding of leaves by the genetic regulation of cyclokinin production. However, it could also be argued that these processes might be more effectively altered through conventional breeding. Many promising properties for coping with drought stress have been introduced for years through conventional plant breeding. These include changing the length of the growing season and the timing of sensitive stages; selecting for small leaves and early stomatal closure to reduce transpiration; selecting for high root activity and deep rooting systems; and selecting for tolerance to salinity. In short, traditional breeding methods and modern methods based on biotechnology should be seen as complementary.

Water Productivity in Rain-fed Agriculture

Eighty per cent of the agricultural land worldwide is rain-fed, with – in developing countries – generally low yield levels and large on-farm water losses during occasional periods of heavy rainfall. This suggests there are significant opportunities for improvements in crop water productivity.

Serraj and his co-authors of the International Crops Research Institute for the Semi-Arid Tropics (ICRISAT) describe in Chapter 8 the complexities of drought management of rain-fed cereal and legume crops in the semi-arid tropics. These cereal and legume crops are characterized by their ability to withstand periods of water scarcity and still produce grain and biomass. Drought stress is a complex issue because of the unpredictability of its occurrence and duration during the growing season, the high evaporative demand on the crop and the low fertility of the soil in which these plants grow. In addition, the effect of drought stress is often compounded by other stress factors, such as infection by root and stalk rot-causing fungi, which can bring about severe lodging and premature death. Current insufficient understanding of the combined effect of all these factors on crop yield complicates the characterization of the physiological traits required for increased water productivity of the crops.

The authors discuss four genetic-enhancement approaches for the improvement of the adaptation of legume and cereal crops to drought-prone environments:

- development of short-duration genotypes that can escape terminal drought;
- conventional breeding of genotypes with superior yield potential in drought-prone areas;
- physiological breeding of drought-resistant genotypes;
- identification of QTL (described in the previous section) for drought tolerance and their use in marker-assisted breeding.

The focus in Chapter 9, written by Rockström, Barron and Fox, is on rain-fed agriculture on smallholder farms in sub-Saharan Africa. The authors present field evidence suggesting that mitigating the effects of intraseasonal dry spells is the key to achieving higher yield levels and higher crop water productivity. As a result of the unpredictability of dry spells, farmers tend to avert risks. For many smallholder farmers in the semi-arid tropics, it is not worth investing in external inputs, including, most importantly, fertilizers, as the risk of total crop failure remains a reality once every 5 years and the risk of severe yield reduction occurs once every 2 years. However, the authors show that, with significant investments in water harvesting, conservation tillage and supplemental irrigation during short dry spells, yields of staple food crops could be more than doubled in many areas of sub-Saharan Africa.

Rockström *et al.* suggest that the best option for increasing crop water productivity lies in combining such practices with management strategies that enhance infiltration of rain, increase the water-holding capacity of the soils and maximize plant water uptake through timeliness of farming operations and soil fertilization. Obviously, upgrading rain-fed production through supplemental irrigation would have site-specific implications for downstream water users. The authors recognize that the socio-economic viability of water-harvesting structures for supplemental irrigation needs to be carefully considered. Preliminary assessment of manually dug farm ponds and sub-surface tanks indicates that the benefit–cost ratio depends on the opportunity cost of labour, which is often low during the dry season in remote rural areas.

Future Cereal Production and Water Productivity

Crop water productivity varies with location, depending on such factors as cropping pattern, climatic conditions, irrigation technology, field water management and infrastructure, and on the labour, fertilizer and machinery inputs. For example, in 1995, water productivity of rice ranged from 0.15 to 0.60 kg m^{-3} and that of other cereals from 0.2 to 2.4 kg m^{-3}. Cai and Rosegrant report in Chapter 10 on an analysis of crop water productivity at the global and regional levels through an integrated water- and food-modelling framework developed at the International Food Policy Research Institute (IFPRI). The authors explored the impact of technology and management improvement on water productivity. Based on the best available information and assuming that water supplies for agriculture will become more and more restricted, they expect that from 1995 to 2025 crop water productivity will increase: the global average water productivity of rice from 0.39 to 0.52 kg m^{-3} and that of the other cereals from 0.67 to 1.01 kg m^{-3}. This increase is predicted to result from increases in crop yield and in water productivity at the basin level, with the major contribution coming from yield increases. One of the conclusions of this study is that investments in agricultural infrastructure and agricultural research may have higher payoffs than investments in new irrigation systems in order to accelerate this increase in water productivity and hence ensure food security in the next 25 years.

Case Studies

Chapters 11–19 contain a number of case studies that illustrate issues discussed in the first group of chapters.

The first case study (Chapter 11), presented by Oweis and Hachum, demonstrates that sustainable increases in crop water productivity can only be achieved through integrated farm-resources management. This approach combines water conservation, supplemental irrigation, better crop selection, improved agronomic practices and political and institutional interventions. The case study is based on experience with cereal and legume production in the West Asia and North Africa (WANA) region, with a specific example from Syria.

The second case study (Chapter 12) describes efficient management of rainwater to achieve higher crop water productivity and increased groundwater recharge. The example, written by Wani, Pathak, Sreedevi, Singh and Singh, is from the semi-arid tropics in northern India. The authors argue in favour of an integrated watershed management approach and identified community participation, capacity building at local level, multidisciplinary technical backstopping, and the use of scientific tools as important elements in efficient rainwater management.

The third case study (Chapter 13), by Ong and Swallow, illustrates the importance of water consumption by trees in irrigated areas and discusses how water productivity can be increased in forestry and agroforestry. The authors describe the differences in relative importance of the various components of the water balance of a tree cover and an agricultural crop. For a tree cover, direct evaporation from the soil is much less than for a crop but evaporative loss through canopy interception is higher. There are also significant differences between agroforestry systems and forests, as the former tend to have a relatively sparse tree density.

In the fourth case study (Chapter 14), Bowen reviews efforts to increase water productivity in potato cultivation. Potato is generally shallow-rooted and sensitive to even mild water deficits. Increasing water productivity in potato was done through a combination of improved germplasm and agronomic practices for potato production in warm tropical environments. The author concluded from the study that there exists a useful range of genetic variability that could be taken advantage of for the development of more drought-tolerant and water-productive genotypes for rain-fed and irrigated potato production.

The fifth case study (Chapter 15) is from the rice–wheat cropping system in south Asia, which covers about 13.5 million ha. Hobbs and Gupta describe how growth in area and yield per unit land has been responsible for continued growth in production for over 30 years. Future growth, however, must come from yield increases and higher crop water productivity. Improved resource-conservation technologies, such as zero tillage (now being widely adopted) and raised beds, are identified as the key to increasing water productivity. The authors also emphasize the importance of partnerships and participatory approaches in the research and adoption of new technologies by farmers.

In the sixth case study (Chapter 16), Hussain, Sakthivadivel and Amarasinghe illustrate the importance of irrigation-water management on crop water productivity in northern India and Pakistan, also with a focus on the wheat–rice production system. The case study refers to systems where the irrigation water is a combination of canal water and pumped groundwater. They found significant variability throughout the season, not only in canal water supply and groundwater use and quality, but also in non-land factors, such as seed variety, sowing dates and weedicide application. The case study indicates that substantial gains in aggregate yield can be obtained by a more equitable distribution of the canal water, which would boost yields in tail reaches without adversely affecting yields elsewhere.

Most of the major water basins in Thailand are closing, while an increasing amount of water is being diverted from agriculture. In the seventh case study (Chapter 17), Molle raises the question of whether water productivity can be increased by economic measures, such as water pricing and market mechanisms for the reallocation of water to other uses. The case study shows that, in the Chao Phraya basin in Thailand, farmers and irrigation administrators have made substantial adjustments to water scarcity in the dry season. Thus, the benefits of such economic measures are much smaller than expected and the transaction costs and political risks outweigh the possible gains.

The eighth case study (Chapter 18) addresses the need for data to monitor the productivity of land and water resources over vast areas. Bastiaanssen, Ahmad and Tahir illustrate how in this study measurements from the National Oceanic and Atmospheric Administration (NOAA) weather satellite were combined with ancillary data, such as canal water supplies and rainfall data, into a geographic information system (GIS). The satellite data were converted to crop yield, actual evapotranspiration and, indirectly, to net groundwater use. The analysis of data for the Indus basin is carried out at various scales. Large variations exist in crop water productivity, which the authors ascribe to variations in the relation between canal water supply and evapotranspiration. However, at a spatial scale of 6 million ha and higher, water productivity becomes constant, because at that scale in the closing Indus basin, all water supplied is depleted. The study reinforces the importance of groundwater recycling in the Indus basin.

In the ninth case study (Chapter 19), Zhang argues, on the basis of crop water-production functions, for the introduction of deficit irrigation in order to increase on-farm water productivity in semi-arid countries. The case study uses data from Syria, the North China Plain and Oregon, USA. Also in this case study, crop water productivity shows significant spatial and temporal variation. The risk of deficit irrigation, according to the author, can be minimized through appropriate irrigation scheduling to avoid water stress during the most sensitive growth stages. One of the conditions for success is that farmers control the timing and amount of the irrigation applications.

Conclusions

This book makes clear that increasing crop water productivity is a challenge at various levels. The first challenge is to continue to enhance the marketable yield of crops without increasing transpiration. The second challenge is at field, farm and system levels to reduce as much as possible all outflows that do not contribute to crop production. These three levels are interlinked and the available water for crop production must be used to its greatest advantage within the basin. This may involve allowing outflow to occur from some fields, knowing that this outflow is not lost for plant production but will be used better at some other location within the basin. The third challenge is to increase the economic productivity of all sources of water, especially rainwater but also waste-water of various qualities and saline (ground) water. Meeting the challenge will require developing methodologies and tools to be used for the collection and interpretation of relevant data and information. Scientific disciplines must work together in the analysis of interactions, synergies and trade-offs.

There are hopeful signs that these challenges will be met. At plant level, traits and genes for drought and salt tolerance have been identified in a number of crops, and lessons learned in some crops will be applied to others, making use of both conventional and molecular breeding techniques. For example, progress in respect of increasing production without a concomitant increase in evapotranspiration through changes in the harvest index and stay-green factor is expected to yield results for some crops within 5 years. At field level, further improvements in crop water productivity are expected from the introduction of supplemental irrigation in rain-fed agriculture and the expansion of drip, trickle and sprinkler irrigation. Further progress is also expected in the adoption and adaptation of water-productivity-enhancing practices when institutions and policies are amended to provide appropriate incentives for farmers. At basin level the importance of an integrated approach to land and water management is recognized, especially in respect of sustainable conjunctive management of groundwater and surface water.

But the task of achieving gains in water productivity is daunting. Technologies and management approaches appropriate for poor rural farmers need development. Incentives that would facilitate the adoption of water-productivity-enhancing field practices are not clearly understood and are lacking. The growing interdependence among water uses and increasing competition among users complicates the search for solutions that will improve the productivity of basin-wide water resources. Institutions and policies that can deal with these complexities and with political realities and yet create an environment for farmer productivity are needed. There is indeed scope for increased emphasis on research and application in all these areas.

We expect that the discussions of the challenges and the hopeful signs will help in understanding not only the limits but also the opportunities for increasing crop water productivity.

References

Burt, C.M., Clemmens, A.J., Strelkoff, T.S., Solomon, K.H., Bliesner, R.D., Hardym, L.A., Howell, T.A., Members ASCE and Eisenhauer, D.E. (1997) Irrigation performance measures: efficiency and uniformity. *Journal of Irrigation and Drainage Engineering* 123(6), 423–442.

Molden, D. (1997) *Accounting for Water Use and Productivity*. SWIM Report No. 1, International Irrigation Management Institute, Colombo, Sri Lanka.

Perry, C.J. (1996) *The IIMI Water Balance Framework: a Model for Project Level Analysis*. Research Report 5, International Irrigation Management Institute, Colombo, Sri Lanka.

Richards, R.A., López-Castañeda, C. and Gomez-Macpherson, G.A.G. (1993) Improving the efficiency of water use by plant breeding and molecular biology. *Irrigation Science* 14, 93–104.

Willardson, L.S., Allen, R.G. and Frederiksen, H.D. (1994) Universal fractions and the elimination of irrigation efficiencies. Paper presented at the 13th Technical Conference, USCID, Denver, Colorado, 19–22 October 1994. Duplicated.

1 A Water-productivity Framework for Understanding and Action

David Molden, Hammond Murray-Rust, R. Sakthivadivel and Ian Makin
International Water Management Institute, Colombo, Sri Lanka

Abstract

Substantially increasing the productivity of water used in agriculture is essential to meet goals of food and environmental security. Achieving these increases requires research that spans scales of analysis and disciplines. In spite of its importance, we do not have a common conceptual framework and language to facilitate research and communication among stakeholders. The objective of this chapter is to propose a common conceptual framework for water productivity. In a broad sense, productivity of water is related to the value or benefit derived from the use of water. Definitions of water productivity differ based on the background of the researcher or stakeholder. For example, obtaining more kilograms per unit of transpiration is an important means of expressing productivity of water when the interest of analysis is crops. At the basin scale, obtaining more value from water used from irrigated and rain-fed crops, forests, fisheries, ecosystems and other uses is of importance. There are several interrelated definitions of water productivity that are important across scales and domains of analyses. We propose in this chapter a set of definitions for water productivity and show how these are related across scales.

As the analysis moves from individual plants to fields, farms, irrigation systems and water basins, different processes and means of analysis are important. Understanding how measures of water productivity scale up and scale down provides the key to how a group of people of diverse disciplines can work together on this topic. For example, crop scientists and breeders may focus on obtaining more mass per unit of transpiration, while planners and economists may consider policies to allocate water and land resources between different uses. To capture the full benefits of improved water productivity at farm level, it is necessary to integrate these with system- and basin-level changes. We provide a framework to show the interrelationship of the work of various disciplines.

Introduction

Increasing the productivity of water in agriculture will play a vital role in easing competition for scarce resources, prevention of environmental degradation and provision of food security. The argument for this statement is simple: by growing more food with less water, more water will be available for other natural and human uses (Molden and Rijsberman, 2001; Rijsberman, 2001).

Increasing productivity of water is particularly important where water is a scarce resource. Physical scarcity, when there is no additional water in a river basin to develop for further use, is common in an increasing number of either dry or intensively developed basins (IWMI, 2000). In these cases, it is likely that increasingly less water will be available for agriculture and that, to sustain production, increases in water productivity will be necessary.

There are other important situations of scarcity. Economic scarcity describes a situation where there is water remaining in nature to be tapped for productive uses but there is extreme difficulty in developing the infrastructure for this water for economic, political or environmental reasons (IWMI, 2000).

A third common situation occurs when water and infrastructure are available and cultivation techniques are known and yet people do not have ready access to water. For example, a lack of water is often not the cause of a head–tail problem. As another example, poor people are excluded from infrastructural development and do not have equal access to the benefits available from a project. This management-induced scarcity has a variety of causes, including poor infrastructural development and maintenance but, often, it finds its roots in inappropriate or ill-functioning policies and institutions.

Water productivity is dependent on several factors, including crop genetic material, water-management practices, agronomic practices and the economic and policy incentives to produce. Corresponding to this, there are many people working in parallel on means to increase the productivity of water but the effort remains disjointed. Part of the reason is that we do not have a common conceptual framework for communicating about water productivity. The purpose of this chapter is to propose a conceptual framework to enable us to work and communicate better together.

After agreeing to work under the banner of 'more crop per drop' or 'producing more with less water' we immediately have to figure out what this means. These terms have different meanings for different people: more kilograms per unit of evapotranspiration (ET) for some, more production per unit of irrigation water delivered for others or more welfare per drop of water consumed in agriculture for others. In this framework, we accept that these are all important and relevant meanings, and our task is to sort out how these concepts relate to one another.

More Crop per Drop: Which Crop and Which Drop?

In a broad sense, productivity of water refers to the benefits derived from a use of water. The numerator then has a physical or economic term expressing the benefit. The denominator is a water term. The expression is most often given in terms of mass of produce, or monetary value, per unit of water. First, consider the denominator, water.

Scale considerations

Consideration of scales helps to untangle the 'which crop/which drop' problem. Water use and management in agriculture cross many scales: crops, fields, farms, delivery systems, basins, nations and the globe. Working with crops, we think of physiological processes: photosynthesis, nutrient uptake and water stress. At a field scale, processes of interest are different: nutrient application, water-conserving soil-tillage practices, bunding of rice-fields, etc. When water is distributed in an irrigation system, important processes include allocation, distribution,[1] conflict resolution and drainage. At the basin scale, allocation and distribution are again important, but to a variety of uses and users of water. At the national and international scale, trade, prices and virtual water all have relevance. Processes between scales are inter-

[1] Allocation and distribution of irrigation water are primarily for irrigation farmers, but they are also to meet the demands of other domestic, industrial, livestock and fisheries uses.

linked. For example, basin-scale allocation practices can set a constraint on how much water a farmer receives and the influence on farm water-management practices.

First, we surmise that issues of scale heavily influence concepts of water productivity. Second, we can differentiate scales of analysis by considering the processes important at each scale. We jump across scales when key processes of consideration change. Thirdly, actions at one scale often influence what happens at a different scale. Fourthly, the definition of water productivity found useful by people is dependent on the scale of analysis they are working at.

Accounting for use and productivity of water

Water accounting provides a means to generalize about water use across scales, and to understand the denominator of the water productivity better (Molden and Sakthivadivel, 1999). Water accounting can be applied at all scales of interest, and requires the definition of a domain bounded in three-dimensional space and time. For example, at the field scale, this could be from the top of the plant canopy to the bottom of the root zone, bounded by the edges of the field, over a growing season. The task in water accounting is to estimate the flows across the boundaries of the domain during the specified time period.

At the field scale, water enters the domain by rain, by subsurface flows and, when irrigation is available, through irrigation supplies. Water is depleted[2] by the processes of growing plants: transpiration and evaporation. The remainder flows out of the domain as surface runoff or subsurface flows or is retained as soil-moisture storage. In estimating water productivity, we are interested in water inflows (rain plus irrigation, or just rainwater in rain-fed agriculture) and water depletion (evaporation and transpiration).

The water-accounting procedure classifies these inflow and outflow components into various water-accounting categories, as shown in Box 1.1. The main process of irrigation is the supply of water for crop transpiration to maintain a healthy environment for growth and production. Depletion by the intended processes of industrial (cooling, cleaning), domestic (washing, drinking) and agricultural uses (transpiration) is referred to as process depletion. The finger diagram in Fig. 1.1 shows the flow of water at the field scale. On the right side, water is depleted by the processes of transpiration and evaporation. In most cases, at field scale we cannot say that the outflow is depleted, as it may be recaptured somewhere downstream or by pumped groundwater use.

In rain-fed agriculture, the partitioning of evaporation and transpiration takes on special significance. Rockström *et al.* (Chapter 9, this volume), for example, argue that much of the evaporation can be transferred into crop transpiration, thus contributing to increased crop yield and increased water productivity.

Within an irrigation system we have the same inputs as in rain-fed agriculture – rainfall and surface and subsurface flows – plus artificial irrigation supplies. Irrigation infrastructure is primarily built to provide water for crop transpiration, but, in many irrigated areas, infrastructure also provides water for domestic and industrial uses and for fishing and livestock. In addition to the intended depletion by crop transpiration, water is also depleted by evaporation from weeds, trees, fallow land and water bodies. Drainage water is sometimes directed to sinks, such as oceans, saline water bodies or saline groundwater. Other outflows can be recaptured for use. However, disentangling these various processes demands the clear analytical framework of water accounting (see example in Fig. 1.2).

Like many intensively irrigated areas, most depletion is through crop ET. Other city and industrial-process depletion exists, but is small compared with crop ET. Non-crop vegetation contributes to non-process, but

[2] Depletion is when water is rendered unavailable for further use in the present hydrological cycle. This happens by evaporation, flows to sinks and incorporation into products. Water can also be considered depleted when it becomes too polluted for further use.

Box 1.1. Water-accounting definitions.

Gross inflow is the total amount of water flowing into the water-balance domain from precipitation and from surface and subsurface sources.
Net inflow is the gross inflow plus any changes in storage.
Water depletion is a use or removal of water from a water basin that renders it unavailable for further use. Water depletion is a key concept for water accounting, as interest is focused mostly on the productivity and the derived benefits per unit of water depleted. It is extremely important to distinguish water depletion from water diverted to a service or use, as not all water diverted to a use is depleted. Water is depleted by four generic processes:

Evaporation: water is vaporized from surfaces or transpired by plants.
Flows to sinks: water flows into a sea, saline groundwater or other location where it is not readily or economically recovered for reuse.
Pollution: water quality gets degraded to an extent where it is unfit for certain uses.
Incorporation into a product: through an industrial or agricultural process, such as bottling water or incorporation of water into plant tissues.

Process consumption is that amount of water diverted and depleted to produce an intended product.
Non-process depletion occurs when water is depleted, but not by the process for which it was intended. Non-process depletion can be either *beneficial* or *non-beneficial.*
Committed water is that part of the outflow from the water-balance domain that is committed to meet other uses, such as downstream environmental requirements or downstream water rights.
Uncommitted outflow is water that is not depleted or committed and is therefore available for a use within the domain, but flows out of the domain due to lack of storage or sufficient operational measures. Uncommitted outflow can be classified as *utilizable* or *non-utilizable.* Outflow is utilizable if by improved management of existing facilities it could be used consumptively. Non-utilizable uncommitted outflow exists when the facilities are not sufficient to capture the otherwise utilizable outflow.
Available water is the net inflow minus both the amount of water set aside for committed uses and the non-utilizable uncommitted outflow. It represents the amount of water available for use at the basin, service or use levels. Available water includes process and non-process depletion plus utilizable outflows.
A *closed basin* is one where all available water is depleted.
An *open basin* is one where there is still some uncommitted utilizable outflow.
In a *fully committed basin,* there are no uncommitted outflows. All inflowing water is committed to various uses.

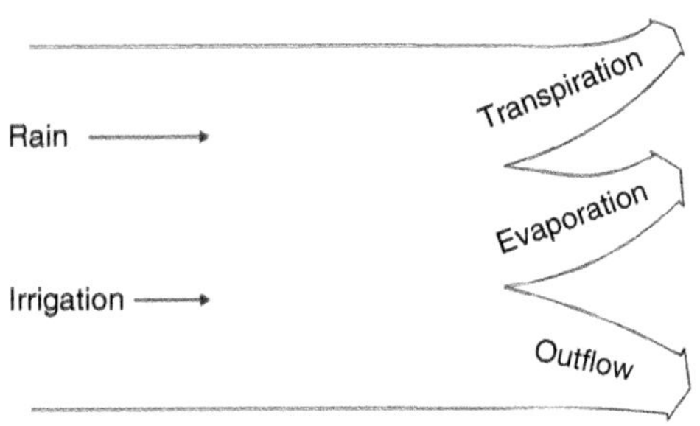

Fig. 1.1. Accounting for water use at field scale. Inflows are from rain and irrigation supplies. Water is depleted by crop transpiration and soil evaporation. The remaining liquid water flows out of the domain.

beneficial, depletion. Irrigation directs flow to drains, which is not reused downstream and is therefore considered depleted. Some of the drainage flow is classified as committed in order to flush salts.

The benefits derived by depletion from trees or water bodies or even flows to saline water bodies can be appreciable. A strict focus on irrigation often leads us to forget that trees have aesthetic and economic value, that water bodies may be important for fisheries or that a sink may really be a wetland providing important ecological services. Unfortunately, some common terminology that we use, such as evaporation or drainage 'loss' or 'wastage', and even 'efficiency' in the way it is commonly defined (see Seckler *et al.*, Chapter 3,

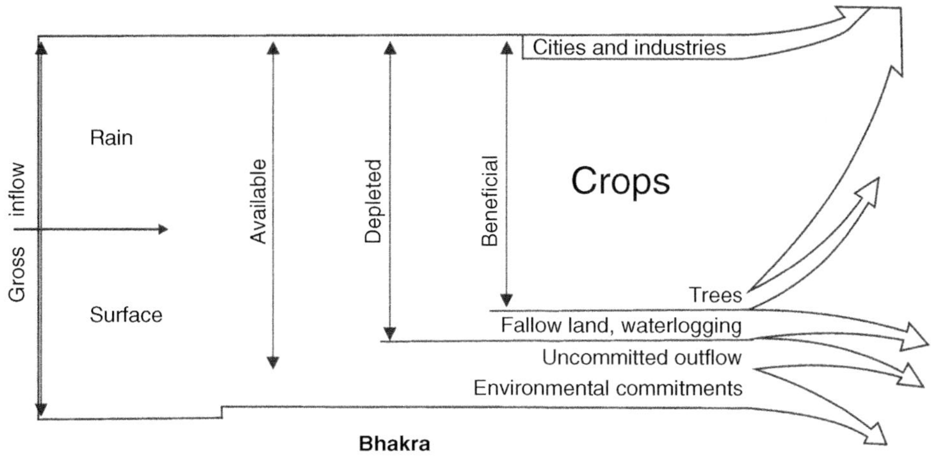

Fig. 1.2. Representation of the Bhakra irrigated area (Molden *et al.*, 2001).

this volume), does not help us to decide what is beneficial and what is not. Figure 1.2 shows a finger diagram for water accounting at an irrigation-system scale, which from top to bottom on the right side shows:

- process depletion (by crops, cities, industries);
- non-process but beneficial depletion, such as evaporation by trees, wetlands within the system or fisheries;
- evaporative depletion of low or negative benefit, such as evaporation from fallow land or from waterlogged areas (that do not have important wetland values);
- flows that are directed by irrigation into sinks, such as seas or inland water bodies that do not add value, and which could have been used within the irrigation system or elsewhere – this is considered as water depleted by irrigation;
- uncommitted flows that are utilizable within the irrigation system or elsewhere;
- committed flows to meet environmental needs or legal or traditional rights of downstream users.

The domain of a basin can be defined by the catchment area of a river system to the salt–freshwater interface.[3] In most cases, the only inflow is precipitation.[4] A useful conceptual advance has been the partitioning of basin water into blue water, which contributes to river runoff, and green water, which evaporates or is transpired (Falkenmark, 2000). By concentrating only on blue water, we omit the benefits derived from the rain and chances of increasing productivity of overall supplies. We agree with this point of view and include precipitation in the water-accounting analysis.[5] Land-use changes are a means of reallocating green water and altering the blue–green balance. Other concerns in water use and management relate to blue water. At the basin scale, other process uses – industrial and domestic, as well as depletion by ecosystems that provide valuable goods and services – are significant.[6]

We can generalize water accounting to other agricultural uses of water. If water is diverted and kept in ponds for fish, the surface evaporation from the pond is accounted for as water depleted by fisheries. If stream

[3] The analysis can be done constructively at sub-basin levels.
[4] Interbasin transfers or subsurface flows can, of course, be significant sources of inflow.
[5] Most irrigation-efficiency calculations start by subtracting effective rainfall.
[6] In many agricultural areas, the quantities of industrial and domestic depletion may be small, but the value derived from the use is significant.

flows are maintained at minimum levels, restricting other uses, the amount of this water should also be considered as depleted by fisheries.[7] In other cases, where fisheries arise because of the development of irrigation reservoirs, value is added to the water without additional depletion.

At the basin scale, outflows require special consideration. Some outflow is required to maintain an environmental balance – to flush out salts and pollutants, prevent saline-water intrusion and supply water and nutrients for coastal fisheries and ecosystems. Floodwater that cannot be captured by existing facilities is considered non-utilizable. The remaining water is considered utilizable for within-basin use.[8]

As we increase the scale of analysis, we tend to add more complexity. But, with increasing competition for water, the types of broad questions we ask are: How can we free water from irrigation for other uses? How can we reduce competition? How can we reduce environmental degradation? All of these require understanding of basin-scale processes.[9] The solutions to these problems most often lie in actions taken at local scales – irrigated and rain-fed fields or within irrigation systems.

Which drop?

So, turning to the fundamental question in water productivity, which cubic metre do we refer to in the denominator of the equation?

Of fundamental concern in agriculture is how much production is derived per unit of crop transpiration.[10] If we could increase the mass per unit of ET all over a basin, production would rise without an increase in water depleted by agriculture. Water productivity in terms of kg per unit of ET, then, seems to be the obvious target that we want to improve.

Water managers, however, tend to be more concerned with the water input. Farmers in rain-fed arid areas, for example, are extremely concerned with capturing and doing the most with limited rainfall. Where an additional supply is available as supplemental irrigation, maximizing the output from a small amount of additional irrigation supply is normally highly productive. For irrigation farmers and system managers, the water supply is the bread and butter of the business. Water supplies, whether rainfall, supplemental irrigation or full irrigation supplies, are candidates for the denominator.

Unfortunately, because scale and environmental factors influence the water supply term, it must be treated with extreme care. For example, where small amounts of irrigation are required in high-rainfall areas and the water-productivity formulation attributes all production to irrigation, productivity per irrigation supply can give high values and thus cannot be compared with areas with low rainfall. Further complicating the matter is that increases in productivity per irrigation supply may not 'scale up' in an intuitive manner (see Box 1.2). It is possible that increasing productivity per unit of supply at field scales may lead to lowering of productivity of supply at larger scales. If, for example, more efficient farm practices are used to grow more crops with the same supply for relatively low-valued uses, thereby reducing supplies to other farmers or uses (especially if they are higher-valued), the overall productivity of basin supplies may be reduced.

In all cases, production per unit of ET remains constant across scales. Increases in productivity of water per unit of ET would lead to increases in productivity of available water. At the farm level, the productivity of diverted water doubles from option I to option II in Box 1.2. Within the system, non-

[7] Retaining minimum flows for fisheries or other ecosystem services could also be classified as an environmental commitment of water.

[8] Utilizable and non-utilizable water is dependent on the degree of infrastructural development. We could define a potentially utilizable flow, which could be depleted within the river basin.

[9] And are also influenced by larger scales, such as trade and virtual water flows.

[10] At scales above farm level, we tend to lump together evaporation and transpiration from agricultural fields as evapotranspiration (ET).

Box 1.2. Production per diverted, depleted and available water.

The following schematic diagram represents two options for an irrigation system to deliver water to farms. In both cases, there are 100 units of production from each farm, 100 units of crop evapotranspiration from each farm and 50 units of rainfall on each farm. At the end of the system shown in the dotted line, 100 units are committed for downstream environmental and agricultural uses.

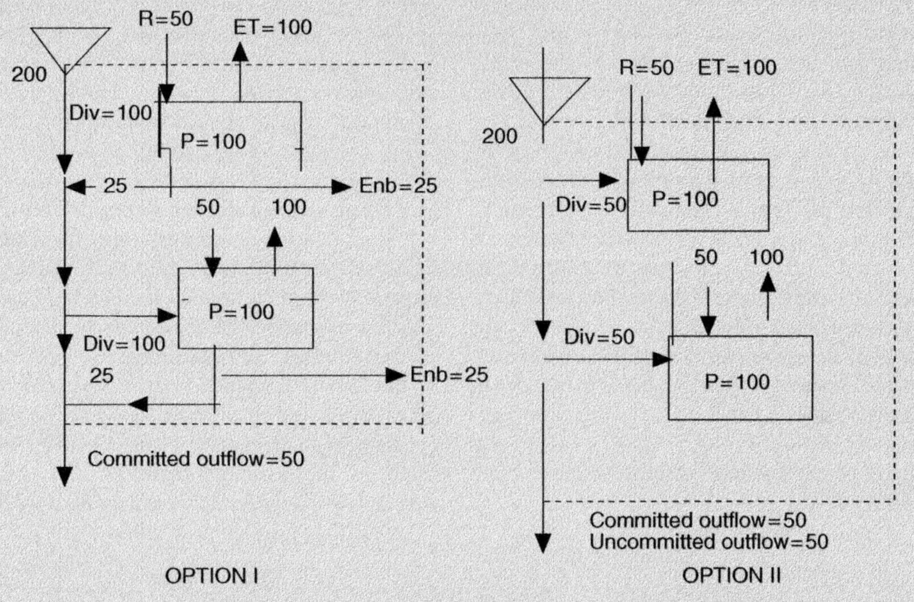

P, production; R, rain; Div, diverted water; ET, evapotranspiration; Enb, non-beneficial evaporation.

The strategies for delivering water are varied. In the first case, 100 units are delivered, while, in the second case, only 50 units are delivered to perfectly match crop evapotranspiration requirements. In the first case, 25 units return to the mainstream, because a waterlogged area has evaporated 25 units, while, in the second case, this waterlogging has been dried up, leaving more water in the mainstream. The following results are obtained:

Option	Diverted	Rain	ET	Available	Production	P/ET	P/Diverted	P/Available
I – farm	100	50	100		100	1.0	1.0	
I – system	200	100	200	250	200	1.0	1.0	0.80
II – farm	50	50	100		100	1.0	2.0	
II – system	100	100	200	250	200	1.0	2.0	0.80

beneficial evaporation is reduced in option II. But this does not lead to increases in productivity of water diverted from the reservoir or water available in the sub-basin. In the second case, the outflow is 100, 50 units more than the commitment of 50 units. In order to realize increased productivity of available supplies, the added 50 units (from decreased non-beneficial evaporation) would have to either be added to increased agriculture within the system or be made available to downstream uses (in which committed water would increase to 100). If an additional 50 units were consumed in agriculture, the productivity of available supplies would have increased to 250/250 = 1.0.

Ideally, we should be able to specify an amount of water available for depletion within any domain. An unambiguous water-management goal would be to increase productivity of available supply. We can define 'available water' at any scale by subtracting the committed and non-utilizable outflow from the net inflow[11] to the domain. (The available water is defined in the finger diagram shown in Fig. 1.3). Increasing the productivity of available water can be achieved by obtaining more per unit of ET and converting non-beneficial depletion to beneficial depletion (water savings) or by reallocating to higher-valued uses. Basin efficiency can be defined by the ratio of beneficial depletion to available water. While available water is an ideal term for water productivity and basin efficiency, it depends on a knowledge of committed flows, which are dependent on allocation rules, water rights and environmental requirements – whose values are unknown or absent in too many situations. We argue strongly for the concept of available supplies and for the need for better definition of rights and requirements within the basin, especially when water is becoming the scarce resource.

Which crop?

The next problem is the numerator of the water-productivity equation. Water productivity can be expressed in physical or economic terms as partial factor productivity (Table 1.1). Physical productivity is defined as the quantity of the product divided by the quantity of the input. Physical production is expressed in terms of mass (kg), or even in monetary terms ($), to compare different crops (Molden *et al.*, 1998). Economic productivity uses valuation techniques to derive the value of water, income derived from water use and benefits derived from water or increased welfare. The valuation discussion requires much more emphasis than we can give here, and we refer readers to Chapter 2 by Barker *et al.* (this volume).

Let us focus the discussion on scale considerations. At the field and crop scale, farmers and researchers are typically interested in the mass of produce. For a farm enterprise, the interest of the farmer shifts to

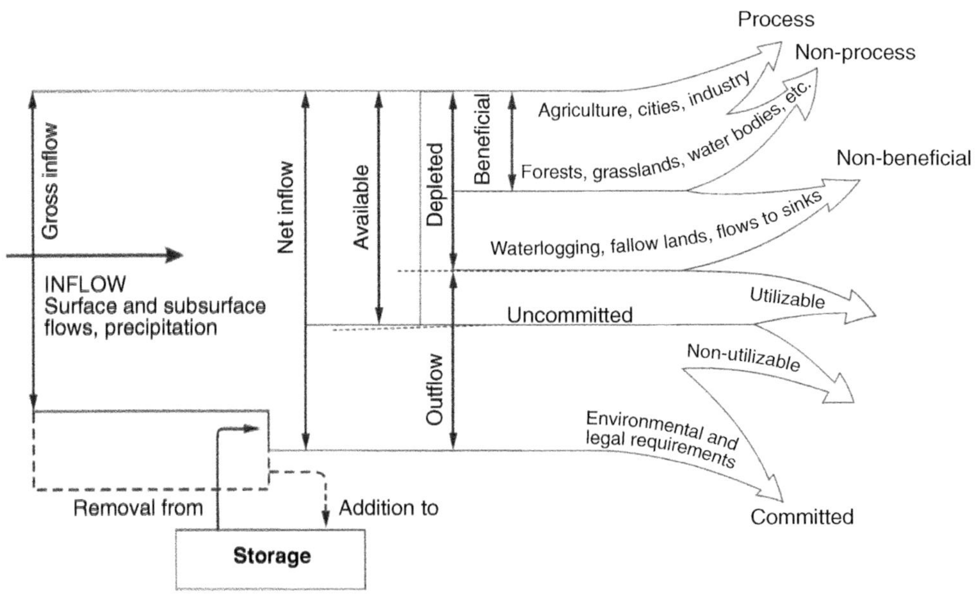

Fig. 1.3. Generalized water-accounting diagram, applicable to basin analysis and analysis at other scales.

[11] Net inflow is the water supply plus rain plus changes in storage. Where there is groundwater depletion, available water should be adjusted to reflect the long-term allowable amount of water that can be withdrawn from groundwater.

Table 1.1. Scale considerations and water productivity.

	Crop	Field	Farm	Irrigation system	Basin
Processes	Water and nutrient uptake and use, photosynthesis, etc.	Tillage, fertilizer application, mulching	Distribution of water to fields, maximizing income	Distribution of water to farms, O&M, fees, drainage	Allocation across uses, regulation of pollution
Scientific interest	Breeders, plant physiologists	Soil scientists, crop scientists	Agricultural engineers, agricultural economists	Irrigation engineers, social scientists	Economists, hydrologists, engineers
Production terms	kg	kg	kg, $	kg, $	$, value
Water terms (cubic metres)	Transpiration	Transpiration, evaporation	Evapotrans-piration, irrigation supply	Irrigation deliveries, depletion, available water	Available water

O&M, operation and maintenance.

income derived or the provision of household food security. In water-scarce situations, though, farmers employ strategies to obtain more mass of production per unit of water supply, such as deficit irrigation (Perry and Narayanamurthy, 1998), supplemental irrigation (Oweis *et al.*, 1999) or water-conservation practices (Rockström *et al.*, Chapter 9, this volume). Water harvesting or employment of drought-resistant crops is also an important strategy for maintaining food security.

Irrigation-system managers may not be concerned about the production derived from irrigation-system water use, as their job tends to be delivery of supplies. However, policy makers, designers and researchers may be keenly interested in the economic output of irrigation systems. It is becoming increasingly apparent that the value of irrigation is not just derived from crops, but rather through the multiple uses of irrigation water (Bakker *et al.*, 1999) or the inadvertent disbenefits produced when irrigation replaces other valuable goods and services. In the Hadejia-Jama'are floodplains in Nigeria, for example, it was found that the value of water in ecosystem services (firewood, fishing, recession agriculture and pastoralism) was found to be much higher than in irrigation (Barbier and Thompson, 1998, as quoted in IUCN, 2000).

At a basin scale, we would like to weigh the benefits of agricultural water uses against other uses. Production derived from industrial or agricultural processes tends to be relatively easier to give a monetary value to than domestic and environmental uses. In water-scarce situations, though, we are forced to consider these trade-offs. Many of these are difficult to quantify and are based on preferences of different members of society. As an example, there is a difficult trade-off between provision of water to poor people for household food security and provision of water for industries. The value of providing access to water for the poor for nutrition may be many times greater than the economic value of the produce, but this must be weighed against the jobs produced in an industrial sector. A first step is to work with stakeholders to highlight which uses are beneficial. Often, choices made are purely political, and the best that can be done is to illustrate and describe the trade-offs at hand.

Achieving Sustainable Increases in the Productivity of Water

In the early stages of river-basin development, we focused on developing and consuming more of the potentially available water by constructing more storage, diversion and distribution facilities. This essentially increases the productivity of potentially available supplies. When we focus on developed water supplies, there are two general pathways for increasing the productivity of water:

- Deplete more available water supply for beneficial purposes, by reducing non-beneficial depletion and converting it to beneficial depletion.[12]
- Produce more output per unit of water depleted.

There are options to increase the productivity of water at each scale of interest, as described below.

Opportunities for increasing water productivity at farm level

Improvements in crop production can only be made at farm level. They result from the deliberate actions of individual farmers increasing production or the value of output with the same volume of available water or maintaining or increasing productivity using less water.

There are a number of different strategies by which farmers can improve water-productivity values, described in detail in Box 1.3. Options include those related to plant physiology, which focus on making transpiration more efficient or productive, agronomic practices, which aim at reducing evaporation, and on-farm agricultural-engineering approaches, which aim at making water application more precise and more efficient.

In practice, many of these different strategies are mixed together because they are complementary. Most programmes to encourage farmers to improve on-farm water management involve combinations of plant physiology, agronomy and agricultural engineering, because there is a synergy involved in applying all three strategies simultaneously.

From the perspective of the Consultative Group on International Agricultural Research and its centers, however, it is important to distinguish carefully between the different strategies, because they require specialist skills that are significantly different. It also helps us in trying to distinguish the level at which we are assessing or measuring water productivity: even at farm level, we get different values of water productivity depending on whether we focus on the plant, the field or the whole farm. We need to know the potential for each component technology rather than the aggregate benefit from the different technologies when applied together.

Farmer motivation to increase water productivity

As discussed elsewhere in this chapter, the need to pay greater attention to water productivity is clearly seen when we look at the relationships between yield and water inputs (Fig. 1.4). With normal plant response functions to water, we always use water less productively in trying to maximize yields. From the so-called 'rational' perspective, if water is scarcer than land then under-irrigation is a logical strategy because it maximizes the scarcer resource.

However, water users are unlikely to use this set of relationships as a major motivation to use water more productively. Under-irrigation is a strategy that can be highly beneficial if there is the prospect of rainfall that will result in big yield increases (e.g. wheat production in Pakistan and north-west India), but it is also a high-risk strategy if irrigation supplies fall below expectations.

Improving productivity of water at irrigation-system and basin scales

While increases in crop production take place on farms, there are a series of water-

[12] In other words, by reducing waste or through real water savings.

Box 1.3. Irrigation systems and basin-level strategies to increase water productivity.

Increasing the productivity per unit of water consumed:

- *Improved water management* – to provide better timing of supplies to reduce stress at critical crop-growth stages, leading to increased yields or, by increasing the reliability of water supply so that farmers invest more in other agricultural inputs, leading to higher output per unit of water.
- *Improving non-water inputs* – in association with irrigation strategies that increase the yield per unit of water consumed; agronomic practices such as land preparation and fertilization can increase the return per unit of water.

Reducing non-beneficial depletion:

- *Lessening of non-beneficial evaporation* – by reducing evaporation from fallow land, by decreasing area of free water surfaces, decreasing non-beneficial or less-beneficial vegetation and controlling weeds.
- *Reducing water flows to sinks* – by interventions that reduce irrecoverable deep percolation and surface runoff.
- *Minimizing salinization of return flows* – by minimizing flows through saline soils or through saline groundwater to reduce pollution caused by the movement of salts into recoverable irrigation return flows.
- *Shunting polluted water to sinks* – to avoid the need to dilute with fresh water, saline or otherwise polluted water should be shunted directly to sinks.
- *Reusing return flows.*

Reallocating water among uses:

- *Reallocating water from lower-value to higher-value uses* – reallocation will generally not result in any direct water savings, but it can dramatically increase the economic productivity of water. Because downstream commitments may change, reallocation of water can have serious legal, equity and other social considerations that must be addressed.

Tapping uncommitted outflows:

- *Improving management of existing facilities* – to obtain more beneficial use from existing water supplies. A number of policy, design, management and institutional interventions may allow for an expansion of irrigated area, increased cropping intensity or increased yields within the service areas. Possible interventions are reducing delivery requirements by improved application efficiency, water pricing and improved allocation and distribution practices.
- *Reusing return flows* – through gravity and pump diversions to increase irrigated area.
- *Adding storage facilities* – so that more water is available for release during drier periods. Storage takes many forms, including reservoir impoundments, groundwater aquifers, small tanks and ponds on farmers' fields.

related actions at irrigation-system and basin scales that influence the basin-scale economic productivity of water. At these scales a series of diversion, distribution and reuse approaches are used either to reduce non-beneficial depletion or to direct water to higher-valued uses.

Within an irrigation system, strategies to reliably distribute water can facilitate the productive use by farmers. These are discussed in detail in the section below. Other strategies are aimed at reducing uncommitted outflows[13] or at mixing water to control pollution loads. These are summarized in Box 1.3.

Basin strategies are typically directed at allocating water supplies and are aimed at shifting water from lower- to higher-valued uses within agriculture and between sectors. But, in addition, typically larger-scale infrastructure or groundwater can be used to promote the reuse of water. For example, the

[13] Not available for downstream use.

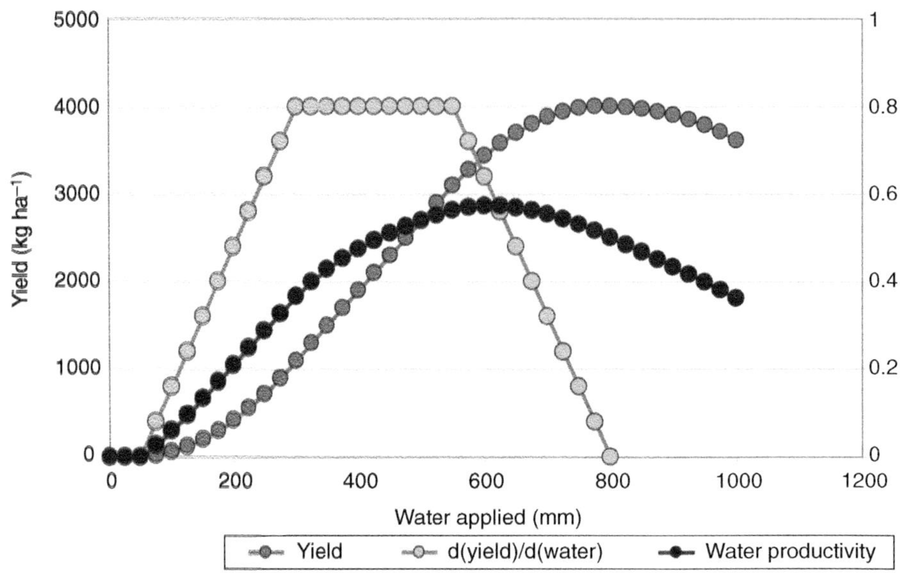

Fig. 1.4. Theoretical relationship between yield and water productivity.

water-diversion structures along the Gediz River (IWMI and GDRS, 2000) or tank cascades (Sakthivadivel *et al.*, 1997) provide a means of recapturing and redistributing water. Pollution control or mixing of high- and low-quality water between sectors has an impact on water productivity. Strategies of land use can be used to increase water productivity.

Linking farm-, irrigation-system- and basin-based water-management strategies to enhance water productivity

To illustrate increases in water productivity, we focus the analysis initially on irrigation systems. A similar analysis could be done for water use by fisheries, rain-fed agriculture or agricultural livestock.

Farm and irrigation-system interactions

There are three possible scenarios that we need to explore to understand how we can maximize the productivity of water at system level.

FARM-LEVEL WATER AVAILABILITY REMAINS THE SAME
If water availability for an individual farm remains the same, then there is no need for any changes at system or subsystem level to increase overall system-level water productivity. All gains will come from a set of uncoordinated actions from some or all of the farmers within the system. The probability is that individual farmers are responding to external factors that encourage them to increase productivity or profitability, independent of their water supply situation. An example of this could be a switch from grain for consumption to production of certified seeds: water consumption is identical but profitability is substantially higher.

ADOPTION OF IMPROVED FARM-LEVEL PRACTICES THAT REDUCE DEMAND FOR WATER Under this scenario there is a reduction in demand for water because of changed farm-level practices. Assuming that farmers cannot increase the area they irrigate, then adoption of new technologies or crops will reduce their overall demand for water. This happened in the Gediz basin in Turkey, where many farmers changed from cotton or grain crops to grapes. Not only are grapes for raisin production more profitable than other crops but they require less water. To take full advantage of this change in demand, system and subsystem canal operations had to change if the saved water was to be used productively elsewhere.

IMPOSITION OF REDUCED WATER DELIVERIES DUE TO REDUCED WATER AVAILABILITY The most common scenario is where there is less water available for agriculture at system or subsystem level. In this case, the pattern of water deliveries is changed and farmers must make specific responses at farm level that will lead to increases in water productivity.

There are several reasons why an individual may try to increase water productivity, and different choices bring with them completely different implications for making water more productive at different scales:

- Reduction in overall water supply.
  - Imposed rationing from a higher level in an irrigation system (equivalent to a reduced water right).
  - Declining groundwater table leading to reduced pumping rates.
- Change in incentives associated with farm-level water management.
  - Imposed incentives to use less water (pricing, withdrawal of subsidies).
  - Declining groundwater resources, which increase operational costs.
- Desires to increase farm-level profitability.
  - Desire to get a better income from limited farm size by increasing production but without changing the basic cropping pattern.
  - Decision to switch to higher-value crops to improve total farm income.

All of these reasons have one thing in common, which is that there have to be commensurate changes in farm-level management of land and water in order to improve water productivity. If such changes do not occur, then the likelihood is that individuals may reduce their cropped area, particularly if there are reductions in water supply, but do not significantly change their irrigation practices. The result will therefore be a decline in overall production without any significant change in water productivity. The yield per depth of water applied will remain the same on the fields that are irrigated, so that no water-productivity gains are made.

With the exception of the relatively rare case where changes at farm level have no impact on demand for water at farm level, we can assume that there is a direct interaction between farm-level and system- or subsystem-level management of water.

The direction of the interaction is an important one. In coping with responses to reductions in demand from individual farmers, the requirement at system and subsystem levels is to manage the additional water so that it can be used productively elsewhere. In cases of reduced water supply, some form of rationing is required that will force farmers to make changes in their farm-level water-management practices. Because these are different management strategies, they are discussed separately.

System-level responses to changes in farm-level water demand

One peculiarity of assessing the productivity of water is that the values are highly dependent on the scale of analysis. A reduction in demand for water at farm level may result in a significant improvement in water productivity when measured at farm level but the values at subsystem level may remain unchanged unless there is a specific effort to utilize the water not used by an individual.

It is because of this peculiarity that, in most cases, we find that an individual working in isolation cannot save much water: it requires the action of others to ensure that the water not used by the individual is used for productive purposes.

Let us assume that an individual farmer refuses the full water entitlement because of the adoption of improved farm-level water-management practices. The refused water will then continue to flow down the canal past the farm gate. If there is no management response, then the refused water may flow to a sink with little or no benefit gained. To ensure that the refused water is used productively requires a clear set of management actions that ensure it is diverted to a productive use elsewhere.

This management action has to be by water users at subsystem level or by staff responsible for system-level operation. In either case, it requires direct communication between managers and water users to continually reassess demand refusal because the refusal is not the same as a permanent reduction in a water right.

This requires that the entire management system becomes demand-responsive and, if the demand response is flexible rather than permanently fixed, then it becomes a complex and extremely difficult task that is likely to stretch the management capacity of most irrigation systems. Most surface irrigation systems are not particularly demand-responsive. They may be able to make changes in long-term allocations through effective seasonal planning, but many find it difficult to be responsive to short-term and ephemeral changes in demand because these require flexible distribution rules.

In those cases where the demand is permanently reduced, it becomes easier to make appropriate changes in system- and subsystem-level rules and water allocations, because, once the change is made at the allocation level, the ensuing water distribution becomes a fixed process.

The situation is made even more complex where groundwater is a substantial source of water. Farmers may use less water by reducing their pumping but it is not always clear whether this results in water saving. That depends on how the aquifer is managed (either formally or informally) by water-user groups or through some form of imposed regulation.

Individual farmers have little incentive to voluntarily refuse water on a permanent basis unless there is some clear tangible benefit. Benefits could be reduced fee payments for water or water services, or deriving a benefit from reducing the adverse effects of getting too much water, such as waterlogging.

We can therefore conclude that, if water-productivity increases at farm level result in individual demand refusals, it will be difficult to capture the unused water in a systematic manner. As a result, we shall not see many system-level changes in water productivity, even though there may be apparent savings at the field and farm levels.

Farm-level responses to changes in system-level water supplies

The much more common scenario is that system-level water availability for irrigated agriculture will decline. This has two distinct differences from the previous discussion of demand-refusal conditions.

First, members of the staff responsible for system-level management have to make reductions in water deliveries. They have several different options open to them, detailed in Box 1.4 and shown graphically in Fig. 1.5. These options are to reduce the irrigated area, reduce discharge per unit area, impose rotational irrigation or shorten the total length of an irrigation season. These options can be combined to result in a series of different scenarios, each of which will have a distinct pattern of water delivery quite different from that which existed before the supply-based reductions were made.

From the perspective of the aim of increasing water productivity at farm level, the option of merely reducing the area irrigated but maintaining the same level of water supply per unit area is unlikely to result in improved water productivity. Individual water users have no external pressure to become more water-productive because there is no change in the relative balance of land and water resources for those farmers who get water.

All other scenarios provide water users with less water, and then they have to make the choice as to whether to adopt the type of measures discussed earlier that can result in improved water productivity at farm level or not. Because their water is now relatively scarcer than land, they can choose to become more water-productive if they so wish.

If water users get less water, then they have a direct incentive to be more water-productive (per unit of irrigation supply) because the benefits are quickly tangible: they can maintain the same cropping inten-

Box 1.4. Management responses to declining water availability at irrigation-system and subsystem levels.

- Reducing area to be irrigated but maintaining the same target discharges and length of irrigation season: this strategy maintains per-hectare water allocation for fewer farmers and is therefore unlikely to encourage adoption of measures to enhance water productivity.
- Reducing per-hectare water supply by reducing target discharges with continuous flow and maintaining total number of days of irrigation supply.
- Reducing per-hectare water supply by reducing target discharges with continuous flow and also reducing total number of days of irrigation supply.
- Reducing water allocation by reducing number of days of irrigation (rotation), maintaining target discharges and maintaining total number of days of irrigation supply.
- Reducing water allocation by reducing number of days of irrigation (rotation), maintaining target discharges and reducing total number of days of irrigation supply.
- Reducing water allocation by reducing number of days of irrigation (rotation), reducing target discharges and maintaining total number of days of irrigation supply.
- Reducing water allocation by reducing number of days of irrigation (rotation), reducing target discharges and reducing total number of days of irrigation supply.

sity as before by using less water per unit area. Whether this involves plant physiology, agronomy or agricultural engineering does not matter: all will mean that production is maintained even though water supplies are less.

Strengthening farm-level and system-level linkages

For water-productivity gains to be substantial and permanent, there must be effective linkages between what happens at farm level and what happens at system level. The choice of a particular technology or approach to farm-level water productivity can be made without regard to system-level management. Similarly, system-level management changes that may try to enhance water productivity can be made without knowledge of what is going on at farm level. However, both of these strategies will be suboptimal.

Let us take a simple example. The single most important innovation to improve the productivity of water has probably been the development and widespread adoption of short-season, high-yielding rice varieties. By reducing the length of life of the plants from, say, 5 to 3.5 months means that the number of days of irrigation drops from about 130 to 85 days without a substantial reduction of yield, if any. Yet the potential saving of some 35% of the total water requirement at field level can only be realized if system-level water issues are reduced by a similar amount.

The same argument applies for all of the technologies: if reduced demand is not matched by reduced supplies, no overall gains will be made at system level. Similarly, if imposed water reductions at system level merely result in fewer farmers getting water at the same level as before, then again there is no overall benefit.

The higher the level of analysis, the greater the importance of recognizing that water-productivity increases require actions not just by an individual water user but by those responsible for water management at system and basin levels.

Linking water-management needs at basin level

In the previous sections, we discussed interactions between water users and service providers. These two-way interactions are relatively easy to identify and describe. In contrast, linkages at basin level are much more difficult to identify and describe, for several reasons.

First, water demands for different sectors are not always quantified with the same pre-

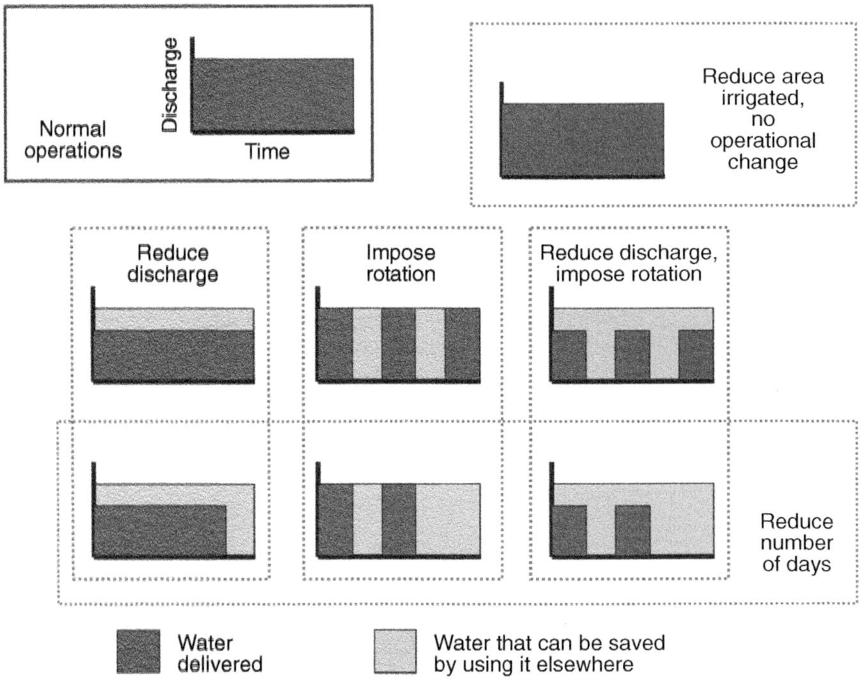

Fig. 1.5. Operational strategies at system or subsystem level that can be used to help save water.

cision. Diversion requirements of urban and industrial sectors are typically better defined than agricultural needs. Return flows are typically not well known and yet represent an important source of water for many people. Other sectors, in particular environmental needs, are less clearly defined. This makes the overall allocation process quite difficult, particularly when water rights do not change in response to changing demands and priorities.

The second complicating factor is that water allocations by sector have to account for multiple uses of the same quantum of water. For example, a single drop of water may serve hydropower, urban, fisheries and then agricultural needs before it is ultimately depleted. Water-management approaches may or may not take advantage of these possible synergies.

Institutional competition adds a third type of complexity. In many countries, some agencies or organizations are reluctant to collaborate with others. Organizations originally charged with large infrastructural construction seem particularly slow to respond to such concerns as environmental protection, water quality and recreation. While the establishment of basin-level management organizations can greatly assist in the process of water allocation between sectors, the actual management of water often rests with individual agencies that still act in a unilateral manner.

Fourthly, land-use decisions or water uses that do not constitute direct stream-flow diversions can have important water-use ramifications. More or less rain-fed agriculture influences movement of both water and salts. For example, in Western Australia, replacement of native forest cover with rain-fed crops resulted in additional recharge and a mobilization of salts (Turral, 1998). The National Land and Water Audit (2000) estimates that total dryland salinity in 2025 will affect 17.5 million ha, dwarfing the impacts of irrigation-induced salinity. Replacing grass with forests or forests with crops influences stream-flow hydrology. Groundwater use or rainwater harvesting may tap into 'blue water' that would otherwise flow into rivers and be available for downstream uses.[14] It is not so obvious that these changes in use may affect other users and the overall productivity of basin water resources.

Clearly, at the basin scale, we are confronted with trade-offs. In closed and closing basins, actions taken on land and water in one part of the basin affect land and water use somewhere else, and in difficult-to-predict ways. These trade-offs are difficult to recognize, much less to quantify. Thus, valuing the productivity of water in its various uses and examining trade-offs require a basin perspective, especially in situations of intensive use, competition and scarcity.

Conclusions and Recommendations

The analysis of water productivity requires a clear understanding of the scale of analysis and the interaction between scales. We have developed expressions of water productivity pertinent to various scales of analysis and have shown the interrelation between scales, means of improving the productivity of water and various actors and disciplines involved in water productivity at various scales.

The productivity of water expressed as mass per unit of water transpired (or ET) is a basic measure of water productivity, valid at any scale. Production per unit of water diverted is an important measure for management, but is not comparable across scales or readily comparable across locations and does not necessarily lead to improvements in the productivity of water diverted at larger scales. Neither measure – productivity per unit of ET or water diverted – provides adequate information about the desires of society to grow more food with less water, to transfer water out of other uses or to use more environmentally sound practices. For this, basin measures related to the amount of water available for agriculture are required.

We have shown that improving the productivity of water in agriculture requires the integrated efforts of many players. For researchers, this does not fall in the domain of one group of specialists but rather requires the efforts of breeders, natural-resource-management specialists, physical scientists, sociologists and more. In practice, it depends on using the synergistic efforts of farmers and water-resource managers at different scales.

While preparing this chapter, three types of issues emerged.

1. We have to be very careful in identifying the scale at which we measure water productivity. The issue of the scale of analysis is fundamental to the improvement of water productivity. As we move from one scale to another, the potential utility of a cubic metre of water changes. Measuring productivity at plant level is relatively simple: the cubic metre refers to the volume of water transpired. But at the basin scale a cubic metre may have many potential uses, each of which values the same water quite differently. Research is required to determine what water productivity really means at different scales within the same basin.

2. There is a need to better understand interactions between scales. Interventions made on a local farm or irrigation scale do not necessarily lead to direct increases in productivity at larger scales, nor do they necessarily free water for higher-value uses, such as cities and, increasingly, the environment. Much more effort is required to understand what impact interventions at one scale have on different scales.

3. We need to examine more closely the trade-offs between different uses of water. One consequence of basin-level analysis of water, particularly in water-scarce basins, is the recognition that each use of water in the basin has impacts on other uses and users. Within agriculture, these trade-offs will involve analysis of water use by fisheries, forests, livestock and field crops. Analysing each water use independently often leads to false conclusions because of these interactions.

Improving water productivity in agriculture will contribute in a major way to the many water problems with which we are

[14] In contrast, converting green-water evaporation to transpiration may not affect other uses of water.

confronted. At present, our combined knowledge is probably sufficient to solve most water-resource problems. What we have not yet done is to combine our knowledge to the maximum effect to address these problems.

References

Bakker, M., Barker, R., Meinzen-Dick, R. and Konradsen, F. (eds) (1999) *Multiple Uses of Water in Irrigated Areas: a Case Study from Sri Lanka.* SWIM Paper 8, International Water Management Institute, Colombo, Sri Lanka.

Barbier, E.B. and Thompson, J.R. (1998) The value of water: floodplain versus large-scale irrigation benefits in Northern Nigeria. *Ambio* 27(6), 434–440.

Falkenmark, M. (2000) Competing freshwater and ecological services in the river basin perspective: an expanded conceptual framework. *Water International* 25(2), 172–177.

IUCN (World Conservation Union) (2000) *Vision for Water and Nature: a World Strategy for Conservation and Sustainable Management of Water Resources in the 21st Century.* Gland, Switzerland.

IWMI (International Water Management Institute) (2000) World water supply and demand 1995 to 2025 (draft). International Water Management Institute, Colombo, Sri Lanka. www.cgiar.org/iwmi/pubs/WWVisn/WWSDOpen.htm.

IWMI and GDRS (General Directorate of Rural Services) (2000) *Irrigation in the Basin Context: the Gediz River Basin Study, Turkey.* International Water Management Institute, Colombo, Sri Lanka.

Molden, D.J. and Rijsberman, F. (2001) Assuring water for food and environmental security. Paper presented at the CGIAR Mid-Term Meeting 2001 in Durban, South Africa, on 26 May. Duplicated.

Molden, D.J. and Sakthivadivel, R. (1999) Water accounting to assess uses and productivity of water. *Water Resources Development* 155(1 & 2), 55–71.

Molden, D.J., Sakthivadivel, R.C., Perry, J., de Fraiture, C. and Kloezen, W.H. (1998) *Indicators for Comparing Performance of Irrigated Agricultural Systems.* Research Report 20, International Irrigation Management Institute, Colombo, Sri Lanka.

Molden, D.J., Sakthivadivel, R. and Habib, Z. (2001) *Basin Use and Productivity of Water: Examples from South Asia.* Research Report 49, International Water Management Institute, Colombo, Sri Lanka.

National Land and Water Audit (2000) *Australian Dryland Salinity Assessment 2000.* Canberra, Australia.

Oweis, T., Hachum, A. and Kijne, J. (1999) *Water Harvesting and Supplemental Irrigation for Improved Water Use Efficiency in Dry Areas.* SWIM Paper 7, International Water Management Institute, Colombo, Sri Lanka.

Perry, C.J. and Narayanamurthy, S.G. (1998) *Farmer Response to Rationed and Uncertain Irrigation Supplies.* Research Report 24, International Irrigation Management Institute, Colombo, Sri Lanka.

Rijsberman, F. (2001) Can the CGIAR solve the world water crisis? Paper presented at the CGIAR Mid-Term Meeting 2001 in Durban, South Africa, on 26 May.

Sakthivadivel, R., Fernando, N. and Brewer, J.D. (1997) *Rehabilitation Planning for Small Tanks in Cascades: a Methodology Based on Rapid Assessment.* Research Report 13, International Irrigation Management Institute, Colombo, Sri Lanka.

Turral, H. (1998) *Hydro Logic? Reform in Water Resources Management in Developed Countries with Major Agricultural Water Use: Lessons for Developing Countries.* Overseas Development Institute, London, UK.

2 Economics of Water Productivity in Managing Water for Agriculture

Randolph Barker,[1] David Dawe[2] and Arlene Inocencio[1]
[1]*International Water Management Institute, Colombo, Sri Lanka;* [2]*International Rice Research Institute, Manila, Philippines*

Abstract

Water is an extremely complex resource. It is both a public and a private good; it has multiple uses; the hydrology requires that we examine potential productivity gains at both the farm and the basin levels; both quantity and quality are important; institutions and policies are typically flawed. For a given situation, economists often disagree on how to value water and on the best strategy for increasing water productivity. This fact notwithstanding, growing scarcity increases the need, if not the demand, for sound economic analyses.

The purpose of this chapter is to lay down some of the concepts and complexities in economic analyses related to increasing water productivity, to provide some examples and to see what this implies regarding the potential for increasing water productivity. We hope that this will help set the stage for productive discussions and the identification of research needs.

The chapter is divided into three main sections. The first section discusses the relationship between efficiency, productivity and sustainability, and emphasizes the confusion in definitions. The second section provides examples at plant, farm, system and basin levels, relating water productivity to both economic efficiency and sustainability. Closely related to this, in the third section we discuss the potentials for increases in water productivity and economic efficiency through incentives created by policy and institutional reforms.

Failure to include the potential for recycling or reuse of water diverted for irrigation in the measurement of irrigation efficiency has led to the widely accepted view that public irrigation systems are poorly managed and that there is considerable scope for increasing water productivity. Water savings do not necessarily lead to higher water productivity and, similarly, higher water productivity does not lead to greater economic efficiency.

A distinction can be made between those measures that increase water productivity by increasing crop yield for a given evapotranspiration (ET) or diversion as opposed to reducing the water-diversion requirements. Measures to increase crop yield for a given ET translate into water-productivity gains at the system and basin levels. However, the management of water to reduce water-diversion requirements is riddled with off-site effects and externalities. Thus, whether water-management practices or technologies designed to increase water productivity and economic efficiency at the farm level translate into water-productivity and economic-efficiency gains at the system or basin level needs to be determined. The basin is a hydrological unit as opposed to an administrative unit. It is only at this level that we can capture and include in our analysis the off-site effects (or, in economic jargon, internalize the externalities).

The growing scarcity and rising value of water in a basin induce farmers to seek ways to increase water productivity and economic efficiency. Recycling or reuse of water is prominent among the prac-

 Water Productivity in Agriculture: Limits and Opportunities for Improvement (eds J.W. Kijne, R. Barker and D. Molden)

tices adopted to increase water productivity, and greater attention needs to be focused on managing surface water and groundwater for conjunctive use. We need a better understanding of biophysical and socio-economic changes in basins over time and improved measures of basin-level efficiencies before we can determine in a given situation the potential for increasing water productivity through policy and institutional reforms.

Finally, as basins become closed, overexploitation of groundwater resources is accompanied by a serious decline in water quality and other problems of environmental degradation. Decisions on basin-level allocations among sectors cannot be based strictly on economic efficiency but they must involve value judgements as to how best to benefit society as a whole. This will include setting priorities in the management of water resources to meet objectives such as ensuring sustainability, meeting food-security needs and providing the poorer segments of society with access to water.

Introduction

Water is an extremely complex resource. It is both a public and a private good; it has multiple uses; the hydrology and externalities require that we examine potential productivity gains at the farm, system and basin levels; both quantity and quality are important in measuring availability and scarcity; and the institutions and policies that govern the use of water are typically flawed.

Given these complexities, it is small wonder that there is little agreement among scientists, practitioners and policy makers as to the most appropriate course of action to be taken to improve the management of water resources for the benefit of society. This fact notwithstanding, the growing scarcity of water increases the need and demand for sound economic analyses.

The purpose of this chapter is to lay down some of the concepts and complexities in economic analyses related to increasing water productivity, to provide some examples and to see what this implies regarding the potential for increasing water productivity. We hope that this will help set the stage for productive discussions and the identification of research needs.

This chapter is divided into three main sections. The first section discusses the relationship between efficiency, productivity and sustainability, emphasizing the confusion in definitions and distinguishing between engineering, biological and economic concepts. The second section provides examples – at the plant, farm, system and basin levels – relating water productivity (WP) to economic efficiency (EE) and to sustainability. Closely related to this, in the third section we discuss the potentials for increases in WP and EE through policy and institutional reforms.

Definitions and Concepts of Efficiency, Productivity and Sustainability

In this section we begin with a discussion of definitions of water-use efficiency (WUE), irrigation-efficiency (IE) and WP. We then define EE and relate EE to IE and WP. We conclude with a brief discussion of WP, EE and sustainability.

Water use and irrigation efficiency

In general terms, we define IE as the ratio of water consumed to water supplied. WP is the ratio of crop output to water either diverted or consumed, the ratio being expressed in either physical or monetary terms or some combination of the two. There are four areas of confusion related to the concept of efficiency.

First, WUE as used in the literature, including the economics literature (e.g. Dinar, 1993) and plant-science literature (e.g. Richards *et al.*, 1993), most commonly refers to what we have defined above as WP: that is to say, it is defined as the ratio of crop output to water input. We believe that in these instances WP is the more appropriate term.

Secondly, the conventional wisdom that irrigation systems in the developing world typically operate at a low level of efficiency

(30–40%) is based on what Seckler *et al.* (Chapter 3, this volume) refer to as classical irrigation efficiency (IEc) or the water consumed divided by the water supplied. IEc is defined in terms of differences between the point of water diversion and the ultimate destination of the water in the root zone of the plant.

IEc = (crop ET − effective rainfall)/(vol. of water delivered − change in root-zone water storage)

IEc at the project level is typically subdivided between conveyance efficiency (water distribution in the main and secondary canals) and field-application efficiency (water distribution to the fields being irrigated). The water diverted but not used for evapotranspiration (ET) includes seepage and percolation, spillover and land preparation, all of which are treated as losses. Classical efficiency decreases as one moves from the field towards the reservoir and conveyance losses are combined with field losses. A high level of IEc may not reflect good management but simply water scarcity. Some scholars prefer to use the term relative water supply (RWS), the inverse of IEc, to avoid the connotation associated with the word 'efficiency'.

Much of the so-called 'losses' in IEc (seepage, percolation and spillovers) can be captured and recycled (for example, by use of tube wells) for use elsewhere in the system. Conversely, many of the so-called 'water savings' practices, such as those that reduce seepage and percolation (e.g. lining canals), are not saving water at all but simply redistributing the water – robbing Peter to pay Paul. The only real losses to the hydrological system are from bare soil and water evaporation (much of which can occur during land preparation) or from flows to the sea or to sinks.

The concepts of neoclassical irrigation efficiency (IEn) or effective irrigation efficiency (Keller and Keller, 1995, 1996; Seckler *et al.*, Chapter 3, this volume) take into account return flows:

IEn = (crop ET − effective rainfall)/(vol. of waters delivered − change in root-zone water storage − vol. of water returned or recycled)

Taking into account return flows results in a higher estimate of IE, which leads to the conclusion that the scope for improving IE is much less than is normally assumed.

Thirdly, we must distinguish between IE and WP at the farm and basin level. To understand this distinction, we need to turn to water-accounting procedures and include non-agricultural water uses (Molden and Sakthivadivel, 1999). This represents another significant step away from the concept of IEc. The operational terms used here (and there are many more) are beneficial depletion and non-beneficial depletion. At the basin level, a potentially wide range of factors can deplete water. Beneficial depletion would include consumption (ET) by the crop being irrigated as well as, for example, beneficial consumption by trees. Non-beneficial depletion includes evaporation and flows to sinks such as the sea. A higher level of efficiency can be achieved by lowering non-beneficial depletion.

Finally, a high efficiency, defined here as a large percentage of beneficial depletion, does not imply a high level of productivity or of economic return. The same degree of beneficial utilization may have substantially different values for the productivity of the water (Seckler *et al.*, Chapter 3, this volume). For example, the same amount of water depleted in the irrigation of cereal crops may have a much higher value in vegetables and fruits or in non-agricultural uses. Furthermore, as water flows through the basin, economists would want to know the benefits and costs associated with various alternatives for reducing diversions and for recycling water.

Economic efficiency and irrigation efficiency

Economic efficiency (EE) takes into account values of output, opportunity costs of inputs and externalities and is achieved when scarce resources are allocated and used such that the net value or net returns (returns minus costs) are maximized. Unlike IE, which is a ratio by definition, EE is a criterion that describes the conditions that must be satisfied to guarantee that resources are being used to generate the largest possible net benefit (Wichelns, 1999).

EE is often consistent with IE. For example, as water becomes scarce and the value of water is high in semi-arid regions, a high IE (although not necessarily the result of improved irrigation management) is consistent with EE. Alternatively, when off-farm impacts can be ignored and water is abundant with low opportunity cost, EE can be achieved even at low IE.

EE in a production setting involves technical and allocative components. A producer is technically efficient when producing the maximum amount of output with a given set of inputs. The producer is allocatively efficient if he/she produces at the point dictated by the prices of outputs and inputs that will maximize returns. A producer is said to be economically efficient if he/she is both technically and allocatively efficient.

Of concern to many economists is the fact that the farm-level price or charge for irrigation water and power for pumping water do not typically reflect the true value of water and would appear to encourage waste. However, farmers and irrigation-system managers will make adjustments in response to water scarcity without price incentives. Furthermore, at the basin level, while analyses based on economic optimization may be useful to policy makers, allocations must take into account the fact that water is a public as well as a private good. Allocations among competing uses involve value judgements as to how to achieve the highest benefit for society as a whole.

Productivity and partial water productivity

The term water productivity (WP) is also defined and used in a variety of ways. There is no single definition that suits all situations. As mentioned previously, in general terms, productivity is a ratio referring to the unit of output(s) per unit of input(s).

The most encompassing measure of productivity used by economists is total factor productivity (TFP), which is defined as the value of all output divided by the value of all inputs. But the concept of partial factor productivity (PFP) is more widely used by economists and non-economists alike. Partial productivity is relatively easy to measure and is commonly used to measure the return to scarce or limited resources, such as land or labour. For example, in the early stages of economic development, agricultural labour is often in surplus and land is the scarce resource. (There are notable exceptions, including many parts of Africa.) Where land is the limiting resource, the greatest economic benefits are achieved by increasing output per unit of land. Therefore, emphasis is placed on technologies that increase yield per hectare (e.g. high-yielding varieties and fertilizer). The change in PFP measured in yield per hectare is a useful indicator of the economic performance of the agricultural sector.

But, as an economy develops, the labour force in agriculture declines and more and more labour is pulled to the non-farm sector. When agricultural labour is in short supply the emphasis shifts to labour-saving technologies (e.g. tractors and mechanical threshers). PFP measured in output per worker is now a better indicator of the economic performance of the agricultural sector.

Until recently, water was not considered a scarce resource. Now, with mounting water shortages and water-quality concerns, there is growing interest in measures to increase WP, which is a specific example in the general class of PFPs. WP is most commonly measured as crop output per cubic metre of water.

Partial water productivity can be expressed in physical or economic terms as follows (Seckler *et al.*, Chapter 3, this volume):

1. Pure physical productivity is defined as the quantity of the product divided by the quantity of the input. Examples include crop yield per hectare or per cubic metre of water either diverted or consumed by the plant. For example, the International Water Management Institute (IWMI) sees as one of its primary objectives 'increasing the crop per drop'.
2. Productivity, combining both physical and economic properties, can be defined in terms of either the gross or the net present value of the product divided by the amount of the water diverted or consumed by the plant.
3. Economic productivity is the gross or net present value of the product divided by the

value of the water either diverted or consumed by the plant, which can be defined in terms of the value or opportunity cost in the highest alternative use.

Economic measures of WP (2 and 3 above) are difficult to estimate. While the net value is more satisfactory than the gross value of the product, the valuation of inputs must be treated in a uniform manner across sites. This can be difficult for land, labour and water (which are also usually the most important inputs). Valuing water is at best a difficult and unsatisfactory process, considering that the marginal value of water varies throughout the season, between seasons, by location, by type of use and by source of water.

There is also the matter of scale or the area over which productivity is measured. Do measures to increase WP at the farm level translate into increases in WP at the system or basin level? Water-accounting procedures that take into account externalities resulting from a farm-level change in water-management practices can be used to measure WP at the system or basin level. Through this process we can determine whether an intervention leads to real water savings (taking into account all return flows, as in IEn). However, at this level, beneficial depletion includes benefits from water use other than for the crop being irrigated, such as water for the environment and other non-agricultural needs.

A distinction can be made between those measures that increase WP by increasing crop yield for a given ET or diversion and those that reduce the water-diversion requirements. In the former case, savings at the plant and field level are realized at the system and basin level. In the latter case, whether increased WP at plant and field level translates into increased productivity at system and basin level needs to be determined. For example, although the water saved in one farming area may be reallocated to higher-value, non-agricultural uses, a reduction in seepage and percolation losses from this area may be at the expense of farmers elsewhere in the system.

However, as the term 'partial' in PFP implies, it tells only part of the story. In general, functions relating output to input (e.g. water, fertilizer) are nearly always concave because the use of higher levels of input is eventually subject to diminishing returns. Under these circumstances, a high WP (or a high IE) in a system or basin may simply reflect a shortage of water rather than good management or EE. In fact, when such a function is purely concave, PFP is maximized by using as little of the input as possible, even when it results in large declines in output (because, as input use declines towards zero, productivity increases towards infinity). Thus, the appropriate goal should be to optimize WP, not maximize it.

Despite the above arguments, many people view higher WP (or higher fertilizer productivity or higher yields) as an inherently good idea. But it is easy to see why measures that show an increase in PFP of water or any other input may provide a misleading result from the perspective of the farmer, as well as from that of the economy as a whole. A technology or management practice that increases water productivity may require the use of more labour and other inputs. For example, a reduction in water application in rice could increase the amount of weeding required. Also, a shift to drip irrigation saves water but also requires capital investment, which might not be cost-effective. Unfortunately, the concept of PFP gives very few guidelines regarding optimization. In fact, without considering the economic and social values of all inputs and outputs, it will be difficult to make progress on this issue. Thus, we now turn to a discussion of the concept of net returns.

Net returns and water productivity

In this section we build on the concept of EE, distinguishing between net private returns and net social returns and relating net returns to WP. Net private returns are defined as the market value of all outputs minus the cost of all inputs, taking into account the opportunity cost of family labour, land and any other inputs that are not purchased on the market. If the net returns to a practice are positive, then it will be beneficial for farmers to adopt the practice. If net returns are negative, it will be disadvantageous for the farmer to adopt the practice and, no matter how large the increase in WP due to the practice, it is unlikely that the farmer will adopt it.

Alternatives for improving net private returns can be categorized as follows (Wallace and Batchelor, 1997):

- agronomic improvements (for example, improved crop husbandry, cropping strategies and crop varieties);
- technical improvements (for example, improved and lower-cost technologies for extracting groundwater);
- managerial improvements (for example, improvements in farm-level resource management or system operation and maintenance (O&M);
- institutional improvements (for example, introduction of water pricing and improvement in water rights).

The first two categories relate to innovations or new technologies that lower costs or increase output per unit of water. The third category, improved management, refers to an increase in technical efficiency or increased output per unit of input with existing levels of technology. The fourth category relates principally to allocative efficiencies encouraged by the creation of market incentives.

Economic theory shows that if a new practice does not have any effects on third parties off the farm (known as technological externalities in the jargon of economics), then the adoption of this practice is advantageous for society as a whole, not just for the farmer. Unfortunately, water management is riddled with externalities, so this theory provides little guidance as to whether or not it is advantageous for society to encourage the adoption of a specific new water-management technology based only on the magnitude of net returns to farmers.

In order to assess whether or not a new technology available to farmers is beneficial to society, one needs to calculate net social returns instead of net private returns. The two concepts are identical, except that net social returns value all inputs and outputs at social prices, not market prices. Social prices are identical to market prices when well-functioning markets exist. When well-functioning markets do not exist, as is almost always the case with water, then one must attach a social value to water, which is defined as the value of the water in the best alternative use (at the margin).

While this opportunity cost is relatively easy to define, it is much harder to measure. For example, one could assign to water a societal value equal to its current value in industrial use. However, if one hypothetically begins to shift water from agriculture to industry, the marginal value of additional water in industry will eventually decline. Thus, in contemplating large transfers of water out of agriculture (as opposed to small, marginal transfers), it is not valid to assume that the per-unit value of the water transferred is equal to the current per-unit value of water in industrial uses.

Furthermore, the concept of net social returns is silent on issues of equity, and most people would agree that equity is important in making decisions on the desirability of implementing policies or technologies that affect WP.

Although it is difficult to measure the net social returns due to the implementation of a policy or technology, it is useful to keep this concept firmly in mind when making judgements about practices that improve WP. At a minimum, this concept reminds us of our ignorance and what specific missing information is desirable for an assessment of new technologies, institutions or policies. Although we shall use the term WP in subsequent discussions, it is always important to bear in mind how much it will cost to increase WP and that not all increases in WP are desirable.

Water productivity, environmental degradation and sustainability

Irrigated agriculture not only competes for water but often contributes to the major degradation of water resources. Consider, for example, those regions of rapidly falling water tables due to groundwater mining or alternatively regions of rising water tables leading to waterlogging and salinity. In the latter case, the social cost may be in the form of environmental degradation or, if corrective measures are taken, the cost to some segments of society may be for appropriate disposal of drainage water. The net social benefit is the difference between returns to the farmer and the cost to society associated with drainage-water pollution (Dinar, 1993).

Ultimately, we must address the issue of sustainability. Unfortunately, there are many definitions of sustainability and sustainable development, ranging from the very broad to the very narrow, which create a potential for misunderstanding (Dixon and Fallon, 1989). We define sustainability as the ability to continue extracting net positive social returns from a resource for an indefinite period of time. Notice that it is not inconsistent with some degree of environmental degradation, i.e. it is not always true for all ecosystems that the optimal rate of degradation is zero, just as it is not always true that the optimal rate of oil extraction from a particular deposit is zero.

One viewpoint in the sustainability debate holds that high-industrial-input agricultural systems are inherently unsustainable (Lynam and Herdt, 1999). Proponents of this view have shifted the debate away from production and income distribution to environmental degradation and input use. The focus on ecosystems by environmentalists and on watersheds by hydrologists has carried the debate substantially above the commodity-based farm and farming-systems level to land, water and other highly valued natural and environmental resources.

Lynam and Herdt (1999) argue that:

> sustainability of common resource systems necessarily incorporates value judgements on multiple criteria over how the community wishes to utilize resources; moreover sustainability of the system will depend more on social institutions controlling access and use than on production technologies.

Relating Water Productivity and Economic Efficiency: Some Examples

Molden *et al.* (2001, Appendix A) provide a comprehensive list of alternatives for increasing WP and Molden *et al.* (Chapter 1, this volume) illustrate how various alternatives can be applied at the crop, farm, system and basin levels. At each of the first three levels, we provide an example illustrating the relationship between WP and EE. At the basin level, we emphasize the relationship between WP and sustainability.

Plant level: increasing water productivity through varietal improvement

The concept of WP used by plant physiologists, molecular biologists and plant breeders refers to the crop output (either grain or biomass) per unit of transpiration by the plant. (This is typically referred to as WUE.) There has been steady improvement in grain yield per hectare through plant breeding in rainfed and, most particularly, in irrigated areas. The development of short-season varieties, reducing the growing time from 5 months to 3.5 to 4, has also been a major source of water savings (more crop per drop per day). The development of water-storage facilities and expansion of the irrigated area in the dry season have allowed these savings to be translated into increases in WP. Thus, there is no question that, over the past three decades, varietal improvement through plant breeding (aided by investments in irrigation and advances in fertilizer technology) has been the major source of increase in WP (Richards *et al.*, 1993).

However, the increase in grain productivity is in some ways deceptive (Richards *et al.*, 1993). In almost all crops, the greater grain yield is not due to an increase in biomass but almost entirely to an improved ratio of grain to biomass (harvest index). As the potential ceiling value for the harvest index is rapidly approaching in many crops, the only way to maintain increases in yield will be to increase biomass (Richards *et al.*, 1993). There appears to be considerable potential for increasing biomass by selecting cultivars for increased WP, defined in this case as the rate at which water lost in transpiration results in the photosynthetic assimilation of carbon in the plant. In many Middle Eastern countries, a very high level of WP has already been achieved. There is thus great hope that research in plant breeding and molecular biology will increase WP in other parts of the world. In other areas, gains in productivity may be achieved through varieties tolerant to saline soil and water conditions.

One of the important features of varietal improvement is that it is relatively less site-specific in terms of potential benefits than most management interventions. Much of the research is funded by international and national agencies. Numerous studies have emphasized the high returns to investment in varietal-improvement research in the past (Evenson *et al.*, 1991; Alston *et al.*, 1995) – although in many instances the benefits ascribed to research may include contributions from irrigation and advances in fertilizer technology. In setting research priorities, a key issue is the size of the geographical area as well as the size of the population upon which the varietal improvement is likely to have an impact. This will determine the benefits of the research relative to its costs. As water scarcity becomes more acute, the potential benefits of this research will increase.

Farm level – adoption of yield-increasing and water-saving technologies: the case of SRI

In promoting the adoption of new technologies, researchers and extension agents often focus on the higher yield potential, ignoring the opportunity cost of family labour and the increased management requirements. This point is illustrated in a draft report on a study of the adoption of the System of Rice Intensification (SRI) in Madagascar (Moser and Barrett, 2003). The paragraphs below are based on this report.

SRI was developed in the early 1990s in Madagascar as a seemingly ideal low-external-input sustainable agriculture (LEISA) technology. The method requires almost no external cash inputs, such as chemical fertilizers, pesticides and seeds. The SRI method involves seeding on dry beds, transplanting younger than 20-day-old seedlings with one seedling per hill, spacing of at least 20 cm × 20 cm, frequent weedings and controlling of the water level to allow aeration of roots during the growth period of the plant. However, the technology requires approximately 50% more labour. Using this method, farmers have repeatedly obtained yields two to three times higher than the 2–3 t ha^{-1} obtained using traditional practices. Owing not only to higher yields but also to the water-saving irrigation practices, the gains in water productivity at the field level could be very high, although water accounting would be required to determine the basin-level impacts of farm-level water savings.

The study undertaken by Moser and Barrett (2003) surveyed 317 households in five villages. Approximately one-third of the farmers adopted SRI but most practised it on only a portion of their land. The adopters tended to have higher education, belong to farmer associations and have higher wealth and income. In contrast, the non-adopters were unskilled agricultural labourers, who, lacking the financial resources to carry them through the 'hungry season', depended on the agricultural wages they received daily. Thus, they cannot afford to spend the extra time required for adopting SRI on their own farms because they are busy working on other people's farms. More importantly, many of those who adopted SRI have since abandoned the technology, often after trying SRI for only one season (Table 2.1). Apparently, the significantly higher yields were not enough to offset the substantially higher labour costs and management requirements.

System level: benefit–cost analysis

We have observed that water savings *per se* may or may not lead to increases in WP. Likewise, an increase in WP may or may not result in higher economic or social benefits. Following the general concepts in our discussion of net returns at the system level, economists assess the merits of an investment by measuring the benefits and costs (B:C) ratio or the internal rate of return (IRR). These are measures of the performance of investments or the productivity of capital. These two terms are defined mathematically as follows:

Table 2.1. SRI adoption and non-adoption patterns in Madagascar, 1993–1999 (from Moser and Barrett, 2003).

	Ambatovaky	Iambara	Torotosy	Anjazafotsy	Manandona	Average[a]
Households trying the method, 1993–1999 (%)	48	16	27	28	21	25
Households using the method in 1999 (%)	26	7	0	13	17	15
Adopters who disadopted (%)	46	53	100	49	19	40

[a] Average is weighted to account for different numbers of households at each site.

B:C ratio:

$$\frac{\sum_{t=1}^{t=n} \frac{B_t}{(1+i)^t}}{\sum_{t=1}^{t=n} \frac{C_t}{(1+i)^t}}$$

The IRR is the discount rate i such that:

$$\sum_{t=1}^{t=n} \frac{B_t - C_t}{(1+i)^t} = 0$$

For the B:C ratio, a social discount interest rate is chosen, typically 10%. If the B:C ratio exceeds 1, then the project has a positive social benefit. If the IRR is greater than the social discount rate (often assumed to be about 10%), then the project has a positive social benefit. While an assessment of environmental costs is now frequently included in the analyses, as with farm-level analyses, this is largely a commodity-oriented approach. Benefits of a given project are typically measured in terms of higher yield and net returns to the farmer for irrigating a specific set of crops.

One of the most well-studied irrigation projects in Sri Lanka is the Gal Oya Water Management Project (Uphoff, 1992; Murray-Rust *et al.*, 1999). A deteriorated irrigation system, the Gal Oya Left Bank Irrigation System, was rehabilitated in the period 1982–1985, using a combination of physical and institutional interventions.

A time-series, impact-assessment model was used to describe the trends and impacts in the system as a whole, as well as in different parts of the system (Amarasinghe *et al.*, 1998; Murray-Rust *et al.*, 1999). The data from 1974 to 1992 covered the period both before and after the rehabilitation. Significant gains have been made in WP for the system as a whole. The tail-end farmers, even though they were less intensively organized, showed the best overall performance in terms of water use, crop production and WP.

Did benefits exceed costs? The project completion report conducted in 1985 estimated an IRR of between 15 and 30% (Project Completion Report, 1985). A subsequent study by Aluwihare and Kikuchi (1991) reported an IRR of 26%. While investment in the construction of new systems in Sri Lanka is no longer profitable, among the major rehabilitation projects conducted in recent years, Gal Oya has had the highest IRR (Table 2.2).

But there are two caveats. First, some of the gains made were at the expense of other water users (D.H. Murray-Rust, personal communication). Prior to rehabilitation, water in the drains was being used by farmers outside of the Left Bank Irrigation System. With this water no longer available, many farmers simply went out of business. We do not know to what degree these 'hidden' costs would lower the IRR. However, this example of off-site effects or externalities emphasizes the need to adopt a basin perspective.

Secondly, although the area irrigated by groundwater is still small, the recent IRR estimates for largely private agro-well and

Table 2.2. Rates of return on irrigation investments in Sri Lanka in recent decades: new irrigation construction and rehabilitation based on 1995 constant prices (Kikuchi *et al.*, 2002).

	C:B ratio	International rate of return (%)
New construction projects[a]		
1980	0.8	12
1985	1.1	9
1990	1.5	7
1995	2.0	5
Major rehabilitation projects[b]		
TIMP 1984	1.04	10
Gal Oya 1987	0.37	26
VIRP 1990	1.09	9
ISMP 1992	0.60	17
MIRP 1994	1.02	10
NIRP 1999	0.88	11

TIMP, Tank Irrigation Modernization Project; VIRP, Village Rehabilitation Project; ISMP, Irrigation System Management Project; MIRP, Major Irrigation Rehabilitation Project; NIRP, National Irrigation Rehabilitation Project.

[a] For the technology level 'New improved varieties, N = 140 kg'.

[b] Years after the names of projects stand for the years when the projects were completed.

pump investments in Sri Lanka are much higher than for public investments in rehabilitation (Kikuchi *et al.*, 2002). But changes in the management of surface water can have a major impact on the groundwater aquifer and overexploitation of groundwater can have negative consequences for both the supply and quality of groundwater. This raises the issue of how best to coordinate the development and management of surface water and groundwater.

Basin level: response to water scarcity and sustainability

As the competition for water increases and river basins become closed for all or part of the year, WP and EE are typically increased by shifting to higher-valued crops, where feasible, and by reallocation of water to industry and domestic uses. Also, water scarcity and the rising value of water can bring forth a response in terms of the development and adoption of new technologies and institutions that can raise water productivity. In economics, these latter changes are explained by the theory of induced innovation (Hayami and Ruttan, 1985). For example, with reference to the Green Revolution, the theory implies that the development of high-yielding, fertilizer-responsive cereal-grain varieties was a response to both rising food-grain and falling fertilizer prices, which made this technology highly profitable. Applying this theory, we see that situations of water shortage and the rising value of water are inducing new techniques, improved management practices and institutional reforms that will raise the productivity of water. The profitability, the feasibility and hence the order of these changes will vary from site to site, depending on local circumstances.

Recent studies of the Gediz basin in Turkey (IWMI and General Directorate of Rural Services, Turkey, 2000), the Chao Phraya basin in Thailand (Molle, Chapter 17, this volume) and the Rio Lerma basin in Mexico (Scott *et al.*, 2001) illustrate the endogenous adjustments that have occurred at both the farm and system levels in response to water shortages.

In the case of the Gediz basin, the adjustments were in response to a prolonged drought from 1989 to 1994. A change was made in the way water was allocated, shifting from a demand- or crop-based system to a supply-based system, with water rationed from the reservoir downward. The result was a significant increase in basin-level irrigation efficiency. To adapt to the dramatically reduced length of the irrigation season, farmers, with the assistance of the government, developed groundwater resources. The shift in cropping pattern over the past decade away from cotton to grapes and orchards is partially explained by the drought, but the entry into the European Customs Union was the overriding factor.

In the Chao Phraya basin, irrigation efficiency has been gradually raised by the use of grating drains, conjunctive use of groundwater, pumped water from ponds and low-lying areas and improved management of dams. Farmers have responded to water shortage and unreliable deliveries in the dry season by sinking tube wells and diversifying crop production and through a spectacular development of inland shrimp farming. This has occurred despite the fact that there are considerable technical constraints and risks in diversification. The centralized water-allocation system has handled the issue of allocation of water to non-agricultural uses relatively well. Basin-level efficiency is high and there appears to be relatively little scope for achieving further productivity gains.

In the Rio Lerma–Chapala basin, water-shortage problems gained prominence with precipitous declines in Lake Chapala (the main source of water for Guadalajara) in the 1980s. IWMI studies have shown the distribution and extent of aquifer depletion (2 m year^{-1}) and growth in agricultural water demand. The Lerma–Chapala Consejo de Cuenca, established in 1993, is the oldest river-basin council in Mexico. It has responsibility for water allocation among users, improving water quality and WUE and conserving the basin ecosystem. However, agricultural, industrial and domestic demand has been rising rapidly, and there is simply not enough water to meet all demand without further overdraft of the aquifer. Water for Lake Chapala and Guadalahara has priority and 240 million m^3 of water formerly used for irrigated agriculture have been reallocated to Lake Chapala. Farmers are beginning to demand that Guadalahara pay for the 240 million m^3.

In summary, in all three basins there has been a response by farmers and irrigation organizations to water shortage that has raised WP and basin-level efficiency and there appears to be relatively little scope for further gains. The non-agricultural demand for water will continue to rise and declining water quality already presents a serious problem. But each of the three basins is at a different stage with respect to basin closure and chronic water shortage. The situation in Mexico is clearly unsustainable. The reduction in irrigated area and, where possible, the shift to high-valued crops on the remaining land can help alleviate the problem.

Allan (1998) has coined the term 'trade in virtual water' to show how international trade can help alleviate water scarcity and increase WP. Mexico provides an interesting example of trade in virtual water (Barker *et al.*, 2000). Over the past 30 years, both fruit and vegetable exports and cereal-grain imports have been increasing rapidly. Figure 2.1 shows that, over the 5 years from 1991 to 1996, the value of fruit and vegetable exports exceeded the value of grain imports by US$1.0–1.5 billion. At the same time, the water saved by the import of cereal grains was about six times the water used for fruit and vegetable production.

Policies and Institutions

There are those who argue that water in large, publicly managed, irrigation systems is being poorly managed and that policy and institutional reforms are needed to create the environment and incentives for saving water and increasing WP. Charges for water or for power for lifting water (if they exist at all) are rarely adequate to cover O&M expenses. As a result, irrigation infrastructure is deteriorating at a rapid rate and overexploitation of groundwater resources is leading to a decline in the water table and in the quality of water.

Others argue that there is much less scope for increasing WP than is commonly believed. Traditional measures of irrigation efficiency are incorrect. Water scarcity, particularly the closing of a basin, creates its own incentive for reforms, leading to changes in

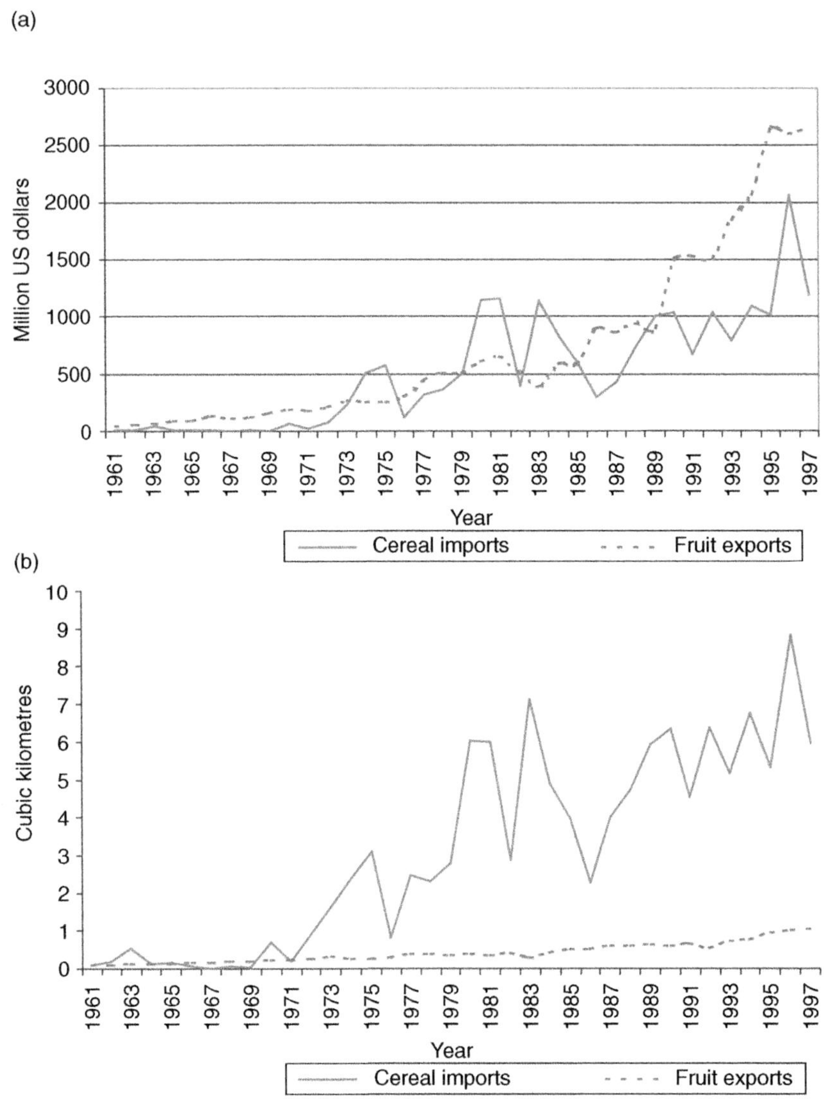

Fig. 2.1. Cereal imports and fruit and vegetable exports in Mexico ((a) US dollars and (b) virtual water, assuming water productivity of 1.2 kg m^{-3} crop evapotranspiration for cereals and 4 kg m^{-3} for fruit and vegetables). Source: Barker *et al.*, 2000.

water-management practices at the farm, system and basin level designed to sustain production. One such example is the spread of pumps and tube wells, largely through private investment, for exploiting groundwater and recycling water from drainage ditches.

There is a strong element of truth on both sides of the argument. As suggested in the previous section, we need much more accurate information on the dynamics of change in water basins over time, noting in particular the changes that occur as water scarcity increases and a basin becomes closed for all or a portion of the year. As competition for water increases, decisions on basin-level allocations among sectors must involve value judgements as to how best to benefit society as a whole. This will include setting priorities in the management of water resources to meet objectives such as ensuring sustainability, meeting food-security needs and providing the poorer segments of society with access to water. These objectives can be incorporated as assumptions or constraints in economic-cum-hydrological optimization models (McKinney *et al.*, 1999).

Faced with growing water shortages, many national policy makers, backed by international experts, have called for improved management of canal irrigation systems. The steps required include: (i) reforms in pricing and charging users for water or water services; (ii) greater participation in the O&M of systems by local user groups; and (iii) the establishment of water rights. In this section, we discuss the first of the two most widely promoted reforms, water-pricing policy and irrigation-management transfer (IMT), and one less publicized area, management for conjunctive use, which appears to offer potential for gains in economic efficiency, equity and WP. We should emphasize that the appropriate policy and institutional reforms will vary depending on the biophysical and socio-economic environment at a given site.

Water-pricing policy

In developed as well as developing countries, there is disagreement regarding the appropriate means by which to price water and the appropriate level of water charges. The pricing of water may involve different objectives, such as cost recovery (who has benefited from the investment in irrigation and who should pay), financing the irrigation agency or reducing wastage of water. Politics also enters heavily into water-pricing decisions. Moreover, many countries lack the tradition, experience and appropriate institutions for pricing irrigation water.

The World Bank has recently undertaken a comprehensive study, 'Guidelines for Pricing Irrigation Water Based on Efficiency, Implementation, and Equity Concerns.' As a part of that study, Johansson (2000) has conducted an exhaustive literature survey on pricing irrigation water. More concise treatment of the issues can be found in Tsur and Dinar (1997) and in Perry (2001). The authors emphasize the fact that water (particularly water used in irrigation) is a complicated natural resource, a complicated economic resource and a complicated political resource. Moreover, while water supplied is a proper measure of service in domestic and industrial uses, water consumed is the appropriate measure in irrigation, and this is particularly difficult to measure.

Tsur and Dinar (1997) discuss several different pricing methods for irrigation water and their implementation costs. These include pricing based on area irrigated, volumetric pricing according to the water used or consumed, output or input pricing, fixed- and variable-rate pricing and water markets. The necessary and sufficient conditions for markets to operate, especially defined and enforceable water rights, are, in most cases, not yet in place. Variable-rate pricing is often suggested in charging for electricity for pumps.

Bos and Wolters (1990) investigated irrigation agencies representing 12.2 million ha of irrigated farms worldwide. They found that water authorities charged on a per-unit area basis in more than 60% of the cases, on a volumetric basis in about 25% of the cases and a combination of area and volumetric methods in 15% of the cases.

Water-pricing methods are most pronounced through their effect on cropping pattern – more so than their effect on water demand for a given crop (Tsur and Dinar, 1997). The various methods differ in terms of amount and type of information and the administrative costs needed in their implementation. The most economically efficient method will depend on physical conditions, such as conveyance structures, water facilities and institutions. If the objective is allocation and not cost recovery, rationing (i.e. assigning water to specific uses) represents an alternative mechanism for coping with water shortages where demand exceeds supply (Perry, 2001).

An example of volumetric-cum-area pricing is found in the Zhanghe irrigation system (ZIS) in Hubei, China (Dong *et al.*, 2001). The province determines the price for water for different uses and water is rationed among sectors when supplies are short. The water-user groups or villages pay the water fee to ZIS on a volumetric basis. The fee for the total volume paid by the group is then divided by the area, and individual farmers are charged according to their irrigated area. Even though farmers pay an area fee, they are well aware that, if they use less water as a group, their fees will be reduced. The savings in water use at the farm level through improved water-management practices, as well as through higher crop yields, have led to an increase over time in the productivity of water for irrigation (Hong *et al.*, 2001). There is also an incentive to save water at the system level. Over the past three decades, water has been diverted to higher-valued, non-agricultural uses, greatly increasing the productivity of ZIS water resources. However, the decrease in water seepage and runoff resulting from water-saving practices (including the lining of canals) may have reduced the water available in downstream tanks within the Zhanghe Irrigation District but outside ZIS, and the negative impact of this is not known.

Participatory irrigation management and irrigation management transfer

In the area of institutional reform, the devolution of management and financial responsibility from irrigation-system managers to local user groups has gained prominence. The popular terms for this are participatory irrigation management (PIM) and IMT. These terms are defined as follows (Groenfeldt and Svendsen, 2000):

- PIM usually refers to the level, mode and intensity of user-group participation that would increase farmer responsibility in the management process.
- IMT is a more specialized term that refers to the process of shifting basic irrigation-management functions from a public agency or state government to a local or private-sector entity.

The interest in transfer of responsibility to user groups rests, in large part, on the desire of many governments to reduce expenditures on irrigation. Among proponents, it is also argued that handing responsibility to local user groups will result in better O&M and increased productivity. PIM/IMT has become one of the cornerstones of the World Bank water-management policy (Groenfeldt and Svendsen, 2000). Recent experience in PIM and IMT seems to suggest that there has been considerably more success in transferring management responsibilities in more advanced countries, such as Turkey and Mexico, than in the developing countries of Asia (Samad, 2001). Where implementation has been successful, government expenditures and the number of agency staff have declined and maintenance has, in some cases, improved, but there is little evidence yet that PIM/IMT has led to an increase in the productivity of irrigation water.

While, under IMT, government responsibility for water management in the lowest level of the irrigation system is being reduced, at the same time water scarcity requires increased government involvement at the highest level of management (Perry, 1999). For example, China has recently centralized control over water diversions from the Yellow River because upstream users were taking so much water that the river often ran dry before reaching the sea. This centralization seems to have increased stream flows in the river. Important areas of centralized management at the basin and sector levels include water allocation among sectors, flood control, drought planning, water-quality regulation and enforcement and groundwater depletion.

Conjunctive use of surface water and groundwater

Historically, the development of the technology of surface-water irrigation preceded that of tube wells, based on compact diesel and electrical power. In fact, the introduction of tube wells in the Indus basin and perhaps in the North China Plain was motivated by concern over the waterlogging and salinization that occurred when canal irrigation caused the water table to rise (O'Mara, 1988). Public drainage wells were installed to lower water tables and reduce waterlogging. A boom followed in tube-well investments for irrigation by individual farmers in south Asia and by communes (and, more recently, by private farms) in north China. Because of the greater convenience and reliability of groundwater, many farmers within surface-irrigation command areas have dug wells or tube wells.

The rate of increase in new areas irrigated by surface water has levelled off. But the irrigated area served by the ever-cheaper tube-well technology has continued to expand to the point where, in India, over half of the area irrigated is from groundwater. The massive investment in tube wells has completely transformed the use of water resources in these regions and has raised problems of resource management that are beyond the grasp of existing irrigation bureaucracies. The overexploitation of groundwater, particularly in the semi-arid areas, is leading to declines in both quantity and quality of water, affecting not only agriculture but also domestic supplies and human health. Often, in many large-scale irrigation systems, tail-end farmers have to supplement surface-water supplies with lower-quality drain water or shallow groundwater (Murray-Rust and Vander Velde, 1994).

One of the greatest potentials for increasing WP lies in the management of surface-water and groundwater resources for conjunctive use, provided this leads to better distribution of water. For example, loss of yield due to salinity could be greatly reduced with improved conjunctive management of surface-water and groundwater resources, especially by better distribution of canal water to maintain optimum levels of water table and salt balances, even in the tail reaches of canal commands (Hussain *et al.*, Chapter 16, this volume). This requires close monitoring of any adverse effects on soil and water quality, as has occurred in irrigation management in the People's Victory Irrigation Canal in the Yellow River basin of China. It has been suggested (M. Wopereis, personal communication, 1998) that farmers in the Senegal River valley, an area with severe soil salinization (e.g. Raes *et al.*, 1996), be equipped to monitor salinity levels themselves. Cheap field conductivity meters can be used for this purpose and such equipment should be within the financial reach of farmers' cooperatives.

Summary and Conclusions

Initially, we addressed the confusion in the definitions of IE, WUE and WP. IE is measured by the ratio of water consumed to water supplied, whereas WP is a ratio of crop output to water either diverted or consumed, measured in either physical or economic terms or some combination of the two. Then we discussed the relationship between WP and EE. Just as water saving does not necessarily result in higher WP, so also higher WP does not necessarily result in higher EE (e.g. the case of SRI).

Measures to increase crop yield for a given ET translate into WP gains at system and basin levels (e.g. through varietal improvements). However, the management of water to reduce water-diversion requirements is riddled with off-site effects or externalities (e.g. the case of Gal Oya). Thus, whether water-management practices or technologies designed to increase WP and EE at farm level result in higher WP and EE at system or basin level needs to be determined. The basin is a hydrological, as opposed to an administrative, unit. It is only at this level that we can capture and include in our analysis the off-site effects (or, in economic jargon, internalize the externalities).

The growing scarcity and rising value of water in a basin induces both farmers and irrigation organizations to seek various ways to increase WP, EE and net returns (e.g. the basin cases in Turkey, Thailand and Mexico). Recycling or reuse of water, particularly through the exploitation of groundwater, is prominent among the practices adopted to increase WP. Greater attention needs to be focused on managing water for conjunctive use.

We need a better understanding of biophysical and socio-economic changes in basins over time and improved measures of basin-level efficiencies before we can determine, in a given situation, the potential for increasing WP through policy and institutional reforms and which reforms are most suitable. Finally, as basins become closed, measures to increase water productivity and exploit groundwater resources are leading to a serious decline in water quality and other problems of environmental degradation. Decisions on basin-level allocations among sectors must involve value judgements as to how best to benefit society as a whole. This will include setting priorities in the management of water resources to meet objectives such as ensuring sustainability, meeting food-security needs and providing the poorer segments of society with access to water.

References

Allan, T. (1998) Moving water to satisfy uneven global needs. 'Trading' water as an alternative to engineering it. *ICID Journal* 47(2), 1–8.

Alston, J.M., Norton, G.W. and Pardey, P.G. (1995) *Science Under Scarcity: Principles and Practice for Agricultural Research Evaluation and Priority Setting*. Cornell University Press, Ithaca, New York.

Aluwihare, P.B. and Kikuchi, M. (1991) *Irrigation Investment Trends in Sri Lanka: New Construction and Beyond*. International Irrigation Management Institute, Colombo, Sri Lanka.

Amarasinghe, U.A., Sakthivadivel, R. and Murray-Rust, H. (1998) *Impact Assessment of Rehabilitation Intervention in the Gal Oya Left Bank*. Research Report 18, International Irrigation Management Institute, Colombo, Sri Lanka.

Barker, R., Scott, C.A., de Fraiture, C. and Amarasinghe, U. (2000) Global water shortages and the challenge facing Mexico. *Water Resources Development* 16(4), 525–542.

Bos, M.G. and Wolters, W. (1990) Water charges and irrigation efficiencies. *Irrigation and Drainage Systems* 4, 267–278.

Dinar, A. (1993) Economic factors and opportunities as determinants of water use efficiency in agriculture. *Irrigation Science* 14, 47–52.

Dixon, J.A. and Fallon, L.A. (1989) The concept of sustainability: origins, extensions, and usefulness for policy. *Society and Natural Resources* 2, 73–84.

Dong, B., Loeve, R., Li, Y.H., Chen, C.D., Deng, L. and Molden, D. (2001) Water productivity in Zhanghe irrigation systems: issues of scale. In: Barker R., Loeve, R., Li, Y.H. and Tuong, T.P. (eds) *Water Saving Irrigation for Rice*. International Water Management Institute, Colombo, Sri Lanka.

Evenson, R.E., Pray, C.E. with the assistance of Ahmed, A. *et al.* (1991) *Research and Productivity in Asian Agriculture*. Cornell University Press, Ithaca, New York.

Groenfeldt, D. and Svendsen, M. (eds) (2000) *Case Studies in Participatory Irrigation Management*. WBI Learning Resources Series, World Bank Institute, Washington, DC, USA, xii, 157pp.

Hayami, Y. and Ruttan, V.W. (1985) *Agricultural Development: an International Perspective*. Johns Hopkins University Press, Baltimore, Maryland.

Hong, L., Li, Y.H., Deng, L., Chen, C.D., Dawe, D., Loeve, R. and Barker, R. (2001) Analysis of changes in water allocations and crop production in the Zhanghe Irrigation System and District, 1966 to 1998. In: Barker, R., Loeve, R., Li, Y.H. and Tuong, T.P. (eds) *Water Saving Irrigation for Rice*. International Water Management Institute, Colombo, Sri Lanka.

IWMI and General Directorate of Rural Services, Turkey (2000) *Irrigation in the Basin Context: the Gediz Study*. International Water Management Institute, Colombo, Sri Lanka, xvii, 124 pp.

Johansson, R.J. (2000) *Pricing Irrigation Water: a Literature Survey*. Working Paper 2449, World Bank, Washington, DC.

Keller, A. and Keller, J. (1995) *Effective Efficiency: a Water Use Concept for Allocating Fresh Water Resources*. Water Resources and Irrigation Division Discussion Paper 22, Winrock International, Arlington, Virginia.

Keller, A. and Keller, J. (1996) *Integrated Water Resources Systems: Theory and Policy Implications*. Research Report 3, International Irrigation Management Institute, Colombo, Sri Lanka.

Kikuchi, M., Barker, R., Weligamage, P. and Samad, M. (2002) *The Irrigation Sector in Sri Lanka: Recent Investment Trends and the Development Path Ahead*. Research Report 62, International Water Management Institute, Colombo, Sri Lanka.

Lynam, J.K. and Herdt, R.W. (1999) Sense and sustainability: sustainability as an objective in international agricultural research. *Agricultural Economics* 3, 381–398.

McKinney, D.C., Cai, X., Rosegrant, M.W., Ringler, C. and Scott, C.A. (1999) *Modeling Water Resources Management at the Basin Level: Review and Future Directions*. SWIM Paper 6, International Water Management Institute, Colombo, Sri Lanka.

Molden, D.J. and Sakthivadivel, R. (1999) Water accounting to assess use and productivity of water. *International Journal of Water Resources Development*, Special Double Issue 15 (1 & 2), 55–71.

Molden, D., Sakthivadivel, R. and Habib, Z. (2001) *Basin-level Use and Productivity of Water: Examples from South Asia*. Research Report 49, International Water Management Institute, Colombo, Sri Lanka.

Moser, C. and Barrett, C.B. (2003) The system of rice intensification in practice: explaining low farmer adoption and high disadoption in Madagascar. In: Bouman, B., Hengsdijk, H., Hardy, B., Bindraban, P., Tuong, T.P., Lafitte, R. and Ladha, J.K. (eds) *Water-wise Rice Production*. IRRI, Los Baños, Philippines.

Murray-Rust, D.H. and Vander Velde, E.J. (1994) Changes in hydraulic performance and comparative costs of lining and desilting of secondary canals in Punjab, Pakistan. *Irrigation and Drainage Systems* 8(3), 137–158.

Murray-Rust, H., Sakthivadivel, R. and Amarasinghe, U.A. (1999) Impact assessment of organizing farmers in the Gal Oya Left Bank Irrigation Scheme, Sri Lanka. *Water Resources Development* 15, 197–217.

O'Mara, G.T. (ed.) (1988) *Efficiency in Irrigation: the Conjunctive Use of Surface and Groundwater Resources*. World Bank, Washington, DC, pp. 128–138.

Perry, C.J. (1999) The IWMI water resources paradigm: definitions and implications. *Agricultural Water Management* 40(1), 45–50.

Perry, C.J. (2001) Water at any price? Issues and options in charging for irrigation water. *Irrigation and Drainage* 50, 1–7.

Project Completion Report (1985) *Final Evaluation Report of Gal Oya*. Planning Resource Corporation (PRC), Englewood, Colorado, pp. 2–17.

Raes, D., Ceuppens, J., Kane, A. and Decke, J. (1996) Water requirements of rice schemes in the Senegal Delta. In: Camp, C.R., Sadler, E.J. and Yoder, R.E. (eds) Evapotranspiration and irrigation scheduling. *Proceedings of the International Conference USA* pp. 87–92.

Richards, R.A., López-Castañeda, C. and Gomez-Macpherson, G.A.G. (1993) Improving the efficiency of water use by plant breeding and molecular biology. *Irrigation Science* 14, 93–104.

Samad, M. (2001) Impact of irrigation management transfer on the performance of irrigation systems: a review of selected Asian experiences. In: *Water Policy Reform: Lessons from Asia and Australia. International Workshop held on 8–9 June 2001, Bangkok, Thailand*. ACIAR Proceedings No. 106.

Scott, C.A., Silva-Ochoa, P., Cruz, V.F. and Wester, P. (2001) Competition for water in the Lerma–Chapala Basin. In: Hansen, A. and van Afferden, M. (eds) *The Lerma–Chapala Watershed: Evaluation and Management*. Kluwer Academic/Plenum Publishers, London.

Tsur, Y. and Dinar, A. (1997) The relative efficiency and implementation costs of alternative methods of pricing irrigation water. *World Bank Economic Review* 11(2), 243–262.

Uphoff, N. (1992) *Learning from Gal Oya*. Cornell University Press, Ithaca, New York.

Wallace, J.S. and Batchelor, C.H. (1997) Managing water resources for crop production. *Philosophical Transactions of the Royal Society London B* 352, 937–947.

Wichelns, D. (1999) Economic efficiency and irrigation water policy with an example from Egypt. *Water Resources Development* 15(4), 543–560.

3 The Concept of Efficiency in Water-resources Management and Policy

David Seckler, David Molden and R. Sakthivadivel[1]
International Water Management Institute, Colombo, Sri Lanka

All science depends on its concepts. These are the ideas which receive names. They determine the questions one asks, and the answers one gets. They are more fundamental than the theories which are stated in terms of them.

(Sir G. Thompson)

Let not even a small quantity of water that comes from the rain go to the sea without being made useful to man.

(King Parakramabahu of Sri Lanka (AD 1153–1186))

Introduction

Many areas of the world are experiencing increasingly severe water scarcity. Recent studies by the International Water Management Institute (IWMI) indicate that one-third of the population of developing countries lives in regions that have absolute water scarcity, in the sense that they do not have sufficient water resources to meet their agricultural, domestic, industrial and environmental needs in the year 2025 (Seckler *et al.*, 1998a,b). An additional 500 million people live in regions of severe economic scarcity; they have a sufficient amount of potential water resources to meet their 2025 needs, but will have to more than double the present utilization of these resources, through large, expensive and possibly environmentally destructive development projects to achieve reasonable amounts of water consumption.

One of the ways of alleviating water scarcity is by increasing the efficiency of water use. Many different ways of increasing water-use efficiency are proposed, ranging from water-saver flushing toilets and low-flow drip-irrigation systems to pricing water to encourage demand reduction and adaptation of water-saving technologies. Indeed, it is sometimes contended that the current efficiency of water use is so low, especially in irrigation, that most, if not all, of future water needs could be met by increased efficiency alone, without development of additional water supplies.

While the potential for saving water through increased efficiency is substantial, it is not as large as might be thought. The reason is that the most commonly used concepts of water-use efficiency systematically underestimate the true efficiency of existing systems by a very large amount. One of the cardinal

[1] The authors are grateful to Vernon W. Ruttan for first suggesting this review.

features of water use is that, when water is used, not all of it is 'used up'. Most of the water remains in the hydrological system, where it is available for reuse or recycling. As water is recycled through the hydrological system, the efficiency of use increases. Thus, while every part of the system may be at low levels of water-use efficiency, the system as a whole can be at high levels of efficiency.

This 'water-efficiency paradox', as it may be called, constitutes the core of this discussion. It is shown how the older, 'classical' concept of efficiency ignored recycling and thus underestimated efficiency. The newer 'neoclassical' concepts represent attempts to integrate water recycling into the concept of water-use efficiency. These conceptual changes have important implications for water-resources management and policy. The classical concept of efficiency often leads to erroneous policies and management systems; the neoclassical concepts point in more effective directions for increasing water-use efficiency. As Willardson *et al.* (1994) rightly observe in the field of irrigation:

> Unless the ideas now associated with irrigation efficiency terms are modified, it will be extremely difficult to properly manage the shrinking supply of freshwater due to the misconceptions and misunderstandings of irrigation efficiency by the engineering, political, and news communities. Yet, much current irrigation literature contains many recommendations to increase irrigation efficiencies in order to create more available water. The economic damage and waste of limited resource management funds caused by such articles and misconceptions are very large.

Indeed, as explained below, these authors and others recommend purging the 'E' word from the literature on irrigation altogether! While we have a good deal of sympathy with this recommendation, we show that the same concept as 'efficiency' is absolutely necessary and, to paraphrase, a rose by any other name remains a rose all the same.

In this chapter, we trace the evolution of the concept of efficiency mainly in the field of irrigation. Irrigation is chosen as the focus because: (i) over 70% of the world's developed water supplies are diverted for irrigation and, therefore, it is especially important to get these concepts right in this field; and (ii) so far as we know, these concepts have been studied more intensely in the field of irrigation than in other areas of water resources. There is a vast and complex literature on the subject of irrigation efficiency. In our opinion, the best comprehensive account of this subject is that of Jensen (1980). In this chapter, we ignore most of the complications and refinement of the subject in order to focus on what we consider to be the central issues pertaining to water-resources policy and management.

The Classical Concept of Irrigation Efficiency

Before the concept of irrigation efficiency was invented, engineers used the irrigation duty to design irrigation systems. The duty is the amount of water that needs to be diverted from a source and applied to the root zone of crops 'to bring the crop to maturity'. (Willardson *et al.*, 1994). The duty is expressed as, say, 0.5 m of depth of water applied per hectare per crop season, or so much per 10-day interval, etc. The irrigation duty is essential for designing the physical structure of water storage and conveyance systems. But there are two problems with irrigation duty. First, it is a rule of thumb, without a clear rationale. Secondly, because of this, it does not indicate whether one is irrigating more or less well. These problems gave rise to the concept of irrigation efficiency.

A major breakthrough was to refine the objective of irrigation in terms of meeting the actual evapotranspiration requirements of crops, or Eta. Once Eta was defined after generations of research, the door opened to the concept of efficiency. Israelsen (1932) is generally credited with rationalizing the duty of irrigation (Willardson *et al.*, 1994) by developing what Keller and Keller (1995) aptly call the 'classical' concept of irrigation efficiency. Israelsen's definition is a direct application of the basic concept of engineering efficiency to the field of irrigation: 'an output divided by an input, both of the same character' (Willardson *et al.*, 1994). Stated simply in contemporary terms, the primary output of irrigation is the amount of water needed to satisfy the crop water requirements (although there are many other uses of irrigation water (noted below)). The crop water requirements are defined here as the Eta requirements of the crop, minus effective precipitation (Pe), (the amount of water that enters the root zone of the crop). This is called net evapotranspiration (NET) (Keller *et al.*, 1996):

$$\text{NET} = \text{Eta} - \text{Pe} \quad (3.1)$$

The input is defined as the amount of water withdrawn or diverted (DIV) from a specific surface-water or groundwater source to achieve NET. Thus classical irrigation efficiency (CE) can be defined as:

$$\text{CE} = \text{NET}/\text{DIV} \quad (3.1a)$$

For example, at the field level, in terms of metres of depth per unit of area:

Eta = 1.0
Pe = 0.2
NET = 0.8
DIV = 2.0
CE = 0.8/2.0 = 40%

Thus, while the duty may be 2.0 m of water, this implies only 40% irrigation efficiency: 60% of the water diverted is not necessary to meet the crop requirement. Put another way, the degree of inefficiency of irrigation in this case is the complement of CE: (1 − CE = 1 − 0.4 =) 60%.

It is commonly said that the water that is not used to satisfy NET is 'wasted' or 'lost'. While this is true from the point of view of meeting the direct objective of satisfying NET, these phrases are a source of endless confusion in the field of water resources. As explained in detail below, only some of this water is truly lost from the hydrological system. Most of it is captured and recycled somewhere else in the system.

In irrigation, there are essentially three sources of classical inefficiency, or putative water 'losses':

- Evaporation from the surfaces of land, water and plants that do not contribute to crop Eta. This includes 'non-beneficial evapotranspiration' by weeds, phreatophytes and other non-beneficial grasses, trees and bushes. However, these uses of water may be of great value in terms of objectives other than irrigation.
- Drainage losses are surface and subsurface losses in the process of delivering the water from the point of diversion to the root zone of crops: leakage from the conveyance system, deep percolation below the root zone of crops and surface drainage from the fields.
- Spillage losses due to mismatches between water supply and demand. When there is more water supplied than demand – for example, irrigation water flowing in canals during heavy rain – the surplus water is spilled into drains.

As discussed further below, the only real water losses to the hydrological system, however, are those from evaporation and flows to 'sinks', such as saline seas. Drainage, spillage and other water flows are losses only in so far as they flow to sinks.

Three important points should be made about the classical concept of irrigation efficiency before proceeding:

1. CE is defined at different scales, in terms of differences between the point of water diversion and the ultimate destination of the water in the root zone of crops (or Eta) (Bos and Nugteren, 1974; Jensen, 1980).

- Application efficiency (Ae) is the ratio of water delivered to the root zone to water delivered to the field.
- Conveyance efficiency (Ec) is the ratio of water delivered to the field to water delivered into the canal from the source.
- Project efficiency (PE) is the overall efficiency of the system, which is also equal to classical efficiency (application efficiency × conveyance efficiency)

For example:

Ae = 0.5
Ec = 0.8
PE = Ae × Ec = 0.40

Thus, CE decreases as the scale of the system increases.

2. As intimated in the discussion of 'non-beneficial evapotranspiration', the output is

not as easy to define as may first appear. Even in the apparently simple case of satisfying Eta, the question of the amount of Eta to be satisfied arises. In cases of water shortage, for example, it may be optimal to practise deficit irrigation, providing less than full Eta to a particular crop area and suffering reduced yields, so that water can be supplied to a greater area (Perry and Narayanamurthy, 1998). Also, irrigation water has multiple uses; in addition to Eta, it is used for moistening land for cultivation and for weed control in paddy irrigation. Because of these optimization and multiple-use complications, some proponents of the concept of CE recommend calling it irrigation 'sagacity' instead (Burt *et al.*, 1997).
3. As noted before, CE ignores the possibility of recycling water 'losses' within the hydrological system. This is the subject of the second part of this discussion.

The concept of CE is used in two important ways. First, it is used as a tool in the design of irrigation and other water-delivery systems. In this example, NET divided by CE (0.8/0.4) is equal to the amount of water (2.0 m) that has to be diverted from the source to satisfy the objective at the destination. Thus, in designing water-delivery systems, engineers explicitly assume a value for CE to size the conveyance system. Secondly, CE is used as a criterion of engineering efficiency. It is generally assumed that the higher the CE the better.

The same concept of engineering efficiency is used in other water sectors. For example, if the objective is to deliver 1 m^3 of water day^{-1} to a household, but 20% of the water is lost in transit because of leakages in the delivery system, the efficiency is 80%. Or, inside the household, if the tap is left on while brushing one's teeth, the (tap to tooth) efficiency may only be 10% because only 10% of the water is beneficially used to meet the objective in this application. The overall efficiency of the domestic water system in this case is only (0.8 × 0.1 =) 8.0%. However, as in irrigation, most of the 92% of the water that represents the inefficiency of household use is not lost to the system as a whole but is captured and recycled.

Influence of classical efficiency

Notwithstanding the problems of the concept of CE it has had enormous influence both within the irrigation profession and in the wider fields of irrigation and water-resources policy and management.

Generations of irrigation engineers have devoted their lives to improving the efficiency of irrigation. Below are some characteristic CEs of various irrigation systems at the farm level (Merriam, 1980; Wolters and Bos, 1990).

1. Conventional gravity = 30–50% (the lower range is mainly in paddy irrigation, to flooded fields).
2. Level basin = 40–70% (the high value is achieved with laser-beam levelling).
3. Sprinkler = 60–75%.
4. Drip = 80–90%.

Since conventional-gravity systems probably comprise 80% or more of the total irrigation systems in the world, shifting from gravity to more efficient forms of irrigation could, theoretically, nearly double the average CE.

This line of thought leads to important effects of Israelsen's (1932) concept outside the irrigation profession. As the concept of irrigation efficiency spread into the realm of water-resources planning, management and policy analysis, it became a commonly accepted fact that irrigation is so inefficient that enormous amounts of water being 'lost' in irrigation could be 'saved' through improved technology and management, and these savings could be used to meet most of the future demands for water by all of the sectors. However, there is a fundamental error in this interpretation of CE, which has led to major mistakes in thinking about irrigation policy and management. Various attempts to solve this error led to the neoclassical revolution in the concept of irrigation efficiency.

The Neoclassical Concept of Irrigation Efficiency

The neoclassical concept of irrigation efficiency developed as a consequence of the evolution of interest in irrigation from the point of view of water-delivery systems to the broader perspective of irrigation management and policy within the context of water resources as a whole, in the entire river basin. It soon became clear that from this perspective the concept of CE was erroneous and misleading. The reason is that the water 'losses' of CE are not necessarily 'real' water losses to the system as a whole – many of these losses are only paper losses – because they are captured and recycled elsewhere in the system. While this problem has probably been in the back of people's heads for a long time (as shown below, it is intimately related to King Parakramabahu's declaration used as an epigraph for this chapter), Wright (1964), Bagley (1965) and Jensen (1967) are the first published references we know that discuss this problem clearly and explicitly. The fact of water recycling set up a 'problem situation', as the philosopher Karl R. Popper (1962) describes it, which evolved through a process of articulation and refinement (or, in Popper's classic phrase, 'conjectures and refutations') to what we call the neoclassical concept of irrigation efficiency.

Net efficiency

This problem was first formally addressed (so far as we know) by Jensen (1977), who proposed revising CE to 'net efficiency' (NE):

$$NE = CE + Er\ (1 - CE) \quad (3.2)$$

where:

CE = classical efficiency (Equation 3.1);
1 − CE = classical inefficiency, i.e. the percentage of the diverted water that is not used to meet the Eta requirements of crops;
Er = the percentage of 1 − CE that is potentially available for recovery, reuse or recycling somewhere in the hydrological system.

Thus, as in the discussion of CE, if 40% of the diversion leaves the system in the form of evapotranspiration and 70% of the remainder is potentially available for reuse, then:

$$NE = 0.40 + 0.7(0.6) = 0.40 + 0.42 = 0.82 \quad (3.2a)$$

Thus, with the same basic parameters, NE is more than twice as high as CE!

Jensen's NE clearly shows the trade-off possibilities between CE in the first term of the equation, and what may be called the 'recycling efficiency' in the second term. For example, assume that it is decided to shift from a surface-irrigation system with a CE of 40% to a sprinkler-irrigation system with a CE of 70% then, following Equation 3.2a, the NE of the sprinkler system is:

$$NE = 0.70 + 0.70(0.30) = 91\% \quad (3.2b)$$

In the shift to sprinkler irrigation, CE increases by (0.70/0.40) − 1 = 75%, but NE increases by only (0.91/0.82) − 1 = 11%. While it might pay to invest in sprinkler irrigation to save water in the first case, it might not pay in the second case – even though the basic water situation is the same in the two cases.

By 1980, with the publication of the state-of-the-art work, *Design and Operation of Farm Irrigation Systems* (Jensen, 1980), NE (or 'effective irrigation efficiency', as it was also called) was the recommended practice. In this volume, Burman *et al.* (1980, p. 220) note that: 'Effective irrigation efficiency ... of a farm, project, or river basin is necessary to estimate or evaluate the *net depletion of water within a river basin* or groundwater system' (writers' italics).

As this discussion proceeds, it will become clear how prescient the statement in italics turned out to be. We shall continue calling Jensen's formulation NE reserving effective efficiency for a later formulation of efficiency within the neoclassical framework discussed in the next section.

A particularly interesting consequence of these neoclassical concepts may be mentioned. It was noted before that CE decreases as the scale or boundary conditions of the system increase, because of increasing water losses. But, in NE or other neoclassical formulations, the opposite is the case: as the scale increases the efficiency generally increases, because of increased water recycling. For example, as discussed further below, studies of the Nile irrigation system in Egypt show that the average CE of irrigation is about 50% but a series of estimations of the neoclassical efficiency of irrigation in the system as a whole has resulted in the latest estimate of 87% (Abu-Zeid and Seckler, 1992; Keller, 1992; Molden *et al.*, 1998).

Effective efficiency

Keller and Keller (1995) developed the concept of effective efficiency (EE), as they called it (see also Keller *et al.*, 1996):[2]

$$EE = NET / I - O(R) \qquad (3.3)$$

where, with the same illustrative quantities as in Equation 3.2:

I = inflows of water from the point of diversion = DIV (= 2.0 m)
NET = 0.8 m
Enb = non-beneficial evaporation = 0.1 m
O = outflows of water from the application = I − (NET + Enb) = 2.0 − 0.9 = 1.1 m
R = the percentage of reusable outflow = 70%
EE = 0.8/{2.0 − 1.1(0.7)} = 0.8/(2.0 − 0.77) = 0.8/1.23 = 65%

The Kellers also incorporate a highly ingenious means of employing pollution (mainly salinity) effects in EE. In brief, they subtract from the outflow the amount of water it would require to dilute to an acceptable level any pollution picked up in the use of the water. This concept pushes the concept of purely physical water efficiency about as far as it is possible to go. Clearly the (negative) value of pollution in the outflow depends on where and how it is reused – for example, rice is more tolerant of salinity than most other crops. But this is a generic problem in the concept of physical efficiency, as noted below, and, within the confines of this concept, it is a major contribution to the theory.

While the NE and EE formulations naturally yield somewhat different values, their substance is clearly the same.

Fractions

A third development in the concept of efficiency is the introduction of 'fractions' to replace concepts of efficiency. As noted before, Willardson *et al.* (1994) extended their critique of the misapplications of (classical) efficiency, quoted above, to the point where they advocated eliminating the word and the concept of efficiency altogether. Instead, they proposed using various fractions in water-resources analysis – especially the consumed fraction (CF) – the ratio of evaporation to the diversion in any given process, such as irrigation.

Since the fractions approach is not, by definition, an efficiency concept, only a few observations will be made about it here. The CF is meant to be used in the context of the water balance of the hydrological system, as discussed in the section on 'Basin Efficiency: the Rate of Beneficial Utilization of Water Resources,' and does not imply a judgement as to whether the water is beneficially consumed or not. For example, in irrigation, the water consumed by both evapotranspiration and Enb is included in CF. Thus, a large CF is no better or worse than a small CF. For this reason, the CF must be considered along with the value of the components of CF, to determine a desirable course of action. This problem is addressed by the use of the term process fraction – i.e. the fraction used by humans for municipal, industrial and agricultural purposes. While the process fraction separates the intended uses from the CF, it does not account for other beneficial uses, such as use by trees, forests and wetlands. CF includes process fraction, non-process fraction and non-beneficial fraction. The excellent discussion of specific water problems in the text of Willardson *et al.* (1994)

[2] This is the same as the Er term used in net efficiency. While the R term was not explicitly used by Keller and Keller (1995) in the original equation for EE, it is clearly implied in the text.

necessarily relies on an implicit evaluation of the CF. The fractions approach has been used in Perry (1996) and Molden (1997) in the context of water productivity, as discussed in the section on 'Basin Efficiency: the Rate of Beneficial Utilization of Water Resources.'

Concluding Observations on the Classical and Neoclassical Concepts of Efficiency

Four important observations should be made before closing the discussion of classical and neoclassical efficiency up to this point.

First, the neoclassical formulations of efficiency are clearly superior in that they include CE only as special cases, where in NE or EE, respectively, Er or R is equal to zero (or is negative). Some of the most important examples of these special cases are as follows:

- Where irrigation is in saline areas and the outflows are too saline to be recycled.
- Where irrigation, or other uses of water, occur next to saline seas, where excess outflows are discharged directly into the sea.
- Where severe mismatches between water supply and demand occur in terms of specific times and places. While the outflows are still in the system, they may be at the wrong place at the wrong time.
- Where, especially in desert areas, outflows go to shallow lakes, where the water is evaporated with little, if any, benefit.

In all of these cases, high CE is called for; but the neoclassical formulations cover these cases as well as all the other cases where the outflows are beneficially recycled.

Secondly, the equation used for CE has an important role to play in the design and management of water-delivery systems, while the neoclassical equations are irrelevant for this purpose. It is best not to use the word 'efficiency' to describe the classical equation, but rather to use another term, such as the 'delivery ratio', as Bos (1997) recommends.

Thirdly, in all the definitions of efficiency up to this point, precipitation only enters the analysis as effective precipitation (Pe). The difference between total precipitation (P) and Pe (P − Pe) – the amount of 'ineffective precipitation', as it were – is lost; it simply vanishes from the system, much like the water 'losses' in CE. This is unacceptable in terms of the water balance of the hydrological system as a whole. Also, as Falkenmark *et al.* (1989) observe, it is important not to neglect 'green water' in concentrating on 'blue water' (diversions). While irrigationists do consider green water, in the form of Pe, in the formulation of NET, it is true that P − Pe is ignored. But irrigationists could reply that hydrologists (excepting Falkenmark *et al.*, 1989) are even worse, because to them 'effective precipitation' is only runoff, and the green water – which does not enter river drainages, but does support most plant life – is treated as a loss! This problem is addressed in the next section.

Fourthly, both the classical and neoclassical formulations of efficiency attempt to stay within the domain of purely physical flows of water, avoiding assignments of values to the flows and quantities of water. But this is an ultimately futile and misleading attempt. Whenever words like efficiency are used, value judgements are necessarily part of the underlying concept and it is best to use them explicitly. At the very least, a distinction must be made between the beneficial and non-beneficial (zero or negatively valued) aspects of water flows. In classical efficiency, this is not a major problem because it is clear that NET is beneficial evaporation. But it becomes a serious problem in the neoclassical formulations, where the outflows can have zero or negative effects – for example, in terms of waterlogging and salinity. Thus it is not just a matter of distinguishing between the amounts of depletion and non-depletion of water in the neoclassical formulations, but a matter of the values of the depletions. For this reason, it would be better to define the Er and O(R) terms of NE and EE as Eb and Ob, where 'b' indicates the amount of beneficial use. This subject leads directly to the concept of the beneficial utilization of water resources discussed in the next section.

Basin Efficiency: the Rate of Beneficial Utilization of Water Resources

The discussion in this part of the chapter remains solidly in the neoclassical tradition. Both the classical and neoclassical concepts of efficiency followed a 'bottom-up' approach, as it were, from the perspective, first, of the individual farmer, through the project level, to intimations of the basin level of analysis. Here, we make the concept of basin efficiency clear and explicit within the overall concept of the beneficial utilization of water resources within river basins. Thus this discussion rather abruptly switches perspective by following a 'top-down' approach. It begins at the level of the river basin as a whole and then, once that is established, extends the analysis down through sub-basin levels to the water sectors, projects and users. In other words, this discussion proceeds from the macro- through the meso- to the micro-level of analysis, rather than in the opposite direction. This change in perspective is important because the whole can be different from the sum of its parts, due to scale and composition effects, and it is important to conduct the analysis of the part within the context of the whole (Keller *et al.*, 1996).

River basins and beneficial depletion[3]

In this discussion, river basins are defined as including the offshore, coastal zone of brackish water formed by the mixture of water from the land and the saline water of external or internal seas and saline aquifers. Also, some basins are interconnected, either by natural or human-made flows, and interbasin transfers between basins must be included in the analysis.

The first and single most important thing to understand about river basins is that, with the exception of usually small and temporary increases of water storage within a river basin, all of the water that annually enters the basin through precipitation, including snow melt or interbasin transfers into the basin, is eventually depleted from the basin. It is depleted either by evaporation, including evapotranspiration, or by discharges to sinks, mainly to inland or external seas and to saline aquifers. Thus, at the river-basin level:

$$P + T - CS = E + S \qquad (3.4)$$

where:

P = total precipitation
T = interbasin transfers (into the basin is positive)
CS = changes in storage in the basin (increase in storage is positive)
E = total evaporation
S = flows to sinks

Since, with the exception of increased CS, all of the water in the basin is depleted, the ultimate question in addressing the utilization of water is not whether the water is depleted or not, but whether it is beneficially depleted or not. The total beneficial depletion (Db) of water in a river basin is:

$$Db = Eb + Sb \qquad (3.4a)$$

where Sb = beneficial flows to sinks.

Eb occurs in such areas as evapotranspiration of valued plants and the perspiration and respiration of animals and in cooling facilities. Some discharges to sinks are also beneficial – for example, in maintaining coastal zones, river flows for navigation, fishing and, in the case of saline aquifers, preventing seawater intrusion. The important aspect of sinks is that, while the water may serve a valuable function in the sink, it is not available for other uses outside the sink.

Non-beneficial depletion through evaporation or discharges to sinks may have either a zero or a negative value. Discharges to seas may have a zero value in water-surplus periods, for example; while discharges to saline aquifers may cause water tables to rise, reducing crop productivity through waterlogging and salinity and polluting domestic water supplies.

[3] For a technical discussion of the material in this section in the context of water accounting, see Molden and Sakthivadivel (1999).

Clearly, it is impossible to 'add up' the sum of the beneficial and non-beneficial uses of water without knowing their exact, positive and negative, values. This fact makes the concept of beneficial utilization an intrinsically qualitative, rather than a quantitative, concept. But it is interesting and important to know, for example, what proportion of the water is being beneficially utilized, even if one does not know the absolute value of beneficial utilization or its net beneficial utilization. For example, if around 87% of the water resources of Egypt are being beneficially utilized, one knows immediately that it will be difficult to increase beneficial use in one area without decreasing it in other areas somewhere else in the system. On the other hand, if only 50% of the water is being beneficially utilized, as is commonly thought of Egypt, then there is large scope for increased beneficial utilization.

Available water supply

Not all of the annual precipitation that enters a basin is available for beneficial use within the basin. The available water supply (AWS) at the basin level is defined as:

$$AWS = (P + T - CS) - N \quad (3.5)$$

where, with P, T and CS as defined in Equation 3.4:

N = non-utilizable water supply, as in discharges of floodwater to sinks.

This term can be defined either as actual N, with existing storage and conveyance facilities, or potential N, with all technically and economically possible water-development facilities.

At the sub-basin level (sb), the AWS term needs to be adjusted by: (i) replacing T by diversions (DIV) from other areas within the basin to the particular area under consideration; and (ii) including committed outflows (C) to other areas from the area under consideration, such as legally or conventionally committed outflows from upper to lower riparian states, or between other subunits within a basin (Molden, 1997):

$$AWS\ (sb) = [(P + DIV) - CS] - (N + C) \quad (3.5a)$$

It could be objected that it is wrong to use total precipitation, including ineffective precipitation (P − Pe), in the definition of irrigation efficiency at the sub-basin level – that this is a 'free' good and that what we want to optimize is diversions, not AWS. But we believe that this traditional approach is mistaken. First, in irrigation, ineffective precipitation can be a partial substitute for diversions by investing in better land- and water-management techniques, such as bunding, field levelling and the like. In other areas, rainwater-collecting devices can serve domestic needs. Secondly, under conditions of water scarcity, there is no free good; water used in one place has an opportunity cost in terms of the value of its use in another place within the system. The concepts of efficiency and productivity need to reflect the values of all the uses and alternative uses within the system.

There is also an important distinction between the amount of AWS that is actually available at a given point in time, with existing water-storage and control facilities, and the amount of AWS that is potentially available in the future with additional facilities. As noted before, this distinction is reflected in the term for non-utilizable supply (N), which is a variable depending on the storage and control facilities up to the ultimate potential AWS at each level.

The basin efficiency (BE) of water resources can now be defined in terms of the ratio of the beneficial utilization of water to AWS at either the basin level or the sub-basin level:

$$BE = (Eb + Sb)/AWS \quad (3.6)$$

BE can be considered in terms of actual AWS, or potential AWS, resulting in either BE (a) or BE (p). At the sub-basin level, the AWS term is replaced by AWS (sb) and the resulting equation is called sub-basin efficiency, BE (sb). Also, if one wished to avoid the nomenclature of efficiency, this equation could be described as the rate of beneficial utilization of water resources (RBU) at the basin and sub-basin levels of analysis.

Types of river basins

River basins can be classified according to the amount of uncommitted discharges to sinks of potentially utilizable water in the dry, low-flow season. For short, we shall call this the discharges of usable water in the dry season.

- An open basin has outflows of usable water in the dry season. In open basins, more water storage could be developed in the dry season and beneficially depleted upstream without diminishing existing uses; in other words, the opportunity cost of additional dry-season depletion is zero.
- A closing basin has no discharges of usable water in the dry season. Therefore, any additional depletion in this season results in a decrease in existing uses. However, closing systems do have discharges of usable water in the wet season. Thus there is at least the possibility that the basin can be reopened through the development of upstream surface and subsurface water storage of wet-season flows for use in the dry season.
- A completely closed basin has no discharges of usable water even in the wet season. In this case, there is no scope for obtaining additional water supplies. Additional water needs can be met only through gains in water productivity – for example, by reducing non-beneficial evaporation or by reallocating water from lower-valued to higher-valued uses.

It is surprising how many closing or closed river basins there are, once one begins looking for them. For example, it is said that such large and important basins as the Indus, the Ganges and the Yellow River basins are closing by this definition, and that the Colorado and the Cauvery River basins are completely closed. Unfortunately, there are few reliable data on discharges of water to sinks for many of the large river basins of the world, much less on the quality of the water. The fact that most water-management agencies do not bother to collect data on what is surely the single most important factor in water management is evidence of the newness of the river-basin perspective, notwithstanding hundreds of years of intuitive understanding of its importance by people like King Parakramabahu.

The productivity of water use

It is important to distinguish between the rate of beneficial utilization, or BE, and the productivity of water use. While the two are related, they are not the same thing. The same degree of beneficial utilization may have substantially different values in terms of the productivity of water. For example, the same amount of water depleted in the irrigation of cereal crops may have a much higher value in vegetable or fruit crops; and it will probably have a higher value in the domestic sectors than in the irrigation sector.

Also, water serves both as an input to the production of a final good, such as irrigation in crop production or wildlife habitats, and as a final good in itself, such as drinking water or the aesthetic value of a beautiful lake. In these and other cases, the value of water is attached to the amount of water diverted to the particular use – for example, the value of so much drinking water supplied to a household – irrespective of the amount of depletion. But it should also be recognized that, if only a small amount of the diversion is depleted, the potential for the outflow being beneficially recycled into other diversions is increased. Repeated reuse of water creates the water multiplier effect, where the sum of the diversions in a river basin can be several times larger than the inflow of water into the basin (Seckler, 1992; Keller *et al.*, 1996). Because of the multiplier effect, the productivity of the water inflow into the basin is often enhanced.

The productivity of water in a given use is defined in terms of the quantity and quality of water diverted or depleted in that use. Given this, there are several different ways of expressing productivity:

- Pure physical productivity is defined as the quantity of the product divided by the quantity of AWS, diverted water or depleted water, expressed as kg m^{-3}. For example, a slogan at IWMI, 'increasing crop per drop', expresses physical productivity.

- Combined physical and economic productivity is defined in terms of the net present value (NPV) of the product divided by the amount of water diverted or depleted. Thus, the quantity of the product is productivity times the amount of AWS or water depleted.
- Economic productivity is the NPV of the product divided by the NPV of the amount of AWS or water diverted or depleted, defined in terms of its value, or opportunity cost, in the highest alternative use.

In estimating the economic value of water, it is more important to understand both the extent and the limitations of what can be rationally accomplished. When water is an input to a final good that has a real market value or shadow price, the marginal value of water, like that of any other input, can be estimated as a derived demand for the input. Obviously, values can also be assigned when water is itself a marketed product, whether a final product, such as drinking water, or an input, such as irrigation water. But, when water or its products are not marketed or when they have non-market values, as in the case of basic needs or ecological imperatives, then it is an abuse of economics to assign real or shadow prices to it as an indicator of its value. All one can rationally do in these cases is to commit agreed-upon quantities of water to these purposes. One can then evaluate the opportunity costs of these commitments in terms of shadow prices in an optimization model. But these shadow prices are costs, not values or benefits. Truly, as Oscar Wilde might have said, 'economists know the cost of everything but the value of nothing'.[4]

The Persistence of Classical Efficiency

It is a remarkable fact that, from the time of their development in 1932 to the present, the neoclassical concepts – whether of NE, BE or EE – have not been widely accepted in the general community of irrigation and water-resource practitioners (see, for example, Clyma and Shafique, 2001). CE prevails, notwithstanding the fact that NE is clearly and demonstrably a more valid concept, developed and recommended by many of the outstanding authorities in the field. For example, in the volume, *Design and Operation of Farm Irrigation Systems*, edited by Jensen (1980) and published by The American Society of Agricultural Engineers, the neoclassical view could not be more clearly articulated (especially in the chapter by Burman *et al.* (1980)). But most professionals remain wedded to the classical view – one, in fact, accused Keller and Keller (1995) of advocating 'sloppy irrigation'!

Indeed, in Jensen (1980, pp. 17–20), reference is made to a debate in 1976 in the USA, where the General Accounting Office published a report on the massive savings of water that could be achieved by increasing CE! The experts in the field used Jensen's NE and water-balance analysis to correct that error. But it is remarkable that now, over 20 years later, the same confusion not only endures but actually predominates in the field of irrigation and water-resources policy and management in the USA and throughout the world.

Because of its importance and interest as an example of the evolution of scientific ideas, we may pause briefly to speculate on the reasons for the persistence of CE. First, there is the matter of training. Most irrigation practitioners were trained before the neoclassical concepts appeared in the later 1970s and went directly into practice – quite properly applying CE to the design of irrigation systems. This imprinted CE in their minds, as it were. Secondly, quite naturally, their professional interests and positions were oriented around CE. Thirdly, a large industry of consulting and construction firms, consultants and donors has been created around the task of rehabilitating and 'modernizing' irrigation systems to increase their CE. Fourthly, CE serves the interest of other professions and groups as well.

[4] For a discussion of these issues in the context of poverty, see Seckler (1966) and, for a brilliant general treatment of the subject, Little (1950).

Economists can use low CE as justification for pricing water and water markets; and environmentalists can use it in their battles against large dams, transbasin diversions and other water-development projects. For all of these reasons, the very idea that old 'sloppy' irrigation systems may already be performing at high degrees of efficiency because they are recycling is hard to accept.

On the other hand, the neoclassical approach has been fully understood and applied by farmers and other practitioners in water law and management in the western states of the USA – and perhaps on a more intuitive basis elsewhere. Western water law explicitly recognizes that one farmer's drainage can be another farmer's irrigation supply, and return flows are zealously protected. In Wyoming, water allotments and charges are made on the basis of the 'consumptive use' of the water (NET), not on the amount of water diverted or applied. Thus it is illegal to increase irrigated land and NET through more (classically) efficient technologies, even if the amount of water diverted is the same.

Indeed, Californians commonly distinguish between 'real' and 'paper' water savings, or what they amusingly refer to as 'wet' vs. 'dry' water savings – depending on whether or not gains in classical efficiency for one user are offset by reduced recycling supplies to another user. An elaborate legal and regulatory framework has been created around water use to apply and enforce these neoclassical concepts (see the interesting case of Colorado in Vissia (1997)). As Burman *et al.* (1980) rightly say, 'The reuse of return flow is one of the main foundations of Western water right management, and its importance is impossible to overestimate.'

Externalities, Regulations and Water Pricing

In terms of economic theory, regulations are a rational response to the problem of externalities – or, as they are also known, aptly for discussion of water, 'spillover effects'. As explained in any standard textbook on economics, externalities occur when the welfare of second parties is affected by the behaviour of first parties without compensation. In the case of external benefits, second parties should compensate first parties for the benefits they receive; in the case of external costs, second parties should be compensated by first parties for the harm they suffer. Compensation is not only a matter of equity; it leads to greater efficiency by 'internalizing the externality'. If the first party has to pay for an external cost, he or she will try to produce less of it and, if paid for an external benefit, will produce more of it. Direct private compensation arrangements between first and second parties are usually impossible in practice. Governments have to intervene to 'internalize the externalities', through taxes, subsidies or regulations. Without government interventions, the market fails to achieve an efficient, much less an equitable, allocation of resources.

Of all the goods and services in the world, water is probably the most externality-ridden. The outflow arising from efficiency Equations 3.1 and 3.1a is an external effect. Typically, it is of the order of 50% in irrigation, and it can be as high as 90% or more in the case of the domestic and industrial sectors. This constitutes a colossal potential source of market failure in water resources.

Given this fact, it is rather amazing to find many economists advocating free-market allocations and pricing of water in the name of efficiency. With externalities, free markets lead to inefficient allocations of resources, as shown in any standard economics text. And, at a practical level, if the developed countries have found it necessary to create elaborate regulatory structures to prevent market failure in water resources, how do advocates of water markets and water pricing imagine that developing countries – with much fewer resources, many more low-volume and poor water users and weak institutional structures – will be able to do this? (See the discussion in Perry *et al.*, 1997.)

Having said this, it should be noted that charging service fees to cover at least the operation and maintenance cost of water-delivery services and even perhaps part of the capital costs is a valid practice that should be implemented everywhere. Also, there are many cases where the outflows create external costs, such as in outflows of water polluted by salinity or other harmful wastes into the water supply. These external costs should be internalized to the polluter either by marginal cost prices or by regulations. And, where the outflows have zero or negative benefits, water prices can and should be used to attain higher efficiencies of water use in the classical sense. In sum, the question of water markets and pricing is not one of either-or but rather of why, where, when and how.

Conclusion

The ultimate goal of water-resources policy and management is to increase the beneficial utilization of water. In the final analysis, there are six basic ways of achieving this goal (Seckler, 1996).

1. Where the AWS at the basin level is underutilized, as in the case of open or closing basins, develop the remaining AWS through additional and improved technical and institutional means.
2. Reduce non-beneficial evaporation and non-beneficial discharges to sinks.
3. Increase the amount of benefits per unit of beneficial evaporation[5] and beneficial discharges to sinks.
4. Reduce water pollution.[6]
5. Reduce waterlogging and flood damage.
6. Reallocate water from lower- to higher-valued uses.

There is a large array of technologies, policies and managerial systems that can be employed for achieving these objectives under specific conditions of time and place. But these systems are a subject beyond the scope of this chapter. We strongly believe that before you act you must think, that sound theory is a necessary (but not sufficient) condition to effective action and that, *per contra*, poor theory can lead to ineffective and even counterproductive actions. Many of the problems of water-resources management today are due to the implementation of false, erroneous or misapplied concepts of efficiency in water-resources policy and management. We hope that the discussion in this chapter will help to resolve that problem.

References

Abu-Zeid, M. and Seckler, D. (eds) (1992) *Roundtable on Egyptian Water Policy, Conference Proceedings.* Water Research Centre, Ministry of Public Works and Water Resources, Cairo, Egypt, and Winrock International, Arlington, Virginia.

Bagley, J.M. (1965) Effects of competition on efficiency of water use. *Journal of Irrigation and Drainage Division of the American Society of Civil Engineers* 91(IR 1), 69–77.

Bos, M.G. (1997) Performance indicators for irrigation and drainage. *Irrigation and Drainage Systems* 11, 119–137.

Bos, M.G. and Nugteren, J. (1974) *On Irrigation Efficiencies.* 1st edn 1974; 2nd edn 1978; 3rd revised edn 1982; 4th edn 1990. Publication 19, International Institute for Land Reclamation and Improvement (ILRI), Wageningen, The Netherlands, 117 pp. 4th edn also published in Farsi with IRANCID, Tehran.

Burman, R.D., Nixon, P.R., Wright, P.R. and Pruitt, W.O. (1980) Water requirements. In: Jensen, M.E. (ed.) *Design and Operation of Farm Irrigation Systems.* Monograph No. 3, American Society of Agricultural Engineers, St Joseph, Michigan, pp. 189–232.

[5] One of the ways of increasing productivity by this means is through 'exogenous' changes in technology, outside water management such as reducing the length of the growing season through plant breeding.
[6] Even a mild degree of pollution reduces the productivity of water – for example, saline water reduces crop yields and polluted drinking water adversely affects health – far short of salinity becoming so severe that it is discharged to sinks.

Burt, C.M., Clemmens, A.J., Strelkoff, T.S., Solomon, K.H., Bliesner, R.D., Hardym, L.A., Howell, T.A., Members ASCE and Eisenhauer, D.E. (1997) Irrigation performance measures: efficiency and uniformity. *Journal of Irrigation and Drainage Engineering* 123(6), 423–442.

Clyma, W. and Shafique, M.S. (2001) *Basin-wide Water Management Concepts for the New Millennium.* American Society of Agricultural Engineers Meeting Paper No. 01–2051, American Society of Agricultural Engineers, Sacramento, California.

Falkenmark, M., Lundqvist, J. and Widstrand, C. (1989) Macro-scale water scarcity requires micro-scale approaches: aspects of vulnerability in semi-arid development. *Natural Resources Forum* 13(4), 258–267.

Israelsen, O.W. (1932) *Irrigation Principles and Practices*, 1st edn. John Wiley & Sons, New York, 422 pp.

Jensen, M.E. (1967) Evaluating irrigation efficiency. *Journal of the Irrigation and Drainage Division, Proceedings of the American Society of Civil Engineers* 93(IR 1), 83–98.

Jensen, M.E. (1977) Water conservation and irrigation systems: climate. In: *Technical Seminar Proceedings,* Columbia, Missouri, pp. 208–250.

Jensen, M.E. (ed.) (1980) *Design and Operation of Farm Irrigation Systems.* Monograph No. 3, American Society of Agricultural Engineers, St Joseph, Michigan, 829 pp.

Keller, A. and Keller, J. (1995) *Effective Efficiency: a Water Use Concept for Allocating Freshwater Resources.* Water Resources and Irrigation Division Discussion Paper 22. Winrock International, Arlington, Virginia.

Keller, A., Keller, J. and Seckler, D. (1996) *Integrated Water Resource Systems: Theory and Policy Implications.* Research Report 3, International Irrigation Management Institute, Colombo, Sri Lanka.

Keller, J. (1992) Implications of improving agricultural water use efficiency on Egypt's water and salinity balances. In: Abu-Zeid, M. and Seckler, D. (eds) *Roundtable on Egyptian Water Policy, Conference Proceedings.* Water Research Centre, Ministry of Public Works and Water Resources, Cairo, Egypt, and Winrock International and Appendix ii, Arlington, Virginia, pp. 16, 721–760.

Little, I.M.D. (1950) *A Critique of Welfare Economics*, 2nd edn. Oxford University Press, Oxford.

Merriam, J.L. (1980) Evaluating irrigation systems and practices. In: Jensen, M.E. (ed.) *Design and Operation of Farm Irrigation Systems.* ASAE Monograph No. 3, American Society of Agricultural Engineers, St Joseph, Michigan, pp. 721–760.

Molden, D. (1997) *Accounting for Water Use and Productivity.* SWIM Report No. 1, International Irrigation Management Institute, Colombo, Sri Lanka.

Molden, D.J. and Sakthivadivel, R. (1999) Water accounting to assess uses and productivity of water. *Water Resources Development* 15(1/2), 55–71.

Molden, D.J., El Kady, M. and Zhu, Z. (1998) Use and productivity of Egypt's Nile. Paper presented at the 14th Technical Conference on Irrigation, Drainage and Flood Control, USCID, Phoenix, Arizona, 3–6 June.

Perry, C.J. (1996) *The IIMI Water Balance Framework: a Model for Project Level Analysis.* Research Report 5, International Irrigation Management Institute, Colombo, Sri Lanka.

Perry, C.J. and Narayanamurthy, S.G. (1998) *Farmer Response to Rationed and Uncertain Irrigation Supplies.* Research Report 24, International Irrigation Management Institute, Colombo, Sri Lanka.

Perry, C.J., Rock, M. and Seckler, D. (1997) *Water as an Economic Good: a Solution, or a Problem?* Research Report 14, International Irrigation Management Institute, Colombo, Sri Lanka.

Popper, K.R. (1962) *Conjectures and Refutations: the Growth of Scientific Knowledge*. Routledge, London.

Seckler, D. (1966) On the uses and abuses of economic science in evaluating public outdoor recreation. *Land Economics* 42(4), 112–123.

Seckler, D. (1992) Irrigation policy, management and monitoring in developing countries. In: Abu-Zeid, M. and Seckler, D. (eds) *Roundtable on Egyptian Water Policy, Conference Proceedings.* Water Research Centre, Ministry of Public Works and Water Resources, Cairo, Egypt, and Winrock International, Arlington, Virginia.

Seckler, D.W., Amarasinghe, U., Molden, D., de Silva, R. and Barker, R. (1998a) *World Water Demand and Supply, 1990–2025: Scenarios and Issues.* Research Report 19, International Water Management Institute, Colombo, Sri Lanka.

Seckler, D., Molden, D. and Barker, R. (1998b) *Water Scarcity in the Twenty-first Century.* International Water Management Institute, Colombo, Sri Lanka.

Vissia, R.J. (1997) Country review of turnover programs, USA. Draft for publication, IWMI.

Willardson, L.S., Allen, R.G. and Frederiksen, H.D. (1994) Universal fractions and the elimination of irrigation efficiencies. Paper presented at the 13th Technical Conference, USCID, Denver, Colorado, 19–22 October 1994. Duplicated.

Wolters, W. and Bos, M.G. (1990) Interrelationships between irrigation efficiency and the reuse of drainage water. In: *Papers Presented at the Symposium on Land Drainage for Salinity Control in Arid and Semi-Arid Regions, Cairo, Egypt, 25 February–2 March 1990*. Vol. 3. *Session III—Session IV–Session V.* Ministry of Public Works and Water Resources, Cairo, Egypt, pp. 237–245.

Wright, K.R. (1964) Water rights engineering under appropriation system. *Journal of the Irrigation and Drainage Division, ASCE* 90(IR 1), 9–15.

4 Rice Production in Water-scarce Environments

T.P. Tuong and B.A.M. Bouman
International Rice Research Institute, Manila, Philippines

Abstract

Rice production in Asia needs to increase to feed a growing population. Though a complete assessment of the level of water scarcity in Asian rice production is still lacking, there are signs that declining quality of water and declining availability of water resources are threatening the sustainability of the irrigated rice-based production system. Drought is one of the main constraints for high yield in rain-fed rice. Exploring ways to produce more rice with less water is essential for food security and sustaining environmental health in Asia. This chapter reviews the International Rice Research Institute (IRRI)'s integrated approach, using genetics, breeding and integrated resource management to increase rice yield and to reduce water demand for rice production. Water-saving irrigation, such as saturated-soil culture and alternate wetting and drying, can drastically cut down the unproductive water outflows and increase water productivity. However, these technologies mostly lead to some yield decline in the current lowland rice varieties. Other new approaches are being researched to increase water productivity without sacrifice in yield. These include the incorporation of the C_4 photosynthetic pathway into rice to increase rice yield per unit water transpired, the use of molecular biotechnology to enhance drought-stress tolerance and the development of 'aerobic rice', to achieve high and sustainable yields in non-flooded soil. Through the adoption of water-saving irrigation technologies, rice land will shift away from being continuously anaerobic to being partly or even completely aerobic. These shifts will have profound changes in water conservation, soil organic-matter turnover, nutrient dynamics, carbon sequestration, soil productivity, weed ecology and greenhouse-gas emissions. Whereas some of these changes can be perceived as positive, e.g. water conservation and decreased methane emission, some are perceived as negative, e.g. release of nitrous oxide from the soil and decline in soil organic matter. The challenge will be to develop effective integrated natural-resource-management interventions, which allow profitable rice cultivation with increased soil aeration, while maintaining the productivity, environmental services and sustainability of rice-based ecosystems.

Introduction

The past years have seen a growing scarcity of water worldwide. The pressure to reduce water use in irrigated agriculture is mounting, especially in Asia, where it accounts for 90% of total diverted fresh water. Rice is an obvious target for water conservation: it is grown on more than 30% of irrigated land and accounts for 50% of irrigation water (Barker *et al.*, 1999). Reducing water input in rice production can have a high societal and

 Water Productivity in Agriculture: Limits and Opportunities for Improvement (eds J.W. Kijne, R. Barker and D. Molden)

environmental impact if the water saved can be diverted to areas where competition is high. A reduction of 10% in water used in irrigated rice would free 150,000 million m^3, corresponding to about 25% of the total fresh water used globally for non-agricultural purposes (Klemm, 1999). However, rice is very sensitive to water stress. Attempts to reduce water in rice production may result in yield reduction and may threaten food security in Asia. Reducing water input for rice will change the soil from submergence to greater aeration. These shifts may have profound – and largely unknown – effects on the sustainability of the lowland rice ecosystem. Our challenge is to develop socially acceptable, economically viable and environmentally sustainable novel rice-based systems that allow rice production to be maintained or increased in the face of declining water availability. This chapter reviews the status of water resources in rice-growing areas and the opportunities and challenges of growing more rice with less water.

Water Resources in Rice-growing Areas

Rice can be grown under irrigated (lowland) or rain-fed (upland or lowland) conditions. Rain-fed rice occupies about 45% of the global rice area and accounts for about 25% of the rice production. Drought has been identified as one of the main constraints for improving yield, which currently averages 2.3 t ha^{-1}. According to Garrity *et al.* (1986), 50% of rain-fed lowland and all rain-fed uplands are drought-prone. Severe and mild droughts often occur in predominantly rain-fed rice areas, such as north-east Thailand, Laos, central Myanmar and east and north-east India (Plate 1).

More than 75% of the rice supply comes from 79 million ha of irrigated lowlands. Rice production in the subtropical regions of north and central China, Pakistan and north-west India mostly depends on wet-season (summer) rainfall, with supplementary irrigation (Plate 2a). Dry-season irrigated rice is concentrated in south China, south and east India and the whole of South-East Asia (Plate 2b). In-depth assessment of the availability of irrigation water in the irrigated rice area is lacking. By overlaying the International Water Management Institute (IWMI)'s water-scarcity atlas (IWMI, 2000) with the International Rice Research Institute (IRRI)'s rice-area maps, it is expected that wet-season irrigated rice areas in north China (2.5 million ha), Pakistan (2.1 million ha) and north and central India (8.4 million ha) will experience 'physical water scarcity' by 2025 (Plate 2a). In addition, about 2 million ha of the dry-season irrigated rice in central India (Plate 2b) will suffer physical scarcity. Most of the approximately 22 million ha dry-season irrigated rice areas in south and South-East Asia fall in the 'economic water scarcity' zone. However, there may be an overestimation of the water availability in the dry season because IWMI's water-scarcity calculations are based on the annual water balance. In principle, water is always scarce in the dry season, when the lack of rainfall makes cropping impossible without irrigation. Thus, there may be rice areas in the 'economic water scarcity' zone affected by 'physical water scarcity' in the dry season.

There is evidence that water scarcity already prevails in rice-growing areas (Plate 3). Consequent overexploitation of groundwater in the last decades has caused serious problems in China and south Asia (Postel, 1997; Sha *et al.*, 2000; Shu Geng *et al.*, 2001). Groundwater tables have dropped, on average, by 1–3 m $year^{-1}$ in the North China Plain, by 0.5–0.7 m $year^{-1}$ in the Indian states of Punjab, Haryana, Rajasthan, Maharashtra, Karnataka and northern Gujarat and by about 1 m $year^{-1}$ in Tamil Nadu and hard-rock southern India. This has led to increased costs of pumping, salinity intrusion, fluoride contamination, land subsidence and the formation of cracks and sink holes (North China Plain). These major groundwater-depletion areas affect rice production in the rice–wheat-growing areas in northern India, Pakistan and China and in the rice-growing areas in Tamil Nadu. In the Ganges delta of Bangladesh, overdrawing of groundwater in the dry season leads to wells falling dry in rice-producing areas, but water levels are restored during the wet season. A specific problem attributed to falling groundwater tables here (and in parts of eastern India) is the appearance of poisonous arsenic.

Heavy upstream water use along some major rivers in Asia is causing severe water shortages downstream. China's Yellow River, which flows for 4600 km through some of Asia's richest farmland, has run dry nearly every year since 1972 (Postel, 1997; Shu Geng *et al.*, 2001). Such is the demand on its water that, in 1997, its final 600 km were dry for more than 4 months. The government of China has prohibited flooded rice cultivation around Beijing (Wang Huaqi *et al.*, 2003). In south Asia, the Ganges and Indus Rivers have little to no outflow to the sea in the dry season. Less dramatic, but more important for rice-growing areas, is the fact that heavy competition for river water between states and different sectors (city, industry) is causing water scarcity for agriculture in southern India's Cauvary delta and in Thailand's Chao Phraya delta (Postel, 1997).

Irrigated rice production is also increasingly facing competition from other sectors. The irrigated rice area in China was reduced by 4 million ha between the 1970s and the 1990s (Barker *et al.*, 1999). Though it is not possible to claim that this reduction in irrigated rice area is entirely due to water scarcity, there is evidence that the reduced area is related to the reduction in the amount of water that is diverted to irrigate rice land. For example, in the 160,000 ha Zhanghe irrigation system (Hubei Province, China), the share of water allocated for irrigation was dominant (about 80%) until the 1980s. Afterwards, Zhanghe reservoir water was increasingly used to meet the growing demand for water by cities and industry and for hydropower generation, and the amount of water allocated for irrigation declined to about 20% in the late 1990s. The irrigated rice area in the 1990s was reduced by about 20% from the level in the 1980s (Fig. 4.1). As a consequence, rice production was also reduced (Dong Bin *et al.*, 2001). Similar examples of increased competition exist elsewhere in Asia. Water from the Angat reservoir in Bulacan Province, the Philippines, is increasingly diverted towards Manila at the expense of downstream water availability for agriculture (Bhuiyan and Tabbal, as cited in Pingali *et al.*, 1997, pp. 196–197). In other areas, water availability is threatened by degrading water quality caused by industrial pollution. Water in the Agno River in the Pangasinan Province is polluted with sediments and chemicals from mining activities upstream (Castañeda and Bhuiyan, 1993). Postel (1997) listed examples of competition between industrial and agricultural uses of water in India.

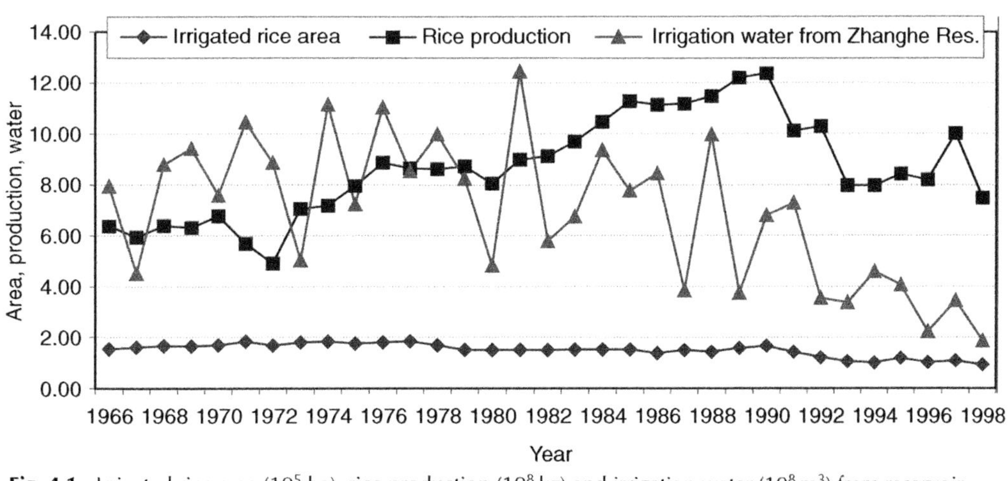

Fig. 4.1. Irrigated rice area (10^5 ha), rice production (10^8 kg) and irrigation water (10^8 m^3) from reservoir (1966–1998), Zhanghe irrigation system, Hubei province, China (from Dong Bin *et al.*, 2001).

Water Productivity in Rice

Rice and water input

Lowland rice in Asia is mostly transplanted or direct (wet)-seeded into puddled, lowland paddy-fields. Land preparation of a paddy consists of soaking, ploughing and puddling. Puddling is mainly done for weed control but it also increases water retention, reduces soil permeability and eases field levelling and transplanting (De Datta, 1981). Soaking is a one-time operation and requires water to bring the topsoil to saturation and to create a ponded water layer. There are often 'idle periods' in between tillage operations and transplanting, prolonging the land preparation period up to 1–2 months in large-scale irrigation systems (Tuong, 1999). The crop growth period runs from transplanting to harvest. During this period, fields are flooded with typically 5–10 cm of water until the final drainage some 10 days before the harvest.

Under flooded conditions, water is required to match outflows (seepage (S) and percolation (P)) to the surroundings and depletions to the atmosphere (evaporation (E) and transpiration (T)). The flow rates of S and P are governed by the water-head (depth of ponded water) on the field and the resistance to water movement in the soil. Because they are difficult to separate in the field, S and P are often taken together as one term, i.e. SP. SP can be as high as 25 mm day^{-1} during land preparation, because soil cracks do not close completely during land soaking (Tuong *et al.*, 1996). Typical SP rates for paddy-fields during the crop growth period vary from 1–5 mm day^{-1} in heavy clay soils to 25–30 mm day^{-1} in sandy and sandy loam soils (Wickham and Singh, 1978; Jha *et al.*, 1981). Only E (from ponded water or moist soil) takes place during land preparation, whereas both E (from soil and water surface between crops) and T occur during the crop growth period. Since it is difficult to separate E and T during crop growth, they are often expressed in one term, evapotranspiration (ET). Typical ET rates of rice in Asia range from 4 to 7 mm day^{-1} (De Datta, 1981; Tuong, 1999).

The water input in paddy-fields depends on the rates of the outflow processes and on the duration of land preparation and crop growth. For a typical 100-day season of modern high-yielding rice, the total water input varies from 700 to 5300 mm, depending on climate, soil characteristics and hydrological conditions (Table 4.1), with 1000–2000 mm as a typical value for many lowland areas. Of all outflows of water from a paddy-field,

Table 4.1. Typical daily rates of water outflows and seasonal water input in lowland rice production in the tropics.

	Daily (mm day^{-1})	Duration (days)	Season (mm)
Land preparation			
Land soaking			100–500
Evaporation	4–6	7–30	28–180
Seepage and percolation	5–30	7–30	35–900
Total land preparation			160–1580
Crop growth period			
Evapotranspiration			
Wet season	4–5	100	400–500
Dry season	6–7	100	600–700
Seepage and percolation			
Heavy clays	1–5	100	100–500
Loamy/sandy soils	15–30	100	1500–3000
Total crop growth			500–3700
Total seasonal water input			660–5280
Typical range of values for total seasonal water input			1000–2000

only T is 'productive' water use, since it leads directly to crop growth and yield formation. Most of the water input to a ricefield, however, is to compensate for E during land preparation and SP during land preparation and the crop growth period. These flows are unproductive as they do not contribute to crop growth and yield formation.

Water productivity

Water productivity is the amount of grain yield obtained per unit water. Depending on the type of water flows considered, water productivity can be defined as grain yield per unit water evapotranspired (WP_{ET}) or grain yield per unit total water input (irrigation plus rainfall) (WP_{IP}). At the field level, WP_{ET} values under typical lowland conditions range from 0.4 to 1.6 g kg^{-1} and WP_{IP} values from 0.20 to 1.1 g kg^{-1} (Tuong, 1999; Bouman and Tuong, 2001). The wide range of WP_{ET} reflects the large variation in rice yield as well as in ET caused by differences in environmental conditions under which rice is grown. Compared with other C_3-type food crops, such as wheat, rice has only slightly lower WP_{ET} values (Table 4.2). However, the WP_{IP} of rice is somewhat less than half that of wheat. The relatively low WP_{IP} of rice is largely due to the high unproductive outflows discussed above (SP and E).

Besides the yield and the size of field-level water outflows, the scale and the boundary of the area over which water productivity is calculated greatly affect its value. This is because the outflow 'losses' by S, P and runoff at a specific location (or field) can be reused at another location within the area under consideration. Data on water productivity across scales are useful parameters to assess whether water outflows upstream are effectively reused downstream. So far, we have found only a few reliable data on the water productivity at different scale levels within irrigation systems (Table 4.3). These limited data suggest that water productivities at scale levels larger than the field level vary widely and are within the variation of water productivities at the field level. The paucity of data on water productivity at scale levels higher than the field level reflects the lack of: (i) data on water flows or yield or both at such scales; and (ii) cooperation between those who work in agriculture (who

Table 4.2. Water productivity of rice, wheat and maize in terms of grain yield (g) per kg of water evapotranspired (WP_{ET}) and per kg of total water (rainfall plus irrigation) input (WP_{IP}) (adapted from Tuong, 1999).

WP_{ET}	WP_{IP}	Source of data used in calculating water productivity	Location
Rice			
	0.05–0.25	Bhatti and Kijne (1992), rainwater not included	Pakistan
1.39–1.61	0.29–0.39	Bhuiyan *et al.* (1995), wet-seeded rice	Philippines
1.1		Sandhu *et al.* (1980)	India
0.88–0.95	0.33–0.58	Kitamura (1990), dry season	Malaysia
0.89		Mishra *et al.* (1990)	India
0.4–0.5		Khepar *et al.* (1997)	India
	0.2–0.4	Bouman and Tuong (2001); 24 data sets	India
	0.3–1.1	Bouman and Tuong (2001); 16 data sets	Philippines
Wheat			
1.0–2.0		Turner (1997)	Australia
1.0–1.5	1.0–1.6	Deju and Lu Jingwen (1993), winter wheat	China
0.65	0.8	Sharma *et al.* (1990)	India
0.87	0.79	Pinter *et al.* (1990)	
Maize			
2.8	2.2–3.9	Stegman (1982)	USA
1.9–2.8	1.9–2.5	Moridis and Alagcan (1992)	Philippines
1.7–2.1	1.6–1.7	Stockle *et al.* (1990)	USA

Table 4.3. Water productivity (g rice kg^{-1} water) in respect of evapotranspiration (WP_{ET}), irrigation (WP_I) and total water input (WP_{IP}) at different scales.

Area (ha)	WP_{ET}	WP_I	WP_{IP}	Location	Source
30–50	0.5–0.6	1–1.5	0.25–0.27	Muda irrigation system, Kendal, Malaysia	Cabangon *et al.* (2002)
287–606	1–1.7	0.4–1	–	Zhanghe irrigation system, Hunan, China	Dong Bin *et al.* (2001)
Over 10^5	–	1–2.5	0.5–1.3		

may have production data) and those who work in the water-management sector (who may have water-flow data).

Strategies for Increasing Water Productivity at the Field Level

Increasing water productivity at the field level can be accomplished by: (i) increasing the yield per unit cumulative ET; (ii) reducing the unproductive water outflows and depletions (SP, E); or (iii) making more effective use of rainfall. The last strategy is important from the economic and environmental points of view, where the water that needs to be provided through irrigation can be offset by that supplied or replaced entirely by rainfall.

Increasing yield per unit ET: germplasm development and agronomic practices

Germplasm development has played an important role in increasing water productivity in rice production. By increasing yield and simultaneously reducing crop duration (and therefore the outflows of ET, S and P), the modern 'IRRI varieties' have about a threefold increase in water productivity compared with the traditional varieties. Most of the increase in WP_{ET}, however, occurred in cultivars released before 1980 (Tuong, 1999). This is because the increase in yield from 1966 to the early 1980s is coupled with a decrease in growth duration, whereas cultivars released after the mid-1980s have a longer duration than those released before 1980 (Peng *et al.*, 1998). Advancement in the development of tropical japonicas (also called the 'new plant type' (IRRI, 1998)) and hybrid rice will enhance water productivity. Peng *et al.* (1998) reported that the ratio of photosynthesis to T was 25–30% higher for the tropical japonica than for the indica type.

In the low-fertility, drought-prone rain-fed environments, breeders have been most successful in manipulating drought escape. Exposure to drought is minimized by reducing crop duration or by minimizing the risk of coincidence of sensitive crop stages with water-deficit periods. The progress in breeding for drought tolerance is less spectacular, and the difficulties encountered are often blamed on the genetic complexity of the trait and its interaction with the environment. Nevertheless, drought-resistant varieties are being bred and released in upland and drought-prone rain-fed lowland areas. Salinity-tolerant varieties, such as Ir51500-AC11-1, allow us to grow rice in areas where salinity problems exclude the cultivation of conventional lowland varieties.

Improved agronomic practices, such as site-specific nutrient management, good weed management and proper land levelling, can increase rice yield significantly without affecting ET and, therefore, may result in increased water productivity (Moody, 1993; Tuong *et al.*, 2000; Hill *et al.*, 2001).

Reducing unproductive water outflows

Large reductions in water input can be potentially realized by reducing the unproductive E and SP flows during land preparation and during the crop growth period (Tuong, 1999; Bouman and Tuong, 2001). There are basically three ways to do so: (i) minimizing the idle periods during land preparation; (ii) increasing the resistance to water flow in the soil; and (iii) decreasing the hydrostatic water pressure.

Minimizing idle periods during land preparation

In transplanted rice, seedlings are usually nurtured in a seedbed for about 2–4 weeks. In irrigation systems that lack tertiary and field channels and with field-to-field irrigation, all the fields surrounding the seedbeds are being tilled (land preparation) and flooded during this period. This land-preparation period can be shortened by the provision of tertiary infrastructure to: (i) supply irrigation water directly to the nurseries without having to submerge the main fields; and (ii) allow farmers to carry out their farming activities independently of the surrounding fields (Tuong, 1999). In the Muda irrigation scheme, Malaysia, increasing the canal and drainage intensity from 10 to 30 m ha^{-1} has enabled farmers to shorten their land preparation by 25 days, resulting in annual water savings of 375 mm in two rice cropping seasons (Abdullah, 1998). In some countries, such as Vietnam and China, specific land areas are set aside for community seedbeds, which can be irrigated independently.

Another way to reduce the idle period during land preparation in irrigation systems without tertiary canals is the use of direct seeding (Bhuiyan *et al.*, 1995). However, the crop growth period in the main field of transplanted rice is shorter than that of direct-seeded rice. Thus, the amount of water saved by direct seeding depends on the balance between the reduction in water use caused by shortened land preparation and the increase in water use caused by prolonged crop growth duration in the main field (after crop establishment (Cabangon *et al.*, 2002)).

Soil management to increase resistance to water flow

The resistance to water flow can be increased by changing the soil physical properties. Cabangon and Tuong (2000) showed the beneficial effects of an additional shallow soil tillage before land preparation to close cracks that cause rapid bypass flow at land soaking. Thorough puddling results in a good compacted plough soil that impedes vertical water flow (De Datta, 1981). Soil compaction using heavy machinery has been shown to decrease soil permeability in north-east Thailand in sandy and loamy soils with at least 5% clay (Sharma *et al.*, 1995). Researchers have even experimented with introducing physical barriers underneath paddy soils, such as bitumen layers and plastic sheets (Garrity *et al.*, 1992). However effective, though, soil compaction and physical barriers are expensive and beyond the financial scope of most farmers.

Water management to reduce hydrostatic pressure

Reducing S and P flows through reduced hydrostatic pressure can be achieved by changed water management (Bouman *et al.*, 1994). Instead of keeping the rice-field continuously flooded with 5–10 cm of water, the floodwater depth can be decreased, the soil can be kept around saturation (saturated soil culture (SSC)) or alternate wetting and drying (AWD) regimes can be imposed. Soil saturation is mostly achieved by irrigating to about 1 cm water depth a day or so after disappearance of standing water. In AWD, irrigation water is applied to obtain 2–5 cm floodwater depth after a larger number of days (ranging from 2 to 7) have passed since the disappearance of ponded water. Wei Zhang and Si-tu Song (1989) reported yield increase under AWD. Our recent work indicates, however, that these are the exception rather than the rule (Bouman and Tuong, 2001; Tabbal *et al.*, 2002b). In most cases, SSC and AWD decrease yield. The level of yield decrease depends largely on the ground water-table depth, the evaporative demand and the drying period in between irrigation events (in the case of AWD). Mostly, however, relative reductions in water input are larger than relative losses in yield, and therefore water productivities in respect of total water input increase (Fig. 4.2). In some cases, AWD even doubled the water productivity compared with conventional flooded irrigation, but with yield reductions up to 30% (e.g. Tabbal *et al.*, 1992).

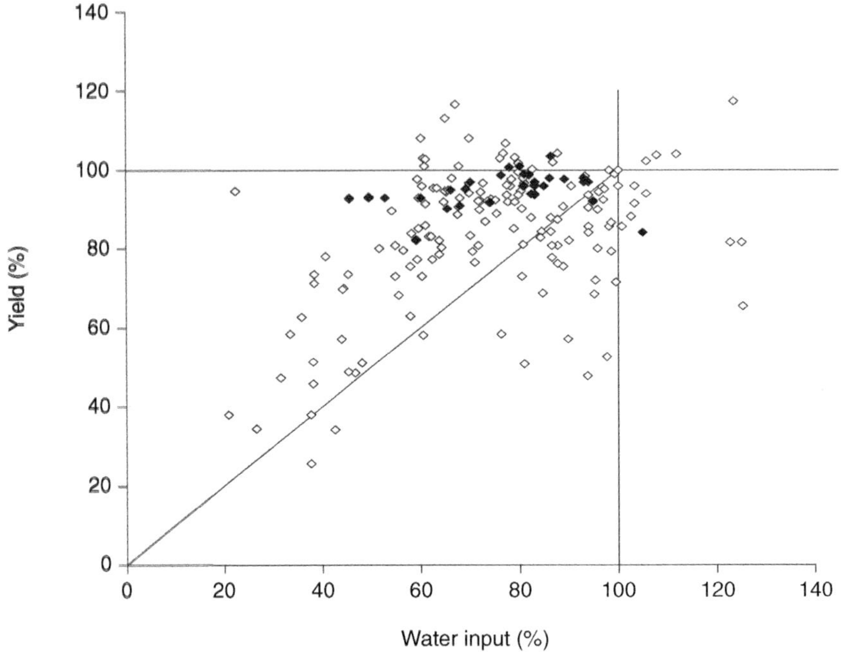

Fig. 4.2. Relative yield versus relative water input. The ◆ markers are data from SSC treatments or having only 1 day no standing water between irrigation turns ($N = 31$); the ◇ markers are from AWD treatments ($N = 149$). Relative yield is calculated as yield in the water-saving treatment over yield of control treatment with ponded water. Relative water input is calculated as the total water input (irrigation plus rainfall) in the water-saving treatment over that in the control treatment. (From Bouman and Tuong, 2001.)

Using Rainfall More Effectively

Dry-seeded rice technology offers a significant opportunity for conserving irrigation water by using rainfall more effectively. In transplanted and wet-seeded rice systems, farmers normally wait for delivery of canal water before they start land soaking. In dry-seeded rice, land preparation is done with dry or moist soil conditions and is started using early monsoonal rainfall. Crop emergence and early growth also occur in the early part of the monsoon, and only later, when canal water is available, is the crop irrigated as needed. Tabbal *et al.* (2002a) demonstrated the feasibility of dry-seeded rice in wet-season irrigated areas in the Philippines. Cabangon *et al.* (2002) reported that dry-seeded rice significantly increased water productivity in respect of irrigation water over wet-seeded and transplanted rice in the Muda irrigation scheme, Malaysia (Table 4.4). However, it was also observed that all three crop-establishment practices had similar total water input and water productivity in respect of total water input. An additional advantage of dry seeding is the early establishment of the crop, which may allow farmers to grow an extra crop after harvest on residual soil moisture (My *et al.*, 1995; Saleh and Bhuiyan, 1995) or using saved irrigation water. In purely rain-fed systems, early establishment and harvest of dry-seeded rice allows the rice plants to escape any late-season drought and hence improve the yield and its reliability.

Emerging Approaches

Raised beds for saturated soil culture

Implementing SSC requires good water control at the field level, and frequent turns of shallow irrigation, which are labour-intensive. Borell *et al.* (1997) experimented with raised

Table 4.4. Mean ± SE of grain yield (t ha^{-1}) and water productivity (g rice kg^{-1} water) in respect of irrigation (WP_I), to total water input (WP_{I+R}), to evapotranspiration from rice area + evaporation from non-rice area (WP_{ET+E}) and to evapotranspiration from rice area (WP_{ET}), in dry-seeded (DS), wet-seeded (WS) and transplanted (TP) irrigation service units (ISU) (from Cabangon *et al.*, 2002).

Parameter	DS ISU	WS ISU	TP ISU
Yield	4.14 ± 0.17[b*]	4.50 ± 0.23[a, b]	4.79 ± 0.23[a]
WP_I	1.48 ± 0.26[a]	0.62 ± 0.30[b]	1.00 ± 0.30[b]
WP_{I+R}	0.27 ± 0.02[a]	0.26 ± 0.02[a]	0.25 ± 0.02[a]
WP_{ET+E}	0.38 ± 0.02[a]	0.42 ± 0.02[a]	0.39 ± 0.02[a]
WP_{ET}	0.48 ± 0.03[b]	0.53 ± 0.04[b]	0.61 ± 0.04[a]

*In a row, mean ± SE followed by the same letter are not significantly different at the 5% level by least significant difference.
SE, standard error.

beds in Australia to facilitate SSC practices. Water in the furrows (30 cm width and 15 cm depth) kept the beds (120 cm wide) at saturation. Compared with flooded rice, water savings were 34% and yield losses 16–34%. Thompson (1999) found that SSC in southern New South Wales, Australia, reduced both irrigation-water input and yield by a bit more than 10%, thus maintaining the irrigation-water productivity. Yield decline due to cold damage is likely for current varieties grown using SSC in that environment. Borell *et al.* (1997) pointed out the need for further research to determine which components of the water balance were responsible for the differences in total water use.

The benefits of growing rice on raised beds with SSC may be extended to a post-rice crop, such as wheat in the rice–wheat system. The productivity of crops sown after rice is often low due to poor soil physical structure and waterlogging from winter rainfall and spring irrigation. A bed system may improve drainage conditions for a post-rice crop.

Aerobic rice

A fundamental approach to reducing water inputs in rice is to grow the crop like an irrigated upland crop, such as wheat or maize. Instead of trying to reduce water input in lowland paddy-fields, the concept of having the field flooded or saturated is abandoned altogether. Upland crops are grown in non-puddled, aerobic soil without standing water. Irrigation is applied to bring the soil water content in the root zone up to field capacity after it has reached a certain lower threshold. The amount of irrigation water should match E from the soil and T by the crop (plus any application inefficiency losses). The potential water savings when rice can be grown as an upland crop are large, especially on soils with high SP rates (Bouman, 2001). Besides cutting down on SP losses, E is also reduced, since there is no standing-water layer.

De Datta *et al.* (1973) experimented with the cultivation of a high-yielding lowland rice variety (IR20) like an upland crop under furrow irrigation. Total water savings were 56% and irrigation water savings 78% compared with growing the crop under flooded conditions. However, the yield was reduced from 7.9 t ha^{-1} to 3.4 t ha^{-1}. Studies on non-flooded irrigated rice using sprinkler irrigation were conducted in Louisiana and Texas, USA (Westcott and Vines, 1986; McCauley, 1990). The experiments used commercial lowland rice cultivars. Irrigation water requirements were 20–50% less than in flooded conditions, depending on soil type, rainfall and water management. The highest-yielding cultivars (producing 7–8 t ha^{-1} under flooded conditions), however, had yield reductions of 20–30% compared with flooded conditions. The most drought-resistant cultivars produced the same under both conditions, but yield levels were much lower (5–6 t ha^{-1}).

New varieties must be developed if the concept of growing rice like an irrigated upland crop is to be successful. Upland rice varieties exist, but have been developed to give stable though low yields in adverse environments where rainfall is low, irrigation is absent, soils are poor or toxic, weed pressure is high and farmers are too poor to supply high inputs. IRRI recently coined the term 'aerobic rice' to refer to high-yielding rice grown in non-puddled, aerobic soil (Bouman, 2001). Aerobic rice has to combine characteristics of both the upland and the high-yielding lowland varieties. Evidence for its feasibility comes from Brazil and northern China. In Brazil, aerobic rice cultivars have come out of a 20-year breeding programme to improve upland rice with yields of 5–7 t ha^{-1} under sprinkler irrigation in farmers' fields (Silveira Pinheiro and Maia de Castro, Los Baños, Philippines, September 2000, personal communication). These varieties are grown commercially on 250,000 ha in the state of Mato Grosso. In north China, aerobic rice cultivars called Han Dao have been developed that yield up to 6–7.5 t ha^{-1} under flash irrigation in bunded fields (Wang Huaqi *et al.*, 2003). In a recent study of farmers testing aerobic rice in north China, it was found that yields of 4.6–6.6 t ha^{-1} were obtained with as little as 476–612 mm of total water input on loamy soils (Bouman *et al.*, 2002). It is estimated that Han Dao varieties are now being pioneered on some 120,000 ha in the North China Plains.

Biotechnology

The recent advances in genomics, the development of advanced analytical tools at the molecular level and genetic engineering provide new avenues for raising the yield potential and enhancing drought-stress tolerance. For example, the incorporation of the C_4 photosynthetic pathway into rice (being a C_3 plant), if achieved, can potentially increase water productivity by 80% (J.E. Sheehy, personal communication). Table 4.2 also indicates that water productivity of maize (a C_4 crop) is significantly higher than that of rice and wheat (C_3 crops).

The currently slow progress in breeding for drought tolerance may be accelerated by the discovery and subsequent manipulation of regulatory genes underlying the complex physiological and biochemical responses of rice plants to water deficit. Common research tools, tolerance mechanisms and breeding solutions are emerging across the evolutionary diversity of crops and plants. The enormous public- and private-sector investments in genomic analysis of *Arabidopsis thaliana*, the cereals and other crops are already contributing greatly to these efforts (Bennett, 2001). Much effort is currently being directed to developing molecular markers for the maximum rooting depth (Champoux *et al.*, 1995), the capacity of roots to penetrate hard pans (Ray *et al.*, 1996) and the capacity of the plant to osmotically adjust to water deficit (Lilley and Ludlow, 1996).

Opportunities and Challenges in the Adoption of Water-saving Practices

Growing rice in continuously flooded fields has been taken for granted for centuries, but the 'looming water crisis' may change the way rice is produced in the future. Water-saving irrigation technologies that were investigated in the early 1970s, such as SSC and AWD, are receiving renewed attention by researchers (Bouman and Tuong, 2001). The basic ingredients of implementing these technologies seem to be in place. But so far, except for China (Li, 2001), the adoption of these technologies has been slow. The challenge is to identify the environmental and socio-economic conditions that encourage farmers to adopt them. In this respect, our research is far from complete. We can, however, identify important factors that affect the farmers' acceptance of water-saving technologies.

Unlike fertilizers and pesticides, water is generally not actively traded on markets in Asia, and government-administered fees for irrigation water are often low or zero. This discourages farmers from treating water as a scarce resource. Farmers have no incentive to adopt water-saving technologies because water conservation does not reduce the farming expenditures nor does it increase income. It can be expected that, when water becomes a real economic good, farmers are more inclined to adopt water-saving technologies. There is evidence that farmers in Asia who are confronted with high costs of water already adopt such technologies. In certain areas in China, where farmers are charged by the volume of water they use, various forms of AWD and reduced floodwater depths have been widely adopted (Li, 2001). Farmers in north-central India (A.K. Singh, Los Baños, Philippines, April 2000, personal communication) who operate pumps to irrigate their fields consciously apply some form of AWD to save pumping costs. Experiences in Australia also show that water trading, by which farmers can sell their water rights to others, encourages farmers to adopt water-conservation measures.

Water-saving technologies that improve productivity and income will be readily accepted by farmers. Dry seeding is widely practised in drought-prone rain-fed systems because of its ability to increase rice yield and its stability and cropping intensity (My *et al.*, 1995; Saleh and Bhuiyan, 1995). In irrigated systems, however, water-saving technologies are mostly associated with some reduction in yield. Technologies that save water for rice and increase productivity of a post-rice crop will be more acceptable to farmers. The prospect of raised beds to increase the total system productivity of the rice–wheat system opens up opportunities to save water. Similarly, farmers may accept dry-seeding technologies in irrigated systems to reduce the labour cost of transplanting and wet land preparation.

All water-saving technologies, from SSC to AWD to dry seeding and aerobic rice, reduce water depth and expose rice-fields to periods without standing water. Poor levelling of rice-fields is common in Asia, leading to heterogeneity in the depth of standing water. This will result in a more competitive and diverse weed flora than in rice under conventional water management. On-farm research has shown that precise land levelling can improve the establishment of direct-seeded rice and increase water productivity (Hill *et al.*, 2001). Improving farmers' knowledge on improved (integrated) weed management will enhance their acceptance of water-saving technologies.

Suitable policies, institutional organization and legislation are needed to promote the adoption of water-saving technologies. The establishment of water-user groups and the implementation of volumetric water charging may be the most important elements behind the successful adoption of AWD in China. New laws prohibiting flooded rice cultivation in parts of the North China Plain and around Beijing are expected to increase farmers' interest in aerobic rice cultivation.

Environmental Impact and Challenges for Sustainable Management of Water-limited Rice Production Systems

Soil submergence is a unique feature of irrigated lowland rice ecosystems. Lowlands producing two or three rice crops per year on submerged soils are highly sustainable, as indicated by sustained nutrient supply capacity, sustained soil carbon levels and sustained trends in rice yields (Buresh *et al.*, 2001). However, the continuous submergence of soil promotes the production of methane, an important greenhouse gas, by the anaerobic decomposition of organic matter. Temporary soil aeration, such as under AWD, can reduce methane emission. Prolonged aeration of soil, such as in aerobic rice, can even reduce methane emission further. Soil aeration, on the other hand, can increase the emission of nitrous oxide, another greenhouse gas. Emissions of methane and nitrous oxide are strongly related to the soil redox potential, a measure of soil oxidation status. Hou *et al.* (2000) suggested that both methane and nitrous oxide emissions could be minimized by maintaining the soil redox potential within a range of −100 to +200 mV. An important research area is to assess whether water-saving technologies can achieve such an intermediate soil redox potential.

Increased soil aeration under AWD and in aerobic rice will also affect the capacity of soil organic matter and the capacity of the soil nutrient supply. The more competitive weed flora associated with water-saving technologies may require a greater reliance on herbicides (Naylor, 1996), which challenges environmental sustainability. Critical issues for water-saving technologies may include how much water and how frequent soil submergence is required for sustaining the productivity and services of rice ecosystems.

The impact of on-farm water saving on the role of water in sustaining environmental health warrants further investigation. In many basins, the drainage and percolation outflows from rice-fields return to the lower reaches of the rivers. They play an important environmental role in sustaining the fresh–saline water balance in estuaries. Reducing the outflows may result in increased salinity intrusion. The reported increased salinity in the Chao Phraya delta, Thailand, is an example. The drying up of the lower reaches of rivers and declining water tables (see examples in section under Water Resources in Rice-growing Areas) indicate that the basin is closing in such areas (Seckler, 1996) and that all the utilizable outflows from upstream have been reused. Water-saving practices that aim at reducing the drainage and percolation outflows from paddies are important options for farmers to maintain rice cultivation in the face of water scarcity, but they may not increase the water availability of the whole basin.

References

Abdullah, N. (1998) Some approaches towards increasing water use efficiency in the Muda irrigation scheme. In: *Proceedings of the Workshop on Increasing Water Productivity and Efficiency in Rice-based Systems, July 1998*. IRRI, Los Baños. Philippines.

Barker, R., Dawe, D., Tuong, T.P., Bhuiyan, S.I. and Guerra, L.C. (1999) The outlook for water resources in the year 2020: challenges for research on water management in rice production. In: *Assessment and Orientation Towards the 21st Century, Proceedings of 19th Session of the International Rice Commission, Cairo, Egypt, 7–9 September 1998*. FAO, Rome, pp. 96–109.

Bennett, J. (2001) Status of breeding for tolerance of water deficit and prospects for using molecular techniques. Paper presented at the International Workshop on Water Productivity, 12–13 November 2001, IWMI, Colombo, Sri Lanka.

Bhatti, M.A. and Kijne, J.W. (1992) Irrigation management potential of paddy/rice production in Punjab of Pakistan. In: Murty, V.V.N. and Koga, K. (eds) *Soil and Water Engineering for Paddy Field Management, Proceedings of the International Workshop on Soil and Water Engineering for Paddy Field Management, 28–30 January 1992*. Asian Institute of Technology, Bangkok, Thailand, pp. 355–366.

Bhuiyan, S.I., Sattar, M.A. and Khan, M.A.K. (1995) Improving water use efficiency in rice irrigation through wet seeding. *Irrigation Science* 16(1), 1–8.

Borell, A., Garside, A. and Shu Fukai (1997) Improving efficiency of water for irrigated rice in a semi-arid tropical environment. *Field Crops Research* 52, 231–248.

Bouman, B.A.M. (2001) *Water-efficient Management Strategies in Rice Production*. Notes 16.2, International Rice Research Institute (IRRI), IRRI, Los Baños, Philippines, pp. 17–22.

Bouman, B.A.M. and Tuong, T.P. (2001) Field water management to save water and increase its productivity in irrigated rice. *Agricultural Water Management* 49(1), 11–30.

Bouman, B.A.M., Wopereis, M.C.S., Kropff, M.J., Berge, H.F.M. and Tuong, T.P. (1994) Understanding the water use efficiency of flooded rice fields. II. Percolation and seepage losses. *Agricultural Water Management* 26, 291–304.

Bouman, B.A.M., Wang Hua Qi, Yang Xiao Guang, Zhao Jun Fang and Wang Chang Gui (2002) Aerobic rice (Han Dao): a new way of growing rice in water-short areas. In: *Proceedings of the 12th International Soil Conservation Organization Conference, 26–31 May 2002, Beijing, China*. Tsinghua University Press, Beijing, China, pp. 175–181.

Buresh, R.J., Dawe, D., Tuong, T.P., Ladha, J.K., Bouman, B.A.M., Lantin, R., Peng, S., Mortimer, M. and Hill, J.E. (2001) Sustainable soil and water management of irrigated rice ecosystems. Paper presented at the Integrated Management for Sustainable Agriculture, Forestry and Fisheries Workshop, 28–31 August 2001, Centro Internacional de Agricultura Tropical, Cali, Colombia.

Cabangon, R.J. and Tuong, T.P. (2000) Management of cracked soils for water saving during land preparation for rice cultivation. *Soil and Tillage Research* 56, 105–116.

Cabangon, R., Tuong, T.P., Tiak, E.B. and Abdullah, N.B. (2002) Increasing water productivity in rice cultivation: impact of large-scale adoption of direct seeding in the Muda irrigation system. In: *Direct Seeding in Asian Rice Systems: Strategic Research Issues and Opportunities. Proceedings of an International Workshop on Direct Seeding in Asia, Bangkok, Thailand, 25–28 January 2000.* IRRI, Makati City, Philippines, pp. 299–313.

Castañeda, A.R. and Bhuiyan, S.I. (1993) Sediment pollution in a gravity irrigation system and its effects on rice production. *Agriculture, Ecosystems and Environment* 45, 195–202.

Champoux, M.C., Wang, G., Sarkarung, S., Mackill, D.J., O'Tool, J.C., Huang N. and McCouch, S.R. (1995) Locating genes associated with root morphology and drought avoidance in rice via linkage to molecular markers. *Theoretical and Applied Genetics* 90, 969–981.

De Datta, S.K. (1981) *Principles and Practices of Rice Production.* IRRI, Los Baños, Philippines, 618 pp.

De Datta, S.K., Krupp, H.K., Alvarez, E.I. and Modgal, S.C. (1973) Water management in flooded rice. In: *Water Management in Philippine Irrigation Systems: Research and Operations.* IRRI, Los Baños, Philippines, pp. 1–18.

Deju, Z. and Lu Jingwen (1993) The water-use efficiency of winter wheat and maize on a salt-affected soil in the Huang Huai river plain of China. *Agricultural Water Management* 23, 67–82.

Dong Bin, Loeve, R., Li, Y., Chen Chongde, Deng Li and Molden, D. (2001) Water productivity in Zhanghe irrigation system: issues of scale. In: Barker, R., Loeve, R., Li, Y. and Tuong, T.P. (eds) *Proceedings of the International Workshop on Water Saving Irrigation for Paddy Rice, 23–25 March 2001, Wuhan.* SWIM Publication, International Water Management Institute, Colombo, Sri Lanka, pp. 97–115.

Garrity, D.P., Oldeman, L.R. and Morris, R.A. (1986) Rainfed lowland rice ecosystems: characterization and distribution. In: *Progress in Rainfed Lowland Rice.* IRRI, Manila, Philippines, pp. 3–23.

Garrity, D.P., Vejpas, C. and Herrera, W. (1992) Percolation barriers increase and stabilize rain-fed lowland rice yields on well drained soils. In: Murty, V.V.N. and Koga, K. (eds) *Proceedings of the International Workshop on Soil and Water Engineering for Paddy Field Management, 28–30 January 1992.* Asian Institute of Technology, Rangsit, Thailand, pp. 413–421.

Hill, J.E., Mortimer, A.M., Namuco, O.S. and Janiya, J.D. (2001) Water and weed management in direct-seeded rice: are we headed in the right direction? In: Peng, S. and Hardy, B. (eds) *Rice Research for Food Security and Poverty Alleviation. Proceedings of the International Rice Research Conference, 31 March–3 April, 2000, Los Baños, Philippines.* IRRI, Los Baños, Philippines, pp. 491–510.

Hou, A.X., Chen, G.X., Wang, Z.P., Van Cleemput, O. and Patrick, W.H., Jr (2000) Methane and nitrous oxide emissions from rice field in relation to soil redox and microbiological processes. *Soil Science Society of America Journal* 64, 2180–2186.

Huke, R.E. and Huke, E.H. (1977) Rice area by type of culture: South, Southeast, and East Asia. A revised and updated database. IRRI, PO Box 933, Manila, Philippines.

IRRI (1998) *Sustaining Food Security Beyond the Year 2000: a Global Partnership for Rice Research, IRRI Rolling Medium-term Plan 1999–2001.* IRRI, Los Baños, Philippines.

IWMI (International Water Management Institute) (2000) *World Water Supply and Demand.* IWMI, Colombo, Sri Lanka.

Jha, K.P., Chandra, D. and Challaiah (1981) Irrigation requirement of high-yielding rice varieties grown on soils having shallow water-table. *Indian Journal of Agricultural Science* 51, 732–737.

Kam, S.P., Fajardo, S.D., Rala, A.B., Hossain, M., Tuong, T.P., Bouman, B.A.M. and Banik, P. (2000) *Multi-scale Drought Risk Analysis of Rainfed Rice Environment.* 30th Annual Scientific Conference of the Crop Science Societies of the Philippines, Batac, Ilocos Norte, Philippines.

Khepar, S.D., Sondi, S.K., Kumar, S. and Singh, K. (1997) *Modelling Effects of Cultural Practices on Water Use in Paddy Fields – a Case Study.* Research Bulletin, Publication No. NP/SWE-1, Punjab Agricultural University, Ludhiana, India.

Kitamura, Y. (1990) *Management of an Irrigation System for Double Cropping Culture in the Tropical Monsoon Area.* Technical Bulletin 27, Tropical Agriculture Research Centre, Ministry of Agriculture, Forestry and Fisheries, Tsukuba, Ibaraki, Japan.

Klemm, W. (1999) Water saving in rice cultivation. In: *Assessment and Orientation Towards the 21st Century. Proceedings of 19th Session of the International Rice Commission, Cairo, Egypt, 7–9 September 1998.* FAO, Rome, pp. 110–117.

Li., Y.H. (2001) Research and practice of water saving irrigation for rice in China. In: *Proceedings of the International Workshop on Water Saving Irrigation for Paddy Rice, 23–25 March 2001, Wuhan, China.* Wuhan University, Wuhan, China, pp. 135–144.

Lilley, J.M. and Ludlow, M.M. (1996) Expression of osmotic adjustment and dehydration tolerance in diverse rice lines. *Field Crops Research* 48, 185–197.

McCauley, G.N. (1990) Sprinkler vs. flooded irrigation in traditional rice production regions of southeast Texas. *Agronomy Journal* 82, 677–683.

Mishra, H.S., Rathore, T.R. and Pant, R.C. (1990) Root growth, water potential, and yield of irrigated rice. *Irrigation Science* 17, 69–75.

Moody, K. (1993) The role of water for weed control in direct-seeded rice. Paper presented at the Workshop on Water and Direct Seeding for Rice, 14–16 June 1993, Muda Agricultural Development Authority, Ampang Jajar, Alor Setar, Malaysia.

Moridis, G.J. and Alagcan, M. (1992) High-frequency basin irrigation design for upland crops in rice lands. *Irrigation and Drainage Engineering* 118, 564–583.

My, T.V., Tuong, T.P., Xuan, V.T. and Nghiep, N.T. (1995) Dry seeding rice for increased cropping intensity in Long An Province, Vietnam. In: Denning, G.L. and Vo-Tong Xuan (eds) *Vietnam and IRRI: a Partnership in Rice Research.* IRRI, Manila, Philippines, and Ministry of Agriculture and Food Industry, Hanoi, Vietnam, pp. 111–122.

Naylor, R. (ed.) (1996) *Herbicides in Asian Rice: Transitions in Weed Management.* Institute for International Studies, Stanford University, Stanford, and IRRI, Palo Alto, California, and Manila, Philippines, 270 pp.

Peng, S., Laza, R.C., Khush, G.S., Sanico, A.L., Visperas, R.M. and Garcia, F.V. (1998) Transpiration efficiencies of indica and improved tropical japonica rice grown under irrigated conditions. *Euphytica* 103, 103–108.

Pingali, P.L., Hossain, M. and Gerpacio, R.V. (1997) *Asian Rice Bowls: the Returning Crisis?* CAB International, Wallingford, UK, in association with IRRI, Los Baños, Philippines, 341 pp.

Pinter, P.J., Zipoli, G., Reginato, R.J., Jackson, R.D., Idso, S.B. and Hohman, J.P. (1990) Canopy temperature as an indicator of differential water use and yield performance among wheat cultivars. *Agricultural Water Management* 18, 35–48.

Postel, S. (1997) *Last Oasis. Facing Water Scarcity.* Norton, New York, 239 pp.

Ray, J.D., Yu, L., McCuch, S.R., Champoux, M., Wang, G. and Nguyen, H.T. (1996) Mapping quantitative trait loci associated with root penetration ability in rice (*Oryza sativa* L.). *Theoretical and Applied Genetics* 92, 627–636

Saleh, A.F.M. and Bhuiyan, S.I. (1995) Crop and rainwater management strategies for increasing productivity of rain-fed rice systems. *Agricultural Systems* 49, 259–276.

Sandhu, B.S., Khera, K.L., Prihar, S.S. and Singh, B. (1980) Irrigation need and yield of rice on a sandy-loam soil as effected by continuous and intermittent submergence. *Indian Journal of Agricultural Science* 50(6), 492–496.

Seckler, D. (1996) *The New Era of Water Resources Management.* Research Report 1, International Irrigation Management Institute, Colombo, Sri Lanka.

Sha, T., Molden, D., Sakthivadivel, R. and Seckler, D. (2000) *The Global Groundwater Situation: Overview of Opportunities and Challenges.* IWMI, Colombo, Sri Lanka.

Sharma, D.K., Kumar, A. and Singh, K.N. (1990) Effect of irrigation scheduling on growth, yield and evapotranspiration of wheat in sodic soils. *Agricultural Water Management* 18, 267–276.

Sharma, P.K., Ingram, K.T. and Harnpichitvitaya, D. (1995) Subsoil compaction to improve water use efficiency and yields of rain-fed lowland rice in coarse-textured soils. *Soil Tillage Research* 36, 33–44.

Shu Geng, Zhou, Y., Zhang, M. and Smallwood, K.S. (2001) A sustainable agro-ecological solution to water shortage in the North China Plain (Huabei Plain). *Journal of Environmental Planning and Management* 44, 345–355.

Stegman, E.C. (1982) Corn grain yield as influenced by timing of evapotranspiration deficits. *Irrigation Science* 3, 75–87.

Stockle, C.O., James, L.G., Bassett, D.L. and Saxton, K.E. (1990) Effect of evapotranspiration underprediction on irrigation scheduling and yield of corn: a simulation study. *Agricultural Water Management* 19, 167–179.

Tabbal, D.F., Lampayan, R.M. and Bhuiyan, S.I. (1992) Water-efficient irrigation technique for rice. In: Murty, V.V.N. and Koga, K. (eds) *Soil and Water Engineering for Paddy Field Management. Proceedings of the International Workshop on Soil and Water Engineering for Paddy Field Management, 28–30 January 1992.* Asian Institute of Technology, Bangkok, Thailand, pp. 146–159.

Tabbal, D.F., Bhuiyan, S.I. and Sibayan, E.B. (2002a) Dry-seeding technique for saving water in irrigated rice production systems. In: *Direct Seeding in Asian Rice Systems: Strategic Research Issues and Opportunities. Proceedings of an International Workshop, on Direct Seeding in Asia, Bangkok, Thailand, 25–28 January 2000.* IRRI, Makati City, Philippines, pp. 281–297.

Tabbal, D.F., Bouman, B.A.M., Bhuiyan, S.I., Sibayan, E.B. and Sattar, M.A. (2002b) On-farm strategies for reducing water input in irrigated rice: case studies in the Philippines. *Agricultural Water Management* 56(2), 93–112.

Thompson, J. (1999) Methods for increasing rice water use efficiency. In: *Rice Water Use Efficiency Workshop Proceedings.* CRC for Sustainable Rice Production, Lecton, pp. 45–46.

Tuong, T.P. (1999) Productive water use in rice production: opportunities and limitations. *Journal of Crop Production* 2(2), 241–264.

Tuong, T.P., Cabangon, R. and Wopereis, M.C.S. (1996) Quantifying flow processes during land soaking of cracked rice soils. *Society of Soil Science America Journal* 60(3), 872–879.

Tuong, T.P., Pablico, P.P., Yamauchi, M., Confesor, R. and Moody, K. (2000) Increasing water productivity and weed suppression of wet seeded rice: effect of water management and rice genotypes. *Journal of Experimental Agriculture* 36, 1–19.

Turner, N.C. (1997) Further progress in crop water relations. *Advanced Agronomy* 58, 293–338.

Wang Huaqi, Bouman, B.A.M., Dule Zhao, Wang Changgui and Moya P.F. (2003) Aerobic rice in northern China – opportunities and challenges. In: Bouman, B.A.M., Hengsdijk, H., Hardy, B., Bidraban, P.S., Tuong, T.P., Lafitte, R. and Ladha, J.K. (eds) *Water-wise Rice Production.* IRRI, Los Baños, Philippines, pp. 143–154.

Wei Zhang and Si-tu Song (1989) Irrigation model of water saving-high yield at lowland paddy field. In: *International Commission on Irrigation and Drainage, Seventh Afro-Asian Regional Conference, 15–25 October 1989,* Vol. I-C. ICID (International Commission for Irrigation and Drainage), Dehli, India, pp. 480–496.

Westcott, M.P. and Vines, K.W. (1986) A comparison of sprinkler and flooded irrigation for rice. *Agronomy* 78, 637–640.

Wickham, T.H. and Singh, V.P. (1978) Water movement through wet soils. In: *Soils and Rice.* IRRI, Los Baños, Philippines, pp. 337–358.

5 Managing Saline and Alkaline Water for Higher Productivity

N.K. Tyagi
Central Soil Salinity Research Institute, Haryana, India

Abstract

Two major approaches to improving and sustaining high agricultural productivity in a saline environment involve: (i) modifying the environment to suit the available plants; and (ii) modifying the plants to suit the existing environment. They could be used separately or together to make possible the productive utilization of poor-quality water without compromising the sustainability of the production resource at different management levels. This chapter discusses the issues arising from the use of these approaches as related to the use of marginal-quality water, at both field and irrigation-system levels.

The results are reviewed of field studies encompassing areas with low to moderate monsoonal rainfall (400–600 mm), underlain by saline/alkaline water and supplemented with deficit canal-water supplies, sufficient only to meet 40–50% of irrigation requirements. Analysis of the results indicates that there are good possibilities of achieving reasonably high water productivity on a sustainable basis by appropriate technological interventions. Some important interventions that have been identified include *in situ* conservation of rainwater in precisely levelled fields; blending saline/alkaline and fresh water to keep the resultant salinity below threshold or to achieve its amelioration; and, if residual sodium carbonate cannot be brought down to acceptable levels, dilution-blending or cyclic application and scheduling irrigation with salty water at less salt-sensitive stages. In high-water-table areas, provision of subsurface drainage facilitates the use of higher-salinity water, reducing the overall irrigation requirement. At higher levels of irrigation systems, it was found that water productivity in saline environments can be improved by a number of measures. These include reallocation of water to higher-value crops with a limited irrigation requirement, spatial reallocation and transfer of water-adopting polices that favour development of water markets and reducing mineralizing of fresh water by minimizing application and conveyance losses that find a path to saline aquifers.

In spite of the technological advances that mitigate salinity damage and the likely economic advantages, there is always a need to exercise caution while practising irrigation with salty water for maintaining sustained productivity.

Introduction

Water productivity in agriculture, which is often used as a criterion for decision-making on crop-production and water-management strategies, is severely constrained by salinity of land as well as of water. Salinity of water is more common than that of the land and it is often the cause of salinity development in soils, largely because of the misuse of salty

 Water Productivity in Agriculture: Limits and Opportunities for Improvement (eds J.W. Kijne, R. Barker and D. Molden)

water for crop production. There are two major approaches to improving and sustaining productivity in a saline environment: modifying the environment to suit the plant and modifying the plant to suit the environment. Both these approaches have been used, either singly or in combination (Tyagi and Sharma, 2000), but the first approach has been used more extensively because it enables the plants to respond better not only to water but also to other production inputs. The development of the management options requires the analysis of sensitivity parameters that affect interaction between salinity and crop yield (Zeng *et al.*, 2001). The sensitivity of crop growth stages often determines management options to minimize yield reductions and to promote the use of salty water. Most management practices aim at keeping salinity in the crop root zone below the threshold salinity of the given crop at the growth stage in consideration. Though the general threshold limits are fairly well established (Maas, 1990), the threshold salinities for different stages are not well defined. The information gap is more serious for alkaline water than for saline water.

Most studies on the effect of salty water on crop yield refer to individual crops, but, in actual practice, the interseasonal salinity balance that actually influences the crop yields is greatly modified by the cropping sequence. The management practices also vary according to the cropping system followed. Therefore, it is important to consider the saline/alkaline water-use practices not only for individual crops but also for the cropping system.

In the past, water productivity has been expressed either in terms of irrigation efficiency (the term mostly used by engineers) or in terms of water-use efficiency (mostly used by agriculturists). The first term has a hydrological basis and can be extended from field to river-basin scale. In other words, the irrigation efficiency can be defined in a system, with one level having a relationship to the other in the irrigation-system hierarchy. This issue is discussed in other chapters in this volume (e.g. by Seckler *et al.*, Chapter 3, and Molden *et al.*, Chapter 1) and is of great importance in planning saline-water use. Most agricultural research has treated saline/alkaline water use in the context of root-zone salinity management, involving the application or withholding of irrigation to maintain an environment favourable to crop production. This approach has enabled the development of management practices at field level without considering their implications and practicability at the farm/irrigation-system/river-basin levels. It should, however, be clearly understood that, just like the water balance, the salinity balance also has to be maintained at field and irrigation-system/basin levels (Tyagi, 2001). Manipulation of water diversions of different qualities and origins can be successfully used as a tool for enhancing water productivity on a sustainable basis (Srinivasulu *et al.*, 1997). Such manipulations would normally involve reallocation and intrasystem/intraseason water transfers, which could be facilitated by development of water markets (Strosser, 1997). This process could begin at the watercourse level, which is the lowest level of large traditional irrigation systems in countries like India and Pakistan, and spread upward in the system hierarchy.

Lastly, productivity should be understood not only in terms of physical outputs, such as grain or biomass yield, but also in economic terms, such as revenue or profit earned per unit of water diverted, at different levels of the irrigation system. Some time ago, much concern was expressed in the state of Haryana (India) when an overall decline in productivity was reported in certain rice-growing areas (Anon., 1998); but, later on, it was discovered that the decline in productivity was due not to any malfunctioning of the system, but to a shift from high-yielding coarse rice varieties to more remunerative basmati rice, which had a lower yield but fetched a far higher price in the market. Incidentally, a salt-tolerant variety of basmati rice (CSR-30) is now available.

Productivity-enhancing measures are discussed that involve the use of saline/alkaline water at field level, such as conjunctive use, water-table management, rainwater conservation in precisely levelled basins and chemical amelioration of alkaline water. Though not exclusive, this discussion of the productivity-enhancing measures is in the context of the rice–wheat system in a monsoonal climate with moderate rainfall (400–600 mm), as prevails in north-west India, where the occurrence of saline/alkaline water is more prevalent (Fig. 5.1). Water reallocation and transfer, water markets and saline-water disposal, which have irrigation-system/basin-level implications, are also briefly presented.

Salinity/Alkalinity Hazards

The most important criterion for evaluating salinity hazards is the total concentration of salts. The quantity of salts dissolved in water is usually expressed in terms of electrical conductivity (EC), mg l^{-1} (p.p.m.) or meq l^{-1}. The cations Na^{+}, Ca^{2+} and Mg^{2+} and the anions Cl^{-}, SO_4^{2-}, HCO_3^{-} and CO_3^{2-} are the major constituents of saline water. Plant growth is adversely affected by saline water, primarily through excessive salts raising the osmotic pressure of the soil solution, resulting in reduced water availability. In field situations, the first reaction of plants to the application of saline water is reduced germination. This reduced initial growth results in smaller plants (lower leaf-area index). Experimental evidence indicates that the interplay of several factors, such as the evaporative demand, salt content, soil type, rainfall, water-table conditions and type of crop and water-management practices, determines salinity build-up in the soil and crop performance resulting from long-term application of saline water.

Some water, when used for the irrigation of crops, has a tendency to produce alkalinity/sodicity hazards, depending upon the absolute and relative concentrations of specific cations and anions. The alkalinity is generally measured in terms of the sodium adsorption ratio (SAR), residual sodium carbonate (RSC) and adjusted SAR. Irrigation

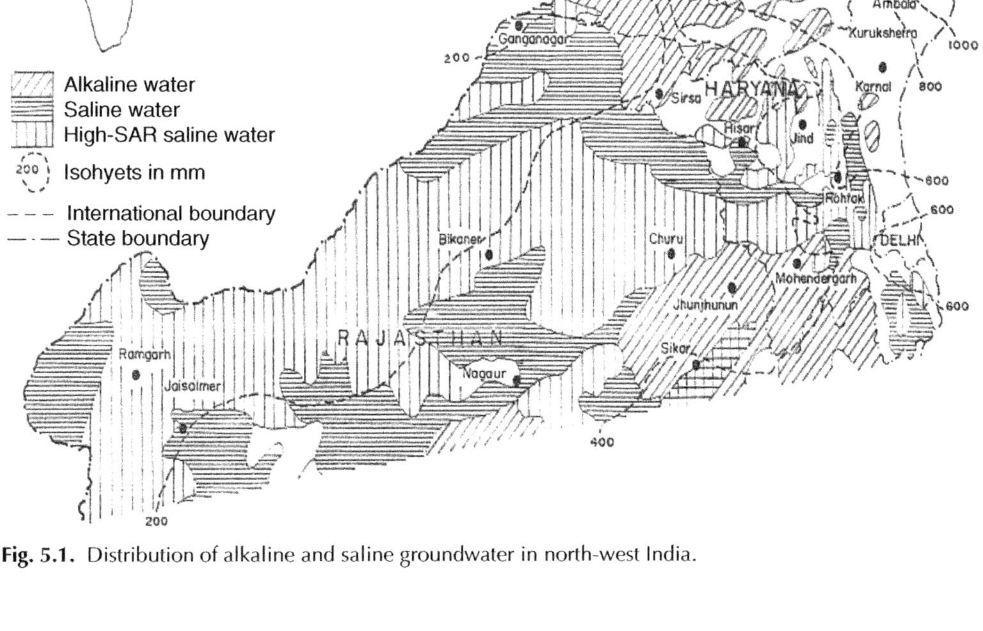

Fig. 5.1. Distribution of alkaline and saline groundwater in north-west India.

with sodic water contaminated with Na^+ relative to Ca^{2+} and Mg^{2+} and high carbonate (CO_3^{2-} and HCO_3^-) leads to an increase in alkalinity and sodium saturation in soils. The increase in exchangeable sodium percentage (ESP) adversely affects soil physical properties, including infiltration and aeration. In the early stages of sodic irrigation, large amounts of divalent cations are released into the soil solution from exchange sites. In a monsoonal climate, alternating irrigation with sodic water and rainwater induces cycles of precipitation and dissolution of salts. Several field observations have shown that, although steady-state conditions are never reached in a monsoonal climate, a quasi-stable salt balance is reached within 4–5 years of sustained sodic irrigation, while a further rise in pH and ESP is very low (Minhas and Tyagi, 1998).

Seasonal Water Balance and Salinization and Desalinization Cycles

In north-west India, the annual weather exhibits three distinct phases, the first of which is the hot and humid season from mid-June to September, when about 80% of the rainfall takes place. This phase covers the growing period of kharif crops, i.e. cotton, pearl millet, maize, sorghum and paddy. The second phase is the the cool and dry season from October to March, which covers the growing period of most rabi crops, including wheat, mustard, gram and barley. The third phase is characterized by hot and dry weather, which prevails from April to mid-June, which covers part of the growing periods of wheat, cotton and maize. A seasonal water-balance analysis shows that, in relative terms, winter and summer months, being dry, are water-deficit periods, whereas the kharif season from mid-June to September has some surplus water (Fig. 5.2). The salinity build-up in the soil is greatly influenced by the weather and the irrigation practice. In waterlogged saline areas, maximum salinity is observed in the pre-monsoonal period in June. This is because, after the first week of April, wheat, which is the dominant irrigated crop, receives no irrigation till its harvest. From mid-April till mid-June, the land remains mostly fallow, when there is no irrigation and there is an upward moisture flux due to high evaporative demand, which results in salinity build-up. With the onset of the monsoon and the planting of crops that receive irrigation, the desalinization of the soil profile takes place, and the salinity reaches a minimum value in

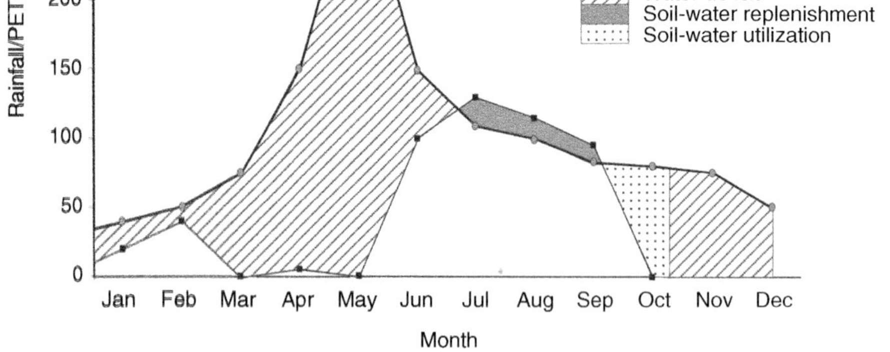

Fig. 5.2. Annual climatic water balance at Karnal. PET, potential evapotranspiration.

October (Fig. 5.3). From November to February, the evaporative demands are low (the value reaches less than 1 mm day^{-1} in December–January) and therefore the upward flux is low. The low initial salinity in the beginning of the rabi season favours saline irrigation, which is further facilitated by low evaporative demands during this season. This limits the rate of salinization in the soil profile due to saline irrigation. By the time the summer season starts, the crops are mature and are able to tolerate higher salinity. The monsoonal water leaches the salts accumulated during the winter and early summer, which is why the limits for the use of saline/sodic water can be higher in this region than recommended elsewhere.

Root-zone Salinity Management

Most research on the use of saline/alkaline water has focused on keeping root-zone salinity under control by various management practices. The important practices include multi-quality water use in different modes, scheduling irrigation with saline water in a manner that avoids its application at sensitive stages, use of chemical amendments, precision levelling and high-frequency irrigation, etc. In situations where high water tables with saline water prevail, subsurface drainage and water-table manipulation are often introduced to promote the use of brackish water.

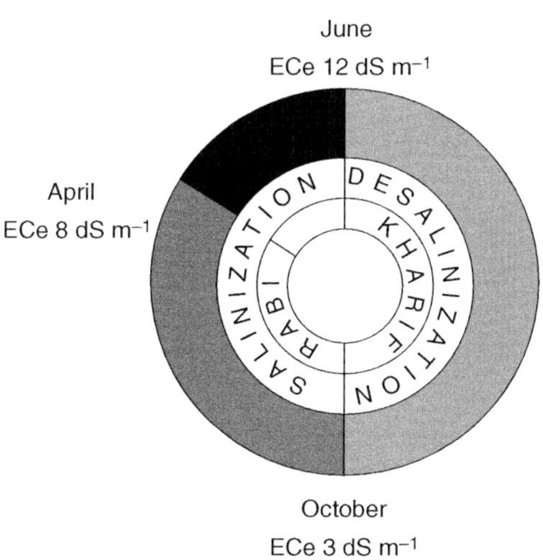

Fig. 5.3. Salinization and desalinization cycle in monsoonal climate. ECe, EC of the soil saturation extract.

Multi-quality irrigation practices

Possible ways of practising multi-quality water use are as shown below. These include direct application of salty water, as well as different modes of blending or cyclic use.

Water-application modes and their impact on productivity

Among the various application modes, direct application of saline water can be practised where salinity of the water is such that a crop can be grown within acceptable yield levels without adversely affecting soil health. It was reported by Boumans *et al.* (1988) that marginal-quality water (EC of 4–6 dS m^{-1}) was being used directly in several locations in Haryana. The average yield depressions for crops, including cotton, millet, mustard and wheat, were less than 20%. When higher-salinity water is used directly, a pre-sowing irrigation, if required, is given with fresh water. To practise joint use of saline and freshwater, the available options are blending and the cyclic mode. Blending is promising in areas where fresh water can be made available in adequate quantities on demand. The potential for blending two different supplies depends on the crops to be grown, salinities and quantities of the two water supplies and the economically acceptable yield reductions. Cyclic use is most common and offers several advantages over blending (Rhoades *et al.*, 1992). In sequential application under the cyclic mode, the use of fresh water and saline water is alternated according to a predesigned schedule. Sometimes, there is inter-seasonal switching, where supplies of fresh water and saline water are applied in different seasons. In a field study, Sharma and Rao (1996) found that saline drainage effluents could be used in different modes without appreciable yield reduction in a wheat crop (Table 5.1).

Table 5.1. Effect of different salinity levels of applied water (blending and cyclic application) over a period of 6 years (1986/87 to 1991/92) on grain yield of wheat.[a]

	Blending		Cyclic application		
EC_{iw} ($dS\ m^{-1}$)	Mean yield ($t\ ha^{-1}$)	Relative yield (%)		Mean yield ($t\ ha^{-1}$)	Relative yield (%)
< 0.6 (FW)	6.0	100	4 FW	6.0	100
6	5.8	96.0	FW + DW	5.8	96.7
9	5.0	80.3	DW: FW	5.6	93.3
12	5.0	80.3	2 FW + 2 DW	5.7	95.0
12 (DW)	4.7	78.3	2 DW + 2 FW	5.4	90.0
			1 FW + 3 DW	5.1	85.0
			4 DW	4.5	75.0

[a]The drainage water had an EC = 12.5–27 $dS\ m^{-1}$ and SAR = 12.3–17.
FW, fresh water; DW, drainage water.

Impact of saline-water use on soil health

The salinity build-up in soil profiles after 6 years of irrigation with different-quality water, in fields provided with subsurface drainage, is shown in Fig. 5.4 (Sharma and Rao, 1996). It can be seen that, for all water with salinity in the range of 0.5–12 $dS\ m^{-1}$, soil salinity at the end of the monsoonal season is reduced to less than 4 $dS\ m^{-1}$.

Several studies have suggested that irrigation water containing salt concentrations exceeding conventional suitability standards can be used successfully on many crops for at least 6–7 years without significant loss in yield. However, uncertainty still exists about the long-term effects of these practices. Long-term effects on soil could include soil dispersion, crusting, reduced water-infiltration capacity and accumulation of toxic elements. The effects on some soil properties (sandy loam soils) of irrigation with high-salinity drainage effluent, as practised in the Sampla drainage area (Haryana), were moni-

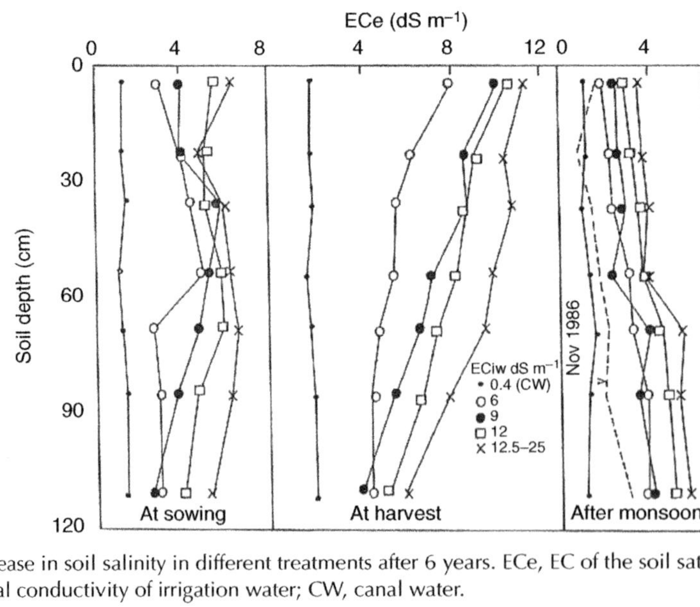

Fig. 5.4. Increase in soil salinity in different treatments after 6 years. ECe, EC of the soil saturation extract; EC_{iw}, electrical conductivity of irrigation water; CW, canal water.

tored for 6 years. Since the SAR of saline drainage water was more (12.3–17.0) than that of canal water (0.7), its use increased soil SAR in all the treatments (Fig. 5.5).

Leaching of salts by monsoonal rains reduced the SAR of the soil saturation extract (SARe) in all the treatments and the remaining SARe values did not constitute any alkaline hazard to the succeeding crops. Similarly, no significant adverse effects were observed on saturated hydraulic conductivity or water-dispersible clay after the monsoonal rains. A slight decrease in hydraulic conductivity after monsoonal leaching will not be a problem during the irrigation season since the negative effect of high SAR of drainage water is offset by the high salinity of the drainage water. The slight variation in water-dispersible clay after 6 years of irrigation with drainage effluent indicates only minimal structural deterioration in soils irrigated with high-salinity drainage effluent. Although no potential adverse effects were observed in these studies at the Sampla farm (Haryana), caution should be exercised when considering the reuse of drainage effluent and the specific conditions should be carefully evaluated.

Use of alkaline water and chemical amelioration

Water having alkalinity/sodicity problems is encountered on a large scale in the rice–wheat-growing areas of Punjab and Haryana in north-west India. Several studies have shown that this water can be used under certain conditions. In a study conducted over a period of 6 years (1981–1987) by Bajwa and Josan (1989), it was found that irrigation with sodic water given after two turns of irrigation with fresh water, to rice as well as to wheat, helped in obtaining yields comparable to those with irrigation with fresh water (Table 5.2). Crop yields even in the case of alternate irrigation with sodic and fresh water were only marginally less than when fresh water alone was used. On average, rice received 18 irrigations, whereas only five turns of irrigation of 6 cm were applied to wheat. In all cases, pre-sowing irrigation was given with fresh water and no amendments to neutralize sodicity were applied. At the end of 6 years, the ESP in plots irrigated entirely with sodic water increased from 3.5 to 46% whereas in alternate irrigation with fresh water and sodic water the ESP

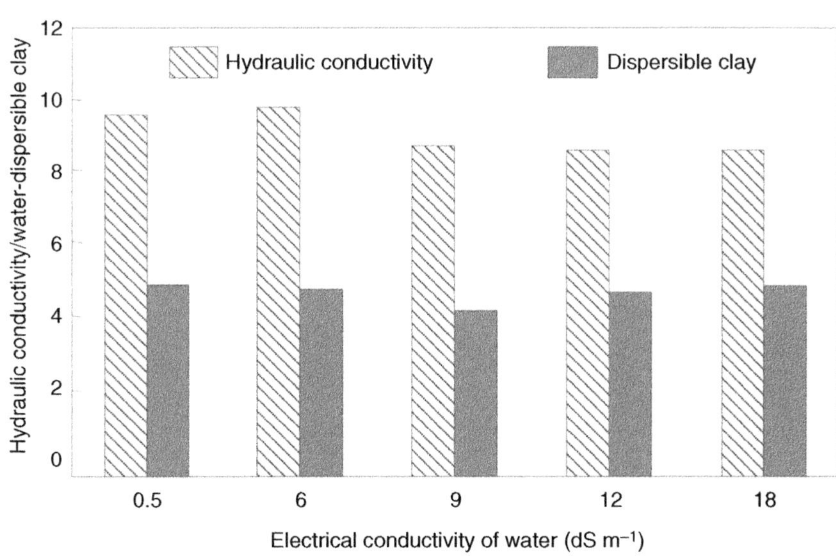

Fig. 5.5. Saturated hydraulic electrical conductivity (mm h^{-1}) of soil saturation extract measured three times during the year, and water-dispersible clay (%) of 0–30 cm layer.

Table 5.2. Average grain yield of rice and wheat as affected by the use of fresh water and alkaline water over a period of 6 years (1981–1986).

		Irrigation-water productivity ($kg\ ha^{-1}\ cm^{-1}$)		
Treatment	Crop yield ($t\ ha^{-1}$)	Rice	Rice–wheat	Wheat
Fresh water (FW)	6.7	5.4	62	180
Alkaline water (AW)	4.2	3.6	39	120
2 FW–AW	62	6.7	5.2	173
FW–AW	58	6.3	5.3	177
FW–2 AW	53	5.7	4.8	160

AW: EC 1.25 dS m^{-1}; SAR = 13.5; RSC = 10 meq l^{-1}.

increased to a level of only 18.2% (Fig. 5.6). The increase in ESP points to the danger involved in the use of these supplies of water.

It should be understood that, when fields are irrigated with poor-quality water, the yields can only be maintained at a lower level than when irrigated with good-quality water if no amendments are applied. The levels at which yields can be sustained depend not only upon the alkalinity of the groundwater but also on the water available from rainfall and canals, etc. Sharma *et al.* (2001), based on a 7-year study (1993–1999), evaluated the sustainable yield index (SYI), which indicates the minimum guaranteed yield as a percentage of the maximum observed yield. The SYI is defined as $(Y - S)/Y_{max}$, where Y is the average yield, S is the standard deviation and Y_{max} the maximum yield (in the study area it was 6 t ha^{-1} for rice and 5 t ha^{-1} for wheat). The SYI ranged from 0.57 to 0.65 in rice and from 0.54 to 0.65 in wheat (Table 5.3) at different doses of applied gypsum. The overall build-up of pH (8.5), SARe (20.7) and EC of the soil saturation extract (ECe) (2.5 dS m^{-1}) in the soil remained below the threshold salinity levels of these crops. This may be due to dilution by rainwater along with the high Ca or Ca + Mg content of the water used. The low level

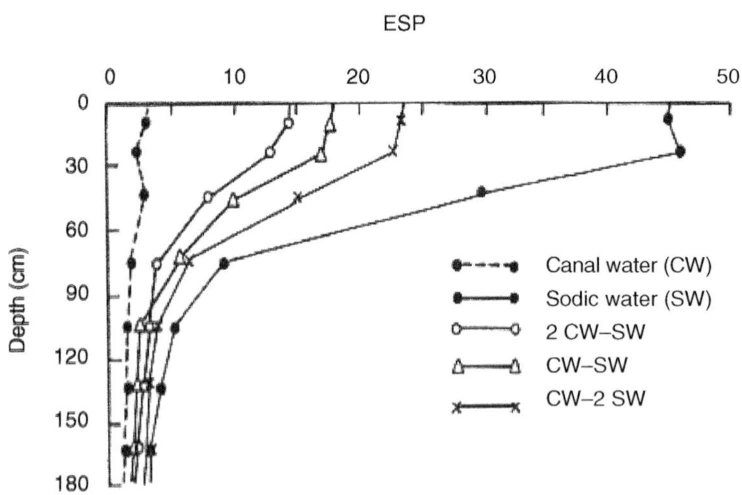

Fig. 5.6. Build-up of exchangeable sodium percentage (ESP) in 0–30 cm soil layer over time (6 years) with sodic water application in different combinations.

Table 5.3. Crop yield and sustainable yield index (SYI) for rice–wheat cropping irrigated with gypsum-amended alkaline water (from Sharma *et al.*, 2001).

Treatment (% GR)	Gypsum applied	Crop yield ($t\ ha^{-1}$)	SYI		
			Rice	Wheat	Rice–wheat
0	0	4.01	3.55	0.57	0.54
12.5	1.24	4.22	3.75	0.60	0.60
25.0	2.50	4.13	3.68	0.60	0.58
50.0	5.00	4.26	3.82	0.61	0.62
75.0	7.50	4.22	3.83	0.62	0.62
100.0	10.00	4.48	3.94	0.62	0.63
Canal water	Nil	4.46	3.85	0.65	0.65

GR, Gypsum requirement for neutralizing completely sodicity.

of sodification could also be attributed to large biological production and dissolution of CO_2 occurring in submerged rice fields. It was concluded that a maximum yield of about 60% in both rice and wheat can be sustained with the use of alkaline water (RSC = 10 meq l^{-1}) if 1.25 t ha^{-1} of gypsum is applied annually to rice–wheat in the medium-rainfall zone (500–600 mm).

Cropping sequence

The irrigation, drainage and agronomic practices vary from crop to crop. Therefore, the crop grown in the previous season greatly influences the production and productivity of the crop in the subsequent season. In a monsoonal climate, crops that favour higher retention and *in situ* conservation of rainwater, which is salt-free, result in lesser salinity/sodicity development in the soil profile at the end of the season, providing a better environment for the next crop. In a 6-year study conducted at the Central Soil Salinity Research Institute (CSSRI) (Sharma *et al.*, 2001), three important cropping sequences (rice–wheat, cotton–wheat and sorghum–wheat) were compared in terms of their productivity when applied with alkaline water. The productivity of the rice–wheat system in kharif and rabi seasons was higher than the sorghum–wheat and cotton–wheat systems (Table 5.4).

Table 5.4. Equivalent rice and wheat yields ($t\ ha^{-1}$) as affected by cropping sequence when irrigated with alkaline water (from Sharma, D.K., 2001, personal communication).

	Equivalent rice yield (kharif)		Equivalent wheat yield (rabi)		Total equivalent yield (wheat)		Soil pH_2	
Cropping sequences	Water quality AW	FW: AW	Water quality AW	FW: AW	Water quality AW	FW: AW	Water quality AW	FW: AW
Sorghum–wheat	2.9	3.5	3.8	4.1	6.22	6.92	9.1	9.0
Rice (basmati)–wheat	4.8	7.0	3.7	4.7	7.62	9.65	9.1	9.0
Cotton–wheat	3.5	4.1	3.5	3.8	6.3	6.66	9.0	9.0
Rice (Jaya)–mustard	4.0	4.3	4.0	4.4	7.27	7.32	9.1	9.0
Rice (Jaya)–berseem (clover)	3.3	4.1	2.7	3.0	5.41	6.31	9.3	9.1

AW, alkaline water; FW, fresh water.

Shallow water-table management

Providing drainage to ensure that the salt concentration does not exceed the level that can be tolerated by crop roots is a requirement for continued productivity. Provision of drainage and leaching over a period of time leads to improvement in the quality of subsoil water in drained fields. The upper few centimetres of subsoil water have very little salinity, and plants could be allowed to use it by manipulating the operation of the drainage system. Thus the plants would meet part of their evapotranspiration needs directly from soil water. The use of groundwater by the crops is related to the water-table depth and the salinity of subsoil water (Chaudhary *et al.*, 1974). Minhas *et al.* (1988) observed that in sandy loam soil with the water table at 1.7 m depth and with groundwater salinity at 8.7 dS m^{-1}, the water table contributed as much as 50% of the requirement when only irrigation was applied.

In another study, a shallow water table at 1.0 m depth with salinity in the range of 3.0 to 5.5 dS m^{-1} gave rise to yield levels equal to the potential yield with good-quality irrigation water, even when the application of surface water was reduced to 50% (Sharma *et al.*, 2001). These fields had been provided with subsurface drainage. The salinity build-up was negligible and the small amount of salt that accumulated was leached in the subsequent monsoonal season. The provision of subsurface drainage also allows the use of higher-salinity water through surface applications (Minhas, 1993; Sharma *et al.*, 2001). The yield reduction with progressively increasing salinity of applied water was much less in fields having a subsurface drainage system than in fields with a deeper water table, which had no need of artificial subsurface drainage. The differences are highly marked at applied water salinities of more than 10 dS m^{-1} (Table 5.5). Relatively higher moisture in the crop root zone in fields with subsurface drainage could be the reason for the higher productivity.

Table 5.5. Relative yield of wheat with saline irrigation under conditions of a deep water table and a high water table but provided with subsurface drainage (from Minhas, 1993; Sharma *et al.*, 1991).

Irrigation-water salinity (dS m^{-1})	Relative yield (%)	
	Deep water table	Shallow saline water table[a]
0.6	95	100
4.0	90	94
8.0	83	86
12.0	60	78
16	42	74[b]

[a]There was provision for subsurface drainage to leach and remove salts.
[b]Salinity varied between 14 and 26.5 dS m^{-1}, the average being 16 dS m^{-1} and the yield varied between 50 and 86%, with an average of 74%.

Improving Economic Efficiency of Water Use

The commonly used definition of water productivity does not take into account the net benefits that accrue from crop production. It should, however, be understood that farmers are interested in increasing water productivity only to the level at which it maximizes their net benefits. The cost of cultivation and the prevailing market price often decide the crop variety that the farmers cultivate, irrespective of the physical water productivity. Growing crops that use less water and have low cost of cultivation but fetch a higher price in the market can enhance economic efficiency. A case in point is the increase in area of basmati rice in several districts of Haryana (Kaithal, Kurukshetra Panipat) in places with marginal-quality water. The yield of basmati rice is only 50% (about 2 t ha^{-1}) of the coarse rice varieties, such as Jaya and IR-8, but its irrigation requirements are about 60–65% of the coarse varieties. Although basmati rice has lower tolerance for sodicity, the supplemental irrigation with alkaline water is also less and its nitrogenous fertilizer demand is only 70% of the coarse variety.

In a field study that involved sequential application of fresh water and alkaline water (FW:AW), the equivalent yield of basmati was 7 t ha^{-1} as compared with only 4.3 t ha^{-1} for Jaya (Table 5.4). The higher economic returns led to its cultivation in a larger area in Haryana, though its physical water productivity may be only half of Jaya or IR-8. In more arid areas, where fresh water during the rabi season is scarce, similar trends are observed with mustard, which replaces wheat because of its much higher salt tolerance and requirement of only one or two post-sowing turns of irrigation compared with four or five turns of irrigation for wheat.

Special Considerations for the Use of Saline/Alkaline Water

The following are the important points that should be considered in developing saline/ alkaline water-use programmes.

Pre-sowing irrigation

Pre-sowing irrigation has a significant influence on crop yields harvested at the end of the season. This is because seed germination and seedling stage are the most sensitive stages. Early salinity stress leads to poor crop stand and considerable yield reduction. The response of wheat to salinity was observed to vary with its growth stage, initial salinity distribution in the soil profile and the modes of saline-water application (irrigation with blended or sequential application) (Sharma *et al.*, 1993). The ECe_{50} (ECe for 50% yield reduction) values increased from 9.3 dS m^{-1} for periods from sowing to crown rooting to 13.2 dS m^{-1} from dough stage to maturity (Fig. 5.7). The effect of pre-sowing irrigation with fresh water and saline water was studied at CSSRI for several crops (Table 5.6). It was observed that one of the most sensitive crops (e.g. mung bean) could sustain irrigation with saline water of 4.7 dS m^{-1} if non-saline water was used at the pre-sowing stage. The water productivity of mung bean, when irrigated with fresh water at pre-sowing and subsequently with saline water (EC_w 4.7), was 41 kg ha^{-1} cm^{-1}, compared with only 12 kg ha^{-1} cm^{-1} when irrigated with saline water throughout the growing period. Though less drastic, a similar trend was observed in mustard. (Note: the values of water productivity are based on water extracted from the soil profile during the growth periods.)

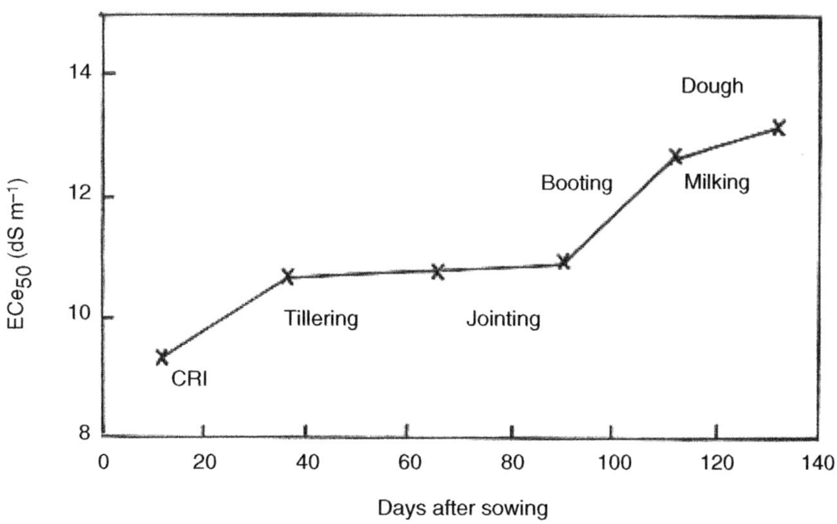

Fig. 5.7. Salinity tolerance of wheat at various growth stages (ECe_{50} denotes ECe for 50% yield reduction). CRI, crown root initation stage.

Table 5.6. Crop yield and water productivity as influenced by irrigation-water salinity and application sequence with different-quality water (from Sharma *et al.*, 1993).

Irrigation-water salinity ($dS\ m^{-1}$)	Water-quality application sequence	Crop yield ($t\ ha^{-1}$)	Water productivity ($kg\ ha^{-1}\ cm^{-1}$)
Mung bean			
0.3	Entire season	2.52	56
4.7	Entire season	0.27	12
4.7	After PI_{FW}	1.56	41
Mustard			
0.3	Entire season	2.32	63
12.3	Entire season	1.05	58
12.3	After PI_{FW}	1.80	64

PI_{FW}, pre-sowing irrigation with fresh water.

Favourable season

Crops grown during the winter season (wheat, mustard and barley) are more tolerant to saline water than those grown during summer (pearl millet, sorghum and groundnut). Also, the soil profile is almost free of salts after the monsoon leaching and has a capacity to receive salts without exceeding critical limits. Added to this is the more favourable evapotranspiration regime of the winter season. Evapotranspiration peaks again after March, when the crop is mature and can tolerate higher salinity.

Crop substitution

Most agricultural crops differ significantly in their tolerance of a concentration of soluble salts in the root zone. It is desirable to choose crops/varieties that can produce satisfactory yields under the conditions resulting from irrigation with saline water. The difference between the tolerance of the least and the most sensitive crops may be eight- to tenfold. This wide range of tolerance allows for considerable use of marginal water supply. The extent by which the tolerance limits for the use of low-quality water are raised governs the greater use of such water, thereby reducing the need for leaching and drainage (Tyagi, 1998). Semi-tolerant to tolerant crops and those with low water requirements should be grown. For example, mustard is salt-tolerant and it requires only one or two turns of irrigation after seeding. Experiments at Sampla (Haryana) indicated that highly saline drainage water can be used for post-planting irrigations of mustard without any substantial loss in yield. Thus mustard can be substituted for wheat in part of the area because it tolerates salinity of up to 6 $dS\ m^{-1}$ for normal yields.

Precision levelling

The use of saline and alkaline water supplies often requires the application of smaller depths at relatively more frequent intervals. In surface-water application methods, the distribution of water and the application depths are greatly influenced by the quality of land levelling. Salinity and non-uniformity in irrigation water have much the same effect on the yield–water response function and both require larger volumes of irrigation water to produce the same yields as can be obtained with non-saline water and uniformly applied water (Howell *et al.*, 1990). In surface irrigation, the uniformity of the soil surface affects the required application depths. In a field study (Tyagi, 1984), it was observed that the system application depth ranged from 40 to 120 mm as the levelling quality decreased

(Fig. 5.8). Higher application depths were associated with lower application efficiencies: with a levelling index (LI) of 0.75 cm, the application efficiency was as high as 90% compared with 45% at an LI of 6.75 cm. The non-uniformity in levelling was reflected in a water-productivity value of 93.1 kg ha^{-1} cm^{-1} at LI = 0.75 cm to 59.1 kg ha^{-1} cm^{-1} at LI = 6.75 cm. The study indicated that to ensure a desired system application depth of 5–6 cm, required to achieve optimum productivity and income, the levelling quality had to be such that the average deviation from the desired depth was less than 3 cm.

Rainwater conservation

Rainwater conservation is the key to the use of poor-quality water as it not only meets part of the irrigation requirements but also facilitates leaching of salt. The quantity of rain that can be conserved within the field depends upon the crop grown during the monsoonal season. Rice paddies offer the most appropriate conditions for retaining rainwater within the field. Raul *et al.* (2001) showed that, in parts of Kalayat and Rajaund administrative blocks in Haryana (India) having alkaline water with an RSC between 5 and 10 meq l^{-1}, rice paddies enabled *in situ* conservation of 95% of monsoonal rains, thereby helping to sustain rice–wheat cropping on 60–70% of the area. In these blocks, between 30 and 40% of the irrigation requirement of rice and over 50% for wheat is met by groundwater mixed with conserved rain, which dilutes the saline/alkaline groundwater to make it usable. Rainwater conservation and the use of gypsum sustain the continued use of these alkaline water supplies in the region.

Enhancing and Sustaining Water Productivity at Irrigation-system Level

One of the options to improve water productivity in physical and economic terms is the transfer of water and spatial reallocation through a change in the water-allocation policies or through a water market. Other options include diversion of water to more productive and profitable uses and reducing salinization of fresh water in areas underlain by saline/alkaline aquifers by improving the on-farm irrigation conveyance efficiency. The sustainability of saline agriculture can be ensured by maintaining the salinity balance within the river basin through evacuation and disposal of salt water to areas outside the basin.

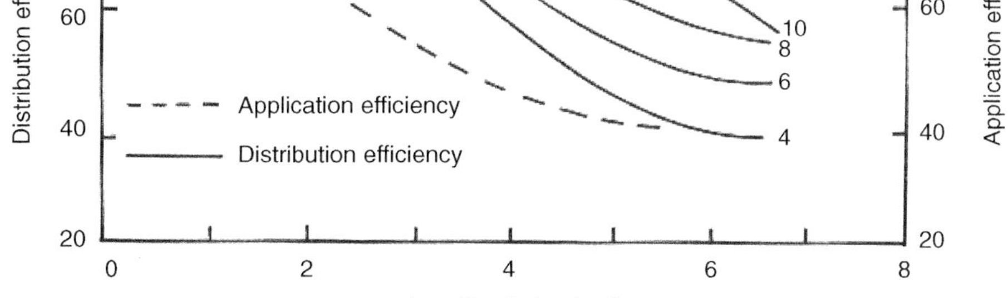

Fig. 5.8. Relationship between levelling index and distribution efficiency at different irrigation depths (from Tyagi, 1984).

Loss in productivity due to salinization of fresh water and its prevention

Fresh water that is lost through seepage and percolation in areas underlain by saline aquifers also becomes saline. Though this water can be reused for irrigation, crop yields will be less. How much less depends on the salt tolerance of the crop, cropping pattern, quantity and quality of applied water and climatic conditions. Obviously, the losses in production and productivity are area-specific. An attempt to estimate the production losses with increasing salinity of groundwater used for irrigation was made for Sirsa and Hisar districts in Haryana and is shown in Fig. 5.9. The financial losses with groundwater salinity of up to 3 dS m^{-1} were within Rs 500 ha^{-1} $year^{-1}$. At higher salinity levels, the losses increased at a very high rate, reaching Rs 8000 ha^{-1} $year^{-1}$ at a groundwater salinity of 10 dS m^{-1}, which has a profound effect on the profitability of the farming enterprise. In areas underlain by saline aquifers, percolation and seepage losses should therefore be reduced as much as possible. Tyagi and Joshi (1996) investigated the techno-economic viability of reducing accretions to groundwater in saline groundwater areas through irrigation-system improvements. Reducing salinization of groundwater by cutting down on up to 75% of the application, distribution and conveyance losses had a high profitability.

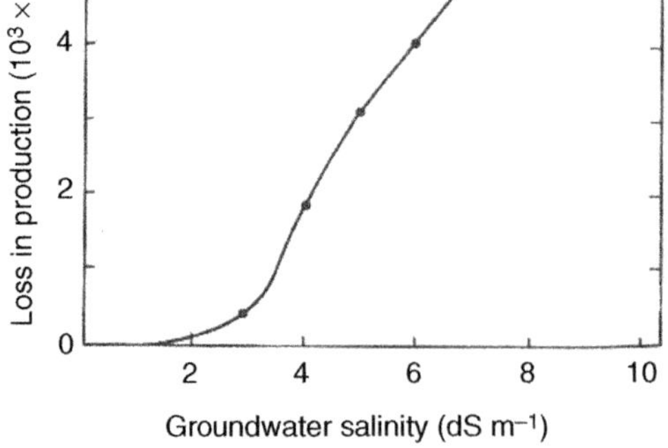

Fig. 5.9. Agricultural production losses as a function of groundwater salinity.

Conjunctive use

Supplies of both fresh water and saline water are limited but the availability of saline groundwater is more dependable. For a given level of canal water and salinity of the groundwater, the farming enterprise will remain profitable until the incremental benefits balance the incremental costs.

A profitability analysis was carried out for wheat irrigated with saline groundwater at a given level of canal-water supply for a watercourse command area in the Kaithal district to see how far the application of saline water would remain economically viable (Anon., 2001). Two levels of canal-water supply (10 and 15 cm ha^{-1}) were considered. It was found that the profit decreased from Rs 12,000 ha^{-1} to Rs 7000 ha^{-1} when the canal-water supply was decreased from 15 cm to 10 cm with a groundwater (EC = 6 dS m^{-1}) use of 15 cm (Fig. 5.10). Since the overall availability of groundwater at system level is also limited, the chance of minimizing productivity losses by applying more groundwater does not appear to be feasible. The only option is to reduce irrigation intensity (irrigated area/cropped area) and to arrive at an optimal mix of irrigated and rain-fed areas.

Productivity increase through the promotion of a groundwater market at watercourse level

The large difference in supply between the head and the tail reaches is a common problem. This problem gets compounded when there is a high overall deficit in canal supplies needed to meet the demand of the culturable command area (CCA) of the canal system. Typical examples are the western Yamuna and Bhakra canal system, where the canal-water supplies are adequate to meet only 30–50% of irrigation demands per crop

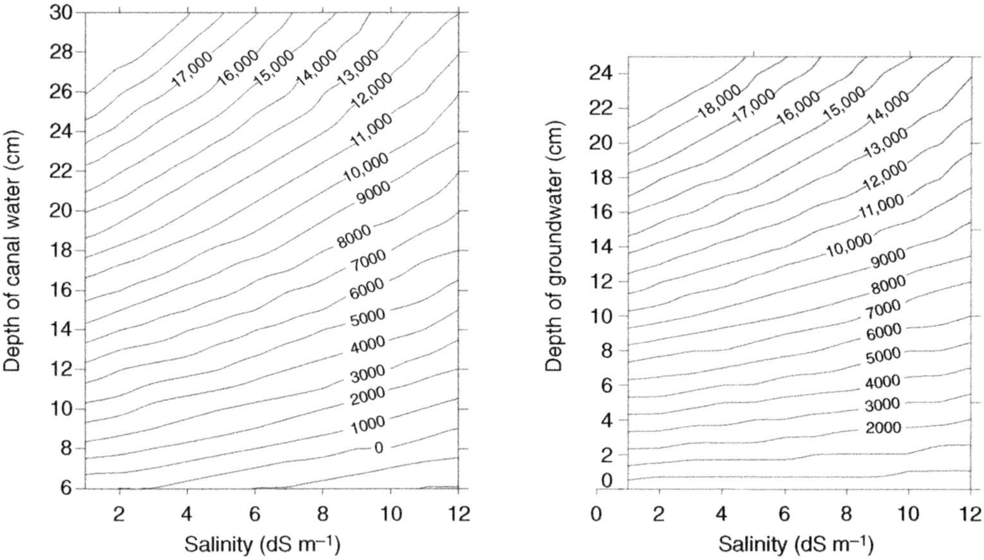

Fig. 5.10. Profitability of conjunctive use of groundwater of varying salinity and canal water at two levels of supply.

season. The water inadequacies at the tail end are further complicated by the progressive decrease in groundwater quality from head to tail reaches. A typical case that has been investigated pertains to the Kaithal circle of Bhakra canal in Haryana. Here the availability of canal water progressively decreased from 25 cm ha^{-1} in the head reach to 8 cm ha^{-1} in the tail reach, with groundwater salinity increasing from 2.5 dS m^{-1} to 6.8 dS m^{-1} (Fig. 5.11).

The water table in the head reach is also substantially higher than in the tail reach. This situation favours the development of

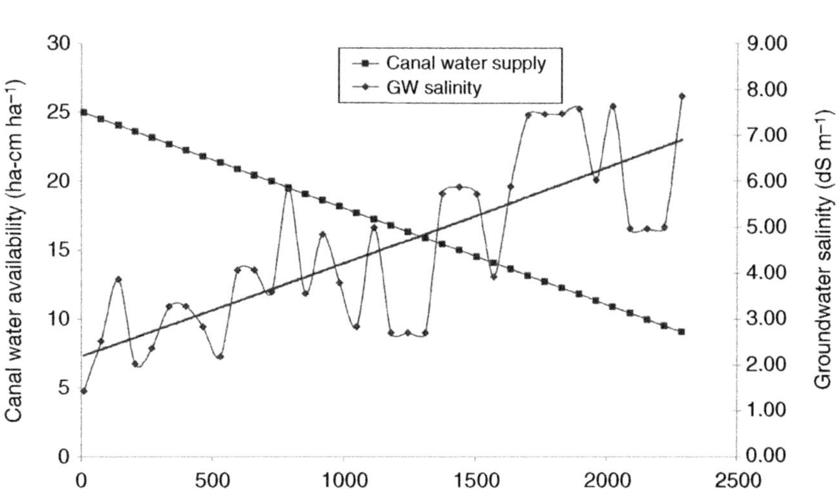

Fig. 5.11. Variation in availability of canal water and salinity of groundwater (GW) from head to tail reach of watercourse no. 25963 L (Batta Minor).

groundwater through shallow tube wells in the head reach and its transfer to the tail reach. Such small-scale water markets are already in existence in Haryana and their existence in the Chistian Subdivision in Punjab (Pakistan) has been investigated by Strosser (1997), who mentioned that the impact of a tube-well water market on farm gross income was significant at 40% of the actual gross income, aggregated for eight sample watercourses. However, he also mentioned that water markets could lead to decreased aquifer recharge and an increase in the soil salinity. The potential increase in relative yield with such groundwater transfer from the head to the tail reach of a watercourse in Batta Minor (Bhakra system) was analysed using the SWAP model (Chandra, 2001). The results indicated that the relative yield would increase from 0.70 to 0.85 in the entire watercourse if 50% of marginal-quality groundwater from the head reach was transferred and used in the tail reach without disturbing canal-water allocation. The relative yield would go up to 0.89 if, instead of blending, the groundwater was used in a cyclic mode (Fig. 5.12).

The state of Haryana has experimented with the transfer of groundwater from freshwater areas with higher rainfall and greater availability of canal water to areas that are less favourably endowed with water. This relieved waterlogging and stabilized the canal water supply in the lower reaches. This practice on a limited scale has been adopted in marginal groundwater areas in the Hisar district by installing shallow tube wells along the branch and distributary canals. Since the projects were state-funded and were not market-oriented, technical and hydrological constraints that operate at higher spatial levels would need to be understood and resolved before promoting saline-water development and use at system level. Particular attention will have to be paid to reduced canal water flow and increased salinity of mixed water as one moves from the head reach of the minors/distributaries/branch canals to their lower reaches.

Balance between saline-water use and disposal

One of the important objectives in groundwater development is to maintain salinity below critical levels for the crops to be grown in the region. Continued recirculation of saline water without any disposal of salts would make the aquifers more saline and ultimately unusable. Therefore, not all saline water can or should be used. How much of it can be used depends upon the supplies of fresh water (canal), rainfall, original salinity of the

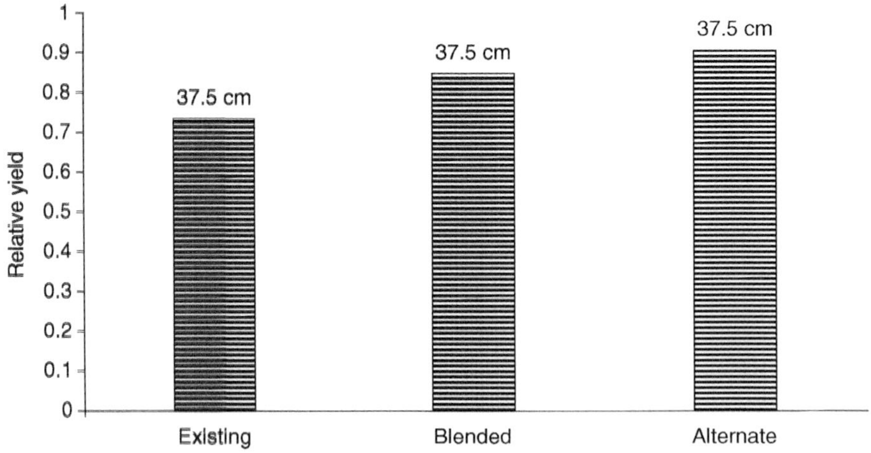

Fig. 5.12. Improvement in water productivity at watercourse (25963L) level by groundwater transfer from head to rail reach.

effluents, soil characteristics, crops and drainage conditions. Srinivasulu *et al.* (1997) have estimated that water equivalent to a minimum of 15% of the annual groundwater recharge with an average EC of 6 dS m^{-1} will have to be disposed of to maintain the salinity balance in groundwater underlying Sirsa and Hisar districts of Haryana. Such a disposal rate would ensure sustainability. Similar estimates will have to be made for other areas.

Extent and Actual Saline Water-use Practices

Irrigation with saline water, developed through shallow tube wells and open wells, is quite extensive. These tube wells were developed primarily for irrigation but have also been providing drainage relief. Studies based on a farm survey conducted in 1983/84 and reported by Boumans *et al.* (1988) estimated that in marginal and saline water zones about 120,000 ha-m were being pumped through more than 68,900 shallow tube wells in 1982/83. It was inferred that the rise in water table was slowed down largely due to these wells. Recent estimates show that 316,000 ha were being irrigated with saline water in the state of Haryana (Manchanda, 1996), of which 75,000 ha were in the region where waterlogging and salinity are either an existing or a potential threat.

Water-use practices

Several water-use practices are in vogue. The survey in the Hisar district (Haryana), mentioned above, also found that saline water pumped by shallow tube wells is, in most cases, used directly without any mixing. Mixing is normally done only if the salinity exceeds 6 dS m^{-1} and, in such cases, the water from the tube well is pumped into a watercourse carrying canal water. Farmers also resort to pumping of groundwater into the canal or watercourse if they perceive that the watercourse discharge is too small to cover the planned irrigation area in the allotted time. Cyclic use of canal and saline water is more common. This is largely because canal water is available for only a few hours after each rotation period of 2–4 weeks' duration and because the opportunity to irrigate with mixed or blended water is small. This constraint could be relaxed if on-farm reservoirs were constructed (Tyagi and Sharma, 2000).

Some farmers do not follow the practice of intraseasonal conjunctive use but reserve a parcel of land for irrigation by saline water only. In that case, they grow salt-tolerant crops, such as mustard, which is not given any pre-sowing irrigation but is sown in residual moisture after the rainy season and is given one or two supplementary turns of irrigation. Since the canal-water charges are levied on an area basis and not on the basis of the number of irrigation turns received from canal water, the farmers save on canal irrigation charges (though the charges are very low) by adopting this practice. The area receiving irrigation exclusively from tube wells with saline water is rotated every season/year to avoid salinization of a particular piece of land. If the tube wells yield water with high RSC, gypsum, which is readily available from the Land Reclamation Corporation outlets, is applied to neutralize the sodicity. Gypsum is either applied to the soil or put into the channel in gunny bags on which water from the tube well falls and slowly dissolves the gypsum. In such cases, gypsum is not powdered but is in the form of big clods. A more scientific way of applying gypsum is through gypsum-dissolving beds, which are specifically constructed for this purpose. Whether applied to the soil or applied with the irrigation water, the basis for computation of the gypsum requirements remains the same. There is, however, a difference in the time of application. In the case of soil-applied gypsum, the entire quantity of gypsum required, estimated on the basis of the amount and quality of the RSC-rich water, is applied all at once. If the sodicity of the soil is already high, the gypsum required to neutralize the RSC of the applied water may have to be applied at the beginning of the season; otherwise, it could be applied before the next crop is planted. In the case of water-applied gypsum, neutralization takes place before its application and there is, therefore, no build-up of sodicity in the soil.

Availability of gypsum is ensured through an organized arrangement with the government.

Epilogue

Saline/alkaline water has been successfully used to augment irrigation supplies and help to raise water productivity in semi-arid regions. This success can be attributed largely to available canal water supplies, which make it possible to plan and practice irrigation with marginal-quality water when it is least harmful and also in diluting the salt concentration in the root zone, keeping it below threshold limits. Monsoonal rainfall, which plays a crucial role in the desalinization cycle, is another factor that regulates the seasonal salt balance in the root zone to permit saline-water use even with traditional irrigation methods. More saline water is used during winter, when it is more productive and least harmful. Similar successes with saline/alkaline water, use have not been achieved in more arid areas, which do not have the benefit of canal irrigation. In those areas, interseasonal fallowing and rain-fed farming with very limited use of saline water applied to salt-tolerant crops continue to be the norm.

In irrigated areas provided with an extensive canal network but with an inadequate water supply, saline groundwater development through shallow tube wells is primarily for irrigation but it also keeps the water table in check. However, continued recirculation and reuse of the marginal-quality water without any disposal of saline water outside the system brings the danger of slowly salinizing both soil and aquifers. In the long run, the practitioners of this technology of using saline/alkaline water, which was initially shown to be successful at the field scale, will have to consider regional salt balances. Simulation studies based on limited data indicate a gradual rise in salinity of both soil and aquifers when the use of saline/alkaline water is extended to larger areas and continued for a long time.

Considering the present situation in respect of saline/alkaline water use, it looks attractive to focus on research that would help develop strategies for the use of this water in areas with only a small and inadequate amount of seasonal rainfall. Harnessing synergetic effects of improved salt-tolerant crop varieties and of improved hydraulic technologies offers a possible approach to enhancing productivity in such areas.

Unlike the crop–water–salinity relationship of saline water, the production functions for alkaline water are not well established. Also, the impact of the use of this water on groundwater aquifers is not well known. Field research and monitoring that would help bridge these gaps in our current understanding deserve our attention.

There are numerous models that help in generating scenarios for the possible consequences of saline-water use on a regional scale. However, models for scenario building at irrigation-system/river-basin scale, where groundwater alkalinity is a problem, are missing. Added to this is the problem of the vast amounts of data that are required but are seldom available for the areas where they are most needed. Therefore, studies aimed at the generation of data to be used in the regional salt- and water-balance model are needed if the sustainability of the technology that improves water productivity at field scale is to be ensured at a higher level of the irrigation system/river basin.

References

Anon. (1998) *Report of XVII Meeting, ICAR Regional Committee No. VI: Agenda Notes and Status of Centre-State Coordination for R&D Linkages in Agricultural Research, Education and Extension.* Karnal, India, p. 270.

Anon. (2001) Draft report: Wheat productivity constraints for rice–wheat cropping pattern of Bhakra Canal System, Central Soil Salinity Research Institute (CSSRI)–IWMI Project, Karnal, India.

Bajwa, M.S. and Josan, M.S. (1989) Effect of alternating sodic and non-sodic irrigation on build up of sodium in soil and crop yield in northern India. *Experimental Agriculture* 25, 199–205.

Boumans, J.H., van Hoorn, J.W., Kruseman, G.P. and Tanwar, B.S. (1988) Water table control, reuse and disposal of drainage water in Haryana. *Agricultural Water Management* 14, 537–545.
Chandra, R. (2001) Development of irrigation water management strategies: a case study. ME thesis, submitted to Maharana Pratap University of Agriculture and Technology, Udaipur. Duplicated.
Chaudhary, T.N., Bhatnagar, V.K. and Prihar, S.S. (1974) Growth response of crops to depth and salinity of water and soil submergence. *Agronomy Journal* 66, 32–35.
Howell, T.A., Cuenca, R.H. and Soloman, K.H. (1990) Crop yield response. In: Hoffman, G.J., Howell, T.A. and Solomon, K.H. (eds) *Management of Farm Irrigation Systems*. American Society of Agricultural Engineers, St Joseph, Michigan.
Maas, E.V. (1990) Crop salt tolerance. In: Tanji, K.K. (ed.) *Agricultural Salinity Assessment and Management*. American Society of Civil Engineers, New York, pp. 262–304.
Manchanda, H.R. (1996) Communication in response to discussion on 'Research on policy issues'. In: *Status Report on Use of Saline and Fresh Waters in Irrigation Commands*. CSSRI, Karnal.
Minhas, P.S. (1993) Modelling crop response to water and salinity stresses. In: Tyagi, N.K., Kamra, S.K., Minhas, P.S. and Sing, N.T. (eds) *Sustainable Irrigation in Saline Environment*. Central Soil Salinity Research Institute, Karnal, pp. 96–109.
Minhas, P.S. and Tyagi, N.K. (1998) *Guidelines for Irrigation with Saline and Alkali Waters*. Bulletin, CSSRI, Karnal.
Minhas, P.S., Sharma, D.R. and Khosla, B.K. (1988) Effect of quality and depth of ground water table on salinization and crop growth. In: *Annual Report*. CSSRI, Karnal, pp. 48–51.
Raul, S., Tyagi, N.K. and Jaiswal, C.S. (2001) Salt and water balance implications of rice–wheat cropping in Haryana. In: Proceedings 1st Asian Regional Conference Held on 16–23 Sepember 2001 in Seoul, Korea. Duplicated.
Rhoades, J.D., Kandiah, A. and Mashali, A.M. (1992) *The Use of Saline Waters for Crop Production*. FAO's Irrigation and Drainage Paper No. 48, FAO, Rome, Italy.
Sharma, D.P. and Rao, K.V.G.K. (1996) Recycling drainage effluent for crop production. In: *Proceedings of Workshop on Waterlogging and Salinity in Irrigated Agriculture, 12–15 March 1996*. CSSRI, Karnal. pp. 134–147.
Sharma, D.P., Singh, K.N., Rao, K.V.G.K. and Kumbhare, P.S. (1991) Irrigation of wheat with saline drainage water on a sandy loam soil. *Agricultural Water Management* 19, 223–233.
Sharma, D.R., Sharma, D.K. and Minhas, P.S. (1993) Feasibility studies on use of saline/sodic waters in conjunction with canal water and amendments. In: Tyagi, N.K., Kamra, S.K., Minhas, P.S. and Singh, N.T. (eds) *Sustainable Irrigation in Saline Environment*. CSSRI, Karnal, pp. 110–117.
Sharma, D.R., Minhas, P.S. and Sharma, D.K. (2001) Response of rice–wheat to sodic irrigation and gypsum application. *Journal of India Society of Soil Science* 49(2), 324–327.
Srinivasulu, A., Tyagi, N.K. and Shukla, K.N. (1997) Conjunctive use of water resources in saline ground water basins: a management model. *ICID Journal* 46(1), 65–89.
Strosser, P. (1997) Analysing alternative policy instruments for the irrigation sector. PhD dissertation, Agricultural University, Wageningen, The Netherlands, 243 pp.
Tyagi, N.K. (1984) Effect of land surface uniformity on irrigation quality and economic parameters in sodic soils under reclamation. *Irrigation Science* 5(3), 151–166.
Tyagi, N.K. (1998) Need, scope and technology for use of marginal quality waters in agriculture. In: *Proceedings of National Seminar on Water Management for Sustainable Agriculture – Problems and Prospects for the 21st Century, 15–17 April 1998*. Indian Agricultural Research Institute (IARI), New Delhi, pp. 175–188.
Tyagi, N.K. (2001) Managing salinity in northwest India: some short and long term options. In: Proceedings International Conference on Agricultural Science and Technology, Beijing, 7–9 September 2001. Duplicated.
Tyagi, N.K. and Joshi, P.K. (1996) Techno-economic analysis of salinity control measures. In: *International Commission on Irrigation and Drainage. Sixteenth ICID Congress, 16–22 September, Cairo, Egypt*, Vol. 1C New Delhi, Q.47.1, pp. 41–53.
Tyagi, N.K. and Sharma, D.P. (2000) Disposal of drainage water: recycling and reuse. In: *Proceedings 8th ICID International Drainage Workshop, New Delhi, 31 January–4 February 2000*, Vol. 3 New Delhi, pp. 199–213.
Zeng, L., Shannon, M.C. and Lesch, S.M. (2001) Timing of salinity stress affects rice growth and yield components. *Agricultural Water Management* 48, 191–206.

6 Water Productivity under Saline Conditions

Jacob W. Kijne
International Water Management Institute, Colombo, Sri Lanka

Abstract

The opportunity for increasing water productivity under saline conditions is contingent on the determination and accurate implementation of the leaching requirement needed to prevent unnecessary percolation below the root zone. The leaching fraction of the applied irrigation water percolates through the root zone to maintain soil salinity at an acceptable level. Crop water use (evapotranspiration) and leaching requirement (LR) together constitute the beneficial depletion of the water resource. Evapotranspiration and leaching are linked through the yield–water-production function. The more the crop growth is affected by salinity, the lower the evapotranspiration and the higher the leaching fraction of the applied irrigation water.

Crops differ in their tolerance for salinity. Under controlled conditions, crops have salinity threshold values below which crop yields are not affected. However, evidence is presented that under field conditions, where plants are subjected to periodic and simultaneous water and salt stress and to non-uniform water application, yields are lowered by salt concentrations below the assumed threshold values. In addition, rather than having one specific seasonal crop salt tolerance (threshold value), crops react differently depending on the timing of the imposed salinity stress.

Irrigation water that is consumed by evapotranspiration leaves the remaining water more concentrated with salts. The leaching requirement increases with the salinity of the water supply and the sensitivity of the crop for salinity. This chapter illustrates how uncertainty about LR, resulting in part from uncertainty about yield–salinity relations, imposes constraints on the possible improvement of water productivity under saline conditions. The chapter points out implications for the successful production of crops with a mixture of saline water and good-quality irrigation water (e.g. conjunctive use of groundwater and canal water).

Introduction

Saline water has been successfully used to grow crops. Saline water can be mixed with better-quality water prior to application, or the two types of water may be applied intermittently. Sensitivity may vary during the growing season, but crops apparently respond to the weighted mean water salinity regardless of the blending method (Letey, 1993). An example of a crop often irrigated with saline water is cotton. Even when irrigated with water of relatively high salinity, the yield of cotton is nearly as much as when irrigated with good-quality water. Cotton is considered a salt-tolerant crop. More sensi-

 Water Productivity in Agriculture: Limits and Opportunities for Improvement (eds J.W. Kijne, R. Barker and D. Molden)

tive crops can also be irrigated with relatively saline water, but they are likely to yield less than when irrigated with good-quality water. Equally high yields, as with the application of non-saline water, can often be obtained by applying more of the saline water. As the salinity of irrigation water increases, its effective quantity decreases (Letey, 1993). The degree by which the quantity is diminished depends on the crop to be grown and the relative yield to be achieved. This relationship is expressed in crop–water–salinity functions.

During the last 100 years, many experiments have been carried out to determine the salt tolerance of crops. Maas and Hoffman (1977) carried out a comprehensive analysis of salt-tolerance data, which was updated by Maas (1990). Based on this analysis, Maas and Hoffman (1977) concluded that crop yield as a function of the average root-zone salinity could be described reasonably well by a piecewise linear response function characterized by a salinity threshold value below which the yield is unaffected by soil salinity and above which yield decreases linearly with salinity. This relationship is found to be variety-specific, and it may also depend on the unique soil conditions, evaporative demand and water-management conditions (van Genuchten and Gupta, 1993).

The threshold–slope model of Maas and Hoffman (1977) has been used widely in a variety of applications in research and water management. Nevertheless, other salinity response functions have been found equally successful in describing the observed data on crop salt tolerance (e.g. van Genuchten and Hoffman, 1984; Dinar *et al.*, 1991). One of the problems with the threshold–slope model in describing experimental data is the rather poor definition of the salinity threshold value for data sets that are poorly defined or erratic or have limited observations. An example of such data is presented in Fig. 6.1 for wheat grown in the Fordwah–Eastern Sadiqia Project of Pakistan – from data reported by Kahlown *et al.* (1998). The relationship between yield and salinity of the applied irrigation water is even more difficult to ascertain, as illustrated in Fig. 6.2, also from Kahlown *et al.* (1998).

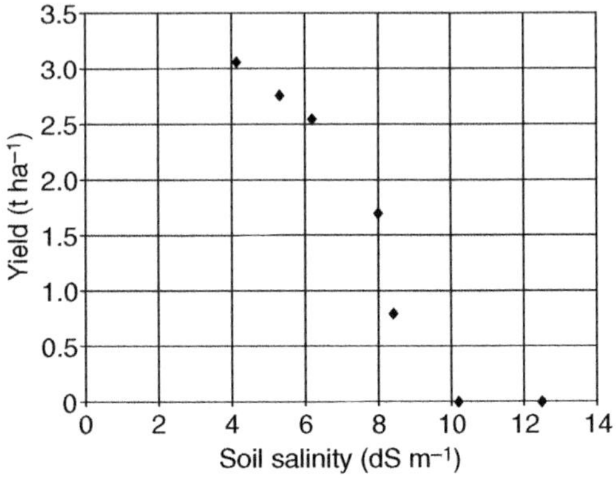

Fig. 6.1. Yield as a function of soil salinity (from Kahlown *et al.*, 1998).

A smooth S-shaped response function, as proposed by van Genuchten and Hoffman (1984), describes the various reported data sets at least as well (see also van Genuchten and Gupta, 1993). The equation for the S-shaped curve is:

$$Y/Y_m = 1/[1 + (c/c_{50})^p] \quad (6.1)$$

In this equation, Y is the yield, Ym yield under non-saline conditions, c is average root-zone salinity, c_{50} is the soil salinity at which the yield is reduced by 50% and p is an empirical constant. The curve shown in Fig. 6.3 is for wheat with an average value of $p = 3$ and $c_{50} = 23.9$ dS m^{-1}. Van Genuchten and Gupta (1993) reported that the value of p in Equation 6.1 is close to 3 for most crops.

Based on lysimeter studies in California, Dinar *et al.* (1991) derived quadratic yield response functions relating yield to the seasonal amount of irrigation water, its average salt concentration and the average soil salinity at the beginning of the season. A major conclusion from this study is that a direct relation between yield and average seasonal salinity does not apply to conditions where several factors are interrelated. For example, when salinity of the soil and the applied water is high and the amount of applied water is not sufficient, average soil salinity itself will not explain yield reduction. One should have relationships between water quantity, water quality, yield, soil salinity and drainage volumes. The quantity of drainage water is likely to increase as more

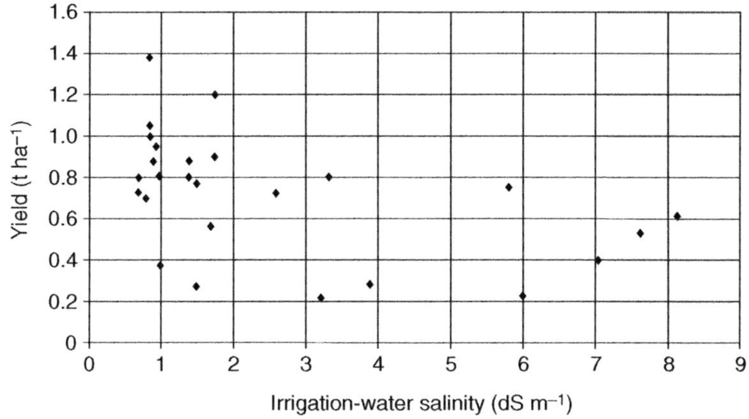

Fig. 6.2. Yield as a function of irrigation-water salinity (from Kahlown *et al.*, 1998).

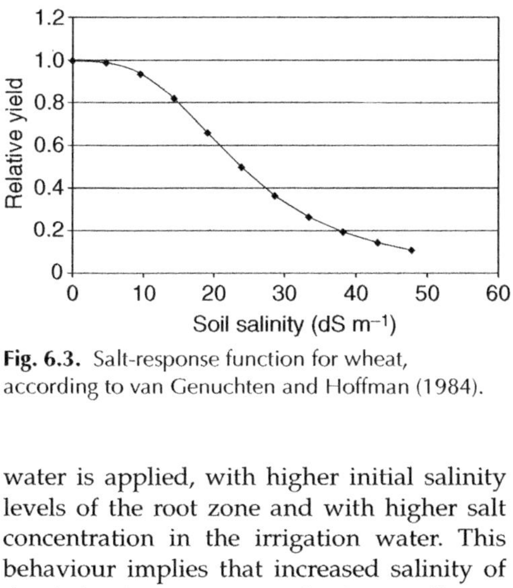

Fig. 6.3. Salt-response function for wheat, according to van Genuchten and Hoffman (1984).

water is applied, with higher initial salinity levels of the root zone and with higher salt concentration in the irrigation water. This behaviour implies that increased salinity of the irrigation water results in smaller or fewer plants with decreased evapotranspiration rates and, hence, in greater deep percolation for a given irrigation application.

When the salinity is mainly the result of sodium salts, the structure of the soil will be adversely affected. High values of the exchangeable sodium percentage (ESP) in the soil can cause the hydraulic parameters, such as percolation rate and infiltration rate, to change significantly. The potential hazard of reduced water infiltration is partly related to the intensity and timing of rainfall. Rainwater has a very low salinity. When it infiltrates the soil, the salinity of surface soil can decrease rapidly, but the soil may remain at almost the same ESP. As a result, the potential of dispersion by rainfall is especially high if the ESP of the soil is high. Rainfall also contributes dispersive energy because of its impact on the soil (Kijne *et al.*, 1998). So far, these effects of sodicity have not been incorporated in any of the salt-response functions. It is to be expected that, with sodic soils, reduced plant growth and, hence, reduced evapotranspiration will not lead to increased percolation for a given irrigation application. Percolation into sodic soils may be so slow that most of the irrigation water will runoff without leaching salts from the root zone.

Apart from the S-shaped relation between yield and soil salinity (Equation 6.1), quadratic yield functions were developed by Dinar *et al.* (1991), quadratic, log–log and linear functions by Datta *et al.* (1998) and a linear function by Lamsal *et al.* (1999). None of these functions show a threshold salinity below which yield is unaffected by salinity. There is now considerable evidence from field observations that yield starts to decline at much lower values of soil salinity than predicted by the threshold–slope functions of Maas and Hoffman (1977). For example, Hussain (1995) reported field data that illustrated this earlier response, and Katerji *et al.* (2000) confirmed this effect in their lysimeter experiments in Bari, Italy. Shalhevet (1994), in a seminal paper on the use of marginal water for crop production, observed that under conditions of high evaporative demand the salinity response function may change so that the threshold salinity decreases and the slope increases, rendering the crop more sensitive to salt. Tyagi (2001) reported a set of empirical relations between relative yield, the amount of water applied as a fraction of pan-evaporation and the salinity of the applied water. These relations were developed at the Central Soil Salinity Research Institute, Karnal, India, for five crops, including wheat, cotton and maize. The curvilinear relations reflect the local conditions and show a gradual decline in yield with an increase in salinity of the irrigation water.

The effect of salinity on yield differs depending on the timing of the salt stress, another factor not considered in salt-response functions. Zeng *et al.* (2001) and Francois *et al.* (1994) reported the importance of timing of salt stress on yield components for rice and wheat, respectively. Shalhevet (1994) hypothesized that the duration of salinization is more significant than sensitivity at a critical growth stage. Zeng *et al.* (2001) argued that this hypothesis can only be tested when the salt-stress periods during the various well-defined growth stages are of equal length, which is the way they designed their experiments. Hence, at least for rice, they repudiated the hypothesis.

In general, yields in farmers' fields tend to be lower for a combination of factors than those predicted on the basis of yields obtained under more controlled conditions (see, for example, Warrick, 1989; Howell *et al.*, 1990; Kijne, 1998). Contributing factors appear to include at least the following: spatial variability of soil structure and fertility, water-application rates, soil salinity, plant density and temporal variability in sensitivity of crops to drought and salt stresses.

The accuracy with which yields can be predicted is relevant in the assessment of leaching requirements. Leaching is a non-productive but beneficial water use. Without maintaining an acceptable salt balance in the root zone, it would not be possible to continue to grow crops in many irrigated areas of the world. But how much water should be allocated to leaching? Guerra *et al.* (1998) report data for seepage and percolation in rice-fields ranging from 1–5 mm day^{-1} in puddled clay soils to as high as 24–29 mm day^{-1} in lighter-textured soils. Seepage occurs in irrigation canals but percolation occurs over the whole area planted with rice. The reported range of values implies that percolation from rice-fields can vary from the same order of magnitude as evapotranspiration up to about eight times as much. The latter is surely excessive in terms of salinity control. In this chapter, the focus will be on leaching requirements for non-rice crops.

In most definitions of irrigation efficiency and water productivity, no allowance is made for leaching as beneficial use of irrigation water (Seckler *et al.*, Chapter 3, this volume). Water-productivity values vary with the geographical scale, as Keller and Keller (1995) illustrated for the Nile valley. A major cause of this variation is the fact that runoff or drainage from one field may be reused on another. However, because of its higher salt content, drainage water is inevitably of lower quality than the applied irrigation water. Even runoff will be degraded if it picks up disease organisms, agricultural chemicals or salt (Solomon and Davidoff, 1999).

Reuse of drainage water (including seepage from canals and percolation from fields) between parts of an irrigation system or within an entire river basin complicates the distinction between consumptive and non-consumptive beneficial use of water (Molden *et al.*, Chapter 1, this volume). To correctly determine the potential for reuse of drainage flows, it is necessary to account for all components of the salt and water balances at the different geographical scales and to know the leaching requirements for the crops to be grown.

High water tables are often associated with irrigated agriculture. They provide a source of water for plant growth through capillary rise of water into the root zone. Substantial contributions from shallow groundwater to crop water requirements have been reported in the literature (e.g. Grismer and Gates, 1991; Letey, 1993). However, when this shallow groundwater is saline, the harmful effects caused by the salt accumulation in the root zone probably outweigh the potential benefits of the groundwater as a source of water for plant production. Usually, the only option for sustaining agricultural production on fields underlain by shallow saline groundwater is to install a subsurface drainage system.

Thorburn *et al.* (1995), studying the uptake of saline groundwater by eucalyptus forests in part of the flood-plains of the Murray River in South Australia, showed that groundwater depth and salinity are the main controls on the uptake of groundwater, while soil properties appear to have a lesser effect. Model studies indicated that uptake of saline groundwater would result in complete salinization of the soil profile within 4 to 30 years at the sites studied, unless salts were leached from the soil by rainfall or floodwaters. However, a relatively small amount of leaching may be sufficient to allow groundwater uptake to continue. Thus groundwater, even when saline, may be an important source of water to salt-tolerant plants and trees in arid and semi-arid areas.

Grismer and Gates (1991) carried out a stochastic simulation study for a salinity-affected area underlain by a shallow water table, representative of conditions in the western San Joaquin valley of California. The model analyses the effects of irrigation–drainage management on water-table depth, salinity, crop yield and net economic returns to the farmer over a 20-year planning period. They found that cotton farming on salinity-affected soils subject to shallow saline groundwater is economically optimal if the application efficiency is 75–80%, which may be attainable with well-managed surface irrigation, and a subsurface drainage system is capable of removing 79–93% of the downward flux. The study illustrates the need to approach management strategies on irrigation and drainage together, from a regional perspective.

Research Data

The data for this chapter were collected at the International Water Management Institute (IWMI)'s research sites in irrigation systems in the Indus River basin of Pakistan between 1988 and 1995. The salt problem of the Indus is formidable. Smedema (2000) reported that the average salt influx by the Indus river water, taken at the rim stations, is estimated at 33 million t, while the outflow to the sea contains only 16.4 million t. Hence, the average annual addition of salts to the land and the groundwater amounts to some 16.6 million t. Most of this accumulation takes place in the Punjab. This is in sharp contrast to Egypt, where a large portion of the irrigated land is underlain by subsurface drains that take the drainage water back to the river. The salts do not stay in the Nile basin but are discharged into the Mediterranean Sea. During part of the year, the salt content in the lower Indus is much lower than in the lower Nile (in the Nile delta) and more salt disposal into the Indus could be accepted. However, during critically low flow periods, such disposals would not be possible. The only option during such periods would be to store the drainage water temporarily for release during high flood periods. Extending the left bank outfall drain, now operating in Sindh, into the Punjab may provide a more permanent (but quite expensive) solution than the present inadequate number of evaporation ponds.

Much of the drainage water from agricultural land in Pakistan's Punjab is being reused, either from surface drains or pumped up from shallow groundwater. The leached salts are therefore returned to the land rather than disposed of to the sea. IWMI's research sites in the Indus basin, the data-collection methodology and data analyses were described by Kijne (1996), Kuper and Kijne (1996) and Kuper (1997).

Specifically, information on the quantity and quality of applied irrigation water at the study sites in Punjab, Pakistan, is obtained from Kijne (1996). The electrical conductivity (EC, i.e. the standard measure of salinity) of canal water was 0.2 dS m^{-1} in most of the experimental sites. The EC of pumped groundwater was obtained from measured values of water quality of tube wells in the sample areas. For the calculations of the salt balance of the study sites, Kijne (1996) used 2.5 dS m^{-1} as a representative value for the salinity of pumped groundwater, ignoring the large variations in water quality that often occur even from pumps close to one another. Average values of the leaching fraction (LF) (the fraction of the infiltrated applied water that passes below the root zone) for the three irrigation systems reported in these studies were between 10 and 15% (Kijne, 1996, Table 2).

Data on LFs for four irrigated fields in the Fordwah–Eastern Sadiqia irrigation system, Chistian subdivision, Punjab, studied in considerable detail, are obtained from Kuper (1997). The latter set of data is summarized in Table 6.1.

ECe is the electrical conductivity of soil water at saturation, the usual parameter for measuring soil salinity in the profile. The value in the third column refers to the linearly averaged electrical conductivity of soil water in the profile down to 1 m. No leaching for field 2 (last column of the table) indicates that there may have been capillary flow from the water table (water table was at 2 m depth).

The spatial and temporal variability of soil salinity is large. Values in columns 4, 5 and 6 give some indication of the vertical spatial variability. Soil salinity increases when the soil dries out between irrigations or in rainfall events, and it varies greatly between upper and lower layers of the root zone. It is generally accepted that plants respond to the average salinity in the root zone and vary their water uptake in the growing season depending on relative values of the osmotic potential in the root zone.

The excessive leaching in field 1 (leaching fraction of 0.65) is blamed on a combination of poor water management by the farmer and the light-textured soil with high permeability. Leaching in the other fields is inadequate for maintaining an average root-zone salinity equivalent to an ECe value of 2 dS m^{-1}. The attainable yield level under these low leaching conditions is less than the maximum.

Leaching Requirement

When more water is applied than is taken up by the plant roots, water flows out of the root zone and carries soluble substances, such as salts and agrochemicals, with it. During this process of downward flow (percolation), soil salinity in the root zone increases with depth. In planning the desired leaching requirement (LR), it is commonly assumed that EC values of the soil extract at the lower root-zone boundary corresponding to 25–50% yield reduction are still acceptable. The weighted average EC value for the entire root zone (weighted according to root distribution) would be much less than at the lower root-zone boundary and the corresponding yield reduction for plants growing in this soil would be less than 25–50%. Such yield reductions are assumed to be economically viable (Smedema and Rycroft, 1988).

The rate of downward flow and leaching varies with the soil water content. It is highest during the first couple of days after irrigation, when the soil water content is still above or near field capacity. Thereafter, leaching continues at a much reduced rate. In many soils, the soil solution at field capacity is about twice as concentrated as when the soil is saturated (shortly after irrigation). When the soil dries out further between irrigations, the soil solution becomes even more concentrated.

Table 6.1. Salinity and leaching fractions in four experimental fields, Chistian subdivision, Punjab, Pakistan (from Kuper, 1997).

	Soil type	ECe (dS m^{-1})	Lowest ECe (dS m^{-1})	Highest ECe (dS m^{-1})	90 cm depth	LF
Field 1	Loamy sand	0.75	0.5	0.8	0.75	0.65
Field 2	Sandy loam	1.75	0.5	2.8	1.8	Nil
Field 3	Loam	2.5	1.3	4.2	2.5	0.07
Field 4	Silt loam	4.75	1.5	8.0	6.0	0.01

Not all downward flow is equally effective in leaching salts from the root zone. The most effective leaching occurs when water moves through the soil mass, rather than through cracks between aggregates. Water moving through cracks and wormholes has been called preferential flow. How much of the percolation occurs as preferential flow depends on the structure and texture of soil and is difficult to determine. As a result, the leaching efficiency of the percolating water is also difficult to assess. In cracking clay soils, initially as much as three-quarters of the applied water may flow through the cracks. Once the soil swells up with moisture, cracks close and the leaching efficiency increases (Smedema and Rycroft, 1988).

In its simplest form, for steady-state conditions, the relation between the LR and the amounts of irrigation and drainage water and their EC reduces to:

$$LR = D_d/D_a = EC_a/EC_d \quad (6.2)$$

where D is depth of water (subscript a for applied water; subscript d for drained water) and EC is the corresponding electrical conductivity. Equation 6.2 states that the amount of salt added in the irrigation water must equal the amount drained to maintain the salt balance. If the actual LF is less than the requirement, salt will accumulate (Hoffman, 1990).

The relationship between the salinity of the applied water, the LF and the resulting soil salinity is an important one. It would be easier to estimate expected yields if it were possible to unambiguously predict the soil salinity likely to result from irrigation applications of known salinity and a specified LF. Table 6.2 presents various relationships between LF and the dimensionless ratio of the average weighted root-zone salinity (Cs) and the average salinity of applied water (Ca).

The values in the table are based on steady-state conditions. However, the relationship between soil and water salinity as governed by leaching is a dynamic one, subject to feedback mechanisms between growth of the crop (hence, evapotranspiration) and leaching of salts (see Dinar *et al.* (1991), referred to earlier). In all cases the salinity-tolerance data are from threshold salinity-response functions. In addition, the leaching equations ignore the effect of sodium salts on the soil structure. The variations among the data in the table are due to the site specificity of the relationship between root-zone salinity and salinity of applied water for any given leaching fraction. A contributing factor is the variability in measured values of the EC of soil-saturation extracts. The coefficient of variation of the EC of soil moisture at saturation is about 50% (Kijne, 1996) (see also Datta *et al.* (1998) and Tedeschi *et al.* (2001), who give similar values).

The various analyses that resulted in the data in Table 6.2 indicate that the ratio of root-zone salinity to irrigation-water salinity is very sensitive to changes in the leaching amount at LF below 0.1. The implication is that a small change in the leaching amount can make a large difference in root-zone salinity. This ratio of root-zone salinity to irrigation-water salinity is less sensitive to changes in the leaching amount at LF values between 0.1 and 0.4, which are most common. Hence, in this range of LF values, root-zone salinity increases about linearly with the salinity of the applied water. Therefore, difficulties in the accurate determination of

Table 6.2. Relationships between leaching fraction and ratio of soil salinity over applied water salinity.

LF	Cs/Ca (Pratt and Suarez, 1990)	Cs/Ca (Rhoades, 1982)	Cs/Ca (Hoffman and van Genuchten, 1983)	Cs/Ca (Prendergast, 1993)
0.05	3	7	4	10.5
0.1	2	5	2.6	5.5
0.2	1.25	3	1.4	3
0.3	1	2.5	1.3	2.15
0.4	0.83	2.35	1	1.75

LF from field data can affect the fit of the leaching equations. The study by Prendergast (1993), in particular, emphasizes the need for local data of the salt- and water-balance parameters.

The leaching equation of Hoffman and van Genuchten (1983) uses a root water-uptake function that is exponential with depth and incorporates some empirical coefficients that can be adjusted according to the local conditions. Of the relations reported in Table 6.2, Hoffman and van Genuchten is probably most commonly used in modelling studies where a relationship between leaching and root-zone salinity is required. It is plotted in Fig. 6.4.

Analysis of Data

Leaching water, as was pointed out before, is a beneficial, non-consumptive use of applied irrigation water. Its benefit is in the removal of salt from the root zone. If a portion of the drainage and runoff water is reused elsewhere in the irrigation system, part of their salt load is reapplied, rather than being removed, and the benefit of the drainage and runoff water is reduced. Solomon and Davidoff (1999) have presented analytical expressions relating irrigation-performance parameters for an irrigation system (called a unit) and its subunits (e.g. watercourse command areas (WCAs)) when drainage water and runoff from one subunit are reused on another. The performance parameters considered are the irrigation consumptive-use coefficient, which is defined as the ratio of irrigation water going to consumptive uses over irrigation water applied, and irrigation efficiency (IE) is defined as irrigation water beneficially used over irrigation water applied. The numerator of IE includes beneficial consumptive use (evapotranspiration), beneficial runoff and beneficial drainage water.

Rather than following this analytical analysis, perhaps the same point can be made by the following simplified example. A series of WCAs of an irrigation system, characteristic of conditions in Pakistan's Punjab, apply a blend of canal water and some drainage water from the upstream command area. The EC of the blend applied to the first WCA is 1.35 dS m^{-1}. All WCAs require 100 units inflow to meet their consumptive-use demand (crop evapotranspiration). According to the relationships of Fig. 6.4, the LR is 0.2 to maintain the root-zone salinity at a level corresponding to an EC of 2 dS m^{-1}. Hence, rather than an inflow of 100 units, $100/(1 - LR) = 125$ units of water need to be applied. The EC of the drainage water issuing from this first WCA is assumed to be 2.5 dS m^{-1}.

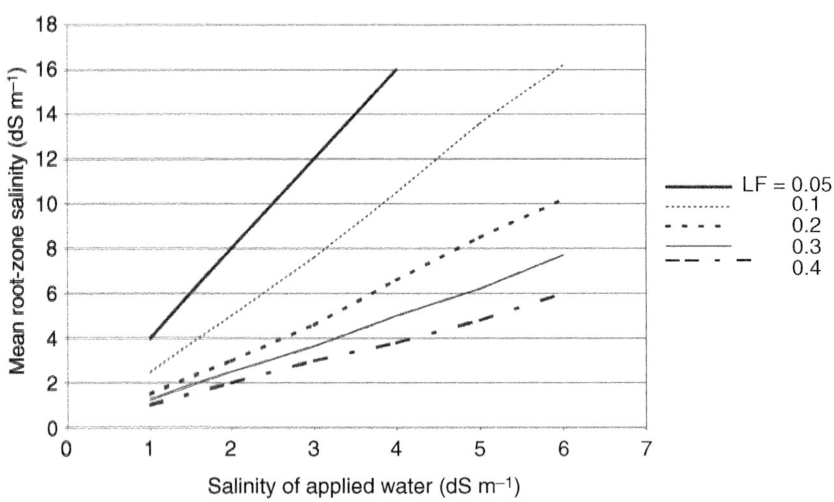

Fig. 6.4. Root-zone salinity as a function of salinity of the applied water and the leaching fraction (Hoffman and van Genuchten, 1983).

In the first example, plotted in Fig. 6.5, the next WCA in line applies a blend consisting of 60% canal water and 40% drainage water from the upstream WCA. The second WCA has as its source of irrigation water a blend of water with an EC of 1.35 dS m^{-1} for the irrigation-water component and an EC of 2.5 dS m^{-1} for the drainage component, resulting in an EC of 1.8 dS m^{-1}. Its LR is 35% and the required inflow is 154 units of water. The drainage water from this second WCA has an EC equal to 2.7 dS m^{-1}. This procedure is repeated for four WCAs. The characteristic values for the fourth WCA are an inflow salinity of 2.5 dS m^{-1}, LR of 45%, inflow of 180 units and drainage salinity of 3.3 dS m^{-1}.

The WCAs of the second example, plotted in Fig. 6.6, take only 10% of their applied water from the upstream drainage flow and 90% from the irrigation supply. In this case, the characteristic values for the fourth WCA are an inflow salinity of 1.74 dS m^{-1}, LR of 36%, inflow of 156 units and drainage salinity of 3.3 dS m^{-1}. The salinization of the water supply is slower when less water is taken from the more saline source. However, the trends are the same: more and more water from the 'good' source needs to be applied to the crop to maintain the root-zone salinity at an acceptable level.

Field 3 in Table 6.1 referred to a farmer's field where the LF was only 0.07. For a water demand of 100 units, this small amount of leaching would bring the inflow to 108 units and, with an EC of 1.35 dS m^{-1}, as in our example, the average EC of the root-zone moisture would be about 10 dS m^{-1}. This level of root-zone salinity would lead to significant production losses of even salt-tolerant crops.

Reuse of drainage flow from another WCA is very common in Pakistan's Punjab. Percolation from one WCA flows to the groundwater and is pumped up by tube wells for reuse elsewhere in the system. In many systems, pumped groundwater makes up between one-half and two-thirds of the irrigation water.

Keller and Keller (1995) used a different method to calculate the leaching requirement:

$$LR = ECa/(5ECe - ECa) \qquad (6.3)$$

where ECa is the EC of the irrigation water and ECe is the EC of the soil-saturation extract for a given crop and a tolerable degree of yield reduction. They assumed an allowable ECe of 1.5 dS m^{-1}. The use of this equation leads to LR values that are almost identical to those obtained in the manner described above.

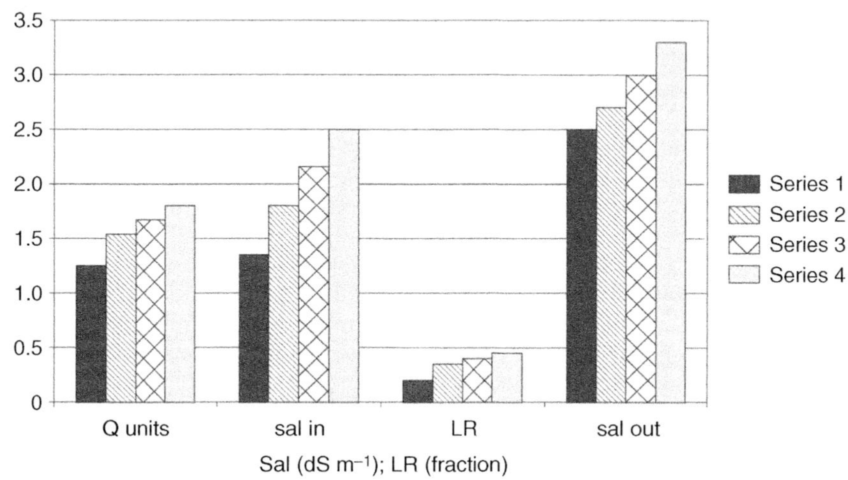

Fig. 6.5. Inflow volume (Q in multiples of 100 units), salinity of inflow and outflow (dS m^{-1}), and leaching requirement (fraction) for four successive reuse cycles, with 40% drainage water blended in.

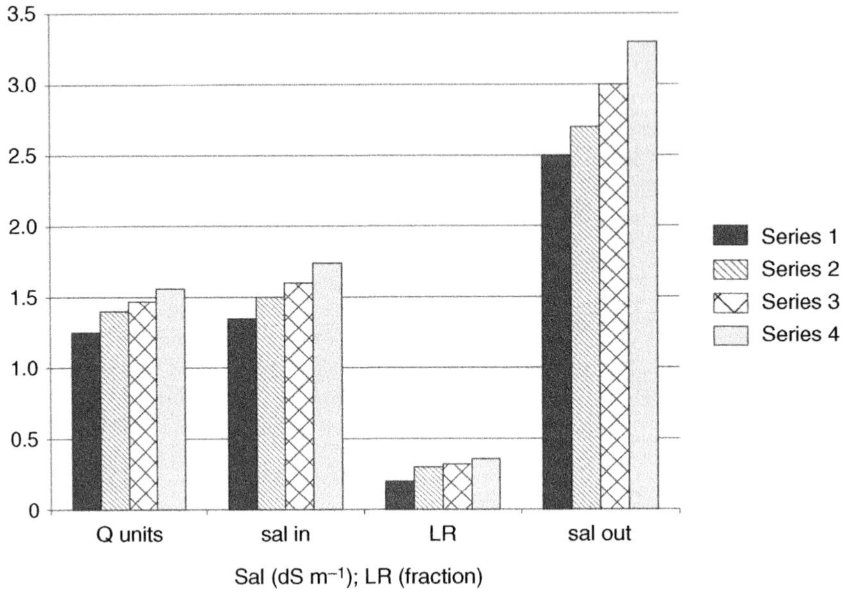

Fig. 6.6. Inflow volume (Q in multiples of 100 units), salinity of inflow and outflow (dS m^{-1}), and leaching requirement (fraction) for four successive reuse cycles, with 10% drainage water blended in.

Discussion

Several factors contributing to the present uncertainty about LRs have been mentioned. The most important ones derive from the inherent complexity of the dynamic plant–soil–water system in terms of its reaction to variations in water quality. Current salt-response functions and leaching equations are valid for static conditions, whereas the system itself is a dynamic one, with seasonal changes in the quality of the applied water, especially where rainfall meets a large part of the crop water demand during one or part of one growing season. Feedback mechanisms in this dynamic system are poorly understood and have rarely been quantified. One example of such a mechanism is the increase in downward flow when crop evapotranspiration declines as a result of salt stress on the crop. Rather than one specific crop, cropping sequences should be considered (see the examples given by Tyagi, Chapter 5, this volume). If the reported threshold values for salt tolerance are too high for most field situations, LR values would be higher than calculated. The effect of this difference is probably small in view of the overall uncertainty in the calculation of leaching requirements. Depth of water table may vary throughout a season or from one season to another, and hence the potential contribution to the evaporative demand of the crop through capillary flow varies as well. The effect of irrigation water rich in sodium salts (alkaline water) on crop production and soil structure is not considered.

Accurate determination of LRs is obviously not easy. Does it matter? It appears that under most conditions more than enough water is applied to the fields to meet the LR. Or, in other words, those low LFs reported in Table 6.2 must surely be exceptions rather than the rule. One gets that impression when considering the values of the relative water supply (the ratio of irrigation supply plus rainfall over water demand) and the relative irrigation supply (irrigation supply over demand) for 26 irrigation systems reported by Molden *et al.* (1998). Relative water supply values varied between 0.8 and 4.0 and half of the systems had values greater than 2.0. The reported variation in relative irrigation supply was between

0.41 and 4.81, while 22 of the 26 systems had values in excess of 1.5. The relative irrigation supply should be near 1 when irrigation supplies tightly fit the gap between demand and rainfall. System-wide values of these two parameters, however, do not tell us where the excess water is applied. In many irrigation systems, subsystems served by a distributary canal in the head reach of a system receive more water per unit land than those located in tail reaches of the same system. This same variation in water distribution is repeated at lower levels of the systems, i.e. between head and tail WCAs within a distributary command area and between farms located in head and tail reaches within the same WCA. The worst salinization often occurs in those tail areas.

A more equitable distribution of water within irrigation systems and better knowledge of LRs would contribute to greater water productivity (yield per unit of water beneficially used for evapotranspiration and leaching of salts) than that presently occurring in many irrigation systems. A condition for such an improvement is more extensive monitoring of the amounts of water and salts applied to and drained from irrigation systems as a whole and especially from their subunits. The data collection should cover all aspects of the water and salt balances at the different levels of irrigation systems. Salemi *et al.* (2000) and Droogers *et al.* (2001) give examples of insights that come from modelling of the water and salt balances in respect of the relation between water application, its salinity and the resulting water productivity for different water application and salinity conditions. The effect of water quality on the attainable water productivity is apparent without explicit knowledge of the LR.

Water productivity in rice cultivation has not been considered in this chapter. Paddy rice is often grown as an ameliorative crop. The high rates of percolation from the fields help reduce the salinity of the root zone for subsequent crops. A drawback of this approach is that rice is often grown on unsuitable light-textured soils that are poorly puddled at the start of the season, leading to excessive percolation rates and rising water tables. Water productivity as low as 0.14 kg m^{-3} of water applied to the rice-fields has been recorded in Pakistan's Punjab. This uncontrolled leaching wastes water.

Kotb *et al.* (2000) describe rice cultivation in salt-affected lands of the northern Nile delta in Egypt. They illustrate that the use of rice paddies to control salinity is faced with a number of constraints, such as periodic water shortages and salinity of supply water, which consists of a blend of fresh water and drainage water. Diversified cropping in the same subsurface drainage system compounds the problems, as rice and the other crops in the cropping system vary in their irrigation and drainage requirements. The authors propose that, to alleviate the problems of water shortage, the rice-cultivation area needs to be reduced by 50% and that rice cultivation in the delta should be consolidated to monitor its extent and to have uniform drainage requirements. Kotb *et al.* (2000) recommend rice cultivation only in saline soils of the delta but perceive that enforcement of such a policy may be difficult to achieve. In addition, long-term changes in the salinity of the delta water resulting from increased drainage-water reuse are not clearly known.

This example is typical in two respects. In many developing countries, the long-term productivity impacts of using saline and sodic irrigation water are unknown and the enforcement of policy measures that would lead to greater equity of distribution is doubtful, at best. A set of measures suggested by Kuper (1997) for a specific command area in Pakistan's Punjab included diversion of good-quality canal water from head to tail reaches to improve the blend of irrigation water available in the tail reaches and thereby curtailing further salinization. The consequence of this measure was that less canal water would be available to head-end farmers, who may object to this measure and compensate for their perceived shortage by pumping more groundwater and hence increasing the likelihood of salinization in the head reaches. The suggested measures were probably not economically viable or enforceable. Because of the current low levels of yield, the expected slight improvements in yield did not raise the economic returns in tail reaches by much (Kijne, 1998).

Unfortunately, few data are available on the economics of salinity-control measures. One complicating factor in the calculation of benefit/cost ratios is that the potential yield level under non-saline conditions is not well known. Yield levels between 4 and 7 t ha^{-1} for wheat and rice irrigated with canal water in India's Punjab (e.g. Tyagi, Chapter 5, this volume, Tables 5.2 and 5.3) are lower than the maximum irrigated yields attained elsewhere when all growth factors are closer to their optimal value.

This chapter has shown that the potential exists for improved water productivity by better-managed leaching practices but is not easily realized. Better knowledge is needed about the magnitude and interaction of the various components of the water and salt balances under field conditions and their changes over time. Those studies are expensive and time-consuming. Modelling studies, such as those discussed by Salemi *et al.* (2000) and Droogers *et al.* (2001), will contribute to our understanding, but they need to be validated in the field. In addition, it should be realized that the recommendations arising from such studies are probably difficult to implement. Reallocation of water supplies to achieve greater equity in access to and quality of water for farmers in different parts of irrigation systems requires greater management inputs and control. Using good-quality water only for high-value crops and poor-quality water for fodder crops and trees is politically unacceptable in a country like Pakistan, where the introduction of such measures would lead to greater poverty and unemployment for those farmers left with the saline groundwater. Reducing cropping intensities or changing cropping patterns to ensure adequate leaching applications is also likely to increase the gap between relatively rich and poor farmers.

In the long term, the installation of subsurface drains in a substantial portion of Pakistan's Punjab and the disposal of saline effluent into salt sinks and ultimately into the sea may be unavoidable. The investments required for this type of work are huge. The recent gradual decline in multilateral infrastructural investments in agriculture gives no reason to think that improved drainage will happen soon. In the meantime, yield levels and water productivity will remain lower than necessary.

References

Datta, K.K., Sharma, V.P. and Sharma, D.P. (1998) Estimation of a production function for wheat under saline conditions. *Agricultural Water Management* 36, 85–94.

Dinar, A., Rhoades, J.P., Nash, P. and Waggoner, B.L. (1991) Production functions relating crop yield, water quality and quantity, soil salinity and drainage volume. *Agricultural Water Management* 19, 51–66.

Droogers, P., Akbari, M., Torabi, M. and Pazira, E. (2001) *Exploring Field Scale Salinity using Simulation Modeling, Example for Rudasht Area, Esfaha Province, Iran.* IAERI–IWMI Research Report 2, Colombo, Sri Lanka.

Francois, L.E., Grieve, C.M., Maas, E.V. and Lesch, S.M. (1994) Time of salt stress affects growth and yield components of irrigated wheat. *Agronomy Journal* 86, 100–107.

van Genuchten, M.T. and Gupta, S.K. (1993) A reassessment of crop tolerance response function. *Journal of the Indian Society of Soil Science* 41, 730–737.

van Genuchten, M.T. and Hoffman, G.J. (1984) Analysis of crop salt tolerance data. In: Shainberg, I. and Shalhevet, J. (eds) *Soil Salinity under Irrigation, Processes and Management.* Ecological Studies No. 51, Springer Verlag, New York, pp. 258–271.

Grismer, M.E. and Gates, T.K. (1991) Hydrological aspects of saline water table management in regional shallow aquifers. In: Dinar, A. and Zilberman, D. (eds) *The Economics and Management of Water and Drainage in Agriculture.* Kluwer Academic Publishers, Boston, Massachusetts.

Guerra, L.C., Bhuiyan, S.I., Tuong, T.P. and Barker, R. (1998) *Producing More Rice with Less Water from Irrigated Systems.* Discussion Paper Series No. 29, International Rice Research Institute, Manila, Philippines.

Hoffman, G.J. (1990) Leaching fraction and root zone salinity control. In: Tanji, K.K. (ed.) *Agricultural Salinity Assessment and Management.* American Society of Civil Engineers, New York.

Hoffman, G.J. and van Genuchten, M.T. (1983) Water management for salinity control. In: Taylor, H., Jordan, W. and Sinclair, T. (eds) *Limitations to Efficient Water Use in Crop Production.* ASA Monograph. American Society of Agronomy, Madison, Wisconsin, pp. 73–85.

Howell, T.A., Cuenca, R.H. and Solomon, K.H. (1990) Crop yield response. In: Hoffman, G.J., Howell, T.A. and Solomon, K.H. (eds) *Management of Farm Irrigation Systems.* American Society of Agricultural Engineers, St Joseph, Michigan.

Hussain, M. (1995) Strategic research program: drainage re-use and economic impacts of salinity in Egypt's irrigation waters. Working Paper Series No. 8–4, Ministry of Public Works, Cairo, Egypt. Duplicated.

Kahlown, M.A., Iqbal, M., Skogerboe, G.V. and Ur-Rehman, S. (1998) *Waterlogging, Salinity and Crop Yield Relationships.* Fordwah–Eastern Sadiqia (South) Irrigation and Drainage Project, Mona Reclamation Experimental Project, WAPDA and International Water Management Institute, Lahore, Pakistan.

Katerji, N., van Hoorn, J.W., Hamdy, A. and Mastrorilli, M. (2000) Salt tolerance classification of crops according to soil salinity and to water stress day index. *Agricultural Water Management* 43, 99–109.

Keller, A.A. and Keller, J. (1995) *Effective Efficiency: a Water Use Efficiency Concept for Allocating Freshwater Resources.* Discussion Paper 22, Center for Economic Policy Studies, Winrock International, Morrilton, Arkansas.

Kijne, J.W. (1996) *Water and Salinity Balances for Irrigated Agriculture in Pakistan.* Research Paper 5, International Irrigation Management Institute, Colombo, Sri Lanka.

Kijne, J.W. (1998) Yield response to moderately saline irrigation water: implications for feasibility of management changes in irrigation systems for salinity control. *Journal of Applied Irrigation Science* 33, 249–277.

Kijne, J.W., Prathapar, S.A., Wopereis, M.C.S. and Sahrawat, K.L. (1998) *How to Manage Salinity in Irrigated Lands: a Selective Review with Particular Reference to Irrigation in Developing Countries.* SWIM Paper 2, International Irrigation Management Institute, Colombo, Sri Lanka.

Kotb, T.H.S., Watanabe, T., Ogino, Y. and Tanji, K.K. (2000) Soil salinization in the Nile delta and related policy issues in Egypt. *Agricultural Water Management* 43, 239–261.

Kuper, M. (1997) Irrigation management strategies for improved salinity and sodicity control. PhD dissertation, Landbouw Universiteit, Wageningen University, Wageningen, The Netherlands.

Kuper, M. and Kijne, J.W. (1996) Advances in the understanding of the impact of irrigation management on salinity, sodicity and soil degradation in Pakistan. Conference Proceedings, IIMI-Pakistan, International Irrigation Management Institute, Lahore, Pakistan. Duplicated.

Lamsal, K., Guna, N., Paudyal, N. and Saeed, M. (1999) Model for assessing impact of salinity on soil water availability and crop yield. *Agricultural Water Management* 41, 57–70.

Letey, J. (1993) Relationship between salinity and efficient water use. *Irrigation Science* 14, 75–84.

Maas, E.V. (1990) Crop salt tolerance. In: Tanji, K.K. (ed.) *Agricultural Salinity Assessment and Management.* American Society of Civil Engineers, New York.

Maas, E.V. and Hoffman, G.J. (1977) Crop salt tolerance – current assessment. *ASCE Journal of Irrigation Drainage Division* 103, 115–134.

Molden, D., Sakthivadivel, R., Perry, C.J., de Fraiture, C. and Kloezen, W. (1998) *Indicators for Comparing Performance of Irrigated Agricultural Systems.* Research Report 20, International Irrigation Management Institute, Colombo, Sri Lanka.

Pratt, F.P. and Suarez, D.L. (1990) Irrigation water quality assessments. In: Tanji, K.K. (ed.) *Agricultural Salinity Assessment and Management,* American Society of Civil Engineers, New York.

Prendergast, J.B. (1993) A model of crop yield response to irrigation water salinity: theory, testing and application. *Irrigation Science* 13, 157–164.

Rhoades, J.D. (1982) Reclamation and management of salt-affected soils after drainage. In: *Proceedings of the First Annual Western Provincial Conference Rationalization of Water and Soil Research and Management.* Lethbridge, Alberta, pp. 123–197.

Salemi, H.R., Mamanpoush, A., Miranzadeh, M., Akbari, M., Torabi, M., Toomanian, N., Murray-Rust, H., Droogers, P., Sally, H. and Gieske, A. (2000) *Water Management for Sustainable Irrigated Agriculture in the Zayandeh Rud Basin, Esfahan Province, Iran.* IAERI–IWMI Research Report 1, International Water Management Institute, Colombo, Sri Lanka.

Shalhevet, J. (1994) Using water of marginal quality for crop production: major issues. *Agricultural Water Management* 25, 233–269.

Smedema, L.K. (2000) *Irrigation-induced River Salinization: Five Major Irrigated Basins in the Arid Zone.* International Water Management Institute, Colombo, Sri Lanka.

Smedema, L.K. and Rycroft, D.W. (1988) *Land Drainage: Planning and Design of Agricultural Drainage Systems*. B.T. Batsford, London.

Solomon, K.H. and Davidoff, R. (1999) Relating unit and subunit irrigation performance. *ASAE Transactions* 42(1), 115–122.

Tedeschi, A., Beltran, A. and Aragues, R. (2001) Irrigation management and hydrosalinity balance in a semi-arid area of the middle Ebro river basin (Spain). *Agricultural Water Management* 49, 31–50.

Thorburn, P.J., Walker, G.R. and Jolly, I.D. (1995) Uptake of saline groundwater by plants: an analytical model for semi-arid and arid areas. *Plant and Soil* 175, 1–11.

Tyagi, N.K. (2001) Application of hydraulic and economic optimization for planning conjunctive use of surface and saline ground water: a case study. In: *IWMI, ICAR, CSSRI, ICAR–IWMI Groundwater Policy Initiative – 2001: Policy Dialogue on Groundwater Management, Held at Central Soil Salinity Research Institute, Karnal, India, 6–7 November 2001.* International Water Management Institute, Colombo, Sri Lanka, pp. 221–237.

Warrick, A.W. (1989) Generalized results from crop yield model with saline waters. *Soil Science Society of America Journal* 53, 1641–1645.

Zeng, L., Shannon, M.C. and Lesch, S.M. (2001) Timing of salinity stress affects rice growth and yield components. *Agricultural Water Management* 48, 191–206.

7 Opportunities for Increasing Water Productivity of CGIAR Crops through Plant Breeding and Molecular Biology

John Bennett
International Rice Research Institute, Manila, Philippines

In regions experiencing an absolute or economic shortage of water, there is an urgent need to increase the water productivity (WP) of crops through breeding or natural-resources management. The options for increasing WP through breeding are: (i) to reduce non-transpirational uses of water; (ii) to reduce transpiration without reducing production; (iii) to increase production without increasing transpiration; and (iv) to enhance tolerance of water-related stresses – drought, salinity and waterlogging or submergence. The Green Revolution achieved substantial increases in WP for rice and wheat by reducing the crop duration and increasing the harvest index. Progress in extending these achievements to other Consultative Group on International Agricultural Research (CGIAR)-mandated crops has been considerable and will accelerate following the recent cloning of several of the underlying genes. The success of breeding for greater WP depends heavily on the use of physiological, molecular and genetic methods to identify useful alleles within the genetic resources held by the CGIAR centres, which, in consequence, have a clear comparative advantage in this important enterprise. The complete sequencing of the *Arabidopsis* and rice genomes will provide crucial help in the discovery of genes for other WP-related traits in all mandated crops. It will soon be possible to enhance WP in popular varieties by DNA-assisted backcrossing, a more efficient breeding strategy than the conventional pedigree method based on phenotypic selection. A major application of molecular breeding is to increasing drought and salt tolerance, particularly at the stress-sensitive flowering stage, with benefits to both water economy and farmers' livelihoods.

The productivity of crops is commonly measured in relation to inputs such as capital, land, energy and labour or fertilizer (Pingali and Heisey, 1999; Socolow, 1999; Ball *et al.*, 2001). In recent years, increasing attention has been given to crop productivity in relation to water consumption (Seckler *et al.*, 1998). WP is an important issue in arid regions that are already experiencing an absolute shortage of water (or are projected to do so in the near future) and in other regions where an economic shortage of water leads to severe competition among water consumers. The magnitude of the problem is illustrated by the fact that agriculture accounts for about 70% of human water use (WMO, 1997). Furthermore, much of the water used in agriculture is lost to the atmosphere as a result of evaporation from the soil and transpiration from leaves, whereas much of the water used for industrial and domestic purposes may be recycled. For every kilogram of grain produced, cereal plants transpire about 1000 kg of water (Tuong, 1999; Tuong and Bouman, 2002, Chapter 4, this volume).

Seckler *et al.* (1998) presented four options for addressing the limitations imposed on crop production by shortages of water. They were: (i) development of additional water resources and water-storage facilities; (ii) increased productivity of existing water supplies; (iii) regional diversion of water; and (iv) increased importation of food. These options can be interpreted entirely in terms of infrastructural development and water engineering, but their implementation would be more effective and sustainable if accompanied by the breeding of crops with greater WP. Such crops would be valuable both in the intensive cropping systems on which global food security rests (Cassman, 1999) and in the stress-affected ecosystems on which 650 million poor people rely for food and livelihood (Alexandratos, 1999). You (2001) proposed that rice cultivation in China should be reduced in favour of more water-efficient crops and that Chinese rice requirements should be met by imports. However, this proposal does not take into account the steps taken in China to increase the WP of rice through breeding and water management (Dong *et al.*, 2001).

This chapter focuses on the opportunities for increasing crop WP through breeding. The first section takes a general look at water use by plants and by farmers and the implications of these uses for WP at the crop level. The second section discusses the opportunities for increasing WP by focusing on four types of trait. The third section focuses on breeding for drought tolerance and the prospects for improving breeding efficiency through genomics.

Water Use and WP at the Crop Level

WP in different cropping systems

The definition of WP differs at different scales of water management. Dong *et al.* (2001) used three definitions of WP in their study of rice cultivation. The WP per unit of evapotranspiration (WP_{ET}) is the mass of crop production divided by the total mass of water transpired by the crop and lost from the soil by evaporation. The WP per unit of irrigation water (WP_I) is the crop production divided by the irrigation flow. The WP per unit of gross inflow (WP_G) is the rice production divided by the rain plus irrigation flow. Two related and widely used physiological concepts are water-use efficiency (WUE), defined as crop production per unit transpiration, and transpiration efficiency (A/T), defined as the ratio of photosynthesis (A) to transpiration (T) (Peng *et al.*, 1998). Scientific disciplines differ in the importance they have given to these parameters.

The most appropriate measure of WP varies with the cropping system. For an entirely rain-fed cropping system, in which no attempt is made to trap rainfall on farm for later use, the goal is to maximize WP_G by planting crops and varieties that utilize rainwater fully and efficiently. WP_{ET} is a valuable parameter by which to judge the water efficiency of different crops, varieties and agronomic practices. A comparison between WP_{ET} and WP_G provides information on the availability of rainwater to the crop, as a result of the root structure of the plant or the structure of the soil. In the absence of irrigation, WP_I is irrelevant.

For a rain-fed cropping system with supplementary on-farm irrigation from surface reservoirs or underground aquifers, WP_I will be a sensitive measure of the timeliness of the irrigation, while WP_{ET} will be a guide to the water efficiency of different crops, varieties or agronomic practices. Cabangon *et al.* (2001) found that WP_{ET} for rice increased from 0.48 g kg^{-1} under dry seeding to 0.53 g kg^{-1} under wet seeding and to 0.61 g kg^{-1} under transplanting. As direct rainfall and on-farm irrigation are difficult to divert to off-farm uses, the value of knowing WP here lies in maximizing production rather than making decisions about water use.

The situation is different for a rain-fed cropping system with supplementary off-farm irrigation or for arid-land farming that is continuously dependent on off-farm water. Here it is likely that water will have to be purchased, possibly in competition with other users. WP_I will then help in decision-making about water allocation. It is important that irrigation water be available at the stages in plant growth (seedling emergence and flowering) when yield is most affected by water shortage. To be sure of supplies at these crucial times, farmers may invest in on-farm storage of off-farm water. Breeders will, of course, wish to develop varieties whose yield is not affected by any water deficit that occurs at these times. They will also wish to make the crop tolerant to low-quality irrigation water (through salt tolerance) and to the waterlogging or flooding that results when rainfall exceeds the sum of evapotranspiration (ET) and seepage.

It is instructive to compare the relevance of WP in the least productive and most productive cropping systems. Poor farmers in rain-fed areas often contend with drought, salinity and waterlogging/flooding. They wish to maximize WP_G, even though they lack the resources to accomplish this objective through water or land management. They look to plant breeding to maximize WP_G through the enhanced tolerance of crops to water-related stresses. One of the most productive cropping systems in the world is the rice–wheat double-cropping system of central and southern China and the Indo-Gangetic plain (Cassman, 1999; Timsina and Connor, 2001). Rice and wheat are grown in summer and winter, respectively. Depending on location, these crops may be largely rain-fed or largely irrigated. Both crops are likely to experience temperature stress and drought at the start and the end of each growing season. In addition, rice may experience submergence in midsummer, and wheat may experience early-season waterlogging and mid-season frost. As provision of irrigation and drainage is expensive, many rice–wheat farmers require cultivars that maximize WP_G through stress tolerance.

WP_{ET} takes into account only water evaporated or transpired and is, therefore, focused on plant behaviour. WP_I and WP_G include not only ET but also water used in other ways for crop production and water that is wasted. In the next two sections, I take a closer look at water use by plants and farmers to gauge opportunities for and limitations on increasing WP.

Water use by plants

Crops and natural vegetation are major users of water. This need arises from four features of plants:

- When plants open the stomata of their leaves to admit atmospheric CO_2 for photosynthesis, they lose water vapour through the same openings, a process known as stomatal transpiration. Many photosynthetic parameters (e.g. electron transport rate, carboxylation efficiency, intrinsic WUE, respiration rate in the light, etc.) are more strongly correlated with stomatal conductance than with water status itself (Medrano *et al.*, 2002).
- Even when stomata are closed, leaves and stems of many species may lose water by transpiration through non-stomatal surfaces.
- Transpiration also serves to cool leaves exposed to high air temperatures, low atmospheric water-vapour pressures or the heating effect of sunlight (Radin *et al.*, 1994).
- Plants use the transpiration stream to transport to the leaves both inorganic nutrients from the soil and a range of chemicals synthesized in the roots, including signal molecules that contribute to whole-plant integration (Peuke *et al.*, 2002).

To satisfy these requirements, plants will transpire in a growing season several hundred times more water than is present in their tissues at any one time.

If plants do not receive enough water to maintain high rates of photosynthesis, total dry-matter accumulation will decline, plant development will be affected and yield will be lost. The extent of yield loss depends on the timing, duration and intensity of the water deficit (Boonjung and Fukai, 1996). It is particularly important that plants gain access to water at the seedling and flowering stages, when yield is most sensitive to water deficit. Poor farmers rely entirely on soil moisture and rainfall to provide water for their crops. They could plant varieties with deep roots (in the uplands) or penetrating roots (in the lowlands) to explore a great volume of soil – a strategy known as drought avoidance. If the rainy season is short and fairly reliable, farmers may plant short-duration varieties that complete flowering before soil moisture declines at the end of the season – the drought-escape strategy. Some lowland varieties can adapt to a slow onset of drought by modifying their chemical constitution to retain as much water as possible through osmotic adjustment and to protect themselves from irreversible damage during stress – the strategy of drought tolerance. A major challenge for breeders is to produce new varieties that display high yield potential under or after drought stress.

Water use by farmers

Farmers with the financial resources to provide supplemental or continuous irrigation achieve much higher yields than farmers in rain-fed environments (Table 7.1), but whether WP_{ET} increases or decreases with irrigation depends on a variety of factors. WP_{ET} may increase if irrigation is supplied at germination and flowering, when water deficit is most damaging, but WP_{ET} may decline if unnecessary amounts of water are supplied at less sensitive stages of the growth cycle or if significant losses occur through evaporation. Farmers may also use large amounts of water for weed control, for growth of a legume crop as green manure, or for flushing salt and other toxic chemicals from the soil. Other uses include provision of water to moderate high or low temperatures through a microclimatic effect, a practice that tends to be limited to rice because of its ability to tolerate flooding of its root system. These practices increase yield but tend to reduce WP. Their elimination can be expected to increase WP, provided yield is maintained by using other approaches to control weeds, supplying N fertilizer, etc.

Plant Traits to Exploit for Increased WP

This section discusses four groups of plant traits that can be exploited to enhance WP. They are: (i) traits that reduce the non-transpirational uses of water in agriculture; (ii) traits that reduce the transpiration of water without affecting productivity; (iii) traits that increase production without increasing transpiration; and (iv) tolerance of three water-related stresses (waterlogging/flooding, salinity and drought). In Table 7.2, these traits are accompanied by an estimate (high, medium, low) of the probability that major progress will be made in the next 5 years. The probability is declared to be high where progress has already been made towards identifying regulatory genes. The probability is considered medium where some of the genes of a relevant pathway have been isolated and, therefore, provide an entry point to the identification of regulatory genes. The probability is described as low where few or no relevant data are available.

Traits that minimize non-transpirational uses of water in agriculture

One of the major water-saving innovations in rice production (measured as WP_G rather than WP_{ET}) is the switch from transplanting to direct seeding (Cabangon *et al.*, 2001), but

Table 7.1. A large minority of rice farmers and consumers depend on rain-fed rice production.

Parameter	Irrigated ecosystem	Rain-fed ecosystem
Production (% of total)	75	25
Land area (% of total)	45	55
Chemical inputs	High	Low
Average yield (t ha^{-1})	>5.0	<2.3
Consumers (billion)	>1.2	>0.8

Table 7.2. Genetic approaches to increasing crop water productivity.

Water-productivity factor	Genetic approach	Probability of major progress in 5 years
Minimize non-transpirational uses of water	Herbicide-resistant crop	Low[a]
	Weed competitiveness	Low
	Heat and cold tolerance at flowering	Medium
	More efficient cooling via evapotranspiration	Medium
	Nitrogen-use efficiency	Medium
	Nitrogen fixation	Low
Reduce transpiration without reducing production	Waxy-cuticle production	Medium
	Rapid stomatal closure	High
	Cooling mechanism for leaves	High
	Rapid canopy closure	Low
	Thicker, more intact Casparian strip	High
	Sustainable production of aerobic rice	Medium
Increase production without increasing transpiration	Short duration, seedling vigour	High
	Higher harvest index	Medium
	C_4 photosynthesis	Medium
	More photosynthesis per unit water transpired	Low
	More dry matter allocated to grain after stress	Medium
	Stay-green flag leaf	Medium
Use cheaper water	Tolerance of salinity	High
Less water management	Tolerance of waterlogging	Medium
	Tolerance of submergence	High

[a]Transgenic mechanism is currently available but its deployment is problematic.

weeds then become a major problem. Herbicide-resistant rice would be a solution, but concerns exist about the spread of herbicide-resistance genes to wild or weedy rice (Rieger *et al.*, 2002). Plants have genetic systems in the nucleus, the mitochondrion and the chloroplast. Insertion of genes for herbicide resistance into the nuclear genome may result in the spread of this trait to wild relatives of crop species, an event that would undermine the value of this trait. As most crop plants show maternal inheritance of plastid DNA ('plastome'), pollen escaping from plants transformed in the plastome will not transmit herbicide-resistance genes to nearby weedy relatives. Daniell *et al.* (1998) reported transformation of the plastome of petunia with the 5-enolpyruvyl Shikimate-3-phospate (EPSP) synthase gene, which confers resistance to the herbicide glyphosate. Of course, it would still be possible for pollen from weedy relatives to fertilize the transgenic plant and thereby create a potentially weedy, herbicide-tolerant hybrid in farmers' fields. Although the rate of this process would be extremely low, given the high self-fertilization rate of elite crop species, further research on eliminating the spread of herbicide resistance will be required before this potentially valuable trait is likely to be widely acceptable. In an alternative approach, the International Rice Research Institute (IRRI) is working with the West Africa Rice Development Association (WARDA) to transfer weed competitiveness from the African cultivated rice *Oryza glaberrima* to the Asian cultivated rice *Oryza sativa*. The basis of this trait appears to be seedling or vegetative vigour.

Green manure is a valuable source of organic nitrogen fertilizer (Ladha and Garrity, 1994). It is derived from short-duration legume crops or the N_2-fixing fern *Azolla*. The use of green manure is frequently advocated but it is often limited by water supply. As water becomes scarcer, it is even less likely that farmers will use green manure, in spite of its benefits. Chemical N fertilizer is widely available but still expensive for many farmers. N-use efficiency is thus a trait that will be highly valued by farmers. A radical alternative approach is to develop N_2 fixation in non-leguminous crops, such as cereals. IRRI pursued this approach until recently. It became clear that rice plants are to some extent '*Rhizobium*-ready' (Kouchi *et al.*, 1999). *Rhizobium* is closely related to *Agrobacterium*, another bacterium that naturally forms relations with dicotyledons but not with cereals. For many years, attempts to achieve *Agrobacterium*-mediated transformation of cereals were of limited success, but, once acetosyringone from potato was introduced as an activator of *Agrobacterium* virulence genes, cereal transformation by this bacterium became routine (Hiei *et al.*, 1997). Is it possible that a breakthrough of similar simplicity will enable cereals to form a symbiotic relationship with *Rhizobium* and fix their own nitrogen?

Australian rice farmers use an extra depth of floodwater (> 20 cm) to create a warmer microclimate during the flowering stage (Williams and Angus, 1994). Without this effect, cold southerly winds cause considerable pollen sterility in high-N plants. Rice is uniquely suited to this use of water because aerenchyma cells in the stem and the root provide a means of overcoming the root anoxia that makes other crops sensitive to waterlogging and flooding (Dennis *et al.*, 2000). This use of water could be dispensed with and WP_I could be significantly enhanced if cold tolerance at the reproductive stage could be increased in rice. Several examples of the enhancement of cold tolerance in rice have been reported (Sakamoto *et al.*, 1998; Saijo *et al.*, 2000). This trait would be equally valuable in other crops that suffer from the effect of cold air during flowering, even if no savings in WP_I were involved.

Traits that reduce the transpiration of water without affecting productivity

Traits that decrease water loss to the air include a waxy cuticle and stomatal closure. It is not clear to what extent the residual water loss from leaves is due to the permeance of the cuticle or to the incomplete closure of stomata (Riederer and Schreiber, 2001). The water permeances of leaf cuticular membranes from 21 plant species were found to cluster according to life form and climate of origin (Schreiber *et al.*, 1996). The lowest water permeances were observed with evergreen leaves from epiphytic or climbing plants growing naturally in a tropical climate. The next group in the order of increasing cuticular permeance comprised xeromorphic plants typically growing in a Mediterranean-type climate. The group with the highest water permeances comprise deciduous plant species with mesomorphic leaves growing in temperate climates. One of the impediments to defining genes that may be useful in decreasing the water permeance of the cuticle is that the chemistry responsible for determining the permeance has not yet been identified, although cutin polymers, their cross-linking, their esterification and their association with epicuticular waxes have all been implicated (Riederer and Schreiber, 2001).

Much more is known about the controls on stomatal closure in response to drought, salinity, soil compaction and other stresses (Luan, 2002; Roberts *et al.*, 2002). Although it is clear that leaf abscisic acid (ABA) promotes stomatal closure, the role of root-derived ABA as the principal long-distance signalling molecule for stress-responsive stomatal closure has recently been challenged by root-grafting experiments (Holbrook *et al.*, 2002). The ABA-dependent events inside guard cells are emerging, including the coordination afforded in *Arabidopsis* by AtRac1, a small guanosine triphosphatase (GTPase) (Lemichez *et al.*, 2001). AtRac1 inactivation by the protein phosphatase abscisic acid-insensitive 1 (ABI1) is the limiting step in the ABA-triggered signalling cascade leading to stomatal closure.

ABA-mediated stomatal closure is not enough to guarantee drought tolerance. Although it reduces water loss and helps to ensure that the soil–root–xylem–leaf–air hydraulic continuum remains intact, it does commit plants to a period without photosynthesis while radiation continues to be absorbed, and ABA may have undesirable side-effects on panicle development (Westgate *et al.*, 1996). It is interesting that the recessive *abh1* mutant of *Arabidopsis* shows ABA hypersensitivity and reduced wilting under drought (Hugouvieux *et al.*, 2001). The *ABH1* gene encodes an mRNA cap-binding protein that acts as a modulator of ABA signalling through alteration of transcript levels for early ABA-signalling elements.

Transpiration is regarded as the only productive water outflow at the field level because it contributes to plant growth by promoting photosynthesis and leaf cooling. However, it is possible that only a fraction of transpiration is actually beneficial and the remainder is wasteful. Peng *et al.* (1998) compared the ratio of photosynthesis (A) to transpiration (T) of seven tropical japonica rice varieties developed by IRRI with those of seven of IRRI's indica varieties. The A/T was determined 1 week after flowering for all varieties and throughout the growing season for one genotype from each type. Both A and T were measured on the topmost fully expanded leaves under saturating light with a portable photosynthesis system. Indica varieties had a higher T than the tropical japonica lines. The differences in A between the two types were relatively small and inconsistent across growth stages and years compared with the differences in T. The A/T was 25–30% higher for the tropical japonica than the indica type over 2 years. A lower carbon isotope ($^{13}C/^{12}C$) discrimination in a tropical japonica line than in an indica variety confirmed that the improved tropical japonica lines had higher A/T values than the indica varieties. These data indicate that significant variation exists in the rice germplasm for the A/T ratio. It would be important to determine whether a higher A/T translates into a higher WP_{ET}.

The role of transpiration in keeping leaves cool is a potential source of difficulty for breeders. As stomatal transpiration and non-stomatal transpiration decline, is it possible that leaf temperature will rise and inhibit production? One possible way of dealing with this issue is to enhance the heat tolerance of crops by non-transpirational means. Two such approaches are through the expression of genes for the heat-shock 101 (HS101) proteins (Queitsch *et al.*, 2000) and ascorbate peroxidase (Shi *et al.*, 2001).

Plants also lose water to drying soil. Since this water eventually contributes to evaporation and thereby reduces WP_{ET}, any trait that reduces such water loss is relevant here. Lignin and suberin form the Casparian strip and play important roles in protecting the stele of roots from water loss (Zeier *et al.*, 1999). Some of the key genes regulating the biosynthesis of these hydrophobic molecules are known. Caffeoyl coenzyme A (CoA) *O*-methyltransferase is a rate-limiting step in the biosynthesis of lignins and suberin (Inoue *et al.*, 1998), and peroxidase activity is essential for later steps in the biosynthesis of these molecules (Roberts and Kolattukudy, 1989). An important step forward will be the identification of transcription factors that upregulate these biosynthetic pathways in response to stress.

Following the lead of agronomists in China and Brazil, IRRI has begun to study the feasibility of establishing a high-input, non-puddled irrigated system for upland rice (Bouman, 2001). Known as 'Han Dao' in China and 'aerobic rice' at IRRI, this system has potential for large water savings, especially on soils with high seepage and percolation rates. Han Dao varieties yield 6–7.5 t ha^{-1} under flash irrigation in bunded fields in north-east China (Wang and Tang, 2000). It will be necessary to breed new varieties that are adapted to this 'aerobic' ecosystem in the tropics.

Traits that increase production without increasing transpiration

One of the most important traits in this category is the harvest index (HI), the proportion of total above-ground dry matter allocated to the harvested organs (e.g. tubers, fruits or seeds). A higher HI was one of the key traits of the high-yielding modern rice and wheat varieties that contributed to the Green Revolution (Khush, 2001). Manipulation of the HI may be achieved at the level of plant architecture and at the level of carbon allocation. At the architectural level, breeders have identified dwarfing genes that reduce vegetative biomass of cereals without affecting grain yield. The *sd1* gene for the semi-dwarf trait of rice has been cloned (Sasaki *et al.*, 2002). It encodes an inactive variant of gibberellic acid 20 (GA_{20}) oxidase-2, an enzyme of gibberellin biosynthesis. The wild-type version or allele of this gene is designated *SD1*. Three different mutant *sd1* alleles from rice germplasm prevent GA biosynthesis and confer the semi-dwarf trait. However, another copy of the gene (GA_{20} oxidase-1) is expressed in flowers and supplies enough GA to promote normal grain filling. The *Rht* dwarfing gene of wheat is a transcription factor in the GA biosynthetic pathway (Peng *et al.*, 1999). As a gain-of-function mutation, it may be useful in reducing the height of other important crop plants and for enhancing WP, because water use is determined by biomass rather than linked to yield.

At the level of carbon allocation, a larger HI implies that a greater proportion of carbohydrate has been deposited in the harvested product, e.g. as starch in cereal grains. The ability of seeds or fruit to accumulate carbohydrate depends on their sink strength. There is a growing consensus that a key step is the unloading of sucrose from the phloem through irreversible hydrolysis by the apoplastic invertase of the sink organ (Fridman *et al.*, 2000; Druart *et al.*, 2001; Nguyen-Quoc and Foyer, 2001). Pollen sterility and embryo abortion in drought-stressed cereals are associated with low levels of apoplastic invertase (Zinselmeier *et al.*, 1999; Saini and Westgate, 2000). Pollen-based male sterility was induced in plants by downregulation of the anther apoplastic invertase by the antisense approach (Goetz *et al.*, 2001). The antisense form of the invertase gene contained part of the gene sequence in the reverse orientation. When this antisense construct was introduced into plants, cells accumulated a novel form of mRNA, which bound strongly to normal invertase mRNA molecules, preventing them from directing the synthesis of invertase molecules. Chopra *et al.* (2000) implicated another sucrose-hydrolysing enzyme, sucrose synthase, in determining the sink strength in mung-bean seeds. Unlike apoplastic invertase, sucrose synthase is cytosolic and catalyses a reversible reaction.

Where the rainy season is reliable but comparatively brief, short duration is a trait that increases WP in crops by permitting drought escape. In such locations, flowering occurs before the onset of terminal drought. In locations with adequate rain throughout the year, short duration can enhance WP by permitting multiple cropping. Although short duration may be selected visually, a molecular understanding is expected to facilitate more precise control of the flowering date. Yamamoto *et al.* (2000) noted that 23 major genes and numerous quantitative trait loci (QTL) for heading date have been reported for rice, a plant with accelerated flowering under short days. A QTL is a genetic locus associated with some fraction of the variance of a quantitative trait, such as heading date. The variance explained by individual QTL varies from large to small. QTL of large effect are few but are readily mapped and potentially useful in breeding. QTL of small effect are numerous and difficult to map with sufficient accuracy for use in breeding. In the case of heading date, however, several QTL of large effect are known. One of such QTL, *Hd1*, was first identified in a segregating population derived from the cross Nipponbare/Kasalath, with the Nipponbare allele reducing the heading date relative to the heading date of Kasalath. *Hd1* has been cloned and shown to encode a zinc-finger transcription factor (Yano *et al.*, 2000). *Hd1* is an allele of the major photoperiod sensitivity gene *Se1*. The Nipponbare allele of this gene, when introduced into rice by transformation, reduced the time to heading under short days from 88 days to 58 days. DNA-based selection may now be used to achieve the same goal. Shorter duration has also been induced in rice by the *Arabidopsis* floral-transcription factor LEAFY (He *et al.*, 2000).

The new varieties of the Green Revolution also enhanced WP. Traditional cultivars are of long duration (150–180 days), with flowering triggered by changes in day length (longer day length for winter wheat and shorter day length for summer maize and rice). The development of photoperiod-insensitive varieties of short to medium duration (90–120 days) enabled crops to increase WP by escaping the late-season drought, which adversely affects flowering and grain development. Shorter duration also permitted double cropping and triple cropping, which make more efficient use of monsoonal floods. By increasing yield and simultaneously reducing crop duration (and therefore the outflows of evapotranspiration, seepage and percolation), the modern varieties of rice have a WP_G that is about threefold higher than that of traditional varieties (Tuong, 1999; Tuong and Bouman, 2002, Chapter 4, this volume).

In most genotypes of sorghum, drought during grain filling hastens leaf senescence, leading to premature death. Stay-green genotypes, in contrast, retain more green leaf area and continue to fill grain normally under drought conditions (Rosenow *et al.*, 1983). Moreover, there is a positive association between stay-green and grain yield under water-limited environments (Borrell and Douglas, 1996). Although the stay-green trait may involve more transpiration as the leaves remain active longer, they appear to give higher yields without requiring supplemental irrigation, increasing WP_G and perhaps even WP_{ET}. QTL *Stg1* and *Stg2* contribute to this trait; they have been mapped but not isolated as yet (Xu, W. *et al.*, 2000).

In 1999, IRRI held a symposium on the theme of achieving C_4 photosynthesis in rice, a C_3 plant (Sheehy *et al.*, 2000). The rationale was to explore the feasibility of achieving in rice the productivity, N-use efficiency and WUE of C_4 plants, such as maize and sorghum, in a plant producing rice grain. Yeo *et al.* (1994) surveyed photosynthetic gas exchange in 22 of the 23 species of the genus *Oryza*. Some species with the highest assimilation rates were assessed for photorespiratory losses, and these were generally around 30%, i.e. similar to those of *O. sativa* varieties. However, a range of *Oryza rufipogon* accessions had photorespiration rates significantly lower than the *O. sativa* genotypes tested. No species in the genus possessed C_4 photosynthetic metabolism, although some did overlap, with compensation concentrations and phosphoenolpyruvate (PEP) carboxylase activities reported for C_3–C_4 intermediate species. Nevertheless, the high level of expression of C_4 maize genes observed in transgenic rice by Ku *et al.* (1999) is indicative of progress using biotechnological tools. Transgenic rice overexpressing maize C_4-type PEP carboxylase, pyruvate-P_i dikinase or both exhibited superior photosynthetic and yield traits (Ku *et al.*, 2000, 2001). However, these superior traits were associated with reduced stomatal resistance and might, therefore, be associated with enhanced transpiration. Measurements of WP_{ET} for these transgenic plants are clearly highly desirable.

Waterlogging and flooding

Waterlogging and flooding are common in rain-fed ecosystems, especially on soils with poor drainage. They can seriously reduce yield (Dennis *et al.*, 2000) and are among the stresses considered by the Food and Agriculture Organization (FAO) and the International Institute for Applied Statistical Research in their estimates of global arable land area and global productivity (Fischer *et al.*, 2001). Roots obtain oxygen for growth and mineral uptake from air pockets in the soil, but, when roots are partially submerged (waterlogged) or completely submerged (flooded), the anoxic conditions prevent root growth and send signals to the rest of the plant to reduce shoot growth and plant productivity. Plants such as rice are tolerant to waterlogging because of their well-developed aerenchyma tissues in the roots and the stem. Some rice varieties, such as FR13A, are even tolerant of 2 weeks of submergence of the entire plant. In a genetic analysis of rice, Xu and Mackill (1996) localized a major gene for submergence tolerance on chromosome 9 (*Sub1*). Nandi *et al.* (1997) additionally localized minor QTL for submergence tolerance on chromosomes 6, 7, 11 and 12. Xu, K. *et al.* (2000) fine-mapped *Sub1* from FR13A, using a very large mapping population derived from a cross between M202 and a derivative of FR13A. Two markers co-segregated with *Sub1* and others were at a distance of 0.2 centiMorgans (cM) on the genetic map of rice (or ~60 kbp at its position on the physical map). The high-resolution map should serve as the basis for map-based cloning of this important locus, as it will permit the identification of bacterial artificial chromosome (BAC) clones (~150 kbp) spanning the region. Sripongpangkul *et al.* (2000) mapped a gene for submergence tolerance from the cultivar IR74; it mapped to the same location as *Sub1* and is presumably allelic with it. Dennis *et al.* (2000) discuss options for achieving waterlogging tolerance in wheat and other crops using genetic engineering.

Salinity

Flooding and waterlogging can lead to the salinization of soil above saline groundwater. Salinity arises also from intrusion of sea water in coastal areas and from the use of irrigation water of low quality. About 10% of the global land area is affected by salinity (Szabolcs, 1989) and about 20% of irrigated land is similarly affected (Yeo *et al.*, 1999). Overlapping sets of QTL for salt tolerance have been detected in tomato at germination and at the seedling stage (Foolad, 1999). A major gene and several QTL for salt tolerance have been reported for rice seedlings (Koyama *et al.*, 2001; Gregorio *et al.*, 2002). Many transgenes have been reported to enhance salt tolerance in *Arabidopsis* or crop plants. One of the most promising approaches is the overexpression of the vacuolar Na^+–H^+ antiporter NHX1 (Apse *et al.*, 1999; Zhang *et al.*, 2001) or of the vacuolar pyrophosphatase (Gaxiola *et al.*, 2001). The pyrophosphatase helps to energize the antiporter by pumping protons in the vacuole and allowing the antiporter to drive Na^+ ions into the vacuole in exchange for the protons. Overexpression of transcription factor *DREB1A* under control of the stress-sensitive *rd29A* promoter increases not only salt tolerance but also drought and cold tolerance in *Arabidopsis* (Kasuga *et al.*, 1999). Bennett and Khush (2002) review the discovery of genes conferring salt tolerance.

The reproductive stage is salt-sensitive in most crops. In the case of rice, this stage lasts for about 70 days, from panicle initiation to grain maturity, with flowering occurring at about the thirtieth day. The most salt-sensitive period is about 7–10 days before flowering (Makihara *et al.* 1999a,b). Few germplasm screens have been conducted to rank accessions according to salt tolerance at the reproductive stage, and yet it is at this stage that an episode of stress has its largest and least-reversible effect on yield. It is unfortunate that most studies on salt tolerance focus on survival of vegetative-stage stress rather than on productivity after a season of realistic stress, including reproductive-stage stress.

Drought

Drought is the most common water-related stress experienced by crops. Across the broad spectrum of research on water, four definitions of drought are in common use (Yevjevich *et al.*, 1978). Meteorological drought is defined as an extended period during which precipitation is below normal. Hydrological drought is an extended period during which stream flow and water levels in lakes and reservoirs are below normal. Socio-economic drought is defined as the meteorological and hydrological condition under which less water is available than anticipated and needed for the normal level of social and economic activity of the region. Agricultural drought is due to a shortage of water in the root zone, such that yield is reduced considerably. It is agricultural drought that is the principal concern here.

Agricultural drought may develop at any time of the cropping season. Its impact is usually most severe at the seedling and flowering stages (Boonjung and Fukai, 1996; Zinselmeier *et al.*, 1999; Saini and Westgate, 2000). It may be prevented by supplemental or continuous irrigation except under conditions of hydrological drought. To help the many farmers who do not have access to irrigation, breeders produce varieties that have an enhanced ability to escape, avoid or tolerate drought (Blum, 1988). The remainder of this chapter focuses on new methods for increasing the efficiency of breeding for drought tolerance.

Breeding for Drought Tolerance

Drought tolerance in CGIAR varieties

The CGIAR centres study a total of 22 mandated crops: six cereals, six food legumes, four forage plants and four tuber crops, along with bananas and plantains. Enhancing drought tolerance is a feature of the breeding programmes for all crops except cowpea, yam, banana and plantain (Table 7.3). These exceptions are grown in humid and subhumid environments and the germplasm collections lack adequate levels of drought tolerance for breeding purposes. The centres often have parallel programmes in crop, water and natural-resources management to ensure that the new lines perform well in their target environments. Farmer participation is often prominent in both the breeding programmes and the management programmes.

A common feature of the breeding programmes is the use of wild relatives of the crop plants as sources of drought tolerance. In its work on the Asian cultivated rice *O. sativa*, IRRI commonly crosses tropical japonica varieties from the uplands and indica varieties from the lowlands in an attempt to produce high-yielding recombinants capable of drought avoidance and drought tolerance (Lafitte *et al.*, 2002). WARDA has introduced genes for drought tolerance into *O. sativa* from the African cultivated rice *O. glaberrima*. The International Center for Agricultural Research in the Dry Areas (ICARDA) has introduced drought tolerance into *Hordeum vulgare* from *Hordeum spontaneum*. *Aegilops tauchii* has been a donor of drought tolerance for *Triticum aestivum* in the wheat programmes of the International Maize and Wheat Improvement Center (CIMMYT) and ICARDA, with *Triticum durum* used as the

Table 7.3. Status of breeding for drought tolerance at CGIAR centres (from http://www.cgiar.org).

Crop	Centre breeding for drought tolerance	Tolerant germplasm
Gramineae		
Rice	IRRI: lowland, upland, released; aerobic WARDA: released	Upland tropical japonicas, *O. rufipogon*, *O. glaberrima*
Maize	CIMMYT: sub-Saharan Africa, released	Ac7643S5
Wheat	CIMMYT, ICARDA: released	*Aegilops tauchii*
Barley	ICARDA: released	*Hordeum spontaneum*
Pearl millet	ICRISAT	PRLT 2/89–33, 863B
Sorghum	ICRISAT	B35, stay-green
Brachiaria	CIAT	Endophytic fungi important
Leguminosae		
Common bean	CIAT: heat	Tepary bean
Chickpea	ICRISAT	ICC4958, large root system
Groundnut	ICRISAT: released	Common
Pigeonpea	ICRISAT	Common
Cowpea	[Needs moist conditions]	None
Soybean	IITA	Northern China germplasm
Lentil	ICARDA: released	Common
Faba bean	ICARDA	Common
Grass-pea	ICARDA: toxin-free variety released	*Lathyrus sativus*
Others		
Potato	CIP	Andean germplasm
Sweet potato	CIP: tolerant late in season	Common
Cassava	IITA	Northern Brazil germplasm
Yam	[Needs high rainfall]	None
Banana	[Needs high rainfall]	*Musa balbisiana* (starchy)
Plantain	[Needs high rainfall]	None

bridging species. The International Crop Research Institute for the Semi-Arid Tropics (ICRISAT) has identified donors of drought tolerance in the germplasm of pearl millet (PRLT 2/89–33 and 863B), sorghum (the stay-green line B35) and chickpea (ICC4958, a line with a large root system). CIMMYT's maize-breeding programme has made considerable use of Ac7643S5 as a donor of drought tolerance. Centro Internacional de la Papa (CIP) is enhancing the drought tolerance of potato, using germplasm from the Andes. The International Institute for Tropical Agriculture (IITA) is using germplasm from northern China and north-eastern Brazil for soybean and cassava, respectively.

Drought-tolerant lines for several mandated crops have already been released by the centres for evaluation by collaborating institutes and farmers. The crops include rice (IRRI, WARDA), maize (CIMMYT), wheat (CIMMYT, ICARDA) and barley (ICARDA), among the cereals, and cowpea (IITA), groundnut (ICRISAT) and lentil (ICARDA), among the food legumes. ICARDA has also released important new lines of grass-pea (*Lathyrus sativus*) for drought-affected areas, but here the main achievement was to use mutation to reduce the neurotoxin levels of already drought-tolerant lines. ICARDA also has a drought-tolerance programme for faba bean. CIP and its collaborators have released drought-tolerant lines of sweet potato.

Several CGIAR-mandated crops have been difficult to improve for drought tolerance because of reproductive barriers or absence of suitable donors in the available germplasm. In its programme to enhance the drought tolerance of the common bean, *Phaseolus vulgaris*, CIAT has shown by grafting that the root system of tepary bean (*Phaseolus acutifolius*) confers considerable tolerance of drought and heat. However, attempts to cross the two species have encountered a reproductive barrier that has not yet been breached. Tissue-culture studies are under way along the lines that have proved successful for interspecific crosses of many other species, including rice. Centro Internacional de Agricultura Tropical (CIAT) is also improving the pasture grasses of the genus *Brachiaria*, especially *B. brizantha*, for which both sexual and apomictic accessions are known. *B. brizantha* is already highly drought-tolerant, but this trait has been attributed in part to the presence of an endophytic fungus, *Acrimonium implicatum*. If the endophyte also proves to be the cause of a serious cattle disease, drought tolerance in this grass might have to be reassessed. Three of IITA's mandated crops – banana and plantain (both with International Network for the Improvement of Banana and Planta (INIBAP)) and yam – need high rainfall. The enhancement of drought tolerance in these crops is not a priority compared with enhancing resistance to biotic stresses. The drought tolerance of banana increases in proportion to the contribution of *Musa balbisiana* to the genome, but the *M. balbisiana* genome also contributes starchiness, rather than sweetness, to the fruit.

Why is breeding for drought tolerance currently inefficient?

In spite of the success of the CGIAR centres in releasing a number of drought-tolerant varieties, breeding for drought tolerance is a slow, painstaking and inefficient process. This situation arises principally from three problems. First, the variability of drought in terms of its timing during the plant growth cycle means that early-season drought, mid-season drought and terminal drought are essentially different challenges. Secondly, for each time of onset of drought, water deficit affects all tissues and involves multiple responses, during both stress and recovery, leading to complex genetic control of drought tolerance. Thirdly, the screening of germplasm collections or breeding materials for drought tolerance is highly sensitive to environmental conditions (soil chemistry, soil texture and weather). These problems have led to the division of drought-breeding programmes into subprogrammes to match the major types of drought-prone environments.

Completely different approaches are taken in these subprogrammes. Where the wet season is brief but reliable, photoperiod-insensitive, short-duration varieties are favoured to allow the crop to pass the very sensitive flowering stage before the onset of water deficit. Where drought is most likely to occur mid-season, photoperiod sensitivity is used to delay flowering until late in the season, when rains are more reliable. Of course, in this situation, the plants must still survive mid-season drought. In upland rain-fed conditions, deep-rooted varieties can continue to grow by tapping the deeper layers for water (Champoux *et al.*, 1995), while in lowland rain-fed conditions plants may require roots with the capacity to penetrate the hardpan, a layer of compacted soil located about 20–25 cm below the soil surface (Ray *et al.*, 1996). In either situation, when water deficit is finally experienced, desired traits include the abilities to: (i) reduce water loss to the soil and the air; (ii) maintain turgor for an extended period; (iii) survive the loss of turgor; and (iv) protect cells against oxidative damage caused by continued absorption of radiation under conditions where stomatal closure prevents photosynthesis. When water is restored, the plants must be able to recover photosynthetic activity and growth, and they must allocate a large fraction of fixed carbon (either stored from before drought stress or synthesized after stress) to grain, fruit and tuber production (Blum, 1998; Richards, 2000).

When water stress is experienced during flowering, pollen sterility and embryo abortion may occur and crop yield may be greatly reduced (Zinselmeier *et al.*, 1999; Saini and Westgate, 2000). It is not entirely clear whether these problems are due to a direct effect of stress on the flowers or whether they arise indirectly from a diminished supply of carbohydrate from the stressed leaves or a premature supply of ABA from stressed leaves and roots (Westgate *et al.*, 1996). There is considerable genetic variation in rice in the capacity to withstand water deficit at flowering and maintain pollen sterility (J.X. Liu and J. Bennett, unpublished data). Proteomic, microarray and genetic approaches are being taken to identify the pathways that permit tolerance of water stress at flowering.

This long list of desirable traits makes the genetics of drought tolerance very complex. The selection for the desired combination of traits is rendered additionally difficult by interactions between the plants and their environment, including soil chemistry, soil texture and weather. To overcome these problems, breeders must conduct large, replicated field trials or use managed environments, such as greenhouses or drip-irrigated fields. Replicated field trials are difficult to conduct early in the breeding programme, when the number of breeding lines is large and the amount of available seed is small. Managed environments are expensive to create and operate and may not fully represent the target environment.

Breeders are focusing increasingly on yield components, rather than on the yield itself. The rationale behind turning to yield components is that each component tends to be determined during a comparatively brief period of the plant growth cycle and is presumed to be controlled by fewer key genes than yield as a whole. The four commonly used yield components for rice are: (i) panicles per square metre; (ii) spikelets per panicle; (iii) spikelet fertility; and (iv) single-grain weight. Each yield component is most sensitive to drought or salinity at a different time in the growth cycle (Boonjung and Fukai, 1996; Makihara *et al.*, 1999a,b). Component i is affected by stress, principally during germination, seedling development and tillering. Components ii, iii and iv are affected by stress, principally during panicle initiation, flowering and grain filling, respectively.

Boonjung and Fukai (1996) exposed rice to water stress for 23–34 days at different growth stages over two growing seasons. They quantified the effect of stress on the four yield components (Fig. 7.1). When drought occurred during tillering, it reduced yield by up to 30% through reductions in the number of panicles per unit area and the number of spikelets per panicle. When drought occurred during panicle development, anthesis was delayed, the number of spikelets per panicle was reduced by 40% and the percentage of filled grains decreased markedly (to zero in 1 year). A decrease in grain yield of > 80%

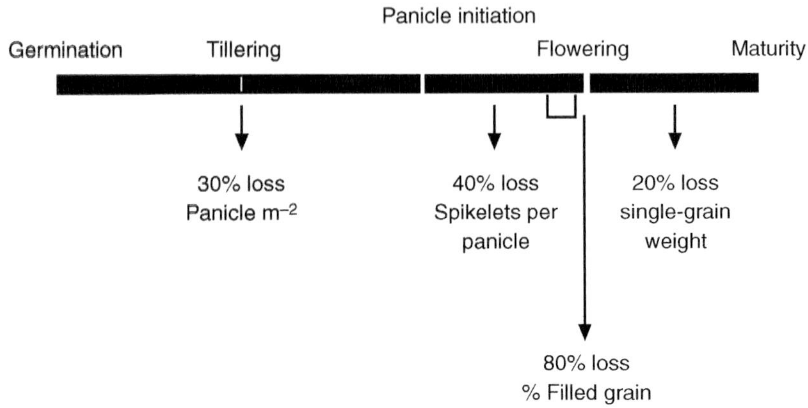

Fig. 7.1. The impact of drought stress on four rice-yield components. Boonjung and Fukai (1996) subjected rice to 3 weeks of water stress at different intervals during the growth cycle and measured the impact on panicles m^{-2}, spikelets per panicle, percentage filled grains and single-grain weight. These components were most severely reduced at tillering, panicle initiation, flowering and grain filling, respectively. The largest impact of stress was recorded at flowering through an 80% reduction in the percentage of fertile grains.

was associated with low dry-matter production during both the drought period and the recovery period. When drought occurred during grain filling, the percentage of filled grains decreased by 60% and individual grain mass decreased by 20%. Makihara *et al.* (1999a,b) obtained similar results for salt stress, but the salt sensitivity of different yield components was genotype-dependent. These observations raise the possibility of simplifying the genetics of drought and salt tolerance by examining the different mechanisms of tolerance at different growth stages for different yield components. The molecular analysis of tolerance would also be made easier because of more precise knowledge of when and from which tissue to extract protein and RNA for analysis. Molecular analysis of stress responsiveness has already begun, using microarrays (Kawasaki *et al.*, 2001; Seki *et al.*, 2001) and proteomics (Moons *et al.*, 1995; Thiellement *et al.*, 1999; Salekdeh *et al.*, 2002a,b). Figure 7.2 illustrates the analytical methods of microarrays and proteomics. Interesting cDNAs are identified by sequencing and interesting proteins by mass spectrometry (Salekdeh *et al.*, 2002a). The genotype × environment interactions of these mechanisms should be greatly reduced compared with yield as a whole; the genetics is simpler and their duration is shorter.

Drought-related Traits

One major approach to understanding and simplifying the genetics of drought tolerance focuses on mapping QTL that condition drought-related physiological traits. Many of the drought-related traits studied at the level of QTL analysis relate to root behaviour. In the case of rice, most mapping populations are derived from intersubspecific crosses between upland tropical japonica cultivars, such as CT9993 and Azucena, and lowland indica cultivars, such as IR62266 and IR64. Key QTL have been mapped for root morphology, root distribution and drought avoidance (Champoux *et al.*, 1995; Price and Tomos, 1997; Yadav *et al.*, 1997; Ali *et al.*, 2000; Courtois *et al.*, 2000; Kamoshita *et al.*, 2002), root penetration ability (Ray *et al.*, 1996; Price *et al.*, 2000; Zheng *et al.*, 2000), osmotic adjustment and dehydration tolerance (Lilley *et al.*, 1996; Zhang *et al.*, 1999), stomatal conductance, leaf rolling and heading date (Price *et al.*, 1997), cell-membrane stability (Tripathy *et al.*, 2000) and ABA accumulation (Quarrie *et al.*, 1997). The contribution of a QTL to the variance of its trait is often quite small, making it difficult to map the QTL with sufficient accuracy for use in marker-assisted selection (MAS) (< 5 cM) or for map-based cloning (< 1 cM) and requiring several QTL to be pyramided to reconstruct the trait to an adequate extent.

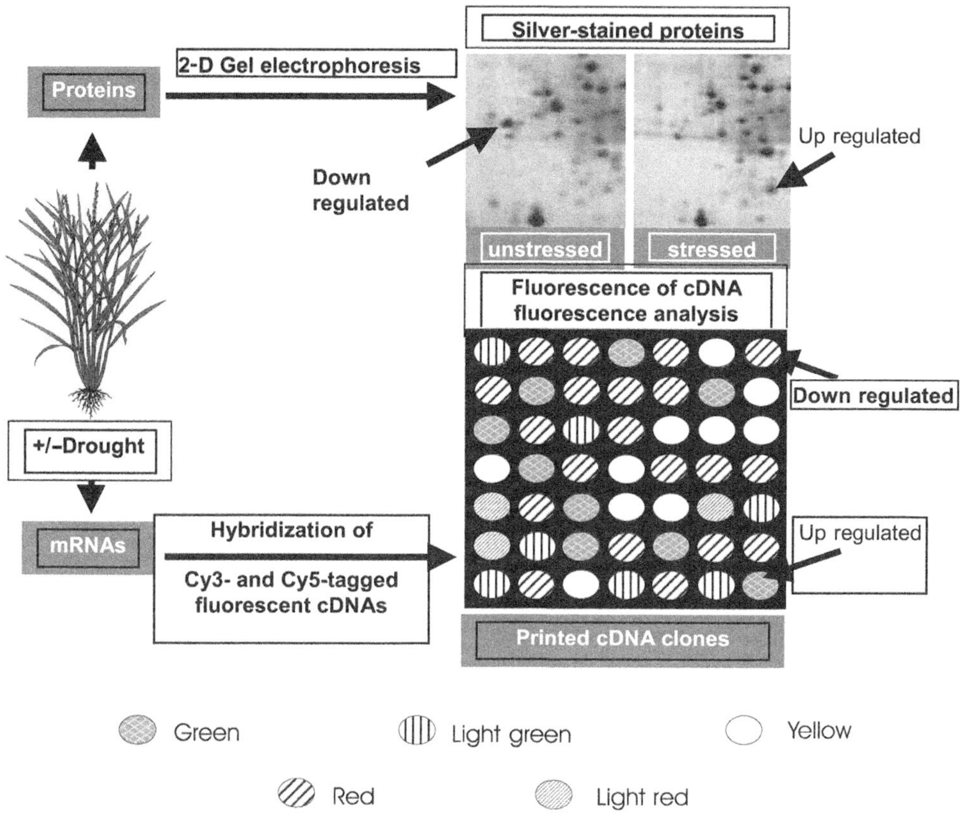

Fig. 7.2. Comparison of gene expression in drought-stressed and unstressed plants through proteomics and microarray analysis. Proteins extracted from plant tissue are separated by two-dimensional electrophoresis (isoelectric focusing, followed by dodecylsulphate-polyacrylamide gel electrophoresis), and then visualized by silver staining and quantified by scanning (Salekdeh *et al.*, 2002a). The most commonly observed changes are changes in abundance, but changes in position as a result of cleavage or phosphorylation are also seen. The mRNA populations extracted from plant tissues are used as a template for cDNA synthesis and concomitant tagging with the fluorophores Cy3 (green, stressed) and Cy5 (red, unstressed). The cDNA populations are hybridized to immobilized arrays of cDNA clones on glass slides, and a fluorescence detector scans over the slide to record hybrid abundance. Both digital and false-colour representations of the hybridization data are recorded (Kawasaki *et al.*, 2001). Green fluorescence shows a cDNA upregulated by stress, red fluorescence is a sign of downregulation by stress and yellow fluorescence indicates that in the cell the cDNA is probably unresponsive to stress.

QTL for drought-related traits have also been reported for other crops, especially cereals. These traits include osmotic adjustment in wheat and barley (Teulat *et al.*, 1998), anthesis silking interval (ASI) in maize (Ribaut *et al.*, 2002) and the stay-green trait in sorghum (Xu, K. *et al.*, 2000). In spite of changes in the chromosome number, the genomes of the cereals display a high degree of synteny, or conservation of gene order, along homologous chromosomes (Gale *et al.*, 2001). As a result, it is sometimes possible to predict the genomic location of a gene in one cereal from its known location in another cereal. The frequency of this sort of prediction will increase as comparative mapping improves. Orthologous genes for osmotic adjustment in barley, rice and wheat have been located in syntenic regions of the respective genomes by QTL mapping (Zhang *et al.*, 1999).

ASI is the time difference in days between anthesis in tassels (male) and silking in ears (female). A longer ASI (due to a longer delay in silking) is strongly associated with greater drought sensitivity (Edmeades *et al.*, 1993). At least six QTL for ASI have been identified and together they account for 47% of the total variance of this trait in the mapping population. QTL for ASI are among several markers being considered by CIMMYT for use in molecular breeding for drought tolerance (Ribaut *et al.*, 2002), but they are not effective by themselves. This situation may arise from the fact that, whereas a long ASI is sufficient to confer drought sensitivity (silks develop too late for efficient pollination), a short ASI is not sufficient to confer drought tolerance. As the separate development of tassel and ear in maize has no counterpart in rice, wheat and barley, it might be thought that ASI has no relevance to these cereals, but, to the extent that ASI is a sign of altered carbohydrate allocation to tassel and ear, it may be relevant to the fundamental question of carbohydrate allocation between organs and between spikelets within a panicle in small-grain cereals.

Candidate genes for drought tolerance

In recent years, many drought-responsive genes have been identified in plants, especially in *Arabidopsis thaliana*. These genes have been identified by several approaches:

- Studies on the anabolic and catabolic pathways for metabolites that accumulate in drought-stressed plants (e.g. proline, glycine, betaine, trehalose, ABA).
- Analysis of other mechanisms of drought tolerance.
- Analysis of protein or mRNA changes in response to drought.
- Analysis of signal-transduction pathways.
- Mapping of QTL for drought tolerance, using segregating populations.

All of these methods are being accelerated as a result of the sequencing of the *Arabidopsis* and rice genomes. A sixth method (isolation of mutants) has so far been less successful with drought than with salinity tolerance.

Alleles and pyramids

All accessions of a particular crop species are expected to contain essentially the same genes. Differences in agricultural performance between accessions are thought to be due to allelic differences within the same gene set. Thus, achieving a high level of drought tolerance depends on finding the most appropriate alleles of key genes and combining or pyramiding them together.

One successful method of screening a germplasm collection for the best alleles is to apply standardized phenotyping protocols to the collection and then to conduct detailed genetic analysis of the best performers. Some of these accessions will owe their superior performance to positive alleles at a considerable number of QTL of small effect and are not of interest here. Other accessions will perform well because they contain positive alleles at one or two major genes or QTL of large effect; these alleles are of great interest in crop improvement. The two types of accessions can be distinguished by advanced backcross analysis.

Desirable alleles may also be recovered from accessions that do not perform particularly well in phenotypic tests. Such alleles might be recognized first only after introgression into an elite genetic background by backcrossing (Tanksley and McCouch, 1997). The genetic basis of these effects is beyond the scope of this chapter, but the recipe for success is easily summarized: hard work, attention to detail and luck in the choice of accessions to be screened. Given the very large size of the germplasm collections in most CGIAR centres, it is impractical to screen a whole collection, but DNA fingerprinting could be exploited to select a core germplasm set of the most divergent accessions for closer analysis. The new techniques of diversity array technology (DaRT) (Jaccoud *et al.*, 2001) and targeting induced local lesions in genomes (TILLING) (McCallum *et al.*, 2000) facilitate these steps.

Desirable alleles might also be the product of a carefully designed programme of random or directed mutagenesis. An exquisite example of directed mutagenesis is the use of recombinant DNA technology to change the promoter on the gene encoding the drought response element building protein (DREB1A) transcription factor to enhance abiotic stress tolerance in *Arabidopsis* (Kasuga *et al.*, 1999). Naturally occurring alleles and mutant alleles may be moved into new genetic backgrounds by marker-assisted backcrossing, but alleles produced by promoter switching and other forms of recombinant DNA technology must be introduced as transgenes, after which they may also be manipulated with the help of markers (Fig. 7.3). Ideally, allele pyramiding will become possible not only for drought but also for other traits that contribute to WP, such as salinity tolerance and waterlogging/flooding tolerance.

In summary, the alleles that will eventually be pyramided to confer a high level of drought tolerance on crop plants may derive from many different genetic sources. They may take several different forms:

- Precisely mapped major genes or QTL of large effect that confer drought tolerance through a specific mechanism but have not yet been cloned and identified.
- Known genes discovered by map-based cloning of major genes or QTL or through some other approach.
- Novel alleles produced from known genes by mutagenesis or recombinant DNA technology and reintroduced into plants by genetic engineering.

The major genes and major QTL will be pyramided using tightly linked flanking markers, whereas alleles of known genes and novel transgenic alleles will be pyramided through the use of allele-specific DNA probes based on the genes themselves.

The discovery of genes for enhancing WP is not an isolated activity. It forms part of a breeding and resources-management programme that begins and ends with farmers (Fig. 7.4). The overall starting-point for such a programme is participatory rural appraisal, involving a wide spectrum of stakeholders. One task of the stakeholders is to determine how much of the responsibility for increasing WP should be shouldered by breeding and how much by natural-resources management, especially water management. Poor farmers will not be able to afford most management options and will look more to breeding for solutions. Intractable traits, such as drought tolerance, require detailed physiological, bio-

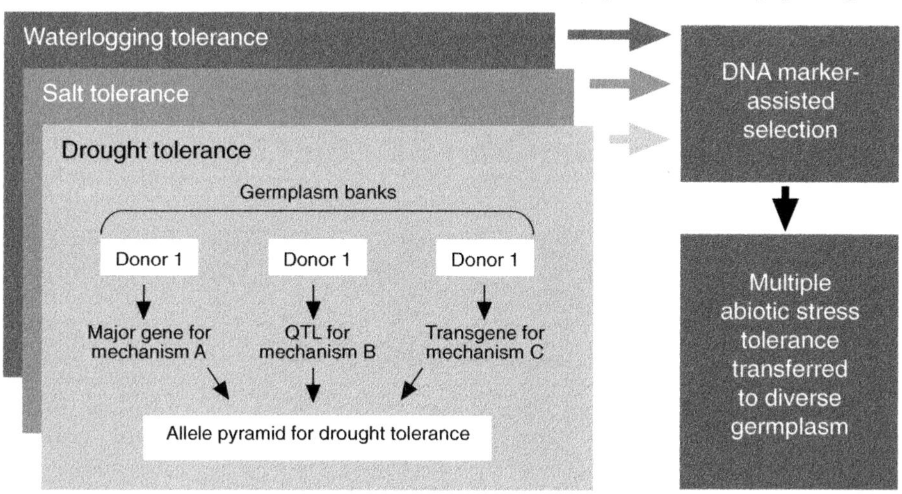

Fig. 7.3. Identification and use of genes conferring tolerance of drought, salt or waterlogging. Tolerance of an abiotic stress involves several distinct molecular or cellular mechanisms. Mapping of the corresponding major genes or major quantitative trait loci (QTL) allows the use of DNA markers to backcross these mechanisms into popular varieties that are sensitive to stress. The sequencing of plant genomes increases the probability that the underlying genes can be isolated and used directly to search for superior alleles. The isolated genes may also be modified (e.g. by promoter switching) to create entirely novel alleles suitable for reintroduction into plants as transgenes. Several different donors may have to contribute genes before an adequate level of tolerance can be assembled through allele pyramiding. The initial assembly of the pyramid and its subsequent transfer to popular varieties both depend on the development of a robust set of molecular markers.

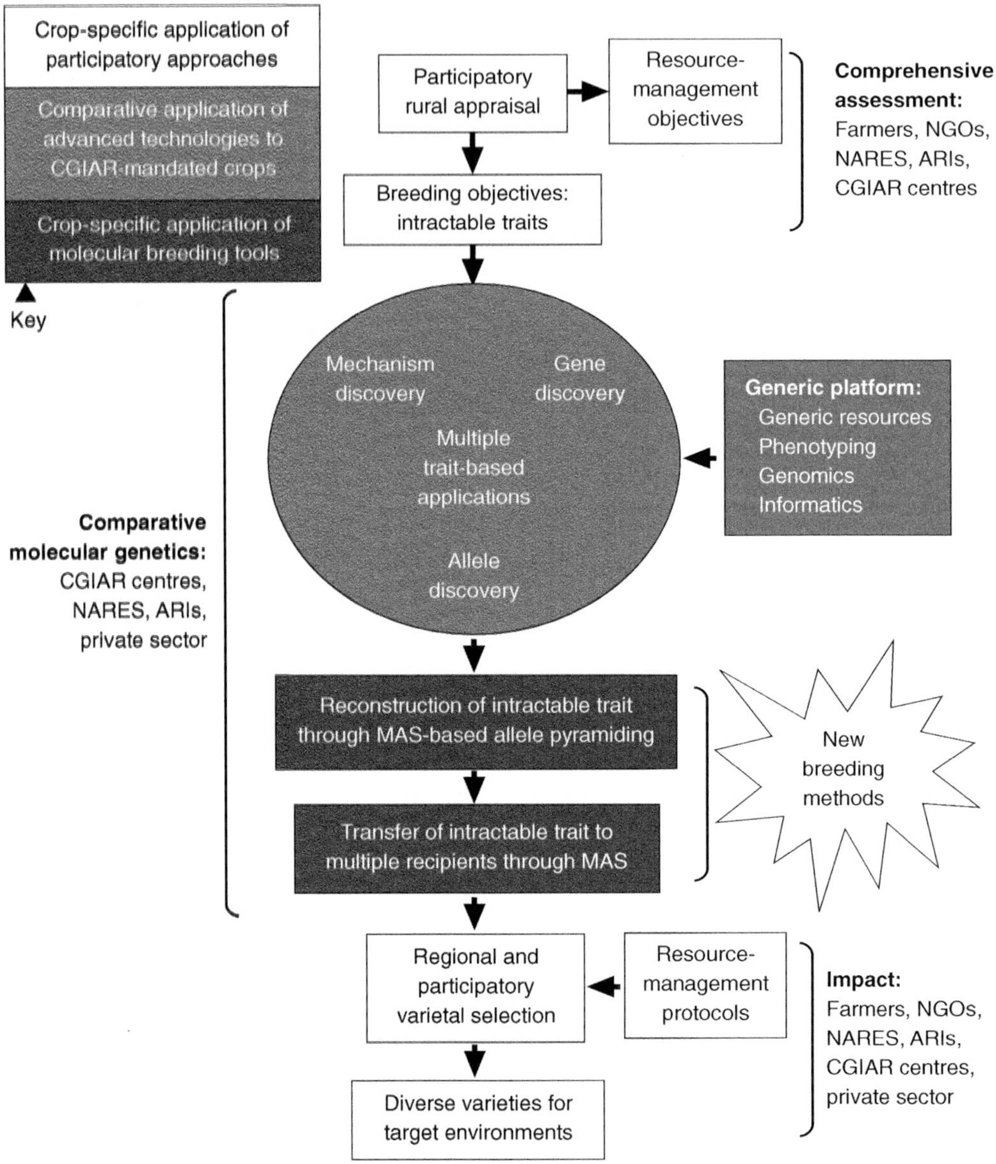

Fig. 7.4. A comprehensive programme of gene and allele discovery to enhance water productivity across the CGIAR-mandated crops. Crop-specific rural appraisal identifies the breeding objectives for key intractable traits, such as drought and salinity, in different target environments. The comparative molecular genetics of those traits helps the discovery process and allows orphan crops to benefit from genomic analysis of the major cereals and *Arabidopsis*. The prerequisite for discovery is a platform consisting of genetic resources (including mutants and mapping populations), standardized phenotyping protocols, genomic resources and tools and bioinformatics. Repeated application of the resource platform allows multiple traits to be analysed, simplified and improved in multiple crops. Marker-assisted selection (MAS) replaces field-based selection and thereby accelerates the enhancement of water productivity. NGOs, non-government organizations; NARES, National Agricultural Research and Extension System; ARIs, Advanced Research Institutes; CGIAR, Consultative Group on International Agricultural Research.

chemical and molecular study, but the order of discovery of mechanisms, genes and alleles will vary depending on the point of entry into the analysis. These steps are therefore depicted as a circle in Fig. 7.4. Feeding into this circular discovery process is a generic technical platform consisting of genetic resources, phenotyping, genomics and informatics. The platform is generic in the sense that all of its components (with the possible exception of phenotyping) are largely independent of whichever traits are under investigation.

The new breeding activities highlighted in Fig. 7.4 were shown in more detail in Fig. 7.3. The primary function of these activities is to show the efficacy of the pyramided set of alleles developed to enhance the intractable trait in question. In this proof-of-concept stage, breeders would use robust molecular markers to backcross the set of alleles into only a limited set of varieties. At the final stages of the process, when a wider range of stakeholders become directly involved again, the set of alleles is backcrossed into a much larger number of locally popular varieties by local breeders. Participatory varietal selection can be included if it can increase the likelihood of uptake of the new variety.

Future Prospects: Linkage between Challenge Programmes

Water will be the most important challenge for agriculture over the next century. Can we breed plants that use water more efficiently? If the answer is no, the increasing demand for water from non-agricultural sectors will leave food security permanently in jeopardy. If the answer is yes, the prospects will be bright for achieving long-term food security, even in resource-poor, highly populous developing countries. The decisive factor will be how quickly the new crop varieties can be generated and released to farmers. The speed of this process will be determined by the strength of the linkage between scientists working on water and scientists working on gene discovery and the determination of these scientists to achieve significant gains in WP over a wide range of crops.

The new challenge programmes of the CGIAR offer a way of linking the water-research community and the genomic community across the major crops. WP is one of the themes of the Water and Food Challenge Program (WFCP), and water-related stresses are high-priority issues for the Genetic Resources Challenge Program (GRCP). Figure 7.4 shows how these two challenge programmes could be linked in relation to intractable problems, such as drought, salinity, waterlogging and submergence. WFCP has a comparative advantage over GRCP in conducting the rural appraisals needed to identify target environments, set objectives for breeding and resource management and evaluate and disseminate new varieties. GRCP has a comparative advantage over WFCP in discovering key genes and alleles in the germplasm collections of the CGIAR centres and combining the alleles to produce unprecedented levels of stress tolerance and greatly enhanced WP for all crops.

This strategy begins and ends with farmer participation. However, more research is needed to find the best ways of integrating participation into breeding programmes. Participatory rural appraisal is essential in defining the environmental adaptability expected of a new variety, the traits required for uptake of the variety by farmers and integrating breeding and natural-resources management to ensure that the variety and its management are consistent with the local environment and the resources of the farmers and their community. The current comprehensive assessment on water resources provides an opportunity to refine appraisal techniques and set the stage for the challenge programmes.

Acknowledgements

Drought research in my laboratory is supported by grants to IRRI from the governments of Germany and Switzerland. I thank Randy Barker and David Mackill for comments on the manuscript and Renee Lafitte, Len Wade, T.P. Tuong, Gary Atlin and Bas Bouman for discussions.

References

Alexandratos, N. (1999) World food and agriculture: outlook for the medium and longer term. *Proceedings of the National Academy of Sciences USA* 96, 5908–5914.

Ali, M.L., Pathan, M.S., Zhang, J., Bai, G., Sarkarung, S. and Nguyen, H.T. (2000) Mapping QTL for root traits in a recombinant inbred population from two indica ecotypes in rice. *Theoretical and Applied Genetics* 101, 756–766.

Apse, M.P., Aharon, G.S., Snedden, W.A. and Blumwald, E. (1999) Salt tolerance conferred by overexpression of a vacuolar Na^+/H^+ antiport in *Arabidopsis*. *Science* 285, 1256–1258.

Ball, V.E., Butault, J.P. and Nehring, R. (2001) *U.S. Agriculture, 1960–96: a Multilateral Comparison of Total Factor Productivity*. ERS Technical Bulletin No. 1895, US Department of Agriculture, Washington, DC, 56 pp.

Bennett, J. and Khush, G.S. (2002) Plant breeding for salt tolerance. *Journal of Crop Production* 7, 1–25.

Blum, A. (1988) *Plant Breeding for Stress Environments*. CRC Press, Boca Raton, Florida, 223 pp.

Blum, A. (1998) Improving wheat grain filling under stress by stem reserve mobilization. *Euphytica* 100, 77–83.

Boonjung, H. and Fukai, S. (1996) Effects of soil water deficit at different growth stages on rice growth and yield under upland conditions. 2. Phenology, biomass production and yield. *Field Crops Research* 48, 47–55.

Borrell, A.K. and Douglas, A.C.L. (1996) Maintaining green leaf area in grain sorghum increases yield in a water-limited environment. In: Foale, M.A., Henzell, R.G. and Kneipp, J.F. (eds) *Proceedings of the Third Australian Sorghum Conference*. Occasional Publication No. 93, Australian Institute of Agricultural Science, Melbourne, Australia, pp. 315–322.

Bouman, B.A.M. (2001) Water-efficient management strategies in rice production. *International Rice Research Notes* 16(2), 17–22.

Cabangon, R.J., Castillo, E.G., Bao, L.X., Lu, G.A., Wang, G.H., Cui, Y.L., Tuong, T.P., Bouman, B.A.M., Li, Y.H., Chen, C.D. and Wang, J.Z. (2001) Impact of alternate wetting and drying irrigation on rice growth and resource-use efficiency. In: Barker, R., Li, Y.H. and Tuong, T.P. (eds) *Water-saving Irrigation for Rice: Proceedings of an International Workshop*. International Water Management Institute, Colombo, Sri Lanka, pp. 55–79.

Cassman, K.G. (1999) Ecological intensification of cereal production systems: yield potential, soil quality, and precision agriculture. *Proceedings of the National Academy of Sciences USA* 96, 5952–5959.

Champoux, M.C., Wang, G., Sarkarung, S., Mackill, D.J., O'Toole, J.C., Huang, N. and McCouch, S.R. (1995) Locating genes associated with root morphology and drought avoidance in rice via linkage to molecular markers. *Theoretical and Applied Genetics* 90, 969–981.

Chopra, J., Kaur, N. and Gupta, A.K. (2000) Ontogenic changes in enzymes of carbon metabolism in relation to carbohydrate status in developing mungbean reproductive structures. *Phytochemistry* 53, 539–548.

Courtois, B., McLaren, G., Sinha, P.K., Prasad, K., Yadav, R. and Shen, L. (2000) Mapping QTL associated with drought avoidance in upland rice. *Molecular Breeding* 6, 55–66.

Daniell, H., Datta, R., Varma, S., Gray, S. and Lee, S.B. (1998) Containment of herbicide resistance through genetic engineering of the chloroplast genome. *Nature Biotechnology* 16, 345–348.

Dennis, E.S., Dolferus, R., Ellis, M., Rahman, M., Wu, Y., Hoeren, F.U., Grover, A., Ismond, K.P., Good, A.G. and Peacock, W.J. (2000) Molecular strategies for improving waterlogging tolerance in plants. *Journal of Experimental Botany* 51, 89–97.

Dong, B., Loeve, R., Li, Y.H., Chen, C.D., Deng, L. and Molden, D. (2001) Water productivity in the Zhanghe irrigation system: issues of scale. In: Barker, R., Li, Y.H. and Tuong, T.P. (eds) *Water-saving Irrigation for Rice*. International Water Management Institute, Colombo, Sri Lanka, pp. 97–115.

Druart, N., De Roover, J., Van den Ende, W., Goupil, P., Van Laere, A. and Rambour, S. (2001) Sucrose assimilation during early developmental stages of chicory (*Cichorium intybus* L.) plants. *Planta* 212, 436–443.

Edmeades, G.O., Bolanos, J., Hernandez, M. and Bello, S. (1993) Causes for silk delay in a lowland tropical maize population. *Crop Science* 33, 1029–1035.

Fischer, G., Shah, M., van Velthuizen, H. and Nachtergaele, F.O. (2001) *Global Agro-ecological Assessment for Agriculture in the 21st Century*. International Institute for Applied Systems Analysis, Laxenburg, Austria, 33 pp.

Foolad, M.R. (1999) Comparison of salt tolerance during seed germination and vegetative growth in tomato by QTL mapping. *Genome* 42, 727–734.

Fridman, E., Pleban, T. and Zamir, D. (2000) A recombination hotspot delimits a wild-species quantitative trait locus for tomato sugar content to 484 bp within an invertase gene. *Proceedings of the National Academy of Sciences USA* 97, 4718–4723.

Gale, M., Moore, G. and Devos, K. (2001) Rice – the pivotal genome in cereal comparative genetics. *Novartis Foundation Symposium* 236, 46–53.

Gaxiola, R.A. *et al.* (2001) Drought- and salt-tolerant plants result from overexpression of the AVP1 H^+-pump. *Proceedings of the National Academy of Sciences USA* 98, 11,444–11,449.

Goetz, M., Godt, D.E., Guivarc'h, A., Kahmann, U., Chriqui, D. and Roitsch, T. (2001) Induction of male sterility in plants by metabolic engineering of the carbohydrate supply. *Proceedings of the National Academy of Sciences USA* 98, 6522–6527.

Gregorio, G.B., Senadhira, D., Mendoza, R.D., Manigbas, N.L., Roxas, N.L. and Guerta, C.Q. (2002) Progress in breeding for salinity tolerance and associated abiotic stresses in rice. *Field Crops Research* 76, 91–101.

He, Z., Zhu, Q., Dabi, T., Li, D., Weigel, D. and Lamb, C. (2000) Transformation of rice with the *Arabidopsis* floral regulator LEAFY causes early heading. *Transgenic Research* 9, 223–227.

Hiei, Y., Komari, T. and Kubo, T. (1997) Transformation of rice mediated by *Agrobacterium tumefaciens*. *Plant Molecular Biology* 35, 205–218.

Holbrook, N.M., Shashidhar, V.R., James, R.A. and Munns, R. (2002) Stomatal control in tomato with ABA-deficient roots: response of grafted plants to soil drying. *Journal of Experimental Botany* 53, 1503–1514.

Hugouvieux, V., Kwak, J.M. and Schroeder, J.I. (2001) An mRNA cap binding protein, ABH1, modulates early abscisic acid signal transduction in *Arabidopsis*. *Cell* 106, 477–487.

Inoue, K., Sewalt, V.J., Murray, G.B., Ni, W., Sturzer, C. and Dixon, R.A. (1998) Developmental expression and substrate specificities of alfalfa caffeic acid 3-*O*-methyltransferase and caffeoyl coenzyme A 3-*O*-methyltransferase in relation to lignification. *Plant Physiology* 117, 761–770.

Jaccoud, D., Peng, K., Feinstein, D. and Kilian, A. (2001) Diversity arrays: a solid state technology for sequence information independent genotyping. *Nucleic Acids Research* 29, E25.

Kamoshita, A., Zhang, J., Sciopongco, J., Sarkarung, S., Nguyen, H.T. and Wade, L.J. (2002) Effects of phenotyping environment on identification of quantitative trait loci for rice root morphology under anaerobic conditions *Crop Science* 42, 255–265.

Kasuga, M., Liu, Q., Miura, S., Yamaguchi-Shinozaki, K. and Shinozaki, K. (1999) Improving plant drought, salt, and freezing tolerance by gene transfer of a single stress-inducible transcription factor. *Nature Biotechnology* 17, 287–291.

Kawasaki, S., Borchert, C., Deyholos, M., Wang, H., Brazille, S., Kawai, K., Galbraith, D. and Bohnert, H.J. (2001) Gene expression profiles during the initial phase of salt stress in rice. *Plant Cell* 13, 889–906

Khush, G.S. (2001) Green revolution: the way forward. *Nature Reviews of Genetics* 2, 815–822.

Kouchi, H., Takane, K., So, R.B., Ladha, J.K. and Reddy, P.M. (1999) Rice ENOD40: isolation and expression analysis in rice and transgenic soybean root nodules. *Plant Journal* 18, 121–129.

Koyama, M.L., Levesley, A., Koebner, R.M.A., Flowers, T.J. and Yeo, A.R. (2001) Quantitative trait loci for component physiological traits determining salt tolerance in rice. *Plant Physiology* 125, 406–422.

Ku, M.S.B., Agarie, S., Nomura, M., Fukayama, H., Tsuchida, K., Ono, K., Hirose, S., Toki, S., Miyao, M. and Matsuoka, M. (1999) High-level expression of maize phosphoenolpyruvate carboxylase in transgenic rice plants. *Nature Biotechnology* 17, 76–80.

Ku, M.S.B., Cho, D., Ranade, U., Hsu, T.-P., Li, X., Jiao, D.-M., Ehleringer, J., Miyao, M. and Matsuoka, M. (2000) Photosynthetic performance of transgenic rice plants overexpressing maize C_4 photosynthesis enzymes. In: Sheehy, J.E., Mitchell, P.L. and Hardy, B. (eds) *Redesigning Rice Photosynthesis to Increase Yield*. International Rice Research Institute, Manila, Philippines, pp. 193–204.

Ku, M.S.B., Cho, D., Li, X., Jiao, D.-M., Pinto, M., Miyao, M. and Matsuoka, M. (2001) Introduction of genes encoding C_4 photsynthesis enzymes into rice plants: physiological consequences, In: Goodie, J.A. and Chadwick, D. (eds) *Rice Biotechnology: Improving Yield, Stress Tolerance and Grain Quality*. John Wiley & Sons, Chichester, pp. 100–116.

Ladha, J.K. and Garrity, D.P. (1994) *Green Manure Production Systems for Asian Lowlands*. International Rice Research Institute, Manila, Philippines, 194 pp.

Lafitte, H.R., Courtois, B. and Atlin, G.N. (2002) The International Rice Research Institute's experience in field screening for drought tolerance and implications for breeding. In: Saxena, N.P. and O'Toole, J.C. (eds) *Field Screening for Drought Tolerance in Crop Plants with Emphasis on Rice*. International Crop Research Institute for the Semi-Arid Tropics, Patancheru, India, pp. 25–40.

Lemichez, E., Wu, Y., Sanchez, J.P., Mettouchi, A., Mathur, J. and Chua, N.H. (2001) Inactivation of AtRac1 by abscisic acid is essential for stomatal closure. *Genes and Development* 15, 1808–1816.

Lilley, J.M., Ludlow, M.M., McCouch, S.R. and O'Toole, J.C. (1996) Locating QTL for osmotic adjustment and dehydration tolerance in rice. *Journal of Experimental Botany* 47, 1427–1436.

Luan, S. (2002) Signaling drought in guard cells. *Plant Cell Environment* 25, 229–237.

McCallum, C.M., Comai, L., Greene, E.A. and Henikoff, S. (2000) Targeting induced local lesions IN genomes (TILLING) for plant functional genomics. *Plant Physiology* 123, 439–442.

Makihara, D., Tsuda, M., Morita, M., Hirai, Y. and Kuroda, T. (1999a) Effect of salinity on the growth and development of rice (*Oryza sativa* L.) varieties. *Japanese Journal of Tropical Agriculture* 43, 285–294.

Makihara, D., Tsuda, M., Hirai, Y. and Kuroda, T. (1999b) Effects of saline irrigation at various growth stages on rice yield. *Japanese Journal of Crop Science* 68, 487–494.

Medrano, H., Escalona, J.M., Bota, J., Gulias, J. and Flexas, J. (2002) Regulation of photosynthesis of C3 plants in response to progressive drought: stomatal conductance as a reference parameter. *Annals of Botany* 89, 895–905.

Moons, A., Bauw, G., Prinsen, E., Van Montagu, M. and Van der Straeten, D. (1995) Molecular and physiological responses to abscisic acid and salts in roots of salt-sensitive and salt-tolerant Indica rice varieties. *Plant Physiology* 107, 177–186.

Nandi, S., Subudhi, P.K., Senadhira, D., Manigbas, N.L., Sen-Mandi, S. and Huang, N. (1997) Mapping QTL for submergence tolerance in rice by AFLP analysis and selective genotyping. *Molecular and General Genetics* 255, 1–8.

Nguyen-Quoc, B. and Foyer, C.H. (2001) A role for 'futile cycles' involving invertase and sucrose synthase in sucrose metabolism of tomato fruit. *Journal of Experimental Botany* 52, 881–889.

Peng, J., Richards, D.E., Hartley, N.M., Murphy, G.P., Devos, K.M., Flintham, J.E., Beales, J., Fish, L.J., Worland, A.J., Pelica, F., Sudhakar, D., Christou, P., Snape, J.W., Gale, M.D. and Harberd, N.P. (1999) 'Green revolution' genes encode mutant gibberellin response modulators. *Nature* 400, 256–261.

Peng, S.B., Laza, R.C., Khush, G.S., Sanico, A.L., Visperas, R.M. and Garcia, F.V. (1998) Transpiration efficiencies of indica and improved tropical japonica rice grown under irrigated conditions. *Euphytica* 103, 103–108.

Peuke, A.D., Jeschke, W.D. and Hartung, W. (2002) Flows of elements, ions and abscisic acid in *Ricinus communis* and site of nitrate reduction under potassium limitation. *Journal of Experimental Botany* 53, 241–250.

Pingali, P.L. and Heisey, P.W. (1999) *Cereal Crop Productivity in Developing Countries: Past Trends and Future Prospects.* Working Paper 99–03, CIMMYT, Mexico D.F., Mexico.

Price, A.H. and Tomos, A.D. (1997) Genetic dissection of root growth in rice (*Oryza sativa* L.). II. Mapping quantitative trait loci using molecular markers. *Theoretical and Applied Genetics* 95, 143–152.

Price, A.H., Young, E.M. and Tomos, A.D. (1997) Quantitative trait loci associated with stomatal conductance, leaf rolling and heading date mapped in upland rice (*Oryza sativa*). *New Phytologist* 137, 83–91.

Price, A.H., Steele, K.A., Moore, B.J., Barraclough, P.B. and Clark, L.J. (2000) A combined RFLP and AFLP linkage map of upland rice (*Oryza sativa* L.) used to identify QTL for root-penetration ability. *Theoretical and Applied Genetics* 100, 49–56.

Quarrie, S.A., Laurie, D.A., Zhu, J., Lebreton, C., Semikhodskii, A., Steed, A., Witsenboer, H. and Calestani, C. (1997) QTL analysis to study the association between leaf size and abscisic acid accumulation in droughted rice leaves and comparisons across cereals. *Plant Molecular Biology* 35, 155–165.

Queitsch, C., Hong, S.W., Vierling, E. and Lindquist S. (2000) Heat shock protein 101 plays a crucial role in thermotolerance in *Arabidopsis*. *Plant Cell* 12, 479–492.

Radin, J.W., Lu, Z., Percy, R.G. and Zeiger, E. (1994) Genetic variability for stomatal conductance in Pima cotton and its relation to improvements of heat adaptation. *Proceedings of the National Academy of Sciences USA* 91, 7217–7221.

Ray, J.D., Yu, L., McCouch, S.R., Champoux, M.C., Wang, G. and Nguyen, H.T. (1996) Mapping quantitative trait loci associated with root penetration ability in rice (*Oryza sativa* L.). *Theoretical and Applied Genetics* 92, 627–636.

Ribaut, J.M., Banziger, M. and Hoisington, D. (2002) Genetic dissection and plant improvement under abiotic stress conditions: drought tolerance in maize as an example. In: *JIRCAS Working Report.* Tankuba, Japan, pp. 85–92.

Richards, R.A. (2000) Selectable traits to increase crop photosynthesis and yield of grain crops. *Journal of Experimental Botany* 51, 447–458.

Riederer, M. and Schreiber, L. (2001) Protecting against water loss: analysis of the barrier properties of plant cuticles. *Journal of Experimental Botany* 52, 2023–2032.

Rieger, M.A., Lamond, M., Preston, C., Powles, S.B. and Roush, R.T. (2002) Pollen-mediated movement of herbicide resistance between commercial canola fields. *Science* 296, 2386–2388.

Roberts, E. and Kolattukudy, P.E. (1989) Molecular cloning, nucleotide sequence and abscisic acid induction of a suberization-associated highly anionic peroxidase. *Molecular and General Genetics* 217, 223–232.

Roberts, J.A., Hussain, A., Taylor, I.B. and Black, C.R. (2002) Use of mutants to study long-distance signaling in response to compacted soil. *Journal of Experimental Botany* 53, 45–50.

Rosenow, D.T., Quisenberry, J.E., Wendt, C.W. and Clark, L.E. (1983) Drought-tolerant sorghum and cotton germplasm. *Agricultural Water Management* 7, 207–222.

Saijo, Y., Hata, S., Kyozuka, J., Shimamoto, K. and Izui, K. (2000) Over-expression of a single Ca^{2+}-dependent protein kinase confers both cold and salt/drought tolerance on rice plants. *Plant Journal* 23, 319–327.

Saini, H.S. and Westgate, M.E. (2000) Reproductive development in grain crops during drought. *Advances in Agronomy* 68, 59–96.

Sakamoto, A., Alia and Murata, N. (1998) Metabolic engineering of rice leading to biosynthesis of glycinebetaine and tolerance to salt and cold. *Plant Molecular Biology* 38, 1011–1019.

Salekdeh, G.H., Siopongco, J., Wade, L.J., Ghareyazie, B. and Bennett, J. (2002a) A proteomic approach to analyzing drought- and salt-responsiveness in rice. *Field Crops Research* 76, 199–219.

Salekdeh, G.H., Siopongco, J., Wade, L.J., Ghareyazie, B. and Bennett, J. (2002b) Proteomic analysis of rice leaves during drought stress and recovery. *Proteomics* 2, 1131–1145.

Sasaki, A., Ashikari, M., Ueguchi-Tanaka, M., Itoh, H., Nishimura, A., Datta, S., Ishiyama, K., Saito, T., Kobayashi, M., Khush, G.S., Kitano, H. and Matsuoka, M. (2002) Green revolution: a mutant gibberellin-synthesis gene in rice. *Nature* 416, 701–702.

Schreiber, L., Kirsch, T. and Riederer, M. (1996) Transport properties of cuticular waxes: ecophysiological relevance for cuticular transpiration. In: Rennenberg, H., Eschrich, W. and Ziegler, H. (eds) *Trees: Contributions to Modern Tree Physiology*. SPB Academic Publishers, Amsterdam, pp. 19–27.

Seckler, D., Upali, A., Molden, D., de Silva, R. and Barker, R. (1998) *World Water Demand and Supply, 1990 to 2025: Scenarios and Issues*. Research Report 19, International Water Management Institute, Colombo, Sri Lanka, 40 pp.

Seki, M., Narusaka, M., Abe, H., Kasuga, M., Yamaguchi-Shinozaki, K., Caminci, P., Hayashizaki, P.Y. and Shinozaki, K. (2001) Monitoring the expression pattern of 1300 *Arabidopsis* genes under drought and cold stresses by using a full-length cDNA microarray. *Plant Cell* 13, 61–72.

Sheehy, J.E., Mitchell, P.L. and Hardy, B. (eds) (2000) *Redesigning Rice Photosynthesis to Increase Yield*. International Rice Research Institute, Manila, Philippines, 293 pp.

Shi, W.M., Muramoto, Y., Ueda, A. and Takabe, T. (2001) Cloning of peroxisomal ascorbate peroxidase gene from barley and enhanced thermotolerance by overexpressing in *Arabidopsis thaliana*. *Gene* 273, 23–27.

Socolow, R.H. (1999) Nitrogen management and the future of food: lessons from the management of energy and carbon. *Proceedings of the National Academy of Sciences USA* 96, 6001– 6008.

Sripongpangkul, K., Posa, G.B.T., Senadhira, D.S., Brar, D., Huang, N., Khush, G.S. and Li, Z.K. (2000) Genes/QTL affecting flood tolerance in rice. *Theoretical and Applied Genetics* 101, 1074–1081.

Szabolcs, I. (1989) *Salt-affected Soils*. CRC Press, Boca Raton, Florida, 274 pp.

Tanksley, S.D. and McCouch, S.R. (1997) Seed banks and molecular maps: unlocking genetic potential from the wild. *Science* 277, 1063–1066.

Teulat, B., This, D., Khairallah, M., Borries, C., Ragot, C., Sourdille, P., Leroy, P., Monneveux, P. and Charrier, A. (1998) Several QTL involved in osmotic-adjustment trait variation in barley (*Hordeum vulgare* L.). *Theoretical and Applied Genetics* 96, 688–698.

Thiellement, H., Bahrman, N., Damerval, C., Plomion, C., Rossignol, M., Santoni, V., de Vienne, D. and Zivy, M. (1999) Proteomics for genetic and physiological studies in plants. *Electrophoresis* 20, 2013–2026.

Timsina, J. and Connor, D.J. (2001) Productivity and management of rice–wheat cropping systems: issues and challenges. *Field Crops Research* 69, 93–132.

Tripathy, J.N., Zhang, J., Robin, S., Nguyen, T.T. and Nguyen, H.T. (2000) QTL for cell-membrane stability mapped in rice (*Oryza sativa* L.) under drought stress. *Theoretical and Applied Genetics* 100, 1197–1202.

Tuong, T.P. (1999) Productive water use in rice production: opportunities and limitations. *Journal of Crop Production* 2, 241–264.

Wang, H.Q. and Tang, S.X. (2000) Upland rice production in China: its past, today and future. Paper presented at the Aerobic Rice Workshop, 7–8 September 2000, IRRI, Los Baños, Philippines.

Westgate, M.E., Passioura, J.B. and Munns, R. (1996) Water status and ABA content of floral organs in drought-stressed wheat. *Australian Journal of Plant Physiology* 23, 763–772.

Williams, R.L. and Angus, J.F. (1994) Deep floodwater protects high-nitrogen rice crops from low-temperature damage. *Australian Journal of Experimental Agriculture* 34, 927.

WMO (1997) *Comprehensive Assessment of the Freshwater Resources of the World.* World Meteorological Organization, Geneva, Switzerland, 52 pp.

Xu, K. and Mackill, D.J. (1996) A major gene for submergence tolerance mapped on rice chromosome 9. *Molecular Breeding* 2, 219–224.

Xu, K., Xu, X., Ronald, P.C. and Mackill, D.J. (2000) A high-resolution linkage map of the vicinity of the rice submergence tolerance locus *Sub1*. *Molecular and General Genetics* 263, 681–689.

Xu, W., Subudhi, P.K., Crasta, O.R., Rosenow, D.T., Mullet, J.E. and Nguyen, H.T. (2000) Molecular mapping of QTL conferring stay-green in grain sorghum (*Sorghum bicolor* L. Moench). *Genome* 43, 461–469.

Yadav, R., Courtois, B., Huang, N. and McLaren, G. (1997) Mapping genes controlling root morphology and root distribution in a doubled-haploid population of rice. *Theoretical and Applied Genetics* 94, 619–632.

Yamamoto, T., Lin, H.X., Sasaki, T. and Yano, M. (2000) Identification of heading date quantitative trait locus *Hd6* and characterization of its epistatic interactions with *Hd2* in rice using advanced backcross progeny. *Genetics* 154, 885–891.

Yano, M., Katayose, Y., Ashikari, M., Yamanouchi, U., Monna, L., Fuse, T., Baba, T., Yamamoto, K., Umehara, Y., Nagamura, Y. and Sasaki, T. (2000) *Hd1*, a major photoperiod sensitivity quantitative trait locus in rice, is closely related to the *Arabidopsis* flowering time gene *CONSTANS*. *Plant Cell* 12, 2473–2484.

Yeo, A.R., Flowers, S.A., Rao, G., Welfare, K., Senanayake, N. and Flowers, T.J. (1999) Silicon reduces sodium uptake in rice (*Oryza sativa* L.) in saline conditions and this is accounted for by a reduction in the transpirational bypass flow. *Plant Cell and Environment* 22, 559–565.

Yeo, M.E., Yeo, A.R. and Flowers, T.J. (1994) Photosynthesis and photorespiration in the genus *Oryza*. *Journal of Experimental Botany* 45, 553–560.

Yevjevich, V., Hall, W.A. and Salas, J.D. (eds) (1978) *Drought Research Needs. Proceedings of Conference on Drought Research Needs. Colorado State University, Fort Collins, Colorado, 12–15 December 1977.* Water Resources Publications, Fort Collins, Colorado, 288 pp.

You, S.C. (2001) Agricultural adaptation of climate change in China. *Journal of Experimental Botany (China)* 13, 192–197.

Zeier, J., Ruel, K., Ryser, U. and Schreiber, L. (1999) Chemical analysis and immunolocalisation of lignin and suberin in endodermal and hypodermal/rhizodermal cell walls of developing maize (*Zea mays* L.) primary roots. *Planta* 209, 1–12.

Zhang, H.X., Hodson, J.N., Williams, J.P. and Blumwald, E. (2001) Engineering salt-tolerant *Brassica* plants: characterization of yield and seed oil quality in transgenic plants with increased vacuolar sodium accumulation. *Proceedings of the National Academy of Sciences USA* 98, 12832–12836.

Zhang, J.X., Nguyen, H.T. and Blum, A. (1999) Genetic analysis of osmotic adjustment in crop plants. *Journal of Experimental Botany* 332, 291–302.

Zheng, H.G., Babu, R.C., Pathan, M.S., Ali, L., Huang, N., Courtois, B. and Nguyen, H.T. (2000) Quantitative trait loci for root-penetration ability and root thickness in rice: comparison of genetic backgrounds. *Genome* 43, 53–61.

Zinselmeier, C., Jeong, B.R. and Boyer, J.S. (1999) Starch and the control of kernel number in maize at low water potentials. *Plant Physiology* 121, 25–35.

8 Management of Drought in ICRISAT Cereal and Legume Mandate Crops

R. Serraj, F.R. Bidinger, Y.S. Chauhan, N. Seetharama, S.N. Nigam and N.P. Saxena

International Crops Research Institute for the Semi-Arid Tropics, Patancheru, India

Abstract

This chapter reviews the International Crops Research Institute for the Semi-Arid Tropics (ICRISAT)'s research achievements in the domain of crop drought tolerance and presents future perspectives in the genetic enhancement of crop water use and drought adaptation in the semi-arid tropics. Exploration of crop genetic variability and genotype–environment interactions has contributed significantly to developing suitable screening methods for specific drought-tolerant traits. Genetic sources of drought tolerance were also identified at ICRISAT for all mandate crops, and some of the associated traits have been well characterized. A large spectrum of genotype duration is now available, from long to short and extra-short duration, and matching genotype duration with likely period of soil water availability is the first strategy used against terminal-drought stress. Identification and genetic mapping of quantitative trait loci for specific drought-tolerant traits using molecular markers are currently receiving greater research focus. This approach provides a powerful tool for dissecting the genetic basis of drought tolerance. If validated with accurate phenotyping and properly integrated in marker-assisted breeding programmes, this approach will accelerate the development of drought-tolerant genotypes. Overall, the progress made at ICRISAT during the last three decades proves that it is realistic to develop varieties that have increased yield under drought-prone conditions. Further multidisciplinary research integrating plant breeding, simulation modelling, physiology and molecular genetics will realize the potential of these approaches and increase the efficiency of crop improvement in drought-prone environments.

Introduction

The International Crops Research Institute for the Semi-Arid Tropics (ICRISAT)'s mandate crops, i.e. pearl millet (*Pennisetum glaucum*), sorghum (*Sorghum bicolor*), chickpea (*Cicer arietinum*), groundnut (*Arachis hypogaea*) and pigeonpea (*Cajanus cajan*), are all known for their relative ability to withstand periods of water-limited conditions and still produce grain and biomass. However, the constant challenge is to reduce yield gaps observed between research plots and farmers' fields under the rain-fed conditions of semi-arid tropics (SAT), in order to ensure sustained food security for the benefit of resource-poor farmers.

Drought stress is a complex syndrome, involving several climatic, edaphic and

 Water Productivity in Agriculture: Limits and Opportunities for Improvement (eds J.W. Kijne, R. Barker and D. Molden)

agronomic factors, and is characterized by three major varying parameters, i.e. timing of occurrence, duration and intensity. The general complexity of drought problems is often aggravated under the SAT conditions, by erratic and unpredictable rainfall and by the occurrence of high temperatures, high levels of solar radiation and poor soil characteristics of the target environments. The high variability in the nature of drought and the insufficient understanding of its complexity have made it generally difficult to characterize the physiological traits required for improved crop performance under drought, consequently limiting plant-breeding efforts to enhance the drought tolerance of crops.

In the agricultural context of SAT and global water challenges, it is critical that both agronomic and genetic management strategies focus on the maximum extraction of available soil moisture and its efficient use in crop establishment, growth, maximum biomass and seed yield. Recent research breakthroughs have revived interest in targeted drought-resistant breeding and the use of new genomics tools to enhance crop water productivity. However, with the fast progress in genomics, a better understanding of the gene functions and drought tolerance physiological mechanisms will also be essential for the progress of genetic enhancement of crop drought tolerance.

It is now well accepted that the complexity of the drought syndrome can only be tackled with a holistic approach, integrating physiological dissection of the resistance traits and molecular genetic tools, together with agronomic practices that lead to better conservation and utilization of soil moisture and matching crop genotypes with the environment.

This chapter reviews the recent progress made at ICRISAT in deciphering the complexity of crop responses to water deficits and developing drought-tolerant varieties of the five mandate crops. The management options for increasing productivity and conserving natural resources adopted by ICRISAT for integrated watershed management are reviewed in a companion chapter (Wani *et al.*, Chapter 12, this volume).

Drought as the Main Challenge for Agriculture in the SAT

The target environments for ICRISAT mandate crops

The agroclimatic and production-system environments of the SAT regions are very diverse and the inherent water constraints that limit crop production are variable. However, it is feasible to broadly characterize the drought patterns of a given environment using long-term water-balance modelling and geographic information system (GIS) tools (Chauhan *et al.*, 2000; Bastiaanssen *et al.*, Chapter 18, this volume). The assessment of the moisture-availability patterns of the target environments is critical for developing genotypes adapted to target environments and to identify environments with similar drought patterns.

Most of the ICRISAT experimental research on drought is accomplished at its centre at Patancheru in peninsular India, characterized by a relatively short growing season in a generally dry semi-arid climate, with high average temperatures and potential evaporation rates (Fig. 8.1). Soils are mainly Alfisols and Vertisols, with low to moderate levels of plant-available water content. In addition, the dry season at this location is generally rain-free, with a high mean air temperature and vapour-pressure deficits, which provide an ideal screening environment to expose plants to controlled drought-stress treatments by managing the timing of irrigation (Bidinger *et al.*, 1987).

The main target environment for ICRISAT work on drought in pearl millet in India is the growing area of the north-western states of Rajasthan, Gujarat and Harayana, where postflowering stress, either alone or in combination with preflowering stress, is a very common feature of the environment (van Oosterom *et al.*, 1996). The focus of pearl-millet research has thus been on terminal drought as it is also the most damaging to grain yield (Mahalakshmi *et al.*, 1987).

Sorghum is one of the most extensively adapted crops; it is grown from 35°S to 45°N of the equator, and the elevation ranges

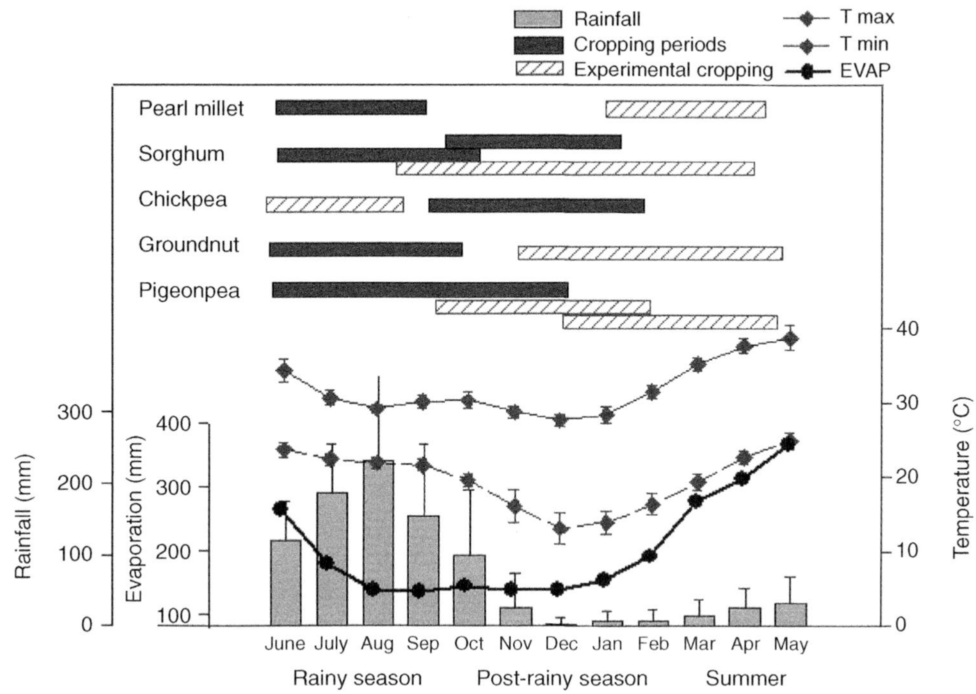

Fig. 8.1. Average climate conditions (1974–2000) and cropping schedule at ICRISAT, Patancheru (17°N 78°E, 542 m).

from sea level to nearly 2000 m a.s.l. The rainfall during the crop season could vary from 300 up to 2000 mm. Terminal-drought stress is most limiting for sorghum production worldwide. In sub-Saharan Africa, drought at both seedling establishment and grain-filling stages is also very common. In India, sorghum is grown during the rainy and the post-rainy seasons. The variable-moisture environment during the rainy season can have a severe impact on grain and biomass yield, affecting both preflowering and postflowering stages.

Characterizing drought in post-rainy-season crops, such as rabi sorghum and chickpea, is simpler, compared with the intermittent drought experienced by rainy-season crops. This is because much of the rainfall is received before the planting of the crop, which is therefore grown almost entirely on stored soil moisture and exposed mostly to progressively increasing (terminal) water deficits. Therefore, the factors governing crop growth and water use in the post-rainy season, i.e. radiation, temperature, vapour pressure and potential evaporation, are relatively stable and predictable, so that simulation modelling of both crop growth and the effects of various crop traits is quite feasible.

Groundnut is an important rainy-season crop in most of the production systems in south Asia and sub-Saharan Africa, where it is grown under varying agroecologies, either as a sole crop or intercropped with sorghum and pigeonpea. Groundnut yields are generally low and unstable under rain-fed conditions, due to unreliable rainfall patterns, with frequent droughts, and to a lack of high-yielding adapted cultivars.

Pigeonpea is grown mainly by resource-poor farmers in India and, to a varying extent, throughout the tropics, usually under rain-fed conditions. Traditionally, medium- to long-duration landraces have been cultivated, with a crop duration of 150–300 days. Pigeonpea can be exposed to intermittent drought stress during dry periods of the rainy season and to terminal-drought stress in the post-rainy season. However, over the last two decades, shorter-duration pigeonpea (SDP) genotypes have been developed, with some genotypes capable of reaching maturity within 90 days (Nam *et al.*, 1993). The introduction of such genotypes has enlarged the scope of pigeonpea cultivation in various, non-traditional cropping systems. However, the developed short-duration genotypes are usually sensitive to intermittent drought.

The Yield Gap of Rain-fed Agriculture in the SAT

Yield losses due to drought are highly variable, depending on timing, intensity and duration, coupled with other location-specific environmental stress factors, such as high irradiance and temperature. Global yield losses due to drought have been estimated to be around 6.7 million t of groundnut, 3.7 million t of chickpea and around 1.8 million t of pigeonpea (Subbarao *et al.*, 1995). It has also been shown that a large proportion of these yield losses can be potentially recovered through efforts in crop improvement (Subbarao *et al.*, 1995).

Drought may cause complete crop failure or a varying amount of reduction in biomass and grain yield. In addition to the direct effect of drought on the yield, the potential beneficial effects of improved crop-management practices, such as fertilizer application or intercropping, are not fully realized in terms of increased production. Drought reduces carbon assimilation through photosynthesis, due to limited gas exchange, and adversely affects symbiotic nitrogen-fixation processes in leguminous crops (Serraj *et al.*, 1999a), resulting in significant reductions in crop yields and soil fertility. Furthermore, the problem of drought is often compounded by related stress factors, such as the infection of roots and stalks, and rot-causing fungi that cause premature death and severe lodging, all of which result in significant yield losses.

Integrated Drought-management Options

Given the increasing scarcity and competition for water resources, irrigation is generally not a possible option for alleviating drought problems in the SAT. For increasing biomass and seed yield, therefore, drought-management strategies, whether agronomic or genetic, need to focus on maximum extraction of available soil moisture and its most efficient use in both crop establishment and maximum crop growth.

The following steps are essential for planning improvement programmes for crop yields for a given target drought-prone area:

- Characterize the major patterns of drought stress and their frequency of occurrence in the target environment.
- Evaluate crop response to the major drought patterns (simulation modelling).
- Match crop phenology (growth period, sowing, flowering, seed filling) with the most favourable period of soil moisture and climatic regimes.
- Develop a strategy for the optimal use of supplementary irrigation, when available.
- Increase the soil water available to crops through agronomic management practices.
- Identify plant traits that would maximize: (i) the use of available soil moisture in transpiration; (ii) the production of biomass per unit water transpired; and (iii) partitioning into seed, thereby conferring enhanced crop water productivity.

Agronomic and genetic options that do not involve the external input of irrigation can only partially alleviate drought effects, because yield is always lower than what can be achieved with irrigation. For example, under drought-prone conditions at ICRISAT, India, chickpea yields higher than 3 t ha^{-1} were obtained in 110 days of crop duration with irrigation in large-plot field trials (ICRISAT, 1982), compared with the average yield at this location of around 1.0 t ha^{-1} in rain-fed conditions in 85 days of crop duration.

Scope for Genetic Enhancement of Yield under Drought

Crop yield and water use

The response of plants to soil-water deficit can be generally described as the sequence of three successive stages of soil dehydration (Fig. 8.2). Stage I occurs at high soil moisture, when water is still freely available from the soil and both stomatal conductance and water-vapour loss are maximal. The transpiration rate during this stage is therefore determined by environmental conditions around the leaves. Stage II starts when the rate of water uptake from the soil cannot match the potential transpiration rate. Stomatal conductance declines, so that keeping the transpiration rate similar to the rate of uptake of soil water results in the maintenance of the water balance of the plant. Finally, stage III begins when the ability of the stomata to adjust to the declining rate of water uptake from the soil has been exhausted and stomatal conductance is at a minimum.

Virtually all major processes contributing to the crop yield, including leaf photosynthetic rate, leaf expansion and growth, are inhibited late in stage I or in stage II of soil drying (Serraj *et al.*, 1999b). At the end of stage II, these growth-supporting processes reach zero and no further growth occurs in the plants. The focus of stage III is survival, and water conservation is essential to allow the plant to endure these severe conditions. Plant survival is a critical trait in natural dryland ecosystems but, for most agricultural situations, stage III has little relevance to questions about increasing crop yield and water productivity, especially in the case of intermittent droughts. Consequently, the amount of water extracted up to the end of stage II determines the cumulative growth by the plants on a particular soil-water reservoir. Not surprisingly, research on soil-water use in crop growth going back more than 100 years has consistently shown an intimate and stable relationship between plant growth and transpirational water use after correcting for variation in atmospheric humidity (Sinclair *et al.*, 1984). Therefore,

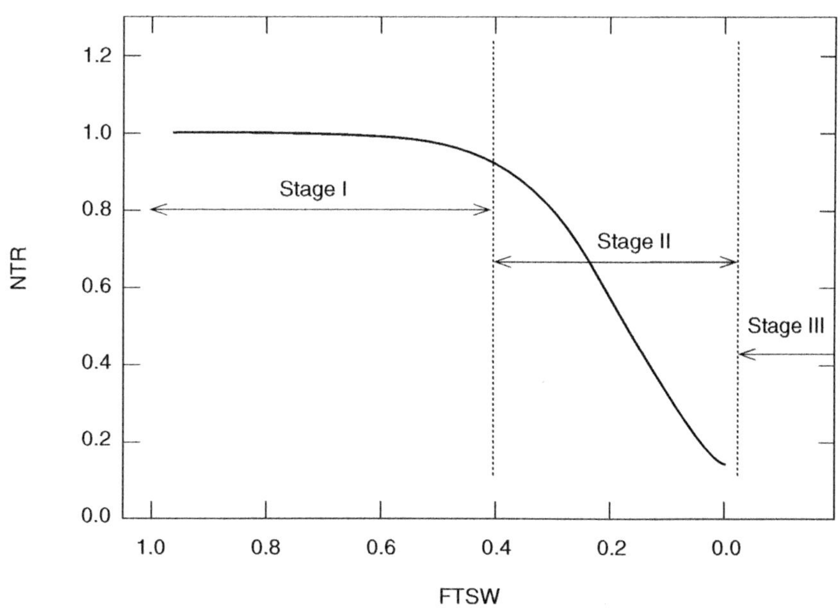

Fig. 8.2. Typical plot of normalized leaf transpiration (NTR) against the fraction of transpirable soil water (FTSW). From data of Sinclair and Ludlow (1986).

options to enhance crop survival do not usually mean an increase in crop yield under drought conditions. Increased crop yields and water productivity require the optimization of the physiological processes involved in the critical stages (mainly stage II) of plant response to soil dehydration.

Genetic-enhancement approaches

Four genetic-enhancement approaches have been implemented at ICRISAT to improve the adaptation of the mandate crops to drought-prone environments. These were:

- the development of short-duration genotypes that can escape terminal drought;
- the development of genotypes with superior yield performance in drought-prone regions following a conventional breeding approach;
- the development of drought-resistant genotypes following the physiological breeding approach;
- the identification of quantitative trait loci (QTL) for drought tolerance and their use in marker-assisted breeding.

A large germplasm collection of the five mandate crops, available in the ICRISAT gene bank, provides the base material for implementing the above four approaches of genetic enhancement in drought resistance.

Development of Short-duration Genotypes

The appropriate crop duration is a compromise of various factors, including the length of the season, the yield potential and the timing of the occurrence of drought stress.

Pearl millet

Drought escape is a major mechanism in pearl millet, determining relative cultivar performance in individual stress environments (Bidinger *et al.*, 1987), and is often a major cause of genotype-by-environment (G × E) interaction in multi-environmental trials (van Oosterom *et al.*, 1996). For example, in the case of the rains ending early, a 1-week difference in time to flowering between two cultivars is equivalent to about 30% of the grain-filling period, which would escape stress in the early-flowering cultivar, but which would be affected by stress in the later-flowering one (Mahalakshmi *et al.*, 1988).

The effects of the timing of the occurrence of single periods of stress before and after flowering provide quantification of the effects of drought escape. For example, an early genotype that flowered 20 days after the onset of terminal-drought stress had about one-quarter of the yield reduction (−12% vs. −51%) of a later-flowering genotype that flowered only 10 days before the onset of the same stress (Mahalakshmi *et al.*, 1987). However, despite the strong effects of drought escape in pearl millet, the scope for using this mechanism in crop improvement under drought conditions still depends upon the predictability of the occurrence of stress (Mahalakshmi *et al.*, 1987).

Sorghum

Breeding for earliness has been a tremendous success, especially for increasing the yield of rainy-season sorghums in India (Seetharama *et al.*, 1982). Such sorghums are also more suited for intercropping with other species. However, this approach also has some disadvantages. For example, reduced vegetative growth of early sorghums results in lower stover yield, which is critical to most resource-poor farmers. Earliness also increases susceptibility to grain moulds, as the grain matures during the end of the season, when it may rain frequently in some years. Earliness is more advantageous during the post-rainy season, although a crop maturing earlier than 3 months may not achieve high yields (Seetharama *et al.*, 1982). In West Africa, phenotypic plasticity derived from photoperiod sensitivity is also an important adaptive trait, useful for matching the crop growth and development with the water-availability period.

Chickpea

Short-duration varieties that mature before the onset of severe terminal drought have proved successful in increasing yield under drought-prone conditions in chickpea (Kumar *et al.*, 1996). However, since seed yield is generally correlated with the length of crop duration under favourable crop-growing conditions, any reduction of crop duration below the optimum would have a penalty in yield (Saxena, 1987). Depending upon the water availability, optimum crop duration for maximum yield would vary. Thus the selection of varieties needs to be matched with the maximum length of the growing period (LGP). Significant progress has been made in developing improved chickpea varieties of short duration that mature in 70–90 days in mild-winter chickpea-growing conditions, as prevailing in peninsular India (Kumar *et al.*, 1996). Even extra-short-duration chickpea varieties, termed super-early, have now been developed (Kumar and van Rheenen, 2000). The development of these new varieties has expanded options to include chickpea as a crop in many prevailing and evolving new production systems, such as rice fallows.

Groundnut

In most of the SAT groundnut-growing regions, the rainfall distribution is erratic and the season length is less than 100 days (Virmani and Singh, 1986). ICRISAT has made considerable progress in shortening the crop duration of groundnut without substantially decreasing the realized yield (Vasudeva Rao *et al.*, 1992). The short-duration varieties developed at ICRISAT have shown 23–411% superior pod yield over local control varieties in the seventh series of international trials across several countries (ICRISAT, unpublished data). However, the early-maturing genotypes usually have shallow root systems, which could make them more susceptible to intermittent dry spells if grown as a rainy-season crop and also result in a reduction of the yield potential. However, genotypic differences in rooting depth have been observed in groundnut (Wright *et al.*, 1991; Nageswara Rao *et al.*, 1993), suggesting scope for combining early maturity with an efficient root system.

Pigeonpea

Traditional long- and medium-duration pigeonpea landraces have evolved under and have apparently adapted to terminal-drought-stress conditions. However, studies in which irrigation has been supplied during the reproductive phase indicate that terminal drought usually reduces grain yield of landraces growing in their typical environment (Chauhan *et al.*, 1992). This is more apparent in the shorter-duration environments closer to the equator, where evapotranspiration is high during the post-rainy season. Thus, in terms of maximizing grain yield, the duration of these landraces seems too long for the common period of soil-moisture availability. However, a large spectrum of genotype duration is now available (Gupta *et al.*, 1989), and matching genotype duration with likely period of soil-water availability is the first line of defence against terminal-drought stress. Further, opting for an SDP cultivar rather than for those traditionally used in a region does not necessarily mean a sacrifice in yield potential, as even extra-short-duration pigeonpea (ESDP) varieties can produce yields above 2.5 t ha^{-1} (Nam *et al.*, 1993).

Screening Tools and Breeding for Drought Tolerance

Screening and selection methodologies

In order to identify sources of drought tolerance, it is necessary to develop screening methods that are simple and reproducible under the target environmental conditions. Therefore, managing drought-screening nurseries requires a careful analysis of likely sources of non-genetic variation among plots, replications and repeated experiments and establishment of procedures for minimizing these factors (Bidinger, 2002).

Several field- and laboratory-screening methods have been used at ICRISAT to screen the mandate crops for drought tolerance, including line-source-sprinkler irrigation, rain-out shelters and measurement of the drought-susceptibility index (DSI).

The line-source-sprinkler irrigation method was first developed by Hanks *et al.* (1976) and further standardized at ICRISAT (Nageswara Rao *et al.*, 1985). This system creates a gradient of drought stress and allows the evaluation of large numbers of genotypes at varying intensities of drought in a given environment. It proved to be useful for screening pearl millet (Mahalakshmi *et al.*, 1990), chickpea (Johansen *et al.*, 1994), groundnut (Nageswara Rao *et al.*, 1985) and pigeonpea (Chauhan *et al.*, 1998). However, where response to applied water is linear, simpler stress/no-stress techniques provide a more efficient means of conducting preliminary evaluations (Mahalakshmi *et al.*, 1990).

When the yield level obtained under stress was not related to the number of days until 50% flowering occurred, the DSI, as proposed by Fisher and Maurer (1978), was calculated from yield under rain-fed conditions and the potential yield under irrigation. The lower the DSI, the greater the drought tolerance of the line. The DSI method was modified by Bidinger *et al.* (1987) to include cases in which yield under stress was related to drought escape and yield potential; it was thus used for screening pearl millet and identifying tolerant genetic material.

Both line-source and DSI methods have been found to be very effective in identifying sources of tolerance to terminal drought in chickpea. Sources of drought tolerance identified by the first method (Saxena, 1987) were further validated by the second (Johansen *et al.*, 1994). More than 1500 chickpea germplasm and released varieties were screened for drought tolerance and evaluated in replicated trials at ICRISAT. Promising drought-tolerant germplasm, such as the line ICC 4958, was used in the drought-improvement programme (ICRISAT, 1992; Saxena *et al.*, 1993).

ICRISAT adopted a holistic approach in screening and selecting groundnut genotypes with superior performance under mid-season and end-of-season drought conditions. To avoid confounding effects of drought incidence with phenology of the crop, the varietal comparisons for drought sensitivity were made within a given taxonomic group. Genotypes resistant to drought have been identified by assessing total dry-matter production and pod yield under a range of drought intensities imposed at critical phases, using a line-source-sprinkler technique. Their ability to recover from mid-season drought has also shown significant genotypic variation (Harris *et al.*, 1988).

The line-source-sprinkler technique was also used to identify several drought-tolerant pigeonpea lines during the rainy seasons of 1986 and 1987. In both seasons, long breaks in rains occurred around the reproductive stage, which facilitated the imposition of drought treatments. Thirty SDP and ESDP advanced breeding lines in the first year and 40 in the second year were sown across a gradient of moisture. Among the pigeonpea hybrids tested, ICPH 8 and ICPH 9 were the most drought-tolerant; a fact which was further confirmed in multilocation trials (Chauhan *et al.*, 1998).

The creation of the rain-out shelter facility at ICRISAT has also significantly improved the precision of drought screening. Thirty-two pigeonpea lines were screened for flowering-stage drought and substantial differences were recorded using the DSI. The line ICPL 88039 showed greater drought tolerance in this screening (Chauhan *et al.*, 1998). Drought screening under the rain-out shelter, though reliable, has limitations of space, and pigeonpea cannot be grown year after year at the same place. To overcome the latter problem, rain-out shelters that can be moved to different places have been designed (Chauhan *et al.*, 1997).

Crop-improvement strategies

Since a strong relationship between the yield potential and the sensitivity of genotypes to end-of-season drought was observed for all ICRISAT mandate crops, a first approach to minimizing yield losses due to terminal drought has been to breed for earliness. But in the case of mid-season drought such a relationship does not hold, as this requires specific genetic-enhancement programmes for drought resistance.

Both conventional and trait-based approaches have been used at ICRISAT in the breeding programmes for drought. The empirical breeding approach is based on the selection for yield and its components under a given drought environment. While such an approach has been partly successful, huge investments in land, labour and capital are required to screen a large number of progenies. In addition, there is evidence of increasingly marginal returns from conventional breeding (Fehr, 1984), suggesting a need to seek more efficient methods for genetic enhancement of drought resistance. On the other hand, associating drought responses with the expression of specific physiological mechanisms can help greatly in establishing screening protocols, which allow better management of G × E interactions. However, it has been argued that a focus on very basic mechanisms is likely to be at the cost of the linkages to final grain yield and increased measurement costs, thereby complicating conventional and molecular breeding for tolerance (Bidinger, 2001).

For pearl millet, it was assumed that grain yield can be improved under water-limited environments if specific traits and responses associated with drought tolerance can be identified and incorporated into elite high-yielding genotypes of appropriate crop duration (Bidinger *et al.*, 2000; Yadav *et al.*, 2002). The QTL-mapping approach is ideal to meet such objectives as it can both identify individual genetic factors associated with a specific response and monitor the incorporation of the identified factors into the breeding programmes. The objective of using mapping-population progenies, based on commercially important hybrid parents, is to improve the drought tolerance of the parents so that the popular hybrids produced by them will have greater tolerance to drought (Yadav *et al.*, 2002).

An ideotype approach was followed for genetic improvement of drought tolerance in chickpea (Saxena and Johansen, 1990). Using ICC 4958 (drought-tolerant parent), Annigeri (a high-yielding parent) and ICC 12237 (a wilt- and root-rot-resistant parent), a three-way cross was made. Following a diversified bulk method of breeding, generations were advanced and nine yield- and root-trait-based selections were made. Yield-based selections were effective in producing varieties with high yield and trait-based selections in producing varieties with a greater degree of drought tolerance (Saxena, 2003). Promising drought-tolerant, *Fusarium*-wilt-resistant lines with high yields are ICCVs 94916–4, 94916–8, 94920–3, 94924–2 and 94924–3 (Saxena, 2003). A backcross programme was also initiated at ICRISAT, with the objective of incorporating drought-tolerant traits in elite cultivars and of combining drought-tolerant traits. Seven varieties that combine the traits of large roots and fewer pinnules were developed (ICCV 98901 to ICCV 98907). A few of these recombinants showed a greater degree of drought tolerance than and a yield similar to that of the high-yielding parent (Saxena, 2003).

The approach followed at ICRISAT for enhancing drought resistance in groundnut has been previously described in detail (Nageswara Rao and Nigam, 2003). An empirical approach was first followed for selection among segregating populations and for evaluation of advanced breeding lines for their sensitivity to mid-season and end-of-season droughts, based on pod and seed yields. While the empirical approach was partly successful, it was concluded that a more efficient breeding approach required the selection of traits associated with drought resistance. There has been significant progress in understanding the physiological basis of genotypic variability of drought response in groundnut, suggesting scope for selecting genotypes with traits contributing to superior performance under water-limited conditions. For instance, substantial genetic variation has been observed in partitioning of dry matter to pods (Nageswara Rao *et al.*, 1993).

Significant genotypic variation in total amount of water transpired (T) and transpiration efficiency (TE) has been shown under field conditions (Wright *et al.*, 1994). Further studies have confirmed large cultivar differences in TE in groundnuts grown in glasshouse and field conditions (Hubick *et al.*, 1988; Wright *et al.*, 1994). These studies made it possible to analyse the yield variation under drought conditions, using the physiological framework proposed by Passioura (1977), where:

$$\text{pod yield} = T \times TE \times \text{harvest index (HI)}$$

Research has also shown that TE and carbon isotope discrimination in leaves (Δ) are well correlated in groundnut (Wright *et al.*, 1988, 1994), suggesting the possibility of using Δ as a rapid, non-destructive tool for selection of TE in groundnut. However, further research has shown that specific leaf area (SLA, $cm^2\ g^{-1}$) is well correlated with Δ and TE in groundnut (Wright *et al.*, 1994). Further studies are currently in progress to compare the efficiency of the trait-based selection approach *vis-à-vis* the empirical approach.

Breeding programmes on pigeonpea in recent years have focused on developing SDP types of 90–150 days of maturity (Gupta *et al.*, 1989). This has made it possible to match phenology with periods of soil-moisture availability, a proven way of combating terminal drought (Chauhan *et al*, 1998). Nevertheless, there is a considerable yield gap, which is largely due to the adverse effect of intermittent droughts in different environments (Nam *et al.*, 1993). The SDP and ESDP cultivars are generally shallow-rooted (Chauhan, 1993), and what they gain by being able to escape drought stress they lose by their inability to extract water from the deeper soil layer. Indeed, it is observed that these genotypes extract water from shallower (< 75 cm) layers, compared with unstressed controls (Nam *et al.*, 2001). In addition, large gaps within the rainy-season rainfall are not unusual in the semi-arid regions, when ESDP and SDP may be forced to grow with limited ability to extract water, owing to their shallow root system. Therefore, more work is needed on screening and selecting pigeonpea cultivars for intermittent-drought tolerance.

Characterization of drought-resistance traits

Identification of simple-to-observe morphological and phenological traits reflective of mechanisms and processes that confer drought tolerance has been a high-priority activity in drought research at ICRISAT. An appropriate screening trait for drought-stress tolerance should fulfil the following criteria: (i) a strong link with higher or more stable grain yield in the target stress environment; (ii) a high level of heritability; and (iii) an easily measurable expression of tolerance, with adequate replication.

The traits associated with some promising drought-tolerant sources for ICRISAT cereal and legume mandate crops are listed in Table 8.1. Most of these characteristics that appear to enhance crop drought resistance are the manifestation of perhaps several individual mechanisms and are most probably under complex genetic control.

The panicle harvest index (PNHI), i.e. the ratio of grain to total panicle weight, has been evaluated as a selection criterion for terminal-stress tolerance in pearl millet in both variety and hybrid-parent breeding (Bidinger *et al.*, 2000). It is also currently used as one of the traits for which QTL are being identified, from a mapping population made from parents that differ in the ability to maintain PNHI under stress. PNHI, however, is readily and inexpensively measured in field experiments and can be readily used as a direct selection criterion. The main potential benefit in identifying a QTL for PNHI would be to allow marker-assisted backcross transfer of improved tolerance of terminal stress to elite lines and varieties, without the requirement for extensive field screening.

In sorghum, delayed senescence or stay-green is considered as a useful trait for plant adaptation to postflowering drought stress, particularly in environments in which the crop depends largely on stored soil moisture for grain filling. To identify superior sources of stay-green, sorghum genotypes have been recently evaluated for patterns of postflowering leaf senescence in replicated field experiments during the 1998/99 and 1999/2000 post-rainy seasons at ICRISAT

Table 8.1. Examples of early-maturing genotypes, putative drought-tolerance traits and genetic sources for the ICRISAT mandate crops.

Crops	Trait	Source	Yield advantage under drought	Reference
Pearl millet	Phenology: early maturing	ICTP 8203,	Yes	Rai *et al.*, 1990
		ICMV 88908	Yes	Witcombe *et al.*, 1995
	PNHI	ICMV 88904	Yes	Witcombe *et al.*, 1997
Sorghum	Phenology: early maturing	S 35	Yes	Rao, 1983
	Stay-green	IS 22380, QL 27, QL 10	ND[a]	Mahalaksmi and Bidinger, 2002
Chickpea	Phenology: extra-short duration	ICCV 2	Yes	Kumar and van Rheenen, 2000
	Large root system	ICC 4958	Yes	Saxena *et al.*, 1993
	Fewer pinnules	ICC 5680	Yes	Saxena and Johansen, 1990
	Small pinnules	ICC 10480	Yes	Saxena and Johansen, 1990
Groundnut	Phenology: short duration	ICGV 92029	Yes	ICRISAT (unpublished)
	Transpiration efficiency	ICGS 76	Yes	Wright and Nageswara Rao, 1994
	Specific leaf area	Tifton 8	Yes	Wright *et al.*, 1994
Pigeonpea	Phenology: extra-short duration	ICPL 87, 83015	Yes	Nam *et al.*, 1993
	Leaf-area maintenance	ICPL 87	Yes	Lopez *et al.*, 1997
	Root and shoot biomass accumulation	ICPH 8, ICPH 9	Yes	Chauhan *et al.*, 1998
	Drought-susceptibility index	ICPL 88039	Yes	Chauhan *et al.*, 1998

[a]Not determined.

(Mahalakshmi and Bidinger, 2002). A collection of 72 stay-green lines was clustered into five groups, based on the percentage of green leaf area at 15, 30 and 45 days after flowering in the 2 years. This work identified several tropically adapted sorghum lines (e.g. IS 22380, QL 27, QL 10, E36 x R16 8/1) with stay-green expression levels equivalent to those of the best temperate lines B 35 and KS 19. The stay-green trait is also currently used for QTL identification in sorghum.

Two important drought-avoidance traits have been characterized and widely used for the genetic enhancement of chickpea: the large root system, which appears to be useful in greater extraction of available soil moisture, and the smaller leaf area, which has been shown to reduce transpirational water loss (ICRISAT, 1992). The chickpea line ICC 4958 has multiple traits of large root size, a rapid rate of root development and extraction of water and a rapid rate of seed development related to its large seed size. Lines ICC 5680 and ICC 10480 have a smaller leaf area, due to either narrow pinnules (ICC 10480) or fewer pinnules (ICC 5680). Recombinants with traits of ICC 4958 and ICC 5680 showed a higher midday leaf relative water content compared with the parents in field trials conducted at ICRISAT (Saxena, 2003).

There is large scope for the genetic improvement of the efficiency of crop water use in groundnut (Wright and Nageswara Rao, 1994). Significant genotypic variation in the total amount of water transpired and TE (defined as the amount of dry matter produced per amount of water transpired) have been found in groundnuts grown in glasshouse and field conditions (Wright *et al.*, 1988, 1994). Groundnut lines ICGS 76, ICGS 44, Tifton 8 and Kadiri 3 were identified as having high TE values (Wright *et al.*, 1994).

Important putative drought-resistance traits for pigeonpea include early vigour, leaf-area maintenance, root and shoot growth rate and development plasticity (Johansen, 2003). Early growth vigour is an important factor in drought resistance as it permits the establishment of a root system that is more effective in extracting water during later drought periods. This is considered to be the main reason why pigeonpea hybrids, such as ICPH 8 and ICPH 9, grow and yield better than the varieties from which they are derived, under both drought and well-watered situations. There are considerable differences in early growth vigour of pigeonpea (Johansen, 2003). Early-maturing genotypes generally show more vigour than later-maturing ones, with hybrids showing most vigour, but there are exploitable differences in this trait within maturity groups.

While reduction in leaf area under drought stress would reduce further transpirational losses and thus enhance survival ability, leaf-area maintenance seems to be an important consideration for pigeonpea under drought (Subbarao *et al.*, 1995). Leaf-area maintenance under intermittent drought stress would involve an integration of several lower-level traits, such as a root system effective in water extraction, dehydration tolerance, leaf movements, etc. Leaf-area maintenance is an easily observable trait, amenable for use in screening segregants of a breeding programme. Pigeonpea shows large genotypic differences for this trait (Lopez *et al.*, 1997). The SDP genotype ICPL 87 performs better than a sister genotype, ICPL 151, which correlates with the greater leaf-area retention in ICPL 87 under drought than in ICPL 151.

QTL and marker-assisted selection (MAS) strategies

Most of the physiological traits associated with drought resistance are quantitative in nature. Using molecular markers, QTL can therefore be detected in an appropriate population of plants. A locus for any quantitative trait can be mapped as long as polymorphism is observed in the segregating populations under analysis and phenotypic information is available for the lines in the population. However, for traits as complex as drought tolerance, the success of the QTL approach is conditioned by the effectiveness of the phenotyping procedure in detecting among recombinant lines repeatable, highly heritable differences that permit the identification of robust QTL. Therefore, a special effort is needed for the conceptualization, design and management of phenotyping programmes for drought tolerance and to maximize the chances of identifying QTL that will be useful in the future improvement of tolerance in the target crop and in the target environment.

A QTL-mapping approach is currently used at ICRISAT to dissect the genetic and physiological basis and apply marker-assisted breeding strategies for several traits linked to drought tolerance, including the PNHI and yield components of pearl millet under terminal drought, root drought-avoidance traits in chickpea and stay-green in sorghum.

For terminal drought in pearl millet, several mapping populations have been developed using restriction-fragment length polymorphism (RFLP) skeleton mapping, trait phenotyping (Hash and Witcombe, 1994) and QTL mapping (Yadav *et al.*, 2002). Test crosses of mapping-population progenies, derived from inbred pollinators and from seed parents differing in their response to drought, were evaluated in a range of managed terminal-drought-stress environments to identify individual QTL associated with drought tolerance. A number of QTL associated with drought tolerance of grain yield and its agronomic and physiological components were identified (Yadav *et al.*, 2002). Some of the identified QTL were common across water-stress environments and genetic backgrounds of the two mapping populations, while others were specific to a particular water-stress environment or genetic background. Interestingly, all the identified QTL contributed to increased drought tolerance through their effect on either increased maintenance of growth or harvest index or both in terminal-drought-stress environments. Programmes for marker-assisted backcross transfer of the identified QTL into the elite parent of these mapping populations have been initiated for the improvement of pearl-millet productivity in water-limited environments (Yadav *et al.*, 2002). The development of near-isogenic lines will also provide an ideal opportunity to further test the effect of the identified QTL and to dissect the associated physiological mechanisms involved in terminal drought.

Phenotyping for chickpea root traits involved in drought avoidance has been carried out in recombinant inbred lines (RILs) of a cross (ICC 4958 × Annigeri) and a wide cross (ICC 4958 × *Cicer reticulatum*) (Saxena and Kumar, 2000). Currently, identification of QTL for the large root system of ICC 4958 for developing the MAS technique is in progress.

Compared with other crops, cultivated groundnut with currently available DNA markers shows limited polymorphic variation, which has made it difficult to construct a genetic map for cultivated groundnut. However, polymorphic variation in DNA has recently been detected in selected germplasm of cultivated groundnut, using molecular markers (He and Prakash, 1997; Subramanian *et al.*, 2000). On the other hand, there is still limited information on the biochemical and molecular basis for variation among genotypes for drought resistance (Nageswara Rao and Nigam, 2003). Further research is necessary to develop linkages between the drought-resistance traits and the molecular markers so that MAS tools can be applied in drought-resistance breeding.

Material dissemination and impact

Most of ICRISAT's genetic-enhancement programmes for the past three decades have focused on increased crop productivity and adaptation in the target semi-arid environments. The overall crop-improvement strategies pursued have, therefore, directly and indirectly contributed to increasing the selection pressure for better adaptation to water-limited conditions. With more than 400 varieties released in 170 countries, ICRISAT's research has contributed significantly to increasing crop productivity and food security in smallholder farming across the SAT. Early-maturing varieties have generally resulted in a reduced risk of crop failure, linked to plants' escape from end-of season drought.

Improvements in pearl-millet and sorghum productivity under water-limited conditions are critical for both national and household food securities in the SAT, especially in sub-Saharan Africa. Two successful examples of the impact of genetic enhancement of crop yield under rain-fed conditions are the releases of pearl-millet variety Okashana 1 in Namibia and the sorghum variety S 35 in Chad and Cameroon.

The pearl-millet variety Okashana 1, developed by ICRISAT and identified for use by the Namibian national programme, is grown on almost 50% of the national pearl-millet area, where the main limitations to crop yield are low rainfall, frequent drought and poor crop management (Rohrbach *et al.*, 1999). Okashana 1 is early maturing, has good terminal-drought tolerance and is generally adapted to marginal environments. The development and dissemination of this variety resulted in a high value of internal rate of return in Namibia.

The S 35 sorghum variety has been described as a non-photoperiod-sensitive, high-yielding, early-maturing and drought-tolerant pure line; it originated from ICRISAT's breeding programme in India and was later advanced and promoted in Cameroon and Chad (Yapi *et al.*, 1999). Its introduction into drought-prone areas of Chad has been very successful, resulting in an estimated yield advantage of about 51% over farmers' local varieties and consequently in a very high internal rate of return (Yapi *et al.*, 1999).

For chickpea, extra-early kabuli genotypes were developed at ICRISAT through the introgression of desi-kabuli, which matured in less than 3 months. Among these lines, the variety ICCV 2 was released in India as 'Swetha' and in Sudan as 'Wad Hamid' (http://grep.icrisat.org/archives/kabuli.htm). This variety has performed well in Egypt, Tanzania and Ethiopia. ICCV 2 is currently the world's shortest-duration kabuli chickpea, able to grow fast on the conserved receding soil moisture and to mature before the moisture is depleted from the deeper soil layers. However, being an extra-short-duration variety, ICCV 2 has a limited yield potential, lower than the traditional desi types (Kumar and van Rheenen, 2000).

The chickpea line ICC 4958 was confirmed as a drought-tolerant source in many field trials, both at ICRISAT and by the Indian national agricultural research system (NARS) partners (Saxena, 2003). It also proved to be the most drought-tolerant in spring-planted chickpea in Mediterranean types of climate. Seven other varieties (ICCV 98901 to ICCV 98907) were developed in order to incorporate drought-tolerant traits in agronomically superior cultivars. Additional promising drought-tolerant, *Fusarium*-wilt-resistant lines developed with high yield are ICCVs 94916–4, 94916–8, 94920–3, 94924–2 and 94924–3 (Saxena, 2003).

Groundnut research on drought in ICRISAT has been mainly targeted on south Asia and West Africa. As the growing season is becoming short, particularly in West Africa, drought-escape and early-maturing genotypes have significantly contributed to maintaining crop productivity under terminal drought. Examples of groundnut varieties released in India are ICGS 11 and ICGS 37, which are tolerant to end-of-season drought. Other cultivars released with tolerance to mid-season drought in India include ICGS 44, ICGS 76 and ICG(FDRS) 10.

In pigeonpea, the adoption of the SDP cultivar ICPL 87 in southern India led to 93% yield gains, in addition to improved soil fertility and reduced production costs (Bantilan and Parthasarathy, 1999). Line UPAS 120, another SDP cultivar, released in northern India, has also shown great drought-tolerant levels (Chauhan *et al.*, 1998). In addition to the short-term economic impact of its cultivation as a second crop in the post-rainy season, the widespread adoption of SDP improves long-term yield stability and system sustainability through the improvement of soil fertility.

Future Thrust

Integrated strategies for crop genetic improvement in drought-prone environments have recently been reviewed (CIMMYT, 2000). A framework for their specific application to ICRISAT mandate crops is outlined in Fig. 8.3.

A systematic characterization of the SAT drought environments where the mandate crops are grown is still to be done, in order to enable adequate targeting of drought-resistant traits, using historical climatic series, GIS tools, water balance and crop-simulation models.

With improved knowledge of probable soil-moisture availability over time, it becomes easier to further exploit the drought-escape option, considering the spectrum of crop durations and germplasm available for all mandate crops.

The ideotype approach for incorporating the relevant drought-resistance traits requires a better knowledge of the physiological mechanisms involved in drought resistance and their genetic control. Simple mechanistic models that can reliably simulate crop growth and yield in different environments can also be used for the assessment of the putative drought-tolerant traits in a wide range of target environments.

Despite the methodological difficulties, genetic enhancement of root systems for more effective water extraction would seem a high-priority effort for rain-fed chickpea and ESDP. Dissection of root traits and development of a screening system relevant to field conditions are therefore needed, in parallel with extensive genotyping and the search for molecular markers. Other promising integrated traits for improving drought resistance and crop water productivity include the PNHI in pearl millet, stay-green in sorghum and transpiration-use efficiency in groundnut. There seems to be much scope for improving such characters, using QTL and molecular breeding techniques, aided by physiological characterization and conventional breeding, in order to significantly improve the ability of the mandate crops to withstand drought stress in defined target environments.

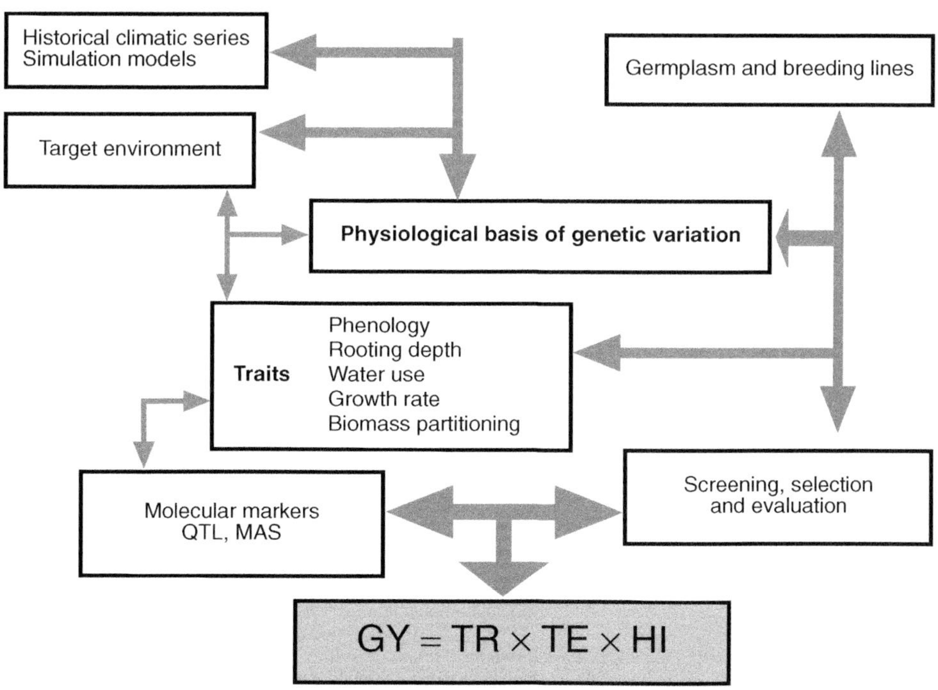

Fig. 8.3. Framework of an integrated strategy for genetic enhancement of crop grain yield (GY) and its components under water-limited conditions at ICRISAT. TR, total plant water transpired; TE, transpiration efficiency; HI, harvest index.

References

Bantilan, M.C.S. and Parthasarathy, D. (1999) *Efficiency and Sustainability Gains from Adoption of Short-duration Pigeonpea in Nonlegume-based Cropping Systems*. ICRISAT Impact Series No. 5, Patancheru, India.

Bidinger, F.R. (2002) Field screening for drought tolerance – principles and illustrations. In: Saxena, N.P. (ed.) *International Workshop on Field Screening for Drought Tolerance in Rice*. International Crops Research Institute for the Semi-Arid Tropics, Patancheru, India, pp. 109–124.

Bidinger, F.R., Mahalakshmi, V. and Rao, G.D.P. (1987) Assessment of drought resistance in pearl millet [*Pennisetum americanum* (L.) Leeke]. I. Factors affecting yield under stress. *Australian Journal of Agricultural Research* 38, 37–48.

Bidinger, F.R., Chandra, S. and Mahalakshmi, V. (2000) Genetic improvement of tolerance to terminal drought stress in pearl millet (*Pennisetum glaucum* (L.) R. Br.). In: Ribaut, J.-M. and Poland, D. (eds) *Molecular Approaches for the Genetic Improvement of Cereals for Stable Production in Water-limited Environments. Proceedings of a Strategic Planning Workshop Held at CIMMYT, El Batan, Mexico, 21–25 June 1999*. CIMMYT, El Batan, Mexico DF, pp. 59–63.

Chauhan, Y.S. (1993) Pigeonpea. In: Salam, A. and Wahid, W.A. (eds) *Rooting Pattern of Tropical Crops*. Tata McGraw Hill, New Delhi, India.

Chauhan, Y.S., Saxena, N.P. and Johansen, C. (1992) Abiotic factors limiting chickpea and pigeonpea production. In: Sachan, J.N. (ed.) *Proceedings of the National Symposium on New Frontiers in Pulses Research and Development, 10–12 November 1989*. Directorate of Pulses Research, Kanpur, India, pp. 111–123.

Chauhan, Y.S., Saxena, N.P., Rao, R.C.N., Johansen, C. and Ravindranath, K. (1997) Portable rainout shelter, a useful tool in drought research. *Food Legume Newsletter* 25, 9.

Chauhan, Y.S., Wallace, D.H., Johansen, C. and Singh, L. (1998) Genotype-by-environment interaction effect on yield and its physiological bases in short-duration pigeonpea. *Field Crops Research* 59, 141–150.

Chauhan, Y.S., Bantilan, F.T., Padmaja, R., Ahmed, I. and Moinuddin, M. (2000) *Spatial and Temporal Distribution of Soil Water Availability in Bangladesh, Indo-Gangetic Plains of India, Pakistan and Sri Lanka*. International Crops Research Institute for the Semi-Arid Tropics, Patancheru, India.

CIMMYT (2000) Molecular approaches for the genetic improvement of cereals for stable production in water-limited environments. In: Ribaut, J.-M. and Poland, D. (eds) *Molecular Approaches for the Genetic Improvement of Cereals for Stable Production in Water-limited Environments. Proceedings of a Strategic Planning Workshop Held at CIMMYT, El Batan, Mexico, 21–25 June 1999*. International Maize and Wheat Improvement Centre, El Batan, Mexico.

Fehr, W.R. (ed.) (1984) *Genetic Contributions to Yield Gains of Five Major Crop Plants*. Special Publication No. 7, Crop Science Society of America (CSSA), Madison, Wisconsin.

Fisher, R.A. and Maurer, R. (1978) Drought resistance in spring wheat cultivars. I. Grain yield responses. *Australian Journal of Agricultural Research* 29, 898–912.

Gupta, S.C., Johansen, C. and Singh, L. (1989) Uniformity of nomenclature in short-duration pigeonpea. *International Pigeonpea Newsletter* 10, 3–6.

Hanks, R.J., Keller, J., Rasmussen, V.P. and Wilson, G.D. (1976) Line source sprinkler for continuous variable studies crop production. *Soil Science Society American Journal* 40, 426–429.

Harris, D., Mathews, R.B., Nageswara Rao, R.C. and Williams, J.H. (1988) The physiological basis for yield differences between four groundnut genotypes in response to drought. III. Developmental processes. *Experimental Agriculture* 24, 215–226.

Hash, C.T. and Witcombe, J.R. (1994) Pearl millet mapping populations at ICRISAT. In: Witcombe, J.R. and Duncan, R.R. (eds) *Use of Molecular Markers in Sorghum and Pearl Millet Breeding for Developing Countries*. Overseas Development Administration, London, pp. 69–75.

He, G. and Prakash, C.S. (1997) Identification of polymorphic DNA markers in cultivated peanut (*Arachis hypogaea* L.). *Euphytica* 97, 143–149.

Hubick, K.T., Farquhar, G.D. and Shorter, R. (1988) Heritability of genotypic × environmental interactions of carbon isotope discrimination and transpiration efficiency in peanut (*Arachis hypogaea* L.). *Australian Journal of Plant Physiology* 15, 779–813.

ICRISAT (1982) Yield potential of chickpea. In: *Annual Report for 1981*. International Crops Research Institute for the Semi-Arid Tropics, Patancheru, India.

ICRISAT (1992) *ICC 4958, a Drought Resistant Chickpea*. Plant Material Description No. 33, International Crops Research Institute for the Semi-Arid Tropics, Patancheru, India.

Johansen, C. (2003) An overview of prospects for genetic enhancement of drought resistance in pigeonpea. In: Saxena, N.P. (ed.) *Management of Agricultural Drought – Agronomic and Genetic Options*. Oxford & IBH Publishing, New Delhi.

Johansen, C., Krishnamurthy, L., Saxena, N.P. and Sethi, S.C. (1994) Genotypic variation in moisture response of chickpea grown under line-source sprinklers in a semi-arid tropical environment. *Field Crops Research* 37, 103–112.

Kumar, J. and van Rheenen, H.A. (2000) A major gene for time of flowering in chickpea. *Journal of Heredity* 91, 67–68.

Kumar, J., Sethi, S.C., Johansen, C., Kelley, T.G., Rahman, M.M. and van Rheenen, H.A. (1996) Potential of short-duration varieties in chickpea. *Indian Journal of Dryland Agriculture Research* 11, 26–29.

Lopez, F.B., Chauhan, Y.S. and Johansen, C. (1997) Effects of drought stress timing on leaf area development and canopy light interception of short-duration pigeonpea. *Journal of Agronomy and Crop Science* 178, 1–7.

Mahalakshmi, V. and Bidinger, F.R. (2002) Evaluation of stay-green sorghum germplasm lines at ICRISAT. *Crop Science* 42, 965–974.

Mahalakshmi, V., Bidinger, F.R. and Raju, D.S. (1987) Effect of timing of water deficit on pearl millet (*Pennisetum americanum*). *Field Crops Research* 15, 327–339.

Mahalakshmi, V., Bidinger, F.R. and Rao, G.D.P. (1988) Timing and intensity of water deficits during flooring and grain-filling in pearl millet. *Agronomy Journal* 80, 130–135.

Mahalakshmi, V., Bidinger, F.R. and Rao, G.D.P. (1990) Line-source vs. irrigated/non irrigated treatments for evaluation of genotype drought response. *Agronomy Journal* 82, 841–844.

Nageswara Rao, R.C. and Nigam, S.N. (2003) Genetic options for drought management in groundnut. In: Saxena, N.P. (ed.) *Management of Agricultural Drought – Agronomic and Genetic Options*. Oxford & IBH Publishing, New Delhi.

Nageswara Rao, R.C., Sardar Singh, Sivakumar, M.V.K., Srivastava, K.L. and Williams, J.H. (1985) Effect of water deficit at different growth phases of peanut. 1. Yield responses. *Agronomy Journal* 77, 782–786.

Nageswara Rao, R.C., Williams, J.H., Wadia, K.D.R., Hubick, K.T. and Farquhar, G.D. (1993) Crop growth, water-use efficiency and carbon isotope discrimination in groundnut (*Arachis hypogaea* L.) genotypes under end-of-season drought conditions. *Annals of Applied Biology* 122, 357–367.

Nam, N.H., Chauhan, Y.S. and Johansen, C. (1993) Comparison of extra-short-duration pigeonpea with short-season legumes under rain-fed conditions on Alfisols. *Experimental Agriculture* 29, 307–316.

Nam, N.H., Chauhan, Y.S. and Johansen, C. (2001) Effects of timing of drought stress on extra-short duration pigeonpea. *Journal of Agricultural Science, Cambridge* 136, 179–189.

Passioura, J.B. (1977) Grain yield, harvest index and water use of wheat. *Journal of Australian Institute of Agricultural Science* 43, 117–120.

Rai, K.N., Kumar, K., Anand, Andrews, D.J., Rao, A.S., Raj, A.G.B. and Witcombe, J.R. (1990) Registration of ICTP 8203 pearl millet. *Crop Science* 30, 959.

Rao, N.G.P. (1983) *Sorghum Breeding Research Report*. ICRISAT/OUA/STRC/SAFGRAD Collaborative Research, Samaru, PMB 1044, International Crops Research Institute for the Semi-Arid Tropics, Zaria, Nigeria.

Rohrbach, D.D., Lechner, W.R., Ipinge, S.A. and Monyo, E.S. (1999) *Impact from Investments in Crop Breeding: the Case of Okashana 1 in Namibia*. ICRISAT Impact Series No. 4, ICRISAT, Patancheru, India.

Saxena, N.P. (1987) Screening for adaptation to drought: case studies with chickpea and pigeonpea. In: *Adaptation of Chickpea and Pigeonpea to Abiotic Stresses: Proceedings of Consultants' Workshop*. International Crops Research Institute for the Semi-Arid Tropics, Patancheru, India, pp. 63–76.

Saxena, N.P. (2003) Management of drought in chickpea – a holistic approach. In: Saxena, N.P. (ed.) *Management of Agricultural Drought – Agronomic and Genetic Options*. Oxford & IBH Publishing, New Delhi.

Saxena, N.P. and Johansen, C. (1990) Chickpea ideotypes or genetic enhancement of yield and yield stability in south Asia. In: *Chickpea in the Nineties: Proceedings of the Second International Workshop on Chickpea Improvement, 4–8 December 1989*. International Crops Research Institute for the Semi-Arid Tropics, Patancheru, India, pp. 81–85.

Saxena, N.P. and Kumar, J. (2000) Phenotyping of RILs, for drought resistant traits, as inputs for identification of QTLs for root traits in chickpea. In: *Archival Report for 1998 and 1999: Genetic Resources and Enhancement Program*. International Crops Research Institute for the Semi-Arid Tropics, Patancheru, India (limited circulation), p. 51.

Saxena, N.P., Johansen, C., Saxena, M.C. and Silim, S.N. (1993) Selection for drought and salinity tolerance in cool-season food legmes. In: Singh, K.B. and Saxena, M.C. (eds) *Breeding for Stress Tolerance in Cool Season Food Legumes.* John Wiley & Sons, Chichester, pp. 245–270.
Seetharama, N., Reddy, B.V.S., Peacock, J.M. and Bidinger, F.R. (1982) Sorghum improvement for drought resistance. In: *Drought Resistance in Crops with Emphasis on Rice.* International Rice Research Institute, Los Baños, Philippines, pp. 317–338.
Serraj, R., Purcell, L.C. and Sinclair, T.R. (1999a) Inhibition of N_2 fixation by drought. A review. *Journal of Experimental Botany* 50, 143–155.
Serraj, R., Allen, L.H. and Sinclair, T.R. (1999b) Soybean leaf growth and gas exchange response to drought under carbon dioxide enrichment. *Global Change Biology* 5, 283–292.
Sinclair, T.R. and Ludlow, M.M. (1986) Influence of soil water supply on the plant water balance of four tropical grain legumes. *Australian Journal of Plant Physiology* 13, 329–341.
Sinclair, T.R., Tanner, C.B. and Bennett, J.M. (1984) Water-use efficiency in crop production. *BioScience* 34, 36–40.
Subbarao, G.V., Johansen, C., Slinkard, A.E., Rao, R.C.N., Saxena, N.P. and Chauhan, Y.S. (1995) Strategies for improving drought resistance in grain legumes. *Critical Reviews in Plant Sciences* 14, 469–523.
Subramanian, V., Gurutu, S., Nageswara Rao, R.C. and Nigam, S.N. (2000) Identification of DNA polymorphism in cultivated groundnut using random amplified polymorphic DNA (RAPD) assay. *Genome* 43, 1–5.
van Oosterom, E.J., Whitaker, M.L. and Weltzien, E.R. (1996) Integrating genotype by environment interaction analysis, characterization of drought patterns, and farmer preferences to identify adaptive plant traits for pearl millet. In: Cooper, M. and Hammer, G.L. (eds) *Plant Adaptation and Crop Improvement.* CAB International, Wallingford, pp. 383–402.
Vasudeva Rao, M.J., Nigam, S.N. and Huda, A.K.S. (1992) The thermal time concept as a selection criterion for earliness in peanut. *Peanut Science* 19, 7–10.
Virmani, S.M. and Singh, P. (1986) Agroclimatological characteristics of the groundnut-growing regions in the semi-arid tropics. In: *Agrometeorology of Groundnut, Proceedings of International Symposium, 21–26 August 1985, ICRISAT Sahelian Centre, Niamey, Niger.* ICRISAT, Patancheru, India.
Witcombe, J.R., Rao, M.N.V.R. and Lechner, W.R. (1995) Registration of ICMV 88908 pearl millet. *Crop Science* 35, 1216–1217.
Witcombe, J.R., Rao, M.N.V.R. and Hash, C.T. (1997) Registration of ICMV 88904 pearl millet. *Crop Science* 37, 1022–1023.
Wright, G.C. and Nageswara Rao, R.C. (1994) Groundnut water relations. In: Smartt, J.E. (ed.) *The Groundnut Crop: a Scientific Basis for Improvement.* Chapman & Hall, London, pp. 281–336.
Wright, G.C., Hubick, K.T. and Farquhar, G.D. (1988) Discrimination in carbon isotopes of leaves, correlates with water use efficiency of field grown peanut cultivars. *Australian Journal of Plant Physiology* 15, 815–825.
Wright, G.C., Hubick, K.T. and Farquhar, G.D. (1991) Physiological analysis of peanut cultivar response to timing and duration of drought stress. *Australian Journal of Agricultural Research* 42, 453–470.
Wright, G.C., Nageswara Rao, R.C. and Farquhar, G.D. (1994) Water-use efficiency and carbon isotope discrimination in peanut under water deficit conditions. *Crop Science* 34, 92–97.
Yadav, R.S., Hash, C.T., Bidinger, F.R., Cavan, G.P. and Howarth, C.J. (2002) Quantitative trait loci associated with traits determining grain and stover yield in pearl millet under terminal drought stress conditions. *Theoretical and Applied Genetics* 104, 67–83.
Yapi, A.M., Debrah, S.K., Dehala, G. and Njomaha, C. (1999) *Impact of Germplasm Research Spillovers: the Case of Sorghum Variety S 35 in Cameroon and Chad.* ICRISAT Impact Series No. 3, ICRISAT, Patancheru, India.

9 Water Productivity in Rain-fed Agriculture: Challenges and Opportunities for Smallholder Farmers in Drought-prone Tropical Agroecosystems*

John Rockström,[1] Jennie Barron[2] and Patrick Fox[2]
[1] *UNESCO-IHE Institute for Water Education, Delft, The Netherlands;* [2] *Department of Systems Ecology, Stockholm University, Stockholm, Sweden*

Abstract

Considering the persistently growing pressure on finite freshwater and soil resources, it becomes increasingly clear that the challenge of feeding tomorrow's world population is, to a large extent, about improved water productivity within present land use.

Rain-fed agriculture plays a critical role in this respect. Eighty per cent of the agricultural land worldwide is under rain-fed agriculture, with generally low yield levels and high on-farm water losses. This suggests a significant window of opportunity for improvements. Ninety-five per cent of current population growth occurs in developing countries and a significant proportion of these people still depend on a predominantly rain-fed-based rural economy.

This chapter presents the agrohydrological rationale for focusing on water productivity in rain-fed agriculture, identifies key management challenges in attempts to upgrade rain-fed agriculture and presents a set of field experiences on system options for increased water productivity in smallholder farming in drought-prone environments. Implications for watershed management are discussed, and the links between water productivity for food and securing an adequate flow of water to sustain ecosystem services are briefly analysed. The focus is on sub-Saharan Africa, which faces the largest food-deficit and water-scarcity challenges.

The chapter shows that there are no agrohydrological limitations to doubling or even quadrupling on-farm staple-food yields, even in drought-prone environments, by producing more 'crop per drop' of rain. Field evidence is presented suggesting that meteorological dry spells are an important cause of low yield levels. It is hypothesized that these dry spells constitute a core driving force behind farmers' risk-aversion strategies. Risk aversion also contributes to the urgent soil-fertility deficits resulting from insignificant investments in fertilizers. For many smallholder farmers in the semi-arid tropics, it is simply not

*The field research in Burkina Faso and Kenya was financed through grants from the Research Department (Sarec) of the Swedish International Development and Cooperation Agency (Sida), and was carried out in collaboration with IRD (Institut de Recherche pour le Dévéloppement) and INERA (Institut National de l'Environnement et de la Recherche Agricole) in Burkina Faso, and in collaboration with KARI (Kenyan Agricultural Research Institute) and the Regional Land Management Unit (RELMA) of Sida in Kenya. We appreciate the use of statistical software INSTAT (v. 6.5), which was provided free of charge by the Statistical Service Centre, University of Reading, UK.

 Water Productivity in Agriculture: Limits and Opportunities for Improvement (eds J.W. Kijne, R. Barker and D. Molden)

worth investing in fertilizers (and other external inputs) so long as the risk for crop failure remains a reality every fifth year and the risk of yield reductions every second year. These high risks are associated with periodic water scarcity during the growing season (i.e. not necessarily cumulative water scarcity).

Results are presented from on-farm agrohydrological field research with innovations in water harvesting and conservation tillage among smallholder farmers in semi-arid rain-fed farming systems. These results indicate that upgrading rain-fed production systems through supplemental irrigation during short dry spells can lead to large increases in water productivity. Downstream implications of increased upstream withdrawals of water for upgrading of rain-fed food production are discussed.

Finally, it is argued that some of the most exciting opportunities for water-productivity enhancements in rain-fed agriculture are found in the realm of integrating components of irrigation management within the context of rain-fed farming, e.g. supplemental or microirrigation for mitigating the effects of dry spells. Combining such practices with management strategies that enhance soil infiltration and improving water-holding capacity and the potential of water uptake of plants can have a strong impact on agricultural water productivity. This suggests that it is probably time to abandon the largely obsolete distinction between irrigated and rain-fed agriculture, and instead focus on integrated rainwater management.

Introduction: a Broadened Water-management Approach

The sheer magnitude of future food needs, to be met by production systems depending on a finite freshwater resource, indicates the necessity to focus on water productivity in both irrigated and rain-fed agriculture. However, there are several reasons why special attention should be given to rain-fed agriculture. Almost all population growth (95%) takes place in tropical developing countries, and it is also there that the bulk of present undernutrition occurs. In sub-Saharan Africa, over 60% of the population depends on rain-based rural economies, generating in the range of 30–40% of the countries' gross domestic product (GDP) (World Bank, 1997). Rain-fed agriculture worldwide is practised on approximately 80% of the agricultural land (the remaining 20% is under irrigated agriculture). This proportion varies substantially between tropical regions, from approximately 95% in sub-Saharan Africa to 65% in Asia (FAOSTAT, 1999). Rain-fed agriculture will remain the dominant source of food production during the foreseeable future (Parr *et al.*, 1990). Yields from rain-fed agriculture are often low, generally around 1 t ha^{-1} in semi-arid tropical agroecosystems (Rockström, 2001), and this fact explains why rain-fed agriculture is estimated to contribute only some 60% of the world crop production (FAO, 2002). There is ample evidence to suggest that the low productivity in rain-fed agriculture is due more to suboptimal performance related to management aspects rather than to low physical potential (Agarwal and Narain, 1997; Benites *et al.*, 1998; Rockström and Falkenmark, 2000; SIWI, 2001). This means that in the developing countries with the most rapid population growth, dependence on rain-fed agriculture operating at suboptimal level is high. Furthermore, it has been estimated that there is limited new land to be put under agriculture (McCalla, 1994; Young, 1999), contrary to the last three decades, when the bulk of increased food production in, for example, sub-Saharan Africa came from expansion of agricultural land. There is thus a growing pressure to increase agricultural productivity through raised yields per unit soil and unit water.

In this chapter, water productivity (WP) broadly signifies the efficiency of water use at the production system or farm level. At this scale, the production of more economic biomass per unit of water is expressed both in terms of more crop per unit evapotranspiration (ET) (which includes a shift from non-productive evaporation to productive transpiration without external hydrological implications) and in terms of more crop per unit rainfall or even per unit harvested water (e.g. rain plus harvested run-on surface flow). The latter involves soil and water management with possible implications for downstream accessibility of water. Downstream access to water as a result of increased water withdrawals upstream is an issue of concern, but it is assumed here that there are overall gains and synergies to be made by maximizing the efficient use of every raindrop where it falls. In other words, WP improvements should be in focus for all water flowing through the landscape, from upstream to downstream, and not as is generally the case once the water has reached a perennial river downstream, usually after an erosive journey. This rationale is in line with Evanari *et al.* (1971), who showed that a larger effective volume of water in a catchment can be generated for productive agricultural use through numerous small water-harvesting structures collecting local surface runoff than by one large storage structure located downstream. Similarly, using the rationale of Seckler *et al.*, (1998) (see also Seckler *et al.*, Chapter 3, this volume), on the erroneous view that all water applied in irrigation is consumed (i.e. a large proportion of the flow can be reused elsewhere), a unit of efficiently used local rainwater does not necessarily mean a unit lost for downstream use. For example, many water-harvesting systems have as both a direct and indirect objective the changing of the partitioning of flow, e.g. from surface to subsurface runoff, rather than increasing consumptive use (Scott and Flores-Lopez, 2003).

This chapter presents the agrohydrological rationale to focus on WP in rain-fed agriculture, and identifies key management challenges in the attempt to upgrade rain-fed agriculture. It presents some field experiences on system options for increased WP in smallholder farming in drought-prone environments. Implications for watershed management are discussed, and the links between WP for food and securing an adequate water flow to sustain ecosystem services are briefly analysed. The focus is on semi-arid and dry subhumid tropical agro-ecosystems, where the increase in WP is most important. Most of the research examples are taken from sub-Saharan Africa, which faces the largest food-deficit and water-scarcity challenges today.

Rainwater Management: the Rationale

A broad approach to WP in land management that covers both irrigated and rain-fed agriculture has implications for water-resources management. Conventionally, the focus of attention regarding global, regional and national freshwater resources and withdrawals has been on the stable and accessible surface and subsurface flow of water in rivers, lakes and groundwater, the so-called blue-water branch in the hydrological cycle (UN, 1997; Cosgrove and Rijsberman, 2000). Blue water is withdrawn not only as direct blue (liquid)-water uses in households, for municipalities, livestock and industry but also as direct withdrawals for irrigated agriculture (of which the consumptive proportion eventually returns to the atmosphere as green vapour or ET flow). Regionally, there are signs of present or predicted near-future physical scarcity of 'blue'-freshwater resources. The International Water Management Institute (IWMI, 2000) estimated that by 2025 30% of the world population might live in regions subject to physical water scarcity (read 'blue'-water scarcity).

The fear of rapidly growing water-scarcity problems, especially in arid and semi-arid tropical regions of the world, is based on analyses comparing blue-water availability with actual blue-water withdrawals, and projections of future withdrawals based on general per capita water requirements. This approach has recently been criticized as it does not include the contribution of rain-fed agriculture in terms of fresh water to cover human water requirements. This has significant implications for water-resources assessments, given the important role of rain-fed food production and that 90% or 1600 m^3 per capita $year^{-1}$ of human freshwater needs are water for food (Rockström and Falkenmark, 2000; Rockström, 2001). However, conventional water-resources assessments highlight the limited possibilities of expansion of direct blue-water withdrawals. Globally, humankind withdraws approximately 4000 Gm^3 $year^{-1}$ (Shiklomanov, 2000), which is projected to reach 5250 Gm^3 $year^{-1}$ in 2025, as a result of population growth and socio-economic development. This is a serious problem in light of the global availability of blue-freshwater flow estimated at 12,500 Gm^3 $year^{-1}$ (Postel *et al.*, 1996). Furthermore, de Fraiture *et al.* (2001) considered that at least 30% of the blue-water flow must be secured as an environmental flow to avoid environmental hazards, such as salt and pollutant build-up and groundwater decline, leaving a utilizable ceiling of 8700 km^3 $year^{-1}$. The increased pressure on finite blue-freshwater resources would suggest limitations in the opportunities to expand the area under irrigation.

This brings our attention to the green-flow branch in the hydrological cycle. Of the global estimated average of 113,000 Gm^3 $year^{-1}$ of precipitation over land areas, 41,000 Gm^3 $year^{-1}$ forms the blue-runoff branch and the remaining 72,000 Gm^3 $year^{-1}$ forms the return flow of green water as ET. Green-water flow sustains rain-fed agriculture, as well as all other water-dependent ecosystems, such as forests, woodlands, grazing lands, grasslands and wetlands.

Partitioning of rainfall in rain-fed agriculture and the biophysical dynamics of green-water flow at plant and production-system level have recently been studied. However, relatively less attention (compared with irrigation efficiency) has been paid to the opportunities at hand to improve agricultural WP within the large (relative to blue-water flow) component of green-water flow in the on-farm water balance and the hydrological cycle at catchment, basin and global levels. In a first global estimate, Rockström *et al.* (1999) calculated global withdrawals of green water to sustain rain-fed agriculture at 4500 Gm^3 $year^{-1}$, compared with some 2500 Gm^3 $year^{-1}$ estimated for irrigated agriculture (Shiklomanov, 2000).

Figure 9.1 shows the geographical distribution of green (rain-fed)- and blue (irrigated)-water withdrawals to produce cereals. Data on blue-water withdrawals for irrigation, as well as data on areas under rain-fed agriculture and estimated grain yields in irrigated and rain-fed farming systems, are taken from IWMI (2000). The green-water withdrawals were calculated assuming a global water productivity in rain-fed grain production of 3000 m^3 t^{-1} grain (ET flow). As seen from Fig. 9.1, the majority of countries (79%) of the world depend predominantly on the return flow of

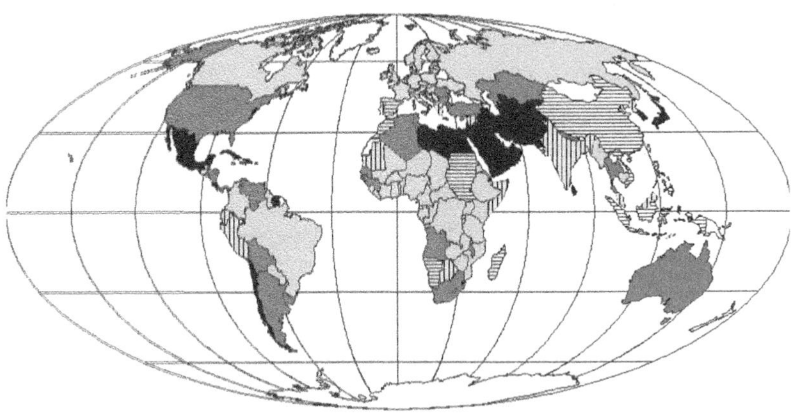

Fig. 9.1. Predominant source of water flow – green or blue – to produce cereals (grain) at country level. ☐: Countries with > 80% green-water dependence, i.e. > 80% of water used to produce cereal foods originates from rain-fed agriculture. ▩: countries where 60–80% originates from green water. ■: countries with > 80% blue-water dependence, i.e. where > 80% of water withdrawals for cereal-food production originates from blue water. ▥: countries where 60–80% originates from blue water. ▤: countries with 40–60% green-water dependence. Countries where some component of source data was lacking are marked in white.

green water in rain-fed agriculture to produce cereals. Not surprisingly, the countries (predominantly in North Africa and the Middle East) that depend primarily on blue-water withdrawals for irrigated grain production correspond to the countries that, in conventional water assessment, are predicted to be facing the most severe water-scarcity problems.

Like all global assessments, the country-scale analysis gives little guidance on challenges and opportunities at the local scale. However, it suggests that there are opportunities to produce more food per drop of water if the focus is changed from the downstream blue-runoff-water resource to the upstream position, where the rainfall enters the soil–plant system. Such a shift towards rainwater management forms a rational entry point for integrated agricultural water management that encompasses both green-rain-fed withdrawals and blue-irrigation withdrawals. Moreover, the shift towards an upstream focus is crucial, especially in respect of resource-poor smallholder farmers, as it opens up the possibility of a kind of water management that will benefit from unutilized gravitational energy.

Upgrading Rain-fed Agriculture: Challenges and Opportunities

Hydroclimatic challenges

Water-related problems in rain-fed agriculture in the water-scarce tropics are often related to high-intensity rainfall with large spatial and temporal variability, rather than to low cumulative volumes of rainfall (Sivakumar and Wallace, 1991; Rockström *et al.*, 1998; Mahoo *et al.*, 1999). Coefficients of variation range from 20 to 40%, increasing as seasonal rainfall averages decrease. The overall result of unpredictable spatial and temporal rainfall patterns indicates a very high risk for meteorological droughts and intraseasonal dry spells. The annual (seasonal) variation in rainfall can typically range from a low of one-third of the long-term average to a high of approximately double the average, meaning that a high-rainfall year can have some six times higher rainfall than a dry year. Generally, the hydroclimatic focus in semi-arid and dry subhumid tropics is on the occurrence of meteorological droughts. Their impact on rain-fed agriculture is complete crop failure, which statistically, for semi-arid lands, occurs about once every 10 years (Stewart, 1988).

Research from several semi-arid tropical regions show that the occurrence of dry spells, i.e. short periods of 2–4 weeks with no rainfall, by far exceeds that of droughts. Stewart (1988), based on research in East Africa, indicated that severe yield reductions due to dry spells occur once or twice in 5 years, and Sivakumar (1992) showed that the frequency of seasonal dry spells lasting 10–15 days was independent of long-term seasonal averages, which range from 200 to 1200 mm in West Africa. Barron *et al.* (2003), studying the frequency of dry spells in semi-arid locations in Kenya and Tanzania, showed a minimum probability (based on statistical rainfall analysis) of 0.2–0.3 for a dry spell lasting more than 10 days at any time of the growing season of a crop, and a probability of 0.7 for such a dry spell to occur during the sensitive flowering stage (maize).

Figure 9.2 shows the probability of dry-spell occurrences based on 21 years of rainfall data (1977–1998) for a site in the semi-arid Machakos district in Kenya. Rainfall is bimodal, with the onset of the long rains in mid-March (day number 75 in Fig. 9.2) and the onset of the short rains in mid-October (day number 288 in Fig. 9.2). The average planting date occurs within the onset windows in Fig. 9.2: on day number 86 (26 March) for the long rains and on day number 304 (30 October) during the short rains. Dry-spell occurrence was also analysed for the same locations, based on water-balance modelling, to assess actual crop water availability. It showed that the maize crop experienced a dry spell exceeding 10 days during 67–80% of the rainy seasons (1977–1998) on a clay soil and during 90–100% of the rainy seasons for a sandy soil.

Obviously, mitigation of intraseasonal dry spells is a key to improving WP in rain-fed agriculture in semi-arid and dry subhumid tropical environments. There are three major avenues to achieve this:

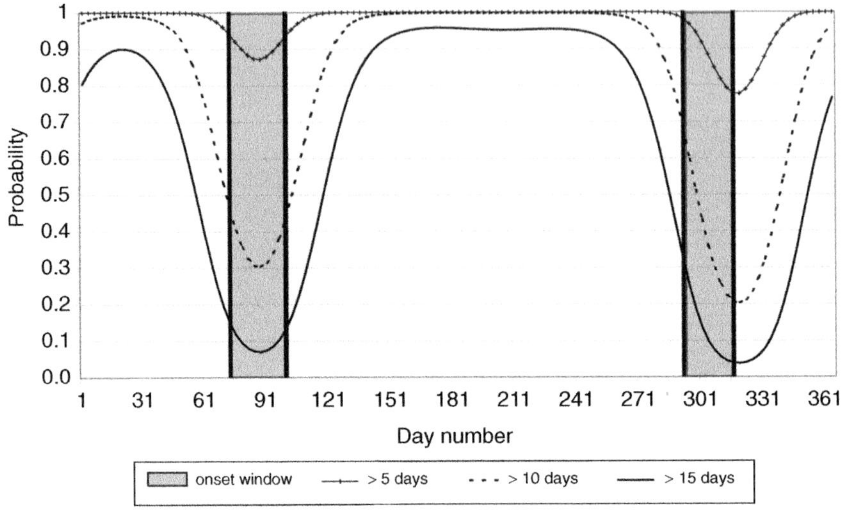

Fig. 9.2. Probability of dry spell exceeding 5, 10 or 15 days. The analysis is based on rainfall data for the period 1977–1998 at Chief Mbiuni Camp, Machakos district, Kenya. Window of planting date for long rains was from day numbers 72 to 100 (12/3 to 11/4) and similarly for short rains from day numbers 291 to 317 (17/10 to 12/11).

- Maximize plant water availability (maximize infiltration of rainfall, minimize unproductive water losses (evaporation), increase soil water-holding capacity and maximize root depth).
- Maximize water-uptake capacity of plants (timeliness of operations, crop management and soil-fertility management).
- Dry-spell mitigation using supplemental irrigation.

Hydroclimatic opportunities

The on-farm water balance can also be analysed for opportunities to improve WP. Despite a general gap in detailed knowledge on rainfall partitioning in rain-fed tropical agriculture, there are several examples of local research, often focusing on specific flow parameters (Sivakumar *et al.*, 1991; Goutorbe *et al.*, 1997; Stephens *et al.*, 1999). Figure 9.3 gives a synthesized overview of the partitioning of rainfall in semi-arid rain-fed agriculture, based on research experiences in sub-Saharan Africa. Soil evaporation generally accounts for 30–50% of rainfall (Cooper *et al.*, 1987; Wallace, 1991), a value that can exceed 50% in sparsely cropped farming systems in semi-arid regions (Allen, 1990). Surface runoff is often reported to account for 10–25% of rainfall (Penning de Vries and Djitèye, 1991; Casenave and Valentin, 1992). Large and intensive rainfall events falling on soils with low water-holding capacities result in significant drainage, amounting to some 10–30% of a rainfall (Klaij and Vachaud, 1992). The result is that productive green-water flow as T is in general reported to account for merely 15–30% of rainfall (J.S. Wallace, Institute of Hydrology, Wallingford, UK, personal communication).

Between 70 and 85% of rainfall can be considered 'lost' to the cropping system as non-productive green-water flow (as soil evaporation) and as blue-water flow (deep percolation and surface runoff). Figure 9.3 thus indicates that there is a high seasonal risk of soil-water scarcity in crop production, in addition to spatial and temporal rainfall variability.

In terms of WP can crop yields in rain-fed agriculture be increased? Rockström and Falkenmark (2000) developed an analytical tool to assess the options available to improve crop yields in semi-arid tropics from a hydrological perspective. In Fig. 9.4, the case of maize cultivated in a semi-arid tropical savan-

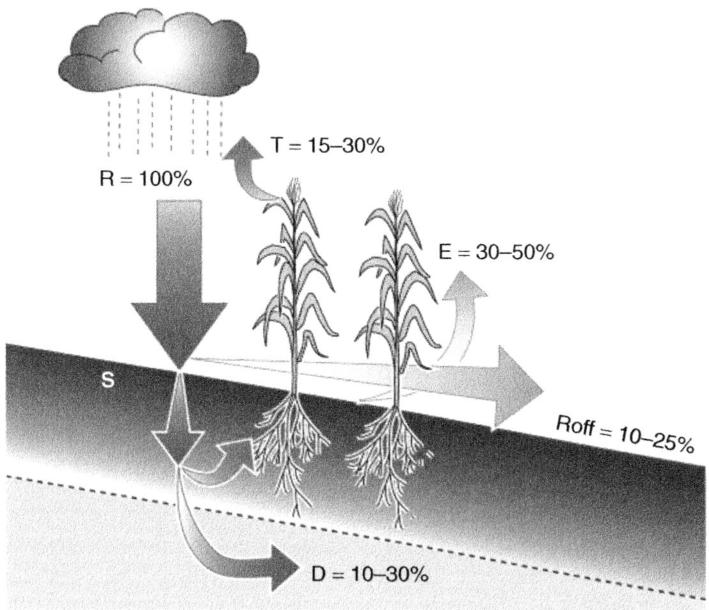

Fig. 9.3. General overview of rainfall partitioning in farming systems in the semi-arid tropics of sub-Saharan Africa. S, soil; R, rainfall; T, transpiration; E, evaporation; Roff, runoff; D, drainage.

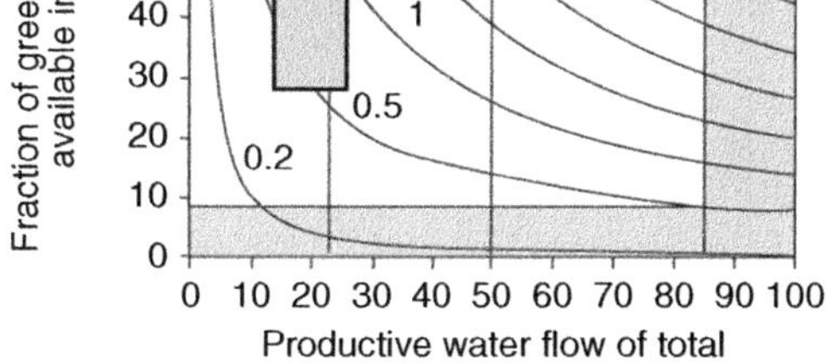

Fig. 9.4. Analysis of the effects of rainfall partitioning and plant water-uptake capacity on maize grain yields under semi-arid conditions. The larger shaded area shows the range of yields experienced on average in sub-Saharan Africa, using the rainfall partitioning range in Fig. 9.3. The smaller shaded area shows the yield range on farmers' degraded fields (Rockström and Falkenmark, 2000).

nah is presented (growing period of 120 days, daily PET = 8 mm day^{-1}, seasonal rainfall = 550 mm). Transpiration water productivity (WP_T) (kg ha^{-1} mm^{-1}) was set at 12.5 kg ha^{-1} mm^{-1} (Falkenmark and Rockström, 1993). The x axis in Fig. 9.4 shows the ratio (%) of productive green water to total green water (ratio of T to ET flow), and is an indicator of the impact of crop management on grain yield (soil fertility, crop species, timing of operations, pest management). The y axis shows the percentage of crop water requirement (CWR) available in the root zone, and is an indicator of the impact of land management on crop yields (basically the percentage of rainfall that infiltrates into the soil and is accessible to the crop).

The concave lines are isolines of equal grain yield (t ha^{-1}), with the lowest-yield line in the lower left corner and the maximum-yield isolines in the upper right-hand corner. The grey border area shows the upper boundary conditions of the model. The attainable yield level in this semi-arid case amounts to 5 t ha^{-1} grain yield. Actual observed yield levels, based on the rainfall partitioning data in Fig. 9.3, are shown by the large square. Poor rainfall partitioning (a vertical drop along the y axis) reduces the possible yields with 1–2.5 t ha^{-1} and poor plant water-uptake capacity reduces yields with 1.5–3 t ha^{-1}. The average actual yield level ranges from 1.5 to 2 t ha^{-1}. The common on-farm reality is shown by the smaller square, with an actual yield range of 0.5–1 t ha^{-1}. In the on-farm case, only 35–55% of the CWR is available in the root zone (due to high runoff, a weak root system and deep percolation) and productive green water amounts to only 15–25% of the total green-water flow (indicating large evaporation losses).

The analysis suggests a large scope for improving yield levels within the available water balance in rain-fed farming systems. It seems that there are no agrohydrological limitations to enabling even a large and stable yield increase from, for example, 0.5 t ha^{-1} to 2 t ha^{-1} (i.e. a quadrupling of yields) in semi-arid environments. The challenges are to maximize infiltration (move up along the *y* axis), to mitigate dry spells (increase the amount of water available in relation to CWR over time) and to improve, primarily, soil-fertility management in order to increase the productive green-water ratio (push the system to the right along the *x* axis).

A note on water productivity

The focus in this chapter is to improve system WP by reducing losses in the on-farm water balance in favour of productive T flow. This is in line with Gregory (1989), who suggested that, because runoff, deep percolation and evaporation can constitute large flows in the water balance, water-use efficiency in semi-arid tropics should be studied in terms of yield per unit rainfall, whereby consideration is given to the impact of management on all water flows. Rainfall water productivity (WP_R) represents a valuable parameter for assessing productivity in semi-arid tropical farming systems (Bennie and Hensley, 2001) as it indicates the extent by which green- and blue-water losses are minimized in favour of productive T flow. Also, management can easily improve WP_R. In contrast, WP_T, which is essentially affected by the atmospheric demand for water and the photosynthetic pathway, i.e. directly linked to the characteristics of the crop species, is relatively difficult to influence within a given cropping system (Sinclair *et al.*, 1984). Instead, from a green-water perspective, WP can more easily be improved by increasing the ratio of evaporation losses from the crop to the evaporation losses from the soil (E_c/E_s). Another option is to convert soil evaporation to plant T, i.e. by increasing yield per unit ET (WP_{ET}).

Water Productivity: System Opportunities

Supplemental irrigation for dry-spell mitigation

An interesting option to increase WP at production-system level is to bridge dry spells through supplemental irrigation of rain-fed crops (Oweis *et al.*, 1999; SIWI, 2001). Supplemental irrigation in smallholder farming systems can be achieved with water-harvesting systems that collect local surface runoff (sheet, rill and gully flow) in small storage structures (100–1000 m^3). Water harvesting, broadly defined as the concentration of surface runoff for productive purposes, has ancient roots and still forms an integral part of many farming systems worldwide (Evanari *et al.*, 1971; Agarwal and Narain, 1997). However, *in situ* systems that aim at water conservation (i.e. maximizing soil infiltration and water-holding capacity) dominate, while storage systems for supplemental irrigation are less common, especially in sub-Saharan Africa (SIWI, 2001).

On-farm research in semi-arid locations in Kenya (Machakos district) and Burkina Faso (Ouagouya) during 1998–2000 indicates a significant scope for improving WP in rain-fed farming through supplemental irrigation, especially if combined with soil-fertility management (Barron *et al.*, 1999; Fox and Rockström, 1999). Surface runoff from small catchments (1–2 ha) was harvested and stored in manually dug farm ponds (100–250 m^3 storage capacity). Simple gravity-fed furrow irrigation was used. During the experimental phase (1998–2000), covering three rainy seasons in Burkina Faso (monomodal rain pattern) and five in Kenya (bimodal rain pattern), supplemental irrigation amounted, on average, to 70 mm per growing season, with a range of 20–220 mm. Seasonal rainfall ranged from 196 to 557 mm in Kenya and from 418 to 667 mm in Burkina Faso. In Kenya, one rainy season was classified as a meteorological drought (short rains of 1998/99), resulting in complete crop failure. One season at each site (long rains 2000 in Kenya and the rainy season 2000 in Burkina Faso) resulted in complete crop failure for most neighbouring farmers, while the water-harvesting system enabled a harvest of an above-average yield (> 1 t ha^{-1}). The seasonal long-term average yield in both areas is approximately 0.5 t ha^{-1}. Grain yields and rain-use efficiencies (kg dry-matter grain per mm rainfall) for the studied water-harvesting system are shown for sorghum in Burkina Faso (Fig. 9.5a) and for maize in Kenya (Fig. 9.5b). Each point represents an average combination of five replicates of water harvesting/fertilizer application for any rainy season. In Burkina Faso, on shallow soil with

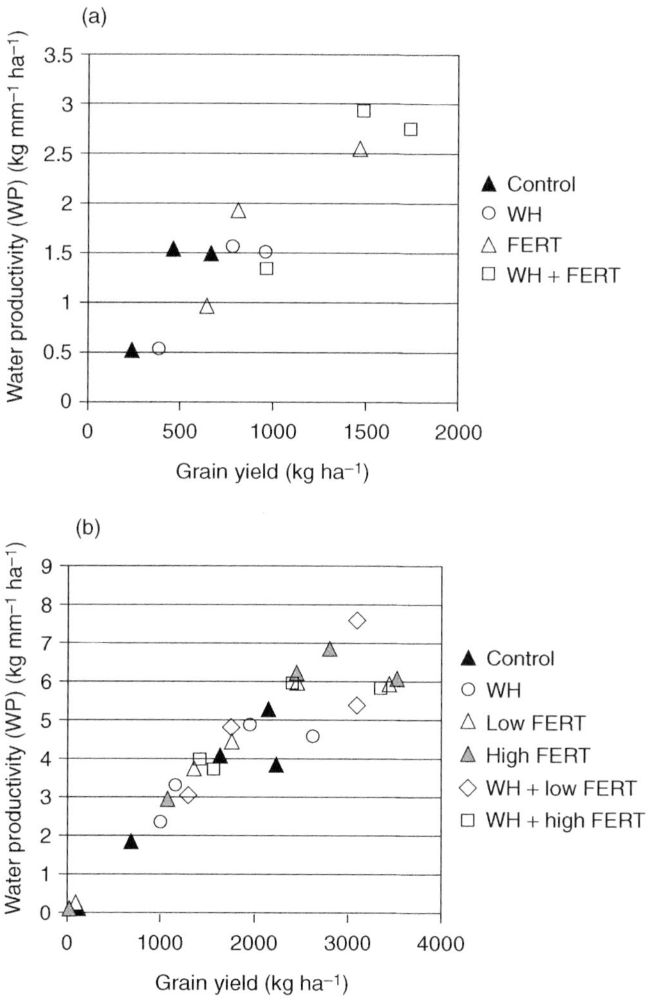

Fig. 9.5. System water productivity (WP) (kg grain per unit rainfall + supplemental irrigation) for sorghum in Burkina Faso (a) and for maize in Kenya (b). Control, traditional farmers' practice with no fertilizer application; WH, supplemental irrigation using water harvesting; FERT, fertilizer application (30 kg ha^{-1} N in Burkina Faso, and two levels in Kenya with low 30 kg N ha^{-1} and high 80 kg N ha^{-1}).

low water-holding capacity, supplemental irrigation alone improved water-use efficiency (rainfall + irrigation) by 37% on average (from 0.9 to 1.2 kg ha^{-1} mm^{-1}) compared with the control (traditional rain-fed practice with manure but no fertilizer). The corresponding ratio for the Kenyan case, on deep soil with high water-holding capacity, was 38% (from 2.2 to 3.1 kg ha^{-1} mm^{-1}).

The largest improvement in yield and WP_R was achieved by a combination of supplemental irrigation and fertilizer application. Interestingly, for both locations, fertilizer application alone (in Kenya with low application of 30 kg N ha^{-1} and high application of 80 kg N ha^{-1}) resulted in a higher average yield and WP_R than supplemental irrigation alone during years with gentle dry spells. For seasons with severe dry spells, e.g. long rains of 2000 in Kenya, non-irrigated crops failed completely, independent of fertilizer level. Nevertheless, the field data indicate that full benefits of water harvesting for supplemental irrigation can only be met by simultaneously addressing soil-fertility management.

An interesting aspect of the observed WP improvements is that WP increases in pace with yield. Assuming a linear relationship between crop yield and T (which is generally the case for a given system), then the WP increase with yield in Figs 9.5a and 9.5b originates from 'crop per drop' improvements as a result of reduction in water losses (evaporation, drainage and runoff). Similar findings of a win–win relationship between WP and yield increase are found by analysing the data of Pandey *et al.* (2000) on maize in the Sahel and of Zhang and Oweis (1999) for wheat in the Mediterranean region. However, the water-use and yield results of Norwood (2000) for limited irrigation of dryland maize do not suggest a linear relation between WP_R and grain yield. The conditions under which WP improvements are achieved as a result of system improvements need further investigation.

The relative contribution to system productivity of supplemental irrigation is assessed by calculating the incremental WP for supplementally irrigated treatments (kg additional grain produced per mm of supplemental irrigation). As seen in Table 9.1 the incremental WP is substantially higher than the seasonal WP (ranging from 2.5 to 7.6 kg ha^{-1} mm^{-1} compared with an overall WP of 0.9–1.2 kg ha^{-1} mm^{-1}). The situation is more complex on soil with greater water-holding capacity and therefore better able to cope with dry spells, as illustrated by the Kenyan case (Table 9.2). Incremental WP improvements are only achieved during rainy seasons with severe dry spells, while during rainy seasons with adequate rain of good distribution (short rains 1999/2000 and 2000/01) the incremental value can be negative.

Table 9.1. Incremental water productivity of supplemental irrigation (Burkina Faso).

Fertilizer application	1998 (kg ha^{-1} mm^{-1})	1999 (kg ha^{-1} mm^{-1})	2000 (kg ha^{-1} mm^{-1})
Non-fertilized	4.9	2.5	3.6
Fertilized	4.6	5.4	7.6

Table 9.2. Incremental water productivity of supplemental irrigation (Kenya).

Fertilizer application	SR 1998/99 (kg ha^{-1} mm^{-1})	LR 1999 (kg ha^{-1} mm^{-1})	SR 1999/2000 (kg ha^{-1} mm^{-1})	LR 2000 (kg ha^{-1} mm^{-1})	SR 2000/01 (kg ha^{-1} mm^{-1})
0 F	6.0	6.3	−9.3	4.2	19.9
30 F	3.5	4.8	32.7	5.5	−17.2
80 F	2.8	4.4	−19.1	7.0	−8.1

Water harvesting for microirrigation

For resource-poor smallholder farmers in water-scarce areas, even small volumes of stored water for supplemental irrigation can significantly improve the household economy. In the Gansu Province in China, small 10–60 m^3 (on average 30 m^3) subsurface storage tanks are promoted on a large scale. These tanks collect surface runoff from small, often treated, catchments (e.g. with asphalt or concrete). Research using these subsurface tanks for supplemental irrigation of wheat in several counties in the Gansu Province (Li *et al.*, 2000) indicates a 20% increase in WP (rain amounting to 420 mm + supplemental irrigation ranging from 35 to 105 mm). WP increased, on average, from 8.7 kg ha^{-1} mm^{-1} for rain-fed wheat to 10.3 kg ha^{-1} mm^{-1} for wheat receiving supplemental irrigation. Incremental WP ranged from 17 to 30 kg ha^{-1} mm^{-1}, indicating the large relative added value of supplemental irrigation. Similar results were observed on maize, with yield increases of 20–88% and incremental WP ranging from 15 to 62 kg ha^{-1} mm^{-1} of supplemental irrigation (Li *et al.*, 2000).

The irrigable land from these subsurface tanks is limited to 400–800 m^2. In many farming communities the tanks are probably only of interest for irrigation of high-value cash crops. A survey in Kenya among smallholder farmers, shows that farmers would rarely consider supplemental irrigation of food crops and would rather use stored water to irrigate cash crops (Jurdell and Svensson, 1998). Inspired by the Chinese subsurface tanks, similar systems are at present being developed and promoted in Kenya and Ethiopia (G. Shone, RELMA/Sida, Nairobi, Kenya, personal communication). In Kenya (Machakos district) these tanks are used to irrigate kitchen gardens, and they enable farmers to diversify sources of income from the land. The microirrigation schemes are promoted together with commercially available low-pressure drip-irrigation systems. Cheap drip kits (e.g. the Chapin bucket kit) save water and labour and are increasingly being adopted by farmers in, for example, Kenya. Combining water harvesting with drip irrigation can result in very significant WP improvements (Ngigi *et al.*, 2000).

Conservation tillage

There is ample evidence indicating that the conventional farming system in the tropics, based on soil inversion using plough and hoe, contributes to soil erosion and soil desiccation. Plough pans impede soil infiltration and root penetration, and frequent soil inversion results in accelerated oxidation of organic matter and soil erosion by wind and rain (Benites *et al.*, 1998). Conservation tillage (CT) covers a spectrum of non-inversion practices from zero-tillage to reduced tillage; it aims to maximize soil infiltration and soil productivity and minimize water losses, while conserving energy and labour. Although CT is not a new concept, the relatively recent successes, e.g. in Brazil (Derpsch, 1998), have inspired research and development efforts in sub-Saharan Africa and Asia. Examples of successful CT systems, where crop yields have been significantly increased, soil erosion reduced and conservation of water improved, can be found in several countries in sub-Saharan Africa, e.g. Ghana, Nigeria, Zimbabwe, Tanzania, South Africa and Zambia (Elwell, 1993; Oldreive, 1993). However, these successes are mostly confined to commercial farmers.

CT has several attractive effects on WP. Traditionally, conservation in agriculture has focused on soil conservation (even though labelled soil and water conservation), with the aim of reducing soil erosion. Even success stories, like the *Fanya juu* terracing in the Machakos district, Kenya (Tiffen *et al.*, 1994), show little or no evidence of actual improvements of WP in agriculture. The recently raised questions of 'what to do between the terraces' in order to increase crop yields and 'how to increase crop per drop of rain' have shifted the focus towards CT, which also enables improved timing of operations, which is crucial in semi-arid rain-fed farming and which has (compared, for example, with storage-water harvesting) the attraction of being applicable on most farmlands.

Table 9.3. Rainwater productivity (kg ha^{-1} mm^{-1}) showing variation (SD) and statistical T-test analyses (comparing CT practices with conventional ploughing with or without fertilizer application).

			Comparative analysis			
			C − FERT		C + FERT	
Treatment	WP_R (kg ha^{-1} mm^{-1})	SD	Statistical significance[a]	Multiplier	Statistical significance[a]	Multiplier
Ripper	10.1	4.9	***	2.8	*	1.4
Ripper + CC	10.6	5.8	***	2.9	**	1.5
Ripper − FERT	8.4	3.7	***	2.4	NS	1.2
Pitting	8.2	4.1	***	2.3	*	1.2
C + FERT	7.0	4.4	***	2.0		
C − FERT	3.6	2.2				

[a]Statistical significance at the 0.001 (***), 0.01 (**) and 0.05 (*) levels of probability.
NS, not significant; Ripper, Magoye ripper; Ripper + CC, Magoye ripper + cover crop (Dolichos lablab); Ripper − FERT, Magoye ripper without fertilizer application; C + FERT, conventional mouldboard plough with fertilizer application; C − FERT, ploughing without fertilizer application.

There are several examples of WP improvements using CT in rain-fed farming. Zero-tillage research using planting drills on wheat in Pakistan show water savings of 15–20% (on average, an estimated 100 mm ha^{-1}) through reduced evaporation, runoff and deep percolation, while increasing yields and saving on fuel (Hobbs *et al.*, 2000).

Over the last decade, promotion of animal- and tractor-drawn CT using rippers and subsoilers among smallholder farmers in the semi-arid Babati district, Tanzania, has resulted in significant WP increases. As seen from Fig. 9.6, the WP of rain was estimated at approximately 1.5 kg ha^{-1} mm^{-1} in the mid-1980s under conventional disc-plough agriculture, compared with a progressively increasing trend from 2 to 4 kg ha^{-1} mm^{-1} during the 1990s after the introduction of mechanized subsoiling (Rockström and Jonsson, 1999).

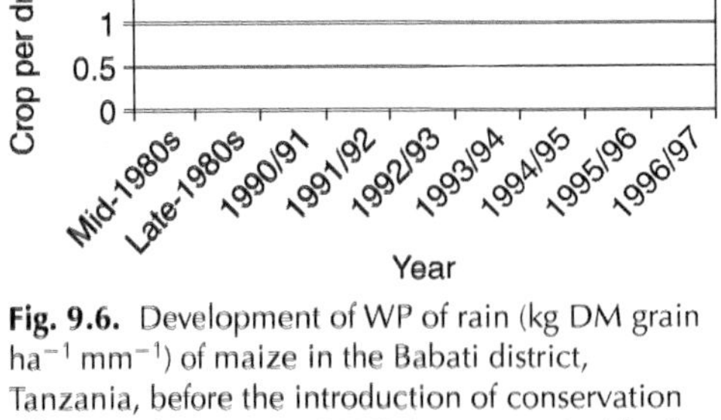

Fig. 9.6. Development of WP of rain (kg DM grain ha^{-1} mm^{-1}) of maize in the Babati district, Tanzania, before the introduction of conservation tillage (mid-1980s to 1990/91) compared with after introduction of conservation tillage (1991/92 onwards). DM, dry matter.

On-farm trials on animal-drawn and manually based CT systems (Magoye ripper and Palabana subsoiler from Zambia developed by IMAG-DLO) in Arusha and Arumeru districts, Tanzania, show similar improvements in WP: WP_R increased when shifting from a mouldboard-plough-based system (C = control) to various CT practices (Table 9.3). The data originate from 2 years (long rains, 1999 and 2000) with six to eight farmers and two replicates per farm. The improved WP is attributed primarily to improved timing of planting, root penetration and soil infiltration.

Watershed management

Upgrading rain-fed agriculture in semi-arid tropical environments will require planning of land management at the watershed scale, rather than having the conventional focus on farm or field level. A shift is needed from the 'soil conservation' approach, where surface runoff entering a farm is seen as a threat to be disposed of (e.g. with graded cut-off drains), to a 'productivity' approach, where surface runoff generated in one part of a watershed is collected and used as a resource for both agricultural and domestic purposes. Such planning is complex among smallholder farmers, since even a small runoff-collecting system will involve multiple landowners. At present, there is little or no attention given to ownership of locally produced surface runoff, but one may expect this to become an issue of importance if runoff is to be successfully managed on a larger scale for local production purposes. That this scenario is not just hypothetical is shown by a recent example from India. Small-scale farmers in the region of Rajastan installed a water-harvesting structure for water retention (a so-called *Johad*) as a strategy to reduce the degradation and livelihood deterioration caused by 3 consecutive years of drought. The Rajastan irrigation district, fearing that the water-harvesting structure would threaten water supply to a dam located downstream, judged the structure illegal (as all rain in the basin is under the authority of the Irrigation Department) and ordered (June 2001) the immediate destruction of the structure (*Down to Earth*, 2001). Many countries lack policies and legislation to manage local initiatives of rainwater management, especially for agricultural purposes (Hartung and Patschull, 2001).

As shown by several hydrological studies at watershed and basin scale, upstream shifts in water-flow partitioning may result in complex and unexpected downstream effects, both negative and positive, in terms of quantity and quality of water (Vertessy *et al.*, 1996). In general, though, increasing the residence time of runoff flow in a watershed, e.g. through practices such as water harvesting and CT, may have positive environmental, as well as hydrological, implications downstream. The hydrological implications at watershed and river-basin level of scaling up system innovations, such as water harvesting, are still unknown and require further research.

System Implications: Balancing Water for Food and Nature

It is estimated that most of the global green-water flow (88%) is at present used to sustain biomass growth in the world's biomes (Rockström *et al.*, 1999). While agriculture (rain-fed and irrigated) accounts for an estimated 7000 Gm^3 $year^{-1}$, forests and woodlands require an estimated average green-water flow of 40,000 Gm^3 $year^{-1}$, grasslands an estimated 15,100 Gm^3 $year^{-1}$ and wetlands an estimated 1100 Gm^3 $year^{-1}$. A doubling of food production over the next 25 years would (without considering WP increases) result in roughly a doubling of water utilization in agriculture. Increased withdrawals of water in rain-fed and irrigated agriculture may have negative implications for water availability to sustain direct human withdrawals and indirect withdrawals to sustain ecosystem services. The expected shifts in water flows in the water balance would affect both nature and economic sectors depending on direct water withdrawals. As suggested in this chapter, a promising avenue for upgrading rain-fed agriculture is through water harvesting, which enables mitigation of dry spells. Such measures would involve the addition of a blue-water component, through storage of surface or subsurface runoff, to the rain-fed system, i.e. developing rain-fed farming into a mixed system with an irrigation component. Carried out on a large scale (e.g. at basin level), water-harvesting promotion may have an impact on downstream blue-water availability. These effects are bound to be site-specific and need to be studied further. However, it is not certain that an increase in ET in rain-fed agriculture upstream automatically results in reduced water availability downstream. Surface runoff generated at the farm level may be lost during its journey through the catchment as evaporation or as blue water of limited use in saline rivers, before reaching a stable surface or subsurface freshwater source. Furthermore, there are large variations in green-water-flow estimates in agriculture, as both rainfall and green-water flow exhibit large spatial and temporal variability. On average, grain crop WP (WP_{ET}) ranges between 3.5 and 10 kg ha^{-1} mm^{-1} (1000–3000 m^3 t^{-1}) for tropical grains. However, as WP_{ET} is affected by biophysical factors as well as management, the actual range is much wider, with WP_{ET} values as low as 1.5 kg ha^{-1} mm^{-1} (6000–7000 m^3 t^{-1}) for degraded and poorly managed systems not being uncommon in rain-fed drylands (Rockström *et al.*, 1998).

Conclusions and Discussion

There is no doubt that the immense challenge of doubling food production over the next 25 years in order to keep pace with population growth requires focus on WP in both rain-fed and irrigated agriculture. As shown in this chapter, even in water-scarce tropical agro-ecosystems, there appear to be no hydrological limitations to doubling or, in many instances, even quadrupling yields of staple food crops in rain-fed smallholder agriculture. Furthermore, evidence suggests that there are several appropriate technologies and methodologies to hand to enable development towards improved soil productivity and WP. In a broad overview of recent projects regarding sustainable agricultural practices and technologies in 52 countries, Pretty and Hine (2001) showed that yield increases as a result of introducing practices such as water-harvesting, CT and drip irrigation amounted to 50–100% on average (with examples of up to 700% increases). The challenge, as pointed out by Pretty and Hine (2001), is to learn from these examples and establish policies that enable their proliferation.

Interestingly, even when focusing on WP in semi-arid rain-fed farming systems, where water is considered a major limiting factor for crop growth, factors other than water are shown to be at least as (if not even more) critical for productivity improvements. The experience with water harvesting for supplemental irrigation in Burkina Faso and Kenya clearly shows that soil-fertility management plays as important a role as water management. In both cases, fertilizer application alone resulted, on average, in higher WP and yields than supplemental irrigation alone. Similarly, for *in situ* water harvesting using CT in Tanzania, water conservation on its own (e.g. ripping and subsoiling) resulted in yields and WP similar to those obtained with improved soil fertility alone in conventionally ploughed systems. However, the water-harvesting studies in Burkina Faso showed that integrated soil-nutrient and water management increased yields threefold, compared with a yield increase of 1.5–2 times over traditional yield levels when either water conservation or better soil fertility was introduced.

Despite these biophysical facts, farmers' investment decisions are strongly influenced by their risk perceptions. Risk of reduced or no return on invested capital in rain-fed semi-arid farming is directly related to the unreliable rainfall distribution. Therefore, as long as farmers 'live at the mercy of rainfall', one should not be surprised at the extremely low level of investments in fertilizers (less than 20 kg ha^{-1} $year^{-1}$ in sub-Saharan Africa), in improved crop varieties and in pest management. To manage water, especially by providing farmers with the means to bridge recurrent dry spells, e.g. through small-scale water harvesting, may be the most sustainable entry point for the improvement of farming systems in general. This form of upgraded rain-fed farming may stimulate further capital and time investment in smallholder rain-fed farming. All evidence suggests that if only crop water access is secured, investments in soil fertility, crop and timing of operations will pay off in terms of substantially increased soil productivity and WP.

This chapter has not considered the social and economic viability of water-harvesting structures for supplemental irrigation among resource-poor farmers. Tentative assessments of manually dug farm ponds and subsurface tanks indicate that the economic viability depends to a large extent on the opportunity cost of labour. With low-value labour (which is often the case during dry seasons in remote rural areas) and considering the dramatic difference a water-harvesting/storage system can play during years of severe dry spells (the difference between total crop failure and having a crop), it is likely that the investment can be readily afforded and quickly recovered. However, there is a need for more detailed studies, which take into account the local environmental, institutional and socio-economic conditions.

The most interesting opportunities for upgrading smallholder rain-fed agriculture may be found in the realm of sectoral and economic system integration and diversification. Reduced risk of crop failures through supplemental irrigation implies the development of a mixed farming system with components of both rain-fed and irrigated agriculture. The time may be ripe for abandoning the sectoral distinction between irrigated and rain-fed agriculture. The implications of such a reform would be substantial. Professionally, there is still a divide between irrigation engineers dealing with irrigation management and agronomists focusing on rain-fed agriculture. Irrigation and rain-fed agriculture generally fall under different ministries (irrigation under 'blue'-water resources ministries and rain-fed agriculture under 'green' ministries of agriculture, natural resources or environment). Integrating the two may result in interesting management and technological advances in the grey zone between the purely blue and purely green food-producing sectors.

Blended upgrading also opens the door to the diversification of farming systems. A smallholder farmer's investment in supplemental irrigation will be an entrepreneurial business step, which will most probably result in a broadened basket of crops produced at farm level. This will reduce farmers' vulnerability to external climatic factors, but will also put increased pressure on the need for functioning markets and infrastructure. Diversification in favour of cash crops with a relatively lower proportion of staple food crops can have interesting virtual water implications. A shift from a staple food crop to a cash crop with similar WP but with a different market value can give rise to virtual water gains. If the same amount of water can be used to generate a higher market price, then the economic gain can be used to buy food grain. This implies a flow of virtual water from regions with a relatively greater (hydrological) comparative advantage for the production of staple foods. In summary, in spite of the wide range of complex biophysical and socio-economic factors affecting WP in rain-fed farming systems in dry subhumid to semi-arid tropics, reducing the risk of crop failure due to water stress may provide the trigger for a much-needed positive spiral of agricultural development in the smallholder sector.

References

Agarwal, A. and Narain, S. (1997) *Dying Wisdom*. Centre for Science and Environment, Thomson Press, Faridabad.

Agnew, C. and Anderson, E. (1992) *Water Resources in the Arid Realm*. Routledge, London, 323 pp.

Allen, S.J. (1990) Measurement and estimation of evaporation from soil under sparse barley crops in northern Syria. *Agricultural and Forest Meteorology* 49, 291–309.

Barron, J., Rockström, J. and Gichuki, F. (1999) Rain water management for dry spell mitigation in semi-arid Kenya. *East African Agriculture and Forestry Journal* 65(1), 57–69.

Barron, J., Rockstrom, J., Hatibu, N. and Gichuki, F. (2003) Dry spell occurrence and maize yields for two locations in semi-arid East Africa. *Agricultural Forest and Meteorology* (in press).

Barrow, C. (1987) *Water Resources and Agricultural Development in the Tropics*. Scientific and Technical, Longman, Harlow.

Benites, J., Chuma, E., Fowle, R., Kienzle, J., Molapong, K., Manu, J., Nyagumbo, I., Steiner, K. and van Veenhuizen, R. (eds) (1998) *Conservation Tillage for Sustainable Agriculture. Proceedings from an International Workshop, Harare, 22–27 June*. Part 1. *Workshop Report*. Deutsche Gesellschaft, GTZ, Eschborn, Germany, 59 pp.

Bennie, A.T.P. and Hensley, M. (2001) Maximizing precipitation utilization in dryland agriculture in South Africa – a review. *Journal of Hydrology* 241, 124–139.

Casenave, A. and Valentin, C. (1992) A runoff classification system based on surface features criteria in semi-arid areas of West Africa. *Journal of Hydrology* 130, 231–249.

Cooper, P.J.M., Gregory, P.J., Tully, D. and Harris, H.C. (1987) Improving water use efficiency of annual crops in the rain-fed farming systems of West Asia and North Africa. *Experimental Agriculture* 23, 113–158.

Cosgrove, W.J. and Rijsberman, F.R. (2000) *World Water Vision – Making Water Everybody's Business – Report to the World Water Council.* EarthScan Publications, London, 64 pp.

de Fraiture, C., Molden, D., Amarasinghe, U. and Makin, I. (2001) PODIUM: projecting water supply and demand for food production in 2025. *Phys. Chem. Earth (B)* 26(11–12), 869–876.

Derpsch, R. (1998) Historical review of no-tillage cultivation of crops. In: Benites, J., Chuma, E., Fowler, R., Kienzle, J., Molapong, K., Manu, J., Nyagumbo, I., Steiner, K. and van Veenhuizen, R. (eds) *Conservation Tillage for Sustainable Agriculture. Proceedings from an International Workshop, Harare, 22–27 June.* Part II. *Annexes.* Deutsche Gesellschaft, GTZ, Eschborn, Germany, pp. 205–218.

Down to Earth (2001) *Down to Earth* 10(5).

Elwell, H.A. (1993) Development and adoption of conservation tillage practices in Zimbabwe. *FAO Soils Bulletin* 69, 129–164.

Evanari, M., Shanan, L. and Tadmor, N.H. (1971) *The Negev, the Challenge of a Desert.* Harvard University Press, Cambridge, Massachusetts.

Falkenmark, M. and Rockström, J. (1993) Curbing rural exodus from tropical drylands. *Ambio* 22(7), 427–437.

FAO (1995) *World Agriculture: Towards 2010. An FAO Study.* Ed. N. Alexandratos. FAO, Rome, Italy, 481 pp.

FAO (2002) *Agriculture: Towards 2015/2030.* Technical Interim Report, Economic and Social Department, FAO, Rome. http://www.fao.org/es/esd/at2015/toc-e.htm

FAOSTAT (1999) http://apps.fao.org/page/collections?subset=agriculture

Fox, P. and Rockström, J. (2000) Water harvesting for supplemental irrigation of cereal crops to overcome intra-seasonal dry-spells in the Sahel. *Phys. Chem. Earth (B)* 25(3), 289–296.

Goutorbe, J.P., Dolman, A.J., Gash, J.H.C., Kerr, Y.H., Lebel, T., Prince, S.D. and Stricker, J.N.M. (eds) (1997) HAPEX-Sahel. *Journal of Hydrology* 188–189, 1–4.

Gregory, P.J. (1989) Water-use efficiency of crops in the semiarid tropics. In: *Soil, Crop and Water Management in the Sudano-Sahelian Zone.* ICRISAT, Patancheru, India, pp. 85–98.

Hartung, H. and Patschull, C. (2001) Legal aspects of rain water harvesting in sub-Saharan Africa. Paper presented at 10th International Conference on Rainwater Catchment Systems (IRCSA), 10–14 September 2001, Mannheim, Germany.

Hobbs, P.R., Dhillon, S., Singh, Y. and Malik, R. (2000) New reduced and zero tillage options for increasing the productivity and sustainability of rice–wheat systems in Indo-Gangetic plains of South Asia. Paper presented at the ISTRO 2000 Conference, 1–6 July 2000, Fort Worth, Texas.

IWMI (2000) *Water Supply and Demand in 2025.* International Water Management Institute, Colombo, Sri Lanka.

Jurdell, F. and Svensson, M. (1998) *Making Blue Water Green: the Viability of Small-scale Earth Dams for Supplementary Irrigation of Cereals in Semiarid Africa (Kenya).* Minor Field Studies No. 42, International Office, Swedish University of Agricultural Sciences, Uppsala.

Klaij, M.C. and Vachaud, G. (1992) Seasonal water balance of a sandy soil in Niger cropped with pearl millet, based on profile moisture measurements. *Agricultural Water Management* 21, 313–330.

Li, F., Cook, S., Geballe, G.T. and Burch, W.R., Jr (2000) Rainwater harvesting agriculture: an integrated system for water management on rain-fed land in China's semi-arid areas. *AMBIO* 29(8), 477–483.

McCalla, A.F. (1994) *Agriculture and Food Needs to 2025: Why We Should be Concerned.* Consultative Group on International Agricultural Research (CGIAR), Washington, DC.

Mahoo, H.F., Young, M.D.B. and Mzirai, O.B. (1999) Rainfall variability and its implications for the transerability of experimental results in semi-arid areas of Tanzania. *Tanzania Journal of Agricultural Science* 2(2), 127–140.

Ngigi, S.N., Thome, J.N., Waweru, D.W. and Blank, H.G. (2000) *Technical Evaluation of Low-head Drip Irrigation Technologies in Kenya.* Research Report, Nairobi University and the International Water Management Institute (IWMI), Nairobi, Kenya, 21 pp.

Norwood, C.A. (2000) Water use and yield of limited-irrigated and dryland corn. *Soil Science Society American Journal* 64(1), 365–370.

Oldreive, B. (1993) *Conservation Farming for Communal, Small-scale, Resettlement and Co-operative Farmers of Zimbabwe. A Farm Management Handbook.* Mazongororo Paper Converters, Zimbabwe, 76 pp.

Oweis, T., Hachum, A. and Kijne, J. (1999) *Water Harvesting and Supplemental Irrigation for Improved Water Use Efficiency.* SWIM Paper No. 7, IWMI, Colombo, Sri Lanka.

Pandey, R.K., Maraville, J.W. and Admou, A. (2000) Deficit irrigation and nitrogen effects on maize in Sahelian environment. I. Grain yield and yield components. *Agricultural Water Management* 46, 1–13.
Parr, J.F., Stewart, B.A., Hornick, S.B. and Singh, R.P. (1990) Improving the sustainability of dryland farming systems: a global perspective. In: Singh, R.P., Parr, J.F. and Stewart, B.A. (eds) *Advances in Soil Science*, Vol. 13, *Dryland Agriculture Strategies for Sustainability*. New York, USA, pp. 1–8.
Penning de Vries, F.W.T. and Djitèye, M.A. (eds) (1991) *La Productivité des pâturages sahéliens. Une étude des sols, des végétations et de l'exploitation de cette resource naturelle*. Pudoc, Wageningen, The Netherlands, 522 pp.
Postel, S.L., Daily, G.C. and Ehlich, P.R. (1996) Human appropriation of renewable fresh water. *Science* 271, 785–788.
Pretty, J. and Hine, R. (2001) *Reducing Food Poverty with Sustainable Agriculture: a Summary of New Evidence. Final Report from the 'SAFE-World' Research Project, February 2001*. University of Essex, Colchester, UK, 133 pp.
Rockström, J. (2001) Green water security for the food makers of tomorrow: windows of opportunity in drought-prone savannahs. *Water Science and Technology* 43(4), 71–78.
Rockström, J. and Falkenmark, M. (2000) Semiarid crop production from a hydrological perspective – gap between potential and actual yields. *Critical Reviews of Plant Science* 19(4), 319–346.
Rockström, J. and Jonsson, L.-O. (1999) Conservation tillage systems for dryland farming: on-farm research and extension experiences. *East African Agriculture and Forestry Journal* 65(1), 101–114.
Rockström, J., Jansson, L.-O. and Barron, J. (1998) Seasonal rainfall partitioning under runon and runoff conditions on sandy soil in Niger. On-farm measurements and water balance modelling. *Journal of Hydrology* 210, 68–92.
Rockstrom, J., Gordon, L., Falkenmark, M., Folke, C. and Engvall, M. (1999) Linkages among water vapour flows, food production, and terrestrial ecosystem services. *Conservation Ecology* 3(2), 5, 1–28. http//www.consecol.org/vol3/iss2/art5
Scott, C.A. and Flores-Lopez, F.J. (2003) Evaporation and infiltration from water bodies in the Lerma–Chapala basin, Mexico. *Journal of the American Water Resources Associaton* (in press).
Seckler, D., Molden, D. and Sakthivadivel, R. (1998) The concept of efficiency in water resource management and policy. International Water Management Institute (IWMI) (mimeo), Colombo, Sri Lanka.
Shiklomanov, I.A. (2000) Appraisal and assessment of world water resources. *Water International* 25(1), 11–32.
Sinclair, T.R., Tanner, C.B. and Bennett, J.M. (1984) Water-use-efficiency in crop production. *BioScience* 34(1), 36–40.
Sivakumar, M.V.K. (1992) Empirical analysis of dry spells for agricultural applications in West Africa. *Journal of Climate* 5, 532–539.
Sivakumar, M.V.K. and Wallace, J.S. (1991) Soil water balance in the Sudano-Sahelian zone. In: Sivakumar, M.V.K., Wallace, J.S., Renard, C. and Giroux, C. (eds) *Soil Water Balance in the Sudano-Sahelian Zone. Proceedings of an International Workshop, Niamey, Niger, 18–23 February 1991*. IAHS Publication No. 199, IAHS Press, Institute of Hydrology, Wallingford, UK, pp. 3–10.
Sivakumar, M.V.K., Wallace, J.S., Renard, C. and Giroux, C. (eds) (1991) *Soil Water Balance in the Sudano-Sahelian Zone. Proceedings of an International Workshop, Niamey, Niger, 18–23 February 1991*. IAHS Publication No. 199, IAHS Press, Institute of Hydrology, Wallingford, UK.
SIWI (2001) *Water Harvesting for Upgrading of Rain-fed Agriculture. Problem Analysis and Research Needs.* SIWI Report 11, Stockholm International Water Institute (SIWI), Stockholm, Sweden, 97 pp.
Stephens, W., Hess, T., Carter, R.C. and Howsam, P. (eds) (1999) Special issue: more from less: improving irrigation water-use efficiency. *Agricultural Water Management* 40(1).
Stewart, J.I. (1988) *Response Farming in Rain-fed Agriculture*. Wharf Foundation Press, Davis, California, 103 pp.
Tiffen, M., Mortimore, M. and Gichuki, F. (1994) *More People, Less Erosion – Environmental Recovery in Kenya*. ACTSPRESS, African Centre for Technology Studies, Nairobi, Kenya, 301 pp.
UN (1997) *Comprehensive Assessment of the Freshwater Resources of the World*. WMO, Geneva, Switzerland.
Vertessy, R.A., Hatton, T.J., Benyon, R.G. and Dawes, W.R. (1996) Long-term growth and water balance predictions for a mountain ash (*Eucalyptus regnans*) forest catchment subject to clear-felling and regeneration. *Tree Physiology* 16(1/2), 221–232.
Wallace, J.S. (1991) The measurement and modelling of evaporation from semi-arid lands. In: Sivakumar, M.V.K., Wallace, J.S., Rénard, C. and Giroux, C. (eds) *Soil Water Balance in the Sudano-Sahelian Zone.*

Proceedings of an International Workshop, Niamey, Niger, 18–23 February 1991. IAHS Publication No. 199, IAHS Press, Institute of Hydrology, Wallingford, UK, pp. 131–148.
World Bank (1997) *World Development Report, 1997*. World Bank, Washington, DC.
Young, A. (1999) Is there really spare land? A critique of estimates of available cultivable land in developing countries. *Environment, Development and Sustainability* 1, 3–18.
Zhang, H. and Oweis, T. (1999) Water–yield relations and optimal irrigation scheduling of wheat in the Mediterranean region. *Agricultural Water Management* 38, 195–211.

10 World Water Productivity: Current Situation and Future Options

Ximing Cai[1,2] and Mark W. Rosegrant[1]
[1] *International Food Policy Research Institute, Washington, DC, USA;*
[2] *International Water Management Institute, Colombo, Sri Lanka*

Abstract

Water productivity is generally defined as crop yield per cubic metre of water consumption, including 'green' water (effective rainfall) for rain-fed areas and both 'green' water and 'blue' water (diverted water from water systems) for irrigated areas. Water productivity defined as above varies from region to region and from field to field, depending on many factors, such as crop patterns and climate patterns (if rainfall fits crop growth), irrigation technology and field water management, land and infrastructure, and input, including labour, fertilizer and machinery. In this chapter, we analyse water productivity at the global and regional levels through a holistic modelling framework, IMPACT-WATER, an integrated water and food model developed at the International Food Policy Research Institute (IFPRI). Scenario analysis is undertaken to explore the impact of technology and management improvement and investment on water productivity and to search for potentials in improving food security through enhancing water productivity. It is found that the water productivity of rice ranged from 0.15 to 0.60 kg m^{-3}, while that of other cereals ranged from 0.2 to 2.4 kg m^{-3} in 1995. From 1995 to 2025, water productivity will increase. The global average water productivity of rice and other cereals will increase from 0.39 kg m^{-3} to 0.52 kg m^{-3} and from 0.67 kg m^{-3} to 1.01 kg m^{-3}, respectively. Both the increase in crop yield and improvement in basin efficiency contribute to the increase in water productivity, but the major contribution comes from increase in the crop yield. Moreover, water productivity of irrigated crops, although higher than that of rain-fed crops in developing countries, is lower in developed countries.

Introduction

Producing enough food and generating adequate income in the developing world to better feed the poor and reduce the number of those suffering will be a great challenge. This challenge is likely to intensify, with a global population that is projected to increase to 7.8 billion in 2025, putting even greater pressure on world food security, especially in developing countries, where more than 80% of the population increase is expected to occur. Irrigated agriculture has been an important contributor to the expansion of national and world food supplies since the 1960s and is expected to play a major role in feeding the growing world population. However, irrigation accounts for about 72% of global and 90% of developing-country water withdrawals; and water availability for irrigation may have to be reduced in many regions in favour of

 Water Productivity in Agriculture: Limits and Opportunities for Improvement (eds J.W. Kijne, R. Barker and D. Molden)

rapidly increasing non-agricultural water uses in industry and households, as well as for environmental purposes. With growing irrigation-water demand and increasing competition across water-using sectors, the world now faces a challenge to produce more food with less water. This goal will be realistic only if appropriate strategies are found for water savings and for more efficient water uses in agriculture.

One important strategy is to increase the productivity of water (Molden, 1997; Molden *et al.*, 2001). Water productivity (WP) is defined as the physical or economic output per unit of water application. In this chapter, using a holistic water–food model, IMPACT-WATER, developed at the International Food Policy Research Institute (IFPRI), we assess the value of WP at both the regional and the global scale in a base year (1995), and project productivity to 2025 under plausible assumptions on food demand and supply and water demand and supply. Food production and consumption are examined simultaneously. The purpose of this chapter is to show how much increase of WP should be achieved between 1995 and 2025 in order to meet demand, and how the increase can be achieved.

Methodology, Data and Assumptions

WP is defined as crop yield per cubic metre of water consumption (WC), including 'green' water (effective rainfall) for rain-fed areas and both 'green' water and 'blue' water (diverted water from water systems) for irrigated areas. WC includes beneficial water consumption (BWC) and non-beneficial water consumption (NBWC). BWC directly contributes to crop growth at the river-basin scale, and NBWC includes distribution and conveyance losses to evaporation and sinks, which are not economically reusable. BWC is characterized by water-use efficiency in agriculture. We use effective efficiency at the river-basin scale (Keller *et al.*, 1996) to represent water-use efficiency, which is the ratio of BWC to WC (in the following it is called basin efficiency, BE, and P is crop production):

$$\mathrm{WP}\ (\mathrm{kg\ m^{-3}}) = \frac{\mathrm{P\ (kg)}}{\mathrm{WC\ (m^3)}} \qquad (10.1)$$

$$\mathrm{WC} = \mathrm{BWC} + \mathrm{NBWC} = \frac{\mathrm{BWC}}{\mathrm{BE}} \qquad (10.2)$$

WP defined as above varies from region to region and from field to field, depending on many factors, such as crop patterns, climate patterns (if rainfall fits crop growth), irrigation technology and field water management, land and infrastructure, and input, including labour, fertilizer and machinery. WP can be increased by either increasing crop yield (i.e. increasing the numerator in Equation 10.1 through other inputs while maintaining a constant water-use level) or reducing WC and maintaining the yield level (i.e. decreasing the denominator), or by both. In this chapter, we compute crop yield and WC through IMPACT-WATER, a modelling framework developed at IFPRI, and then compute WP as crop yield (kg) per cubic metre of WC.

IMPACT-WATER combines an extension of IFPRI's International Model for Policy Analysis of Agricultural Commodities and Trade (IMPACT) to include water in the agricultural supply functions with a newly developed water simulation model (WSM). IMPACT simulates food demand, supply and trade in the global scope (for a detailed description, see Rosegrant *et al.*, 2001a). Crop area and yield are functions of BWC for crop growth under the condition of crop evapotranspiration requirement, as well as of investment in crop and input prices in agricultural research.

$$P = A \times Y \qquad (10.3)$$

$$A = A\ (\mathrm{BWC|ETC,\ crop\ prices,\ irrigation\ investment}) \qquad (10.4)$$

$$Y = Y\ (\mathrm{BWC|ETC,\ crop\ prices,\ input\ prices,\ agricultural\ research\ investment}) \qquad (10.5)$$

where A is the crop harvested area, Y is the crop yield and ETC is the crop evapotranspiration requirement.

Beneficial crop WC depends on effective water availability, including effective rainfall and effective irrigation-water supply.

Effective rainfall is calculated based on total rainfall, crop evapotranspiration requirement and soil characteristics (USDA, 1967). Effective irrigation-water supply is simulated by WSM, taking into account total renewable water, non-agricultural water demand, water-supply infrastructure and economic and environmental policies at the basin, country or regional levels (Fig. 10.1). A detailed description of effective irrigation-water supply within the model can be found in Rosegrant and Cai (2001).

IMPACT-WATER allows an exploration of the relationships between water availability and food production at various spatial scales, from river basins, countries or regions to the global level, over a 30-year time horizon (e.g. 1995–2025). Water availability is treated as a stochastic variable with observable probability distributions, in order to examine the impact of droughts on food supply, demand and prices. China, India and the USA, which together account for about 60% of global grain production, have been disaggregated into several basins. Other countries and regions are aggregated in 33 spatial units. In each unit, eight food crops are considered in detail: rice, wheat, maize, other coarse grains, soybean, potato, sweet potato, and cassava and other roots and tubers. Irrigation requirements for all other crops are also projected.

The starting-point for the analysis is a baseline scenario that incorporates our best estimates of the policy, investment, technological and behavioural parameters driving the food and water sectors. On the food side, total cereal demand is projected to grow by 758 million t between 1995 and 2025, of which 84% of the projected increase will be in developing countries. Expansion in area will contribute very little to future production growth, with a total increase in cereal crop area of only 54 million ha by 2021–2025, from 688 million ha in 1995. The slow growth in crop area places the burden of meeting future cereal demand on growth in crop yield. Although yield growth will vary considerably by commodity and country, in the aggregate and in most countries it will continue to slow down. The global growth rate of yield for all cereals is expected to decline from 1.5% $year^{-1}$ during 1982–1995 to 1.0% $year^{-1}$ during 1995–2020; and, in developing countries, average growth of crop yield will decline from 1.9% $year^{-1}$ to 1.2% $year^{-1}$.

In the water component, the model utilizes hydrological data (precipitation, evapotranspiration and runoff) that re-create the hydrological regime of 1961–1991 (Alcamo, 2000). Non-irrigation water uses, including domestic, industrial and livestock water uses, are projected to grow rapidly. Total non-irrigation water consumption in the world is projected to increase from 370 km^3 in 1995 to 620 km^3 in 2025, an increase of 68%. The largest increase of about 85% is

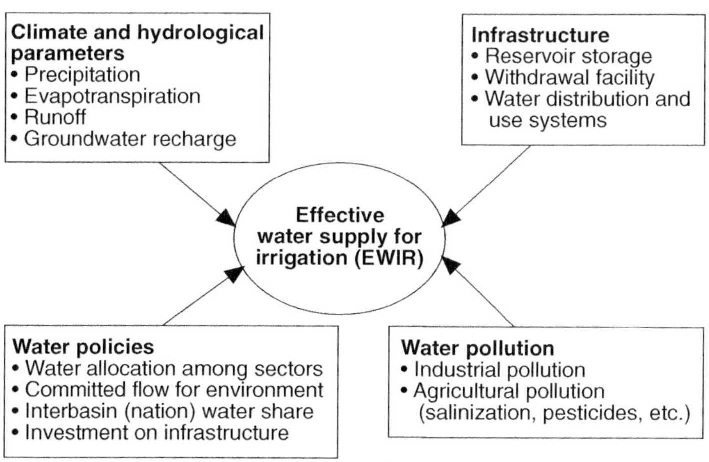

Fig. 10.1. Processes involved in simulating effective water supply for irrigation.

projected for developing countries. Moreover, in-stream and environmental water demand is accounted as committed flow that is unavailable for other uses, and ranges from 15% to 50% of the runoff, depending on runoff availability and relative demands of the in-stream uses in different basins. Irrigation-water demand is estimated and projected, based on crop evapotranspiration and effective rainfall (estimated on a monthly basis), irrigated area and water-use efficiency. Globally, irrigated harvested area for cereals is estimated to be 21 million ha in 1995, and growth is projected to be slow, with a total increase of 24 million ha for irrigated cereals by 2025. The global potential irrigation-water demand is 1758 billion m^3 in 1995 and 1992 billion m^3 in 2021–2025, increasing by 13.4%. The developing world is projected to have much higher growth in potential irrigation-water demand than the developed world between 1995 and 2021–2025, with potential consumptive demand in the developing world rising from 1445 billion m^3 in 1995 to 1673 billion m^3 (average) in 2021–2025, or 15.8%. However, as will be seen below, the effective increase in consumptive use of water for irrigation worldwide is only 3.9%, considerably lower than the growth in potential demand, due to constraints in water supply.

We assume moderate increases in water-withdrawal capacity, reservoir storage and water-management efficiency, based on estimates of current investment plans and the pace of water-management reform. The water outcomes are briefly summarized below:

- *Total maximum allowed water withdrawals.* The total global water withdrawals were 3722 km^3 in 1995, representing 7.8% of global renewable water resources. Water withdrawals for the base year 1995 are estimated as 2795 km^3 in developing countries and 926 km^3 in developed countries. Groundwater pumping in 1995 is up from 817 km^3 (21.9% of total water withdrawals). Total global water withdrawals are projected to increase to 23% between 1995 and 2025. Projected withdrawals increase by 28% in developing countries. Global consumptive use of water will increase by 16%, and the vast majority of the increase will be in developing countries, where consumptive use across all sectors will increase by 18%.
- *Reservoir storage.* The total global reservoir storage for irrigation and water supply is estimated at 3428 km^3 in 1995 (47% of total reservoir storage for all purposes), and is projected to reach 4118 km^3 by 2025, representing a net increase of 690 km^3 over the next 25 years.
- *Effective rainfall use.* It is assumed that effective rainfall use for rain-fed crops will increase by 3–5% in the baseline scenario, due to improvements in water harvesting and on-farm water management and varietal improvement that shifts crop growth periods to better utilize rainfall. This is approximately the equivalent of increasing crop evapotranspiration by 150 km^3.
- *BE.* The average BE for the base year 1995 is assessed at 0.56 globally (0.53 in developing countries and 0.64 in developed countries). Relatively large increases in BE are assumed under the baseline scenario for developed and developing countries where renewable water-supply infrastructure is highly developed (e.g. India, China, and west Asia and North Africa (WANA)). For other regions, such as in sub-Saharan Africa and South-East Asia, where water supply facilities are still fairly underdeveloped, only smaller increases in BE are projected. Based on the above assumptions, the average BE is projected to reach 0.61 worldwide, 0.59 in developing countries and 0.69 in developed countries by 2025. On a global basis, with the improvement in water-use efficiency in the baseline scenario, the global WC demand is 8% lower by 2025 relative to what it would be if effective efficiency remained constant.

Results

Although WP as defined above can be calculated for each of the crops in each spatial unit considered in IMPACT-WATER, without loss of generality, this chapter will focus on the results for rice and total cereals except for rice, mostly at an aggregated spatial scale, i.e. the developing world and the developed world, with some results shown at the region or basin level too. Results from two alternative scenarios are also presented, showing the impact of water-use efficiency and environmental water conservation.

Water productivity in 1995

Figure 10.2 shows a global map of WP of irrigated rice and Fig. 10.3 a similar map of total irrigated cereals excluding rice. The basic elements of these maps are the 36 countries and aggregated regions used in IMPACT (Rosegrant *et al.*, 2001a). Since rice usually consumes more water than other crops, the WP of rice is significantly lower than that of other cereals. Figures 10.2 and 10.3 show that the WP of rice ranges from 0.15 to 0.60 kg m^{-3}, while that of other cereals ranges from 0.2 to 2.4 kg m^{-3}. For both rice and other cereals, WP in sub-Saharan Africa is the lowest in the world. The WP of rice is 0.10–0.25 kg m^{-3} in this region, with an average yield of 1.4 t ha^{-1} and WC ha^{-1} is close to 9500 m^3. For other cereals in sub-Saharan Africa, the average yield is 2.4 t ha^{-1}, the WC is 7700 m^3 ha^{-1} and the average WP is 0.3 kg m^{-3} (ranging from 0.1 to 0.6 kg m^{-3}). Among developing countries, China and some South-East Asian countries have a higher WP of rice, ranging from 0.4 to 0.6 kg m^{-3}; however, the average of the developed world, 0.47 kg m^{-3} (yield, 4.7 t ha^{-1}; WC 10,000 m^3 ha^{-1}), is higher than the 0.39 kg m^{-3} of the developing world (yield, 3.3 t ha^{-1}; WC 8600 m^3). For other cereals, WP is lower than 0.4 kg m^{-3} in south Asia, central Asia, northern and central sub-Saharan Africa; it is 1.0–1.7 kg m^{-3} in China, the USA and Brazil; and 1.7–2.4 kg m^{-3} in Western European countries. The average WP of other cereals in the developed world is 1.0 kg m^{-3} (yield, 4.4 t ha^{-1}; WC 4500 m^3 ha^{-1}); in the developing world it is 0.56 kg m^{-3} (yield, 3.2 t ha^{-1}; WC, 5600 m^3 ha^{-1}).

It should be noted that, because of the level of aggregation, the values shown on these maps do not show the variation of WP within individual countries. Within

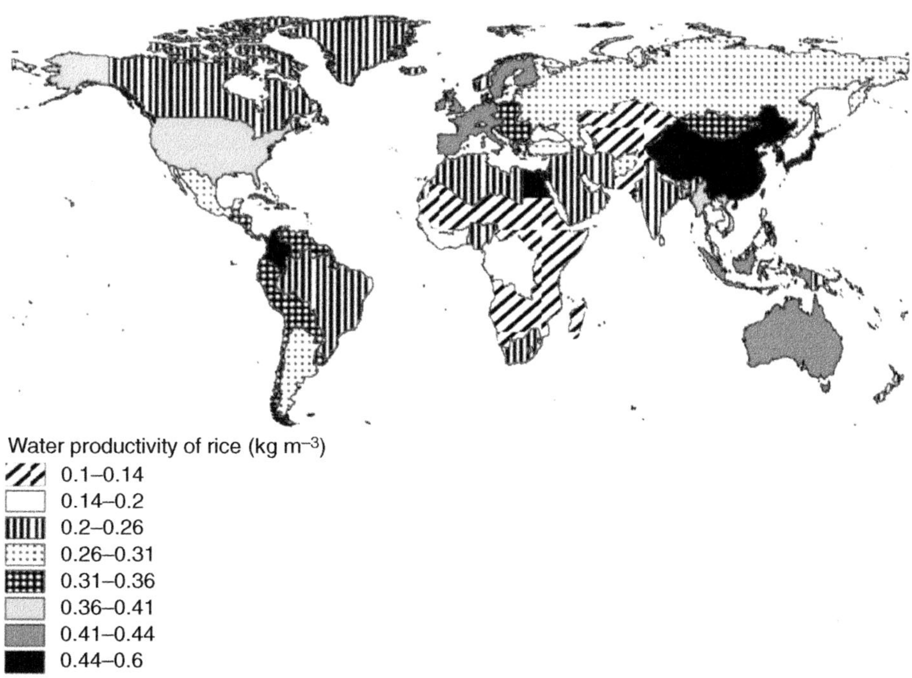

Fig. 10.2. Water productivity of rice in 1995.

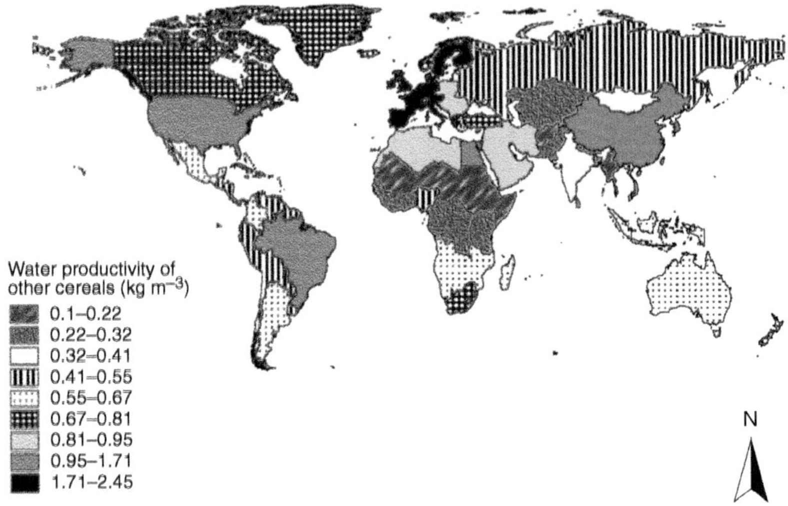

Fig. 10.3. Water productivity of total cereals, excluding rice, in 1995.

some large countries, WP varies significantly. Figure 10.4 shows the WP of total cereal excluding rice in major river basins in China, India and the USA. In China, WP for non-rice cereals ranges from 0.4 to 1.4 kg m^{-3}, with higher WP in the Yangtze River basin and north-east China (the Song-Liao river basin). Crop yields in these areas are relatively higher and water availability is relatively less restricted. However, in India, where non-rice cereal productivity ranges from 0.2 to 0.7 kg m^{-3}, higher WP occurs in northern India (0.4–0.7 kg m^{-3}), where crop yield is higher but water availability is more restricted than in other areas. In the USA, WP ranges from 0.9 to 1.9 kg m^{-3}, with higher values in the north than in the south and the highest in the north-western regions.

Changes in water productivity between 1995 and 2025

IMPACT-WATER simulates crop production and water use from 1995 to 2025, based on which WP year by year is calculated during the period. Figure 10.5 shows a projection of WP of irrigated rice in developing countries, developed countries and the world, from 1995 to 2025, and Fig. 10.6 shows the curves for other cereals. First, we can see that WPs vary from year to year due to variability in climate, which shows that the latter affects water availability and then WP. Secondly, based on our assumption on area and yield growth and on water supply enhancement, WPs are going to increase significantly between 1995 and 2025. For example, WP of other cereals will increase from 1.0 to 1.4 kg m^{-3} in developed countries, from 0.6 to 1.0 kg m^{-3} in developing countries and from 0.7 to 1.1 kg m^{-3} in the world. Figures 10.7 and 10.8 further compare WPs in several regions between 1995 and the average of 2021–2025, for rice and other cereals, respectively.

What is the major reason for the increase in WP from 1995 to 2025, the increase in yield or improvement in water efficiency that decreases WC per hectare? Figure 10.9 compares crop yield and WC for rice between 1995 and the average of 2021–2025. Figure 10.10 shows the same comparisons for other cereals. As can be seen, crop yield increases and WC per hectare decreases, except for a slight increase in WC for other cereals in sub-Saharan Africa. WC per hectare depends on the change of total consumption and the change of crop area. IMPACT-WATER projects a relatively small increase in irrigated cereal crop area, only 24 million ha or 10% from 1995 to 2025 for total irrigated cereals in the world. On the other hand, total realized

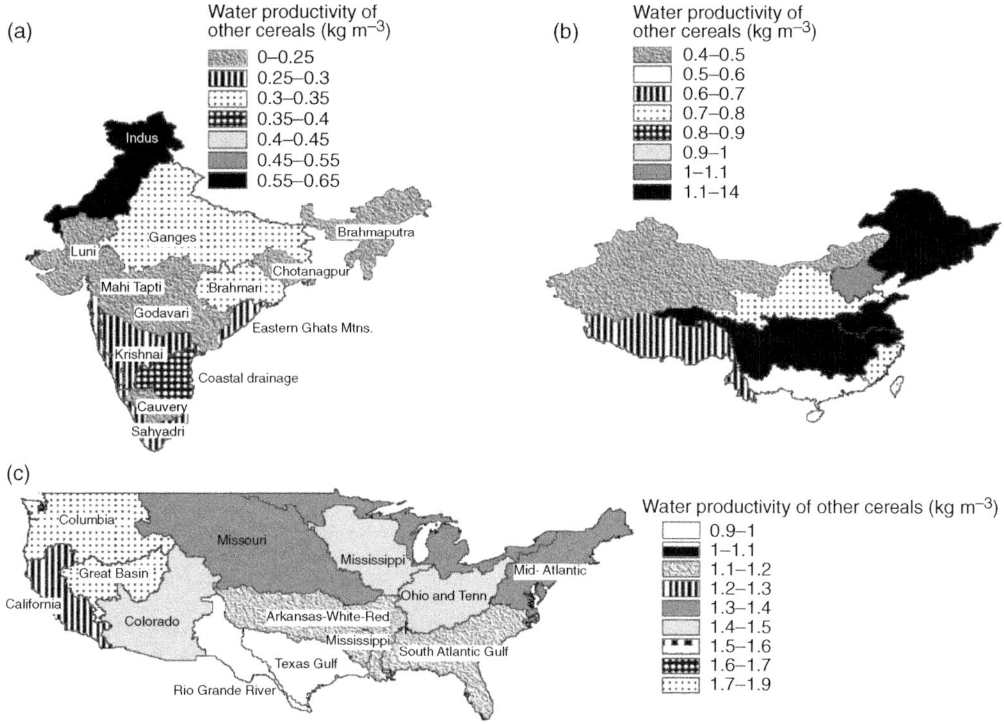

Fig. 10.4. Water productivity of total cereals, excluding rice, in 1995 in river basins in (a) India, (b) China and (c) the USA.

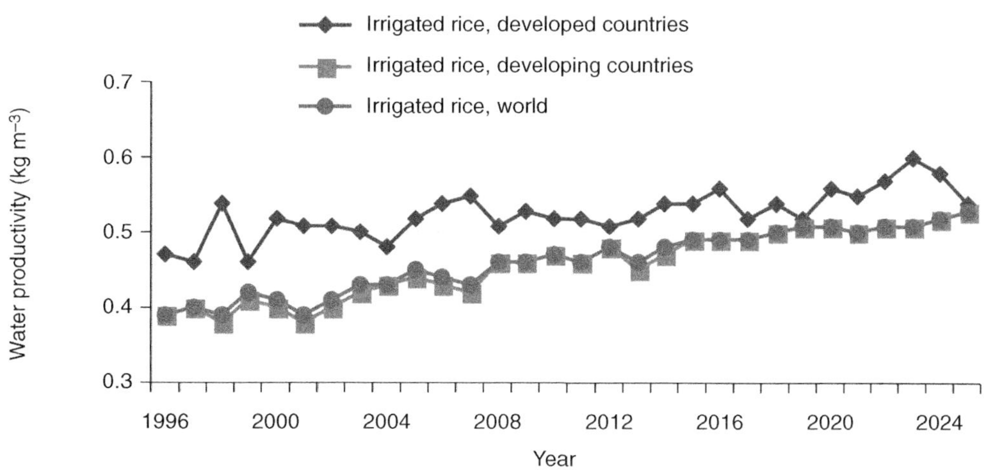

Fig. 10.5. Water productivity of irrigated rice.

crop WC is further determined by the change of water-withdrawal capacity, BE, the change of rainfall harvest and the change of the crop consumption requirements, as well as the amount of water taken by the non-irrigation sectors. Under the baseline scenario, total

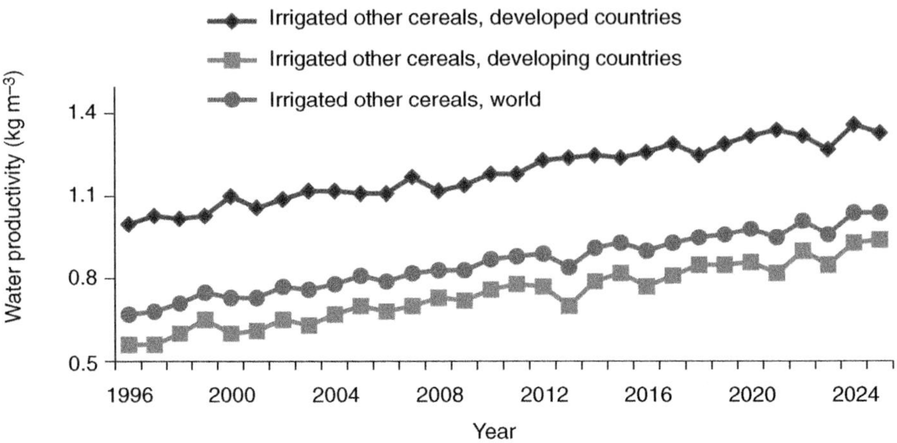

Fig. 10.6. Water productivity of irrigated other cereals.

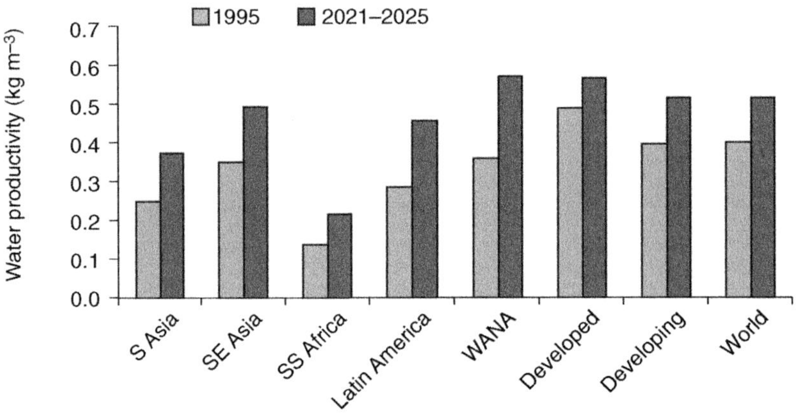

Fig. 10.7. Water productivity of rice in several regions in 1995 and 2021–2025. SS, Sub-Saharan.

global water withdrawals are projected to increase by 23% from 1995 to 2025, with the increase mainly used for non-irrigation sectors (increasing by 62% worldwide from 1995 to 2025). These will increase the total consumption. However, WC can be reduced, because the projected increase in effective river-basin water-use efficiency will decrease the crop consumption demand. All of these factors result in a 3.9% increase in consumptive use of water for irrigation worldwide. Overall, as can be seen in Figs 10.9 and 10.10, the change of WC per hectare is small compared with the change of crop yield. The increase in WP mainly results from the increase in crop yield.

What about the WP of rain-fed crops? Is it comparable to the WP of irrigated crops? Figures 10.11 and 10.12 show WP of rice and other cereals, respectively, from 1996 to 2025 in developing countries. WP for irrigated crops is higher than that of rain-fed crops, at a level of 0.15–0.2 kg m^{-3} for rice and 0.1–0.4 kg m^{-3} for other cereals. The difference becomes larger from 1996 to 2025, due to the higher rate of increase in irrigated yield and the increase in water-use efficiency over time. However, WP of irrigated crops is not higher than that of rain-fed crops everywhere in the world. This can be seen from Figs 10.13 and 10.14, which show the WP of rice and other cereals, respectively, from

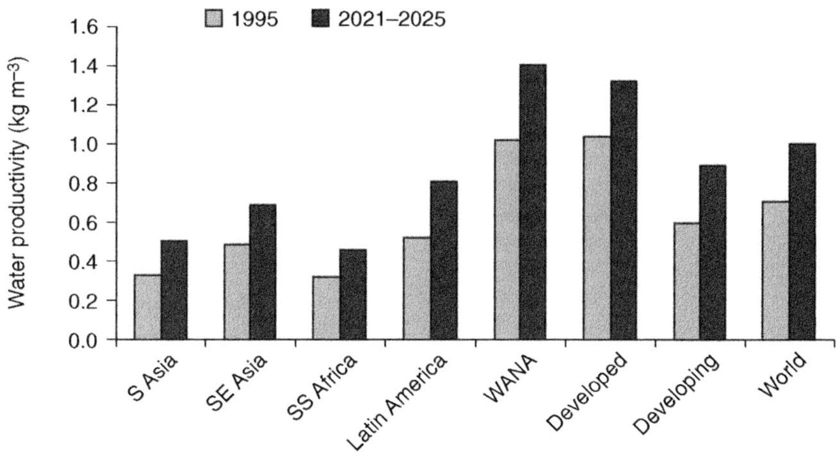

Fig. 10.8. Water productivity of other cereals in several regions in 1995 and 2021–2025. SS, Sub-Saharan.

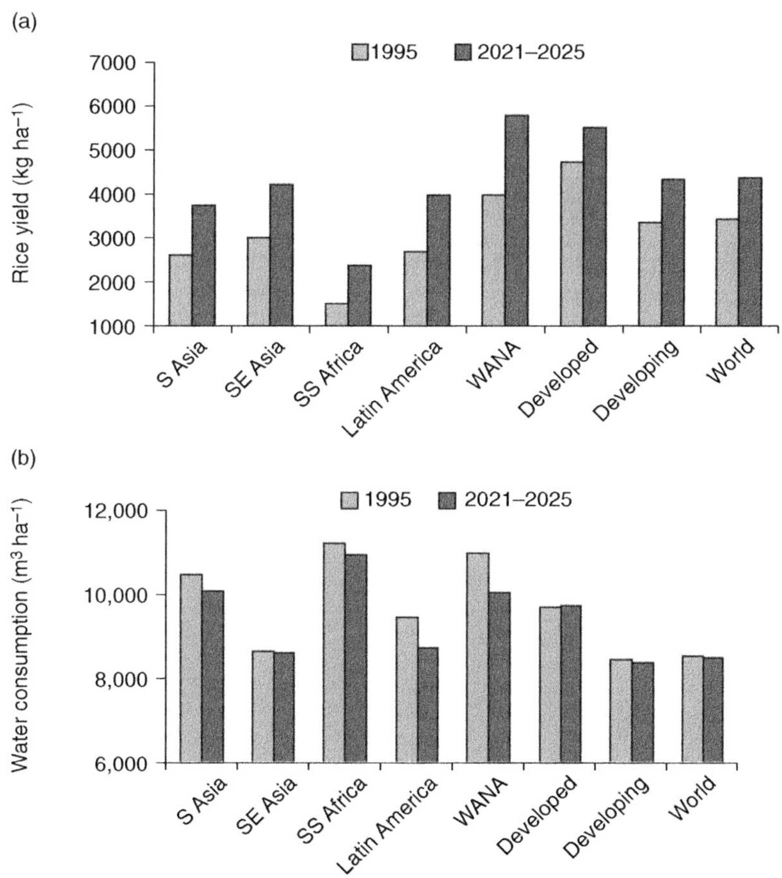

Fig. 10.9. Crop yield (a) and water consumption per hectare (b) of rice in several regions in 1995 and 2021–2025. SS, Sub-Saharan.

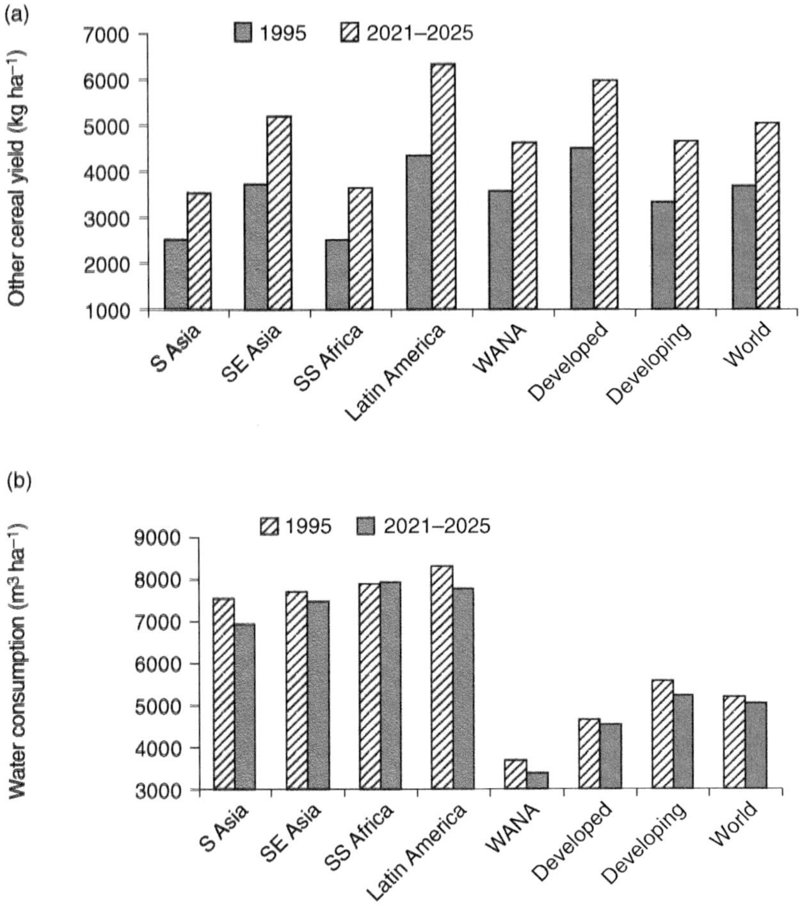

Fig. 10.10. Crop yield (a) and water consumption per hectare (b) of other cereals in several regions in 1995 and 2021–2025. SS, Sub-Saharan.

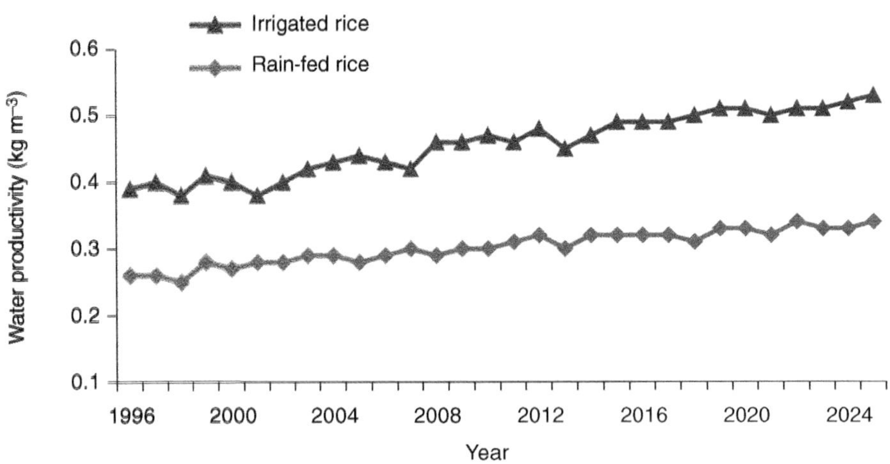

Fig. 10.11. Water productivity of irrigated and rain-fed rice in developing countries.

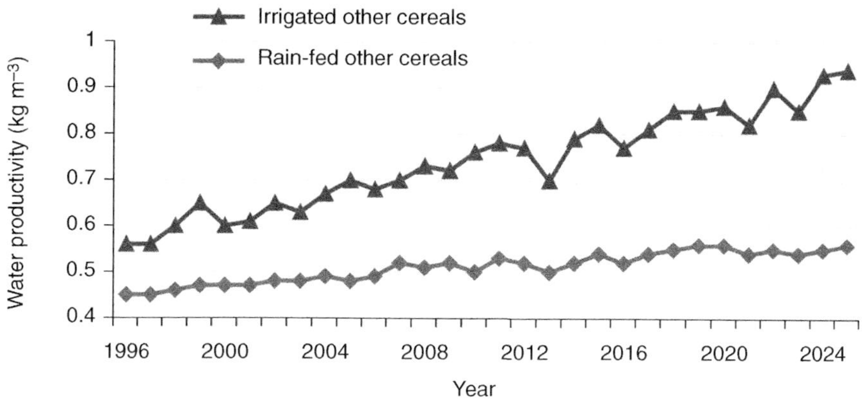

Fig. 10.12. Water productivity of irrigated and rain-fed other cereals in developing countries.

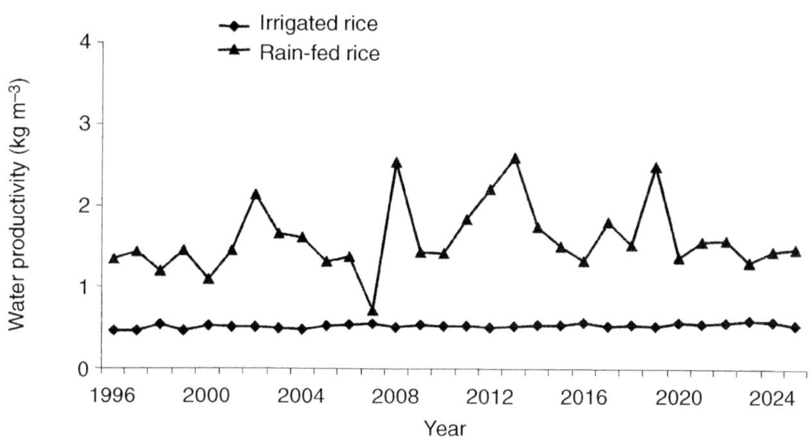

Fig. 10.13. Water productivity of irrigated and rain-fed rice in developed countries.

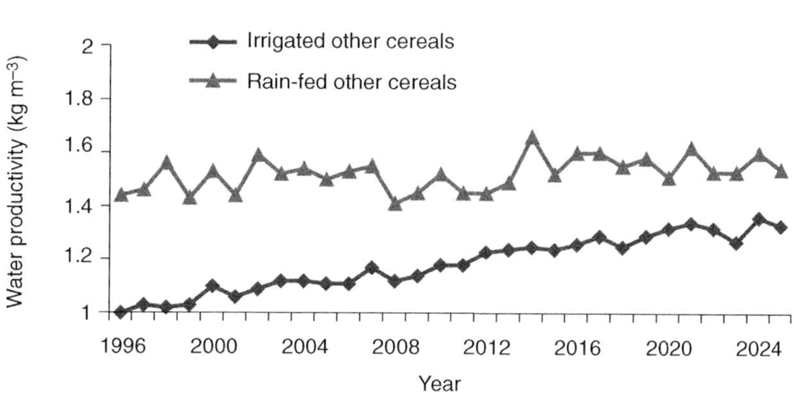

Fig. 10.14. Water productivity of irrigated and rain-fed other cereals in developed countries.

1996 to 2025 in developed countries. The curve of irrigated crops is below the curve of rain-fed crops. This indicates the relatively favourable rainfall conditions for crop growth and high rain-fed crop yields associated with infrastructure and other inputs to rain-fed crops in developed countries, compared with those in developing countries.

Alternative Scenarios

For irrigated crops, water-use efficiency is a key factor in WP. In one scenario, we assume higher basin efficiency (HBE) around the world in the next 25 years, and assess the impact of this on WP. An alternative scenario is defined, based on the increasing concern for environmental reservation of water in the world. This scenario tests the possibility of maintaining the baseline food outputs with larger improvement in effective agricultural water-use efficiency (the same as assumed in the first scenario), but with lower water withdrawal (HBE-LW) so that more water is left for environmental purposes. Assumptions under the two alternatives and the baseline are illustrated in Table 10.1, showing BE, water withdrawal and irrigation consumption under the three scenarios. The two alternative scenarios have HBE and lower water withdrawal and irrigation consumption in both developed and developing countries in 2021–2025. For example, compared with the baseline, HBE-LW results in 443 km^3 or 13% lower withdrawal in developing countries, and 94 km^3 or 7% lower withdrawal in developed countries; and 231 km^3 or 19% less irrigation consumption in developing countries, and 47 km^3 or 17% less irrigation consumption in developed countries.

Table 10.2 compares the WP of rice and other cereals under the three scenarios. Compared with the baseline, higher WPs result from the two alternative scenarios. The highest WP occurs under HBE-LW, which implies that, with HBE, restricting water withdrawals even more will lead to still higher WP. Water use per hectare decreases correspondingly under HBE, and HBE-LW. For example, compared with the baseline, water use per hectare of other cereals is reduced by 4% and 10% under HBE and HBE-LW, respectively, in developing countries, and 6% and 12% under HBE and HBE-LW, respectively, in developed countries. For developing countries, Figs 10.15 and 10.16 show water productivity of rice and other cereals, respectively, during 1996–2025 under the three scenarios. Figures 10.17 and 10.18 show water consumption per hectare of rice and other cereals, respectively, during 1996–2025 under the three scenarios. These curves show that the difference between the baseline and the alternative scenarios contin-

Table 10.1. Estimated and projected values of basin efficiency, water withdrawal and irrigation consumptive use.

		Projected, 2021–2025 average		
	Estimated 1995	Baseline	High basin efficiency	High basin efficiency and low withdrawal
Basin efficiency (%)				
Developing countries	0.54	0.59	0.77	0.77
Developed countries	0.64	0.69	0.81	0.81
Water withdrawal (km^3)				
Developing countries	2764	3486	3347	3043
Developed countries	1144	1277	1228	1183
Irrigation consumptive use (km^3)				
Developing countries	1162	1214	1135	983
Developed countries	268	274	250	227

Table 10.2. Estimated and projected values of water productivity and water use for rice and other cereals.

		Projected, 2021–2025 average		
	Estimated 1995	Baseline	High basin efficiency	High basin efficiency and low withdrawal
Rice				
Water productivity (kg m^{-3})				
Developing countries	0.39	0.53	0.56	0.58
Developed countries	0.47	0.57	0.61	0.63
Water use ha^{-1} (m^3 ha^{-1})				
Developing countries	8,580	8,445	8,040	7,510
Developed countries	10,200	9,730	9,100	8,710
Other cereals				
Water productivity (kg m^{-3})				
Developing countries	0.56	0.94	1.01	1.03
Developed countries	1.00	1.32	1.45	1.5
Water use ha^{-1} (m^3 ha^{-1})				
Developing countries	5,720	5,260	5,040	4,760
Developed countries	4,430	4,530	4,275	3,980

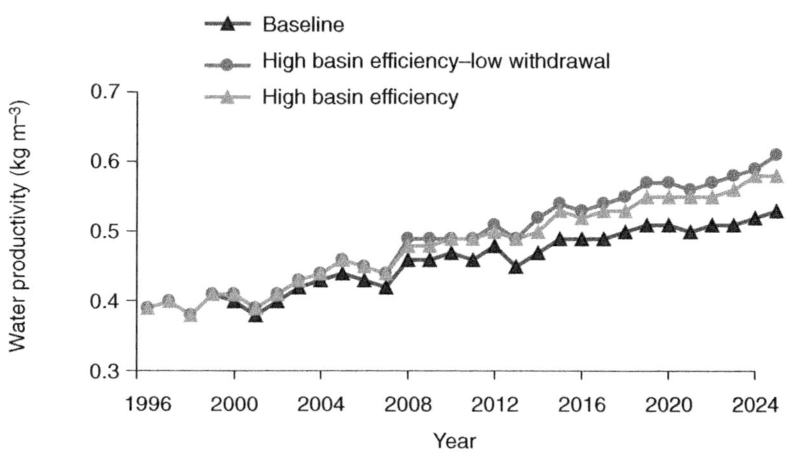

Fig. 10.15. Water productivity of rice under three scenarios.

ues to grow with time, corresponding to higher growth rate of BE and lower growth rate of water withdrawal.

HBE results in significantly higher crop yields and irrigation production during 2021–2025 than in the baseline scenario (except for rice yield in developed countries), which reduces the world price of rice by 15% for rice and 12% for other cereals, and reduces imports of cereals for developing countries from 235 to 213 million tons (Table 10.4). It should be noted that rice yield in developed countries under HBE declines slightly due to less economic incentives (crop prices), since lower prices tend to reduce crop yield. This effect for rice yield in developed countries is stronger than the impact of higher basin efficiency. As designed, HBE-LW compensates for the effect of lower water withdrawal by using larger improvements in BE, so that the baseline food production and demand balance will be maintained and the results come out as expected. Crop yield and production and crop prices under HBE-LW are close to those under

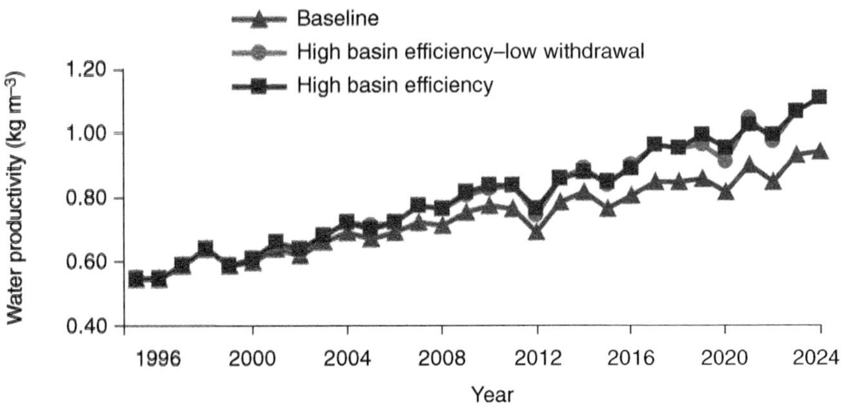

Fig. 10.16. Water productivity of other cereals under three scenarios.

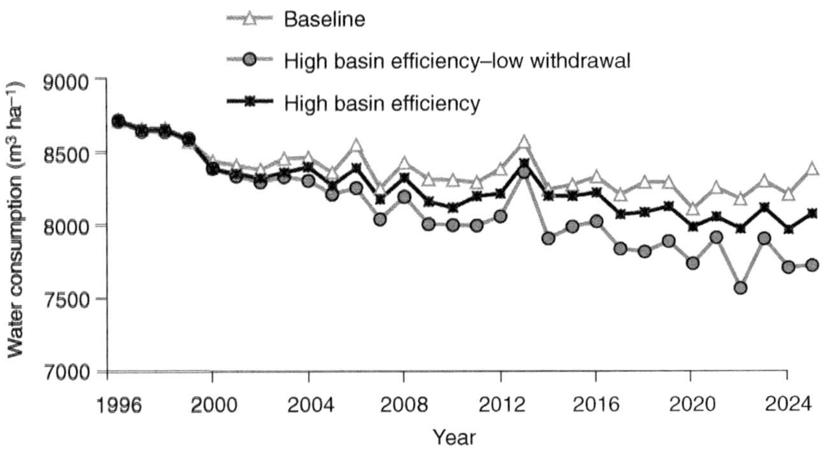

Fig. 10.17. Water consumed per hectare of irrigated rice under three scenarios.

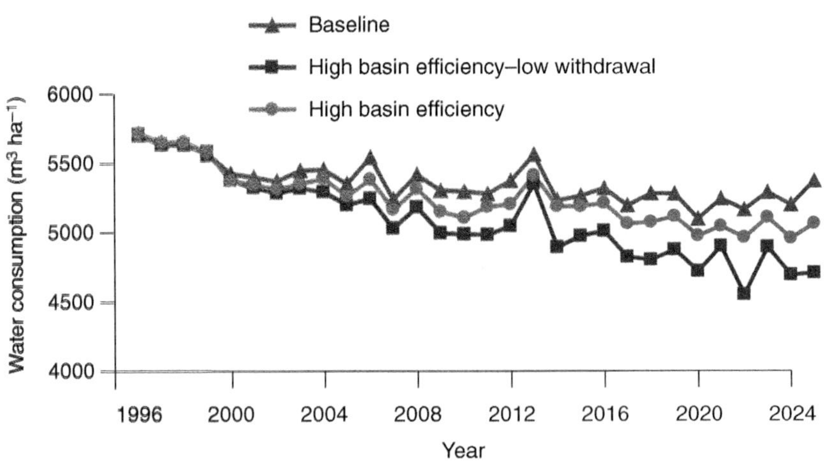

Fig. 10.18. Water consumed per hectare of irrigated other cereals under three scenarios.

Table 10.3. Estimated and projected values of rice and other cereal yields, and irrigated cereal production.

	Estimated 1995	Projected, 2021–2025 average		
		Baseline	High basin efficiency	High basin efficiency and low withdrawal
Rice				
Crop yield (kg ha^{-1})				
Developing countries	3310	4330	4530	4360
Developed countries	4790	5520	5505	5455
Other cereals				
Crop yield (kg ha^{-1})				
Developing countries	3185	4670	5165	4835
Developed countries	4410	6000	6180	5980
Irrigated cereal production (million t)				
Developing countries	557	867	938	880
Developed countries	186	269	274	267

Table 10.4. Estimated and projected world price of rice and other cereals, and developing-country cereal imports.

	Estimated 1995	Projected, 2021–2025 average		
		Baseline	High basin efficiency	High basin efficiency and low withdrawal
World price (US$ t^{-1})				
Rice	285	236	201	239
Other cereals	114	108	95	110
Cereal imports (million t) by developing countries	107	235	213	220

the baseline scenario. A small switch occurs between developed and developing countries: crop yields under HBE-LW are slightly lower than those under the baseline scenario in developed countries, while the opposite is true in the developing countries. This results in slightly lower cereal import to developing countries, as shown in Table 10.4. The reason behind the switch is a relatively more restrictive water-withdrawal condition for developed countries than for developing countries under the baseline scenario. Thus, the further restriction of water-withdrawal under HBE-LW results in a larger effect on irrigated crop production in developed countries.

Conclusions

WP is defined as crop yield per unit of WC (kg m^{-3}) and is computed through an integrated water and food modelling framework, IMPACT-WATER, for individual crops in each spatial unit (individual or aggregated basins) in the global scope during a period of 30 years (1995–2025). It was found that WP of rice ranged from 0.15 to 0.60 kg m^{-3}, while that of other cereals ranged from 0.2 to 2.4 kg m^{-3} in 1995. WP is relatively low in sub-Saharan Africa and high in developed countries. China and South-East Asian countries have higher WP for rice than other countries, mainly because of higher crop yields. WP will increase from 1995 to 2025: the global average WP of rice will increase from 0.39 kg m^{-3} to 0.52 kg m^{-3}, and the global average WP of other cereals will increase from 0.67 to 1.01 kg m^{-3}. Both the increase in crop yield and reduction in WC through improvement in BE contribute to the increase in WP, but the major contribution comes from the increase in crop yield. Therefore, investments in agricultural infrastructure and agricultural research might have higher pay-offs than investments in new irrigation, in order to increase WP and ensure food security in the next 25 years (see also Fan *et al.*, 1999). This conclusion is based on our assumption that water supply is becoming more and more restricted due to source availability and environmental and financial constraints. However, as shown by the HBE alternative, large improvements in BE would significantly increase WP and reduce water-withdrawal constraints (alternative scenario HBE-LW). The technical and financial feasibility for greatly improving BE needs more research (Cai *et al.*, 2001).

We also find that WP of irrigated crops is higher than that of rain-fed crops in developing countries, but lower in developed countries. This shows that, in developing countries, irrigated agriculture is more efficient in resource utilization and food production than rain-fed agriculture; but this also points to the untapped potential to increase WP of rain-fed crops through research and infrastructural investment (Rosegrant *et al.*, 2001b).

References

Alcamo, J.P. (2000) *Personal Communications on Global Change and Global Scenarios of Water Use and Availability: an Application of Water GAP.* Centre for Environmental System Research (CESR) University of Kassel, Kassel, Germany.

Cai, X., Ringler, C. and Rosegrant, M. (2001) *Does Efficient Water Management Matter? Physical and Economic Efficiency of Water Use in the River Basin.* EPTD Discussion Paper 72, International Food Policy Research Institute, Washington, DC.

Fan, S., Hazell, P.B.R. and Thorat, S. (1999) *Linkages between Government Spending, Growth, and Poverty in Rural India.* International Food Policy Research Institute, Washington, DC.

Keller, A.A., Keller, J. and Seckler, D. (1996) *Integrated Water Resources Systems: Theory and Policy Implication.* Research Report No. 3, International Irrigation Management Institute, Colombo, Sri Lanka.

Molden, D.J. (1997) *Accounting for Water Use and Productivity.* SWIM Paper 1, International Irrigation Management Institute, Colombo, Sri Lanka.

Molden, D.J., Sakthivadivel, R. and Habib, Z. (2001) *Basin-level Use and Productivity of Water: Examples from South Asia.* Research Report 49, International Water Management Institute, Colombo, Sri Lanka.

Rosegrant, W.M. and Cai, X. (2001) *Modeling Water Availability and Food Security – a Global Perspective: the IMPACT-Water Model.* Working Paper, International Food Policy Research Institute, Washington, DC.

Rosegrant, M., Cai, X., Cline, S. and Nakagawa, N. (2001) The role of rain-fed agriculture in the future of global food production. Invited background research paper for International Freshwater Conference, Bonn, Germany.

USDA (United States Department of Agriculture, Soil Conservation Service (SCS) (1967) *Irrigation Water Requirement.* Technical Release No. 21, Engineering Division, SCS, Washington, DC.

11 Improving Water Productivity in the Dry Areas of West Asia and North Africa

Theib Y. Oweis[1] and Ahmed Y. Hachum[2]
[1]*International Center for Agricultural Research in the Dry Areas, Aleppo, Syria;* [2]*Water Management and Irrigation Engineering, University of Mosul, Iraq*

Abstract

In the dry areas, water, not land, is the limiting factor in improving agricultural production. Maximizing water productivity, and not yield per unit of land, is therefore a better strategy for on-farm water management under such conditions.

This chapter highlights the major research findings at the International Center for Agricultural Research in the Dry Areas (ICARDA) regarding improving the water productivity of its mandate crops of wheat, barley, lentils, chickpea and faba bean. It is shown that substantial and sustainable improvements in water productivity can only be achieved through integrated farm-resources management. On-farm water-productive techniques, if coupled with improved irrigation-management options, better crop selection and appropriate cultural practices, improved genetic make-up and timely socio-economic interventions, will help to achieve this objective. Conventional water-management guidelines, designed to maximize yield per unit area, need to be revised for achieving maximum water productivity instead. A case study from Syria shows the applicability of this option. It illustrates that, when water is scarce, higher farm incomes may be obtained by maximizing water productivity than by maximizing land productivity.

Introduction

In the dry areas, agriculture accounts for about 80% of the total consumption of water. Water is rapidly becoming scarcer in west Asia and North Africa (WANA) and the competition for its use is growing more intense. In these areas, water is the factor that limits agricultural production. Most of the conventional sources of fresh water in the region have already been developed and the tendency to overexploit the natural resources is growing. Therefore, the only option left, in addition to developing some non-conventional sources, is to feed the ever-increasing population of the region using the same amount (or less) of water. Hence, the efficiency of water use in agriculture needs to increase in a sustainable manner, i.e. food production (quantitatively and qualitatively) per unit of water used has to be raised.

The International Center for Agricultural Research in the Dry Areas (ICARDA) aims to contribute to poverty alleviation in the dry areas by productivity improvement through sustainable natural-resources management. The ultimate goal is to improve the welfare of people in the dry areas of the developing world by increasing productivity and nutrition, while preserving and enhancing the

 Water Productivity in Agriculture: Limits and Opportunities for Improvement (eds J.W. Kijne, R. Barker and D. Molden)

natural-resources base of water, land and biodiversity. The challenge is to coordinate water and land management with the use of improved cultivars in viable cropping systems to increase water productivity (WP) and economic output. If agricultural production and livelihoods in the dry areas are to be sustained, even at current levels, greater priority must be given to improving WP and enhancing the efficiency of water use.

Options potentially available for coping with the consequences of water scarcity in agriculture in the dry areas include the development of additional sources of water and improving the management of all water uses.

Development of additional sources of water

Desalination

Desalination is gaining more importance as advances in the technologies are made. However, it is still an expensive process and hence is currently mainly used in areas where an affordable energy source is available, as in the oil-producing countries. Part of the desalinized water is used for irrigation. Breakthroughs in the cost of desalination would open up great opportunities for several countries of the region.

Marginal-quality water

Marginal-quality water offers good opportunities in many water-scarce areas. Potential sources include natural brackish water, agricultural drainage water and treated effluent. Research shows that substantial amounts of brackish water exist in dry areas, which can be either utilized directly in agriculture or desalinated at low cost for human and industrial consumption. Treated sewage effluent is an important source of water for agriculture in areas of extreme scarcity, such as Jordan and Tunisia. It is, however, a great environmental issue in other countries. The proper reuse of drainage water in agricultural production is also becoming attractive in many countries. By treating the marginal water as a resource rather than as a waste, it is possible to help the growers as well as to contribute to the alleviation of water scarcity and the sustainability of agricultural production systems.

Water transfers

Water transfers between water basins and across national borders have been extensively discussed in the region over the last two decades. Importation of water is under active consideration in the Middle East. The two most relevant options are to transport water by pipeline (Turkey's proposed peace pipeline) and by ship or barrage (big tanks or 'Medusa' bags). Both suggestions are subject to economic, political and environmental considerations, which are yet to be examined. Attempts to transfer water by balloons and tankers have been made but the cost is still high for agricultural purposes.

A project to transfer water by pipelines from Turkey to the Middle East countries was unsuccessful for economic and political reasons. Potential for such projects can only be realized with good regional cooperation and trust between the various parties.

Effective water management

Improved farm water management could have the greatest impact on water availability in dry areas. It is, however, a complex matter and also involves social, economic, organizational and policy issues, in addition to the technical ones. Research has demonstrated that proper management can more than double the return from water (Oweis, 1997). The major areas contributing to improved water management are discussed here.

Cost recovery of water

Although water is extremely valuable and essential in this region, it is generally supplied free or at low and highly subsidized cost. It is widely accepted that water pricing would improve efficiency and ensure better investment levels in water projects. However, the concept is seriously challenged in many countries of the region. The reasons are mostly sociopolitical and one cannot ignore these concerns as they are real and culturally determined. Innovative solutions are needed to put a real value on water for improving the efficiency of its use, while at the same time finding ways from within the local culture to protect the right of people to access water for their basic needs (El Beltagy, 2000).

Existing improved technologies

If properly applied, these technologies may at least double the amount of food produced from present water resources. Implementing precision irrigation, such as trickle and sprinkler systems, laser levelling and other techniques contribute to substantial improvement in water application and distribution efficiency. Although water lost during conveyance and on-farm application may not be a real loss from the basin perspective, its quality is likely to deteriorate and its recovery comes at a cost. To recapture and reuse water lost in this way is easier in large irrigation systems, but their construction comes at a high cost. There is a need to provide farmers with economic alternatives to current practices that are leading to wastage of water, and with incentives that can bring about the needed change.

Improved water productivity

Supplemental irrigation with a limited amount of water, if applied to rain-fed crops during critical stages, can result in substantial improvement in yield and WP. Application of water to satisfy less than the full water requirement of crops was found to increase WP and spare water for irrigating new lands. It has now been widely recognized that optimum WP is achieved by under-irrigating the crop. Adoption of this strategy requires an immediate adjustment to the conventional guidelines on irrigation in the region.

Optimizing agronomic practices and inputs, such as appropriate cropping patterns and fertilization, can also increase WP.

Using both Mendelian breeding techniques and modern genetic engineering, new crop varieties that can increase water-use efficiency while maintaining or even increasing yield levels can be developed. However, more work is needed to integrate all the above-mentioned approaches in practical packages to achieve the largest return from the limited water available.

Participation of all concerned in the management of scarce water resources is the key to successfully implementing more effective measures of water management. Players include the public and private sectors but, most importantly, representatives of the water users, particularly farmers and pastoralists, should be involved in the decision-making on water-management issues. Without appropriate policies, users cannot achieve the objectives of effective water management, but the inadequacy of current policies is the main constraint on improved water use in the region.

Water Scarcity and Mismanagement in the Dry Areas

The extent of the scarcity problem

The dry areas of WANA are characterized by low rainfall with limited renewable water resources. The average annual per capita renewable supplies of water in WANA countries are now below 1500 m^3, well below the world average of about 7000 m^3. This level has fallen from 3500 m^3 in 1960 and is expected to fall to less than 700 m^3 by the year 2025. In 1990, only eight of the 23 WANA countries had per capita water availability of more than 1000 m^3, the threshold for the water-poverty level. In fact, the 1000 m^3 level looks ample for countries like Jordan, where the annual per capita share has dropped to less than 200 m^3 (Margat and Vallae, 1999). Mining groundwater is now a common practice in the region, risking both water reserves and quality. In many countries, securing basic human water needs for domestic use is becoming an issue of concern, let alone the needs for agriculture, industry and the environment.

The current water supplies will not be sufficient for economic growth in all the countries of the region, except Turkey and Iran. Water scarcity has already hampered development in all countries of the Arabian Peninsula, Jordan, Palestine, Egypt, Tunisia and Morocco. Other countries of the region, such as Syria, Iraq, Algeria and Lebanon, are also increasingly affected as scarcity continues to get worse.

It is estimated that nearly one billion people live in the dry areas. About half of the workforce earns its living from agriculture, and water scarcity adds to their misery. At present, the average income of an estimated 690 million people is less than US$2.00 day^{-1}; the average income of 142 million of the 690 million is less than US$1.00 day^{-1} (Rodriguez and Thomas, 1999). Rural women and children suffer the most from poverty and its social and physical deprivations, which include malnutrition and high rates of infant mortality.

For most countries of WANA, almost half the crops of this dry region are grown under irrigation, and agriculture accounts for over 75% of the total consumption of water. With rapid industrialization, urbanization and population growth (double the world average), economic realities seem certain to reallocate water increasingly away from agriculture to other sectors. Moreover, opportunities for large captures of new water are few, if any. Most river systems suitable for large-scale irrigation have already been developed. It is becoming increasingly difficult to avoid unacceptable depletion of the flow to downstream users. Likewise, few major resources of renewable groundwater remain untapped. The tendency is now to overexploit existing sources, which, of course, is unsustainable.

The scarcity of water in some countries of the WANA region has reached the point where freshwater supplies are sufficient only for domestic and industrial use, which have priority. Very soon, the only water available for agriculture in these countries will be either saline or sewage effluent. This situation prevails already in the Gulf countries and will reach other countries, such as Jordan, in the next decade. Nevertheless, despite its scarcity, water continues to be misused. New technologies have made it possible for farmers to deplete aquifers to exhaustion. Desertification or land degradation is another challenge in the WANA region, closely related to water. Several international conferences and conventions, most recently the Convention to Combat Desertification, have brought these issues to the forefront of global concerns. Climatic variation and change, mainly as a result of human activities, are leading to depletion of the vegetative cover, loss of biophysical and economic productivity through exposure of the soil surface to wind erosion and shifting sands, water erosion, salinization of land and waterlogging. Although these are global problems, they are especially severe in the dry areas of WANA.

Compounding these problems is the expanding population. Population growth rates in WANA range up to 3.6%. The total population in WANA is expected to more than double, approaching 930 million by 2020. This will affect food supplies: the grain gap is projected to increase from 51 million t in 1995 to 109 million t by 2020 in the 23 countries of the region (Nordblom and Shomo, 1995). This is a conservative estimate that assumes no growth in per capita consumption. Assuming grain would continue to be priced at around US$130 t^{-1}, importing 109 million t of grain would cost US$14.2 billion!

It is projected that the vast majority of the 23 WANA countries will reach severe water poverty by the year 2025; ten of them are already below that level (Seckler *et al.*, 1999). The increasing pressure on water resources will, unless seriously tackled, escalate conflicts and seriously damage an already fragile environment. This is particularly relevant in respect of countries with shared water resources. In WANA about one-third of the renewable water supplies are provided by rivers flowing from outside the region (Ahmad, 1996). Under the prevailing conditions, regionally integrated water-resources management is obviously the best way to manage the shared water at the basin level. However, considering the importance attached to national sovereignty and the fact that international laws on shared water resources are still inadequate, conflicts between several countries of the region will continue to occur.

The concept of water productivity

Seckler *et al.* (Chapter 3, this volume) have discussed the various concepts of water-use efficiency and/or WP as used in the literature on irrigated agriculture and, therefore, it is unnecessary to repeat them here. WP is shown in Fig. 11.1 in its dependence on crop yield (Y), transpiration (T), evaporation (E), evapotranspiration (ET) and irrigation water (I). From the diagram, transpiration WP (WP_T) and evapotranspiration WP (WP_{ET}) are defined as:

$$WP_T = Y/T$$
$$WP_{ET} = Y/(T + E)$$

It is evident that $WP_{ET} < WP_T$. However, if E = 0, then $WP_{ET} = WP_T$. In this chapter, reference to WP is to WP_{ET}. Furthermore, the irrigation water productivity, WP_I, is defined as follows:

$$WP_I = \Delta Y / \Delta ET$$

in which ΔY and ΔET are the increase in yield and evapotranspiration due to irrigation, respectively.

Integrated approach to on-farm water management

Newly developed water-resources management strategies have become more integrated in the sense of considering all aspects of water scarcity simultaneously. Current policies of water-resources management look at the whole set of technical, institutional, managerial, legal and operational activities required to plan, develop, operate and manage the water-resources system at all scales, i.e. farm, project, basin and national scale, while considering all sectors of the economy that depend on water. Sustainability is a major objective of these policies, wherein it is stipulated that the utilization of resources by future generations should in no way be limited by the use of current generations.

Fundamental to the successful integration of water-resources development and management is the involvement of all stakeholders to the greatest possible extent in the various management activities. Decisions regarding the best use of water must be made by evaluating the economic, social and environmental costs and benefits of alternatives. Integrated water-resources development also means looking at the impacts of policies on the social, economic and environmental aspects of the system.

Economic constraints are particularly important in developing sustainable water-management options. A sustainable-development path can only be secured if development policies consider economic

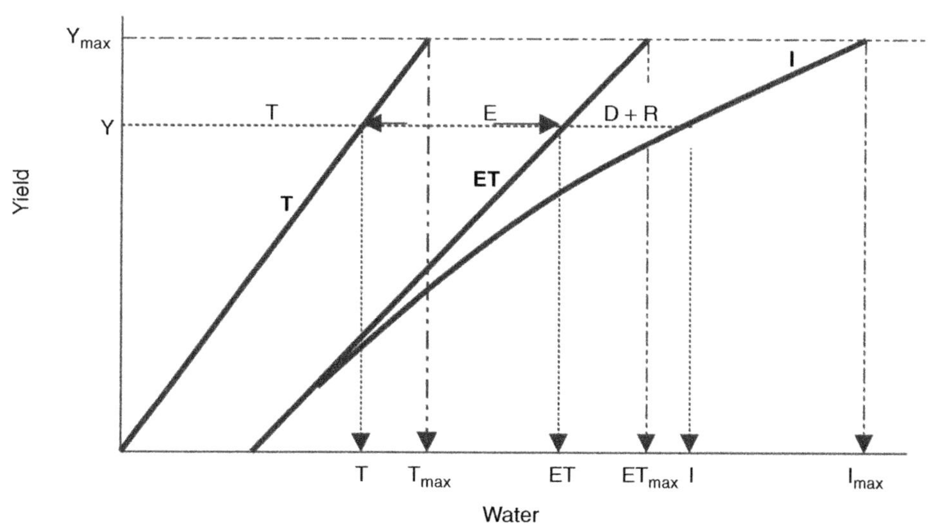

Fig. 11.1. A diagram for water-productivity terms and components: transpiration (T), evaporation (E), irrigation (I), evapotranspiration (ET), drainage (D) (deep percolation) and runoff (R).

aspects, such as costs and benefits to the society and individuals. This means that sustainable development and use of water resources should be compatible with the principles of sustainable economic activities. That is why many past irrigation schemes have failed or were much less successful than expected at the planning stage.

The strategy of any integrated approach for natural-resources management in the dry areas considers water as the central issue and water is accorded the highest priority. Utilization of soil and vegetation is closely linked to water and subject to climatic conditions. This strategy responds to the urgent need for improved productivity using less water by doing research on improved and sustainable WP at the farm level. Research in central and west Asia and North Africa (CWANA), which is the mandate dry region of ICARDA, focuses on the following five areas:

1. In rain-fed areas, optimizing supplemental irrigation, using the limited water available from renewable resources.
2. In drier environments (steppe), promoting efficient water harvesting for improved farmer income and environment.
3. In fully irrigated areas, developing on-farm packages for increased WP and soil and water quality.
4. In all the environments, developing strategies, methods and techniques for the safe and sustainable use of marginal water and treated sewage effluent in agriculture.
5. Developing strategies for the conservation and sustainable utilization of renewable groundwater resources.

Improving crop WP requires exploiting the genetic diversity of landraces and wild relatives for developing improved germplasm suited for stress-inducing environments. Germplasm improvement includes any of the following: increasing crop yield, disease resistance, heat and drought tolerance and, most importantly, the efficiency with which the crop uses water. The following two main strategies are pursued:

1. Selection for increased performance and WP by improving cultivars while maintaining current management conditions and water availability. This is done through improving crop cultivars adapted to marginal conditions through selection for performance, mainly yield and WP, directly in the target environment (Ceccarelli *et al.*, 1998). This is increasingly done through participatory plant breeding, involving farmers, to maximize the selection response. This strategy also focuses on the identification of morphological, physiological and agronomic criteria or traits that are related to increased performance under dry conditions. Such traits may then form the basis for indirect selection for yield and water productivity in dry environments. A new method involves employing marker-assisted selection for quantitative trait loci to identify breeding material with improved performance under dry conditions and higher water productivity.
2. Changing both management practices and crop cultivars concurrently. Different approaches in plant breeding and the numerous aspects of crop management are combined and integrated to develop viable strategies and sets of recommendations for productive, efficient and sustainable production systems. This combination of improved management practices and the crop plants or varieties themselves yields the greatest improvement in crop WP and can result in a quantum jump in both crop productivity and WP (Duivenbooden *et al.*, 1999).

To ensure generalization and transferability of the research results among dry regions, the concept of 'integrated research sites' was implemented together with work on agro-ecological characterization and modelling. A number of carefully selected integrated research sites was identified. Scientists from all disciplines work together on the most important issues in dry-area agriculture – that is, the need for more efficient, sustainable and water-efficient production systems. The strategies and technology packages developed and tested in these integrated research sites are then transferred or extended to other or larger areas, using bio-economic modelling to adapt them to the specific sites and situations with their specific biophysical and socio-economic conditions.

Major Research Achievements

Water-use-efficient techniques

In dry areas, moisture availability to the growing crops is the most significant single factor limiting production. Accordingly, this production factor must receive high priority. Technologies for improving yield, stabilizing production and providing conditions suitable for using higher technology are important, not only for improved yields but also for better WP.

Supplemental irrigation for rain-fed farming

The rain-fed areas play an important role in the production of food in many countries of the region and the world. They cover more than 80% of the land area used for cropping throughout the world and produce some 60% of the total production (Harris, 1991). In the Mediterranean-type climate, rainfall is characterized by its variability in both space and time. In general, rainfall amounts in this zone are lower than seasonal crop water requirements; moreover, its distribution is rarely in a pattern that satisfies the crop needs for water. Periods of severe moisture stress are very common and, in most of the locations, these coincide with the growth stages that are most sensitive to moisture stress. Soil-moisture shortages at some stages result in very low yields. Average wheat-grain yields in WANA range between 0.6 and 1.5 t ha^{-1}, depending on the amount and distribution of seasonal precipitation.

It was found, however, that yields and WP are greatly enhanced by the conjunctive use of rainfall and limited irrigation water. Research results from ICARDA and others, as well as harvests from farmers, showed substantial increases in crop yield in response to the application of relatively small amounts of supplemental irrigation (SI). This increase occurs in areas having low as well as high annual rainfall. Table 11.1 shows substantial increases in wheat-grain yields under low, average and high rainfall in northern Syria with application of limited amounts of SI. Applying 212, 150 and 75 mm of additional water to rain-fed crops increased yields by 350, 140 and 30% over that of rain-fed crops receiving an annual rainfall of 234, 316 and 504 mm, respectively. In addition to yield increases, SI also stabilized wheat production from one year to the next. The coefficient of variation was reduced from 100% to 20% in rain-fed fields that received SI.

The impact of SI is not only on yield but also, more importantly, on WP. The productivity of both irrigation water and rainwater is improved when they are used conjunctively. The average rainwater productivity of wheat grains in the dry areas is about 0.35 kg m^{-3}. However, it may increase to as much as 1.0 kg m^{-3} with improved management and favourable rainfall distribution. It was found that 1 m^3 of water applied as SI at the proper time might produce more than 2.0 kg of wheat grain over that using only rainfall. Data from a 5-year experiment (1991/92–1995/96) at ICARDA's research station in northern Syria (Table 11.2) show

Table 11.1. Yield and water productivity for wheat grains under rain-fed and supplemental irrigation (SI) in dry, average and wet seasons at Tel Hadya, northern Syria (from Oweis, 1997).

Season/annual rainfall (mm)	Rain-fed yield (t ha^{-1})	Rainfall WP (kg m^{-3})	Irrigation amount (mm)	Total yield (t ha^{-1})	Yield increase due to SI (t ha^{-1})	WP_I[a] (kg m^{-3})
Dry (234)	0.74	0.32	212	3.38	3.10	1.46
Average (316)	2.30	0.73	150	5.60	3.30	2.20
Wet (504)	5.00	0.99	75	6.44	1.44	1.92

[a]No surface runoff and drainage occur in the field.
WP_I, irrigation water productivity.

Table 11.2. Rainwater productivity (WP_R), combined rain- and irrigation-water productivity (WP_{R+I}) and irrigation-water productivity (WP_I) of bread-wheat grains in northern Syria.

Year	Rain (mm)	WP_R (kg m^{-3})	SI (mm)	WP_{R+I} (kg m^{-3})	WP_I (kg m^{-3})
1991/92	351	1.04	165	1.16	1.46
1992/93	287	0.70	203	1.23	2.12
1993/94	358	1.08	175	1.17	1.43
1994/95	318	1.09	238	1.08	1.06
1995/96	395	0.91	100	0.90	0.73
Mean water productivity		0.96		1.11	1.36

such an improvement in WP. The amount of water added by SI is not sufficient on its own to support any crop production. However, when supplementing rainfall by irrigation, the rainwater productivity (WP_R) increased in most of the years, particularly in the driest year (1992/93). On average, it increased from 0.96 to 1.11 kg m^{-3}. The last column in the table presents marginal irrigation WP (ratio of increase in yield to increase in evapotranspiration due to irrigation) with an average value of 1.36 kg m^{-3}.

The high WP of SI water is mainly attributed to alleviating moisture stress during the most sensitive stages of crop growth. Moisture stress during wheat flowering and grain filling usually causes a collapse in the crop seed filling and reduces the yields substantially. When SI water is applied before the occurrence of stresses, the plant may produce its potential yield.

Furthermore, using irrigation water conjunctively with rainwater was found to produce more wheat per unit of water than if used alone in fully irrigated areas where rainfall is negligible. In fully irrigated areas, wheat yield under improved management is about 6.0 t ha, using about 800 m^3 ha^{-1} of irrigation water. Thus, WP will be about 0.75 kg m^{-3}, one-third of that achieved with SI. This difference should encourage allocation of limited water resources to the more efficient practice (Oweis, 1997).

Unlike in full (or conventional) irrigation, the time of SI application cannot be determined in advance. When possible, and for rational allocation of limited water supplies, SI should be scheduled at the moisture-sensitive stages of plant growth. For example, for rain-fed cereals in the WANA region, the three most sensitive growth stages are seedling, anthesis and grain filling. Scheduling of SI should coincide with these sensitive periods to make certain that root-zone moisture does not limit growth.

Rainwater harvesting for the drier environments

The drier environments of WANA, the so-called *badia* or steppe, cover most of this region. The steppe receives inadequate annual rainfall for economic dry-farming production. The timing of precipitation in these areas is highly erratic. Most of this limited rainfall comes in sporadic, intense and unpredictable storms, usually on crusted soils with low infiltration rates, resulting in surface runoff and uncontrolled rill and gully water flow. Thus, the land is deprived of its share of rainfall and the growing crops endure severe moisture-stress periods, which significantly reduce yield, if any is produced. Therefore, a large part of the rainfall evaporates directly from the soil surface. Even some of the rain that infiltrates the soil to a shallow depth evaporates again. The rain that runs off usually joins streams and, if not intercepted, flows into a depression and loses its good quality and evaporates. The overall result is that the vast majority of precipitation water is lost as evaporation to the atmosphere without benefits; in other words, rainwater productivity is extremely low. Intervention in these areas is needed.

Water harvesting is one option for making rainwater more available to the crops in dry areas. It increases the amount of water per unit cropped area, reduces the severity of droughts and increases the productivity of rainwater.

Throughout history, water harvesting has shown good potential for increasing the efficiency of rainwater by concentrating it on a smaller area and thus ensuring enough moisture in the root zone of the plants. Indigenous systems, such as *jessour* and *meskat* in Tunisia, *tabia* in Libya, cisterns in north Egypt, *hafaer* in Jordan, Syria and Sudan and many other techniques, are still in use (Oweis *et al.*, 2001). Unfavourable socio-economic conditions over the last decades have caused a decline in the use of these systems, but recently increased water scarcity in the dry areas is favouring the revival of these systems.

Small basin microcatchments in the Muaqqar area of Jordan (mean annual rainfall of 125 mm) have supported almond-trees now for over 15 years without irrigation, despite several years of drought in which annual rainfall dropped below 60 mm. In the same area, small farm reservoirs were able to collect water every year with sufficient amounts to justify profitable agricultural development (Oweis and Taimeh, 1996). In the Mehasseh steppe of Syria (120 mm annual rainfall), rain-fed shrubs have a less than 10% survival rate, while those grown under microcatchments had an over 90% survival rate (Table 11.3). Shrub survival rate can be improved from 10 to 90% with the introduction of water-harvesting interventions (semicircular bunds), even during 3 drought years after 1 relatively normal year. In north-west Egypt (130 mm annual rainfall), small water-harvesting basins with 200 m^2 catchment support olive trees, and harvesting rainwater from the roofs of greenhouses provided about 50% of the water required by the vegetables grown within them (Oweis *et al.*, 2001).

These experiences and many others show that the productivity of rain in the drier environments can be substantially increased when a proper water-harvesting technique is implemented. In large-scale areas, methodologies for using remotely sensed data and ground information in a geographic information system (GIS) framework are often used to identify suitable areas and appropriate methods for water harvesting (Oweis *et al.*, 1998b). It was estimated that 30–50% of the rain in these environments might be utilized if water harvesting is practised, thus improving current rainwater productivity several-fold.

Successfully and sustainably integrating water-harvesting techniques within the agricultural systems in the dry areas is not an easy task. Several limitations exist, including socio-economic, technical and policy-related ones. Unclear landownership and lack of capacity of the farmers to implement these techniques are among the most important constraints.

Efficient on-farm water management

Optimum scheduling of irrigation is by far the most important means for improving crop WP and the key questions in irrigation scheduling are when to irrigate and how much water to apply.

Table 11.3. Shrub survival rate (%) in the Mehasseh steppe (120 mm annual rainfall), Syria, under semicircular microcatchment water harvesting with different sizes.

Year	Rainfall (mm)	Without water harvesting	Diameter of the semicircle (m)		
			2	4	6
1997/98	174	20	96	98	97
1998/99	36	7	92	95	93
1999/00	42	2	92	93	89
Mean		9.7	93.3	95.3	93.0

Deficit irrigation

Deficit irrigation is an optimizing strategy under which crops are deliberately allowed to sustain some degree of water deficit and yield reduction (English *et al.*, 1990). The adoption of deficit irrigation implies appropriate knowledge of crop water use and responses to water deficits, including the identification of critical crop growth periods and of the economic impacts of yield-reduction strategies. Figure 11.2 shows typical results on wheat, obtained from field trials conducted in a Mediterranean climate in northern Syria. The results show significant improvement in SI WP at lower application rates than at full irrigation. The highest WP of applied water was obtained at rates between one-third and two-thirds of full SI requirements, in addition to rainfall. The application of nitrogen improved WP, but, with deficit SI, lower nitrogen levels were needed (Fig. 11.2). This shows that, under deficit-irrigation practice, other cultural practices may also need to be adjusted. Planting dates, for example, interact significantly with the level of irrigation applied. Optimum levels of irrigation to maximize WP need to consider all these factors (Oweis *et al.*, 1998a).

WP is a good indicator for identifying the best irrigation-scheduling strategies with deficit SI of cereals (Zhang and Oweis, 1999), in analysing the water-saving performance of irrigation systems and management practices (Ayars *et al.*, 1999) and to compare different irrigation systems, including deficit irrigation. Experience from Syria showed that applying only 50% of the SI requirement to rain-fed wheat reduces full SI yield by less than 15% only (Oweis *et al.* 2000).

Strategies for optimal deficit SI in rain-fed areas require knowledge of rainfall amounts and distribution, in addition to the sensitivity to moisture stress during the various crop growth stages. Zhang and Oweis (1998) developed and used a quadratic wheat-production function to determine the levels of irrigation water for maximum yield and net profit. They also determined the yields for several levels of under-irrigation that would not reduce the farmer's income below that which the farmer would earn with full irrigation and limited water resources. For sustainable utilization of limited water resources and higher WP, the analysis indicates that a sound strategy would involve maximizing profit.

Analysis of 4 years' data (1996–2000) of SI with winter-sown food legumes on ICARDA's experimental fields, northern Syria, has shown similar trends in water-management options. Table 11.4 shows that, for chickpea, the optimal water management is to under-irrigate the crop by supplying one-third of its full water requirements. For lentil, deficit irrigation with two-thirds of its full water requirement seems to be the best choice. It can be seen that lentil and faba bean are more responsive to irrigation than chickpea.

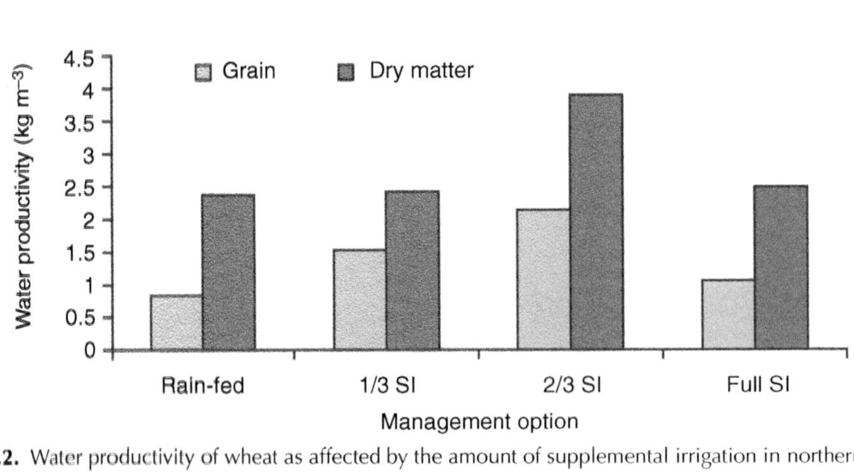

Fig. 11.2. Water productivity of wheat as affected by the amount of supplemental irrigation in northern Syria.

Table 11.4. Mean values of water productivity of four seasons (1996–2000) of lentil, faba bean and chickpea under rain-fed and supplemental irrigation at Tel Hadya, Aleppo, Syria.

Water-management option	Water productivity (production/water (kg m^{-3}))					
	Grain			Biomass		
	WP_R	WP_{R+I}	WP_I	WP_R	WP_{R+I}	WP_I
Lentil						
Rain-fed	0.57			2.14		
One-third SI		0.61	1.13		2.25	3.08
Two-thirds SI		0.69	1.34		2.44	3.56
Full SI		0.71	1.36		2.46	2.70
Chickpea						
Rain-fed	0.61			1.52		
One-third SI		0.68	1.23		1.77	2.70
Two-thirds SI		0.68	0.75		1.53	2.20
Full SI		0.55	0.34		1.48	1.60
Faba bean						
Rain-fed	0.51			1.40		
One-third SI		0.58	1.16		1.40	1.80
Two-thirds SI		0.62	1.53		1.53	2.40
Full SI		0.65	0.98		1.69	1.80

WP_R, rainwater productivity; WP_{R+I}, combined rain- and irrigation-water productivity; WP_I, irrigation-water productivity.

The decision on optimal strategies under varying conditions is a complex one, especially in rain-fed areas where year-to-year amount and distribution vary much. For example, it was found that sowing of rain-fed wheat spread out over the 3 months, November–January, substantially reduces the peak water demand during the critical SI period in the spring (Oweis and Hachum, 2001). This reduction is even greater when deficit irrigation is applied. The analysis was conducted using the simplified optimization model mentioned above. The results showed that a multisowing-date strategy reduced the peak farm water-demand rate by more than 20% thus potentially allowing a reduction in irrigation-system capacity and/or size. Also, the water demand of a larger area can be met with the same water supply. However, optimal sowing dates that minimize farm water demand do not always maximize total farm production and/or WP. The outcome depends on the crop water requirements and yield for each sowing date. Furthermore, selection of the optimal strategy is greatly influenced by the level of water scarcity.

The relationship between yield and water deficit has to be well known when planning deficit irrigation. The existing literature on this subject does not provide firm and ready-to-use information and, hence, there is a great need for research in this area. To determine when to irrigate and how much water to apply, suitable water-stress indicators should be used. These indicators may refer to the depletion of soil water, soil water potential and plant water potential or canopy temperature. For practical reasons, the most widely used indicators are soil water content and soil water potential. However, the spatial variability of the soil and irrigation depth gives rise to highly variable soil water content and/or potential data when these are obtained as point measurements.

There are different ways to manage deficit irrigation. The irrigator can reduce the irrigation depth, refilling only part of the root-zone soil-water capacity, or reduce the irrigation frequency by increasing the time interval between successive irrigations. In surface irrigation, wetting furrows alternately or placing them further apart is one way to implement deficit irrigation.

Cropping pattern and cultural practices

Among the management factors of the more productive farming systems are the use of suitable crop varieties, improved crop rotation, sowing dates, crop density, soil-fertility management, weed control, pests and diseases control, water-conservation measures, irrigation scheduling, water-quality monitoring and drainage. Integration of livestock into the farming system is important for nutrient cycling and fertilization of the soil. The challenge in WANA is to devise relevant practical solutions to the very low yield and WP in the region and implement them in the context of both local biophysical and socio-economic constraints.

The identification of appropriate crops and cultivars with optimum physiology, morphology and phenology to suit local environmental conditions is one of the important areas of research within cropping-system management for improved WP. Plant breeders aim at well-adapted cultivars with higher yield potential, tailored to the specific agroclimatic conditions. The breeding programme seeks improvement of crops so that they are tolerant to cold, drought and heat and resistant to diseases and insects and have vigorous early growth to reduce evaporation losses from the soil surface (Zhang *et al.*, 1998). A seasonal shifting, i.e. development of crop varieties that can be grown or sown in winter (instead of spring) under a lower evaporative demand, represents an additional challenge for breeders aiming at using scarce water more efficiently. However, traits such as winter-hardiness and disease resistance of the cultivars have to be improved. The development of crop varieties for early-growth vigour has been a major concern of ICARDA's winter-cereal breeders for many years.

In the winter-rainfall environment of the WANA region, despite temperature limitations on growth, it pays to sow early (late autumn) so that as much as possible of the crop's growth cycle is completed within the cool, rainy winter and early spring period. Delaying the sowing date prevents crop germination and the establishment of seedlings, because of the rapid drop in air temperature starting generally in November. In the lowlands of the Mediterranean region, where continuous cropping of pure cereal or cereal–legume rotations prevail, mid-November was found to be the optimum sowing time for cereals. Every week's delay after this time results in a yield decrease of 200–250 kg ha^{-1}. If the onset of seasonal rain is delayed, early sowing can be realized by SI.

Soil fertility is another critical factor in WP in WANA's agriculture. Water plays a significant role in fertilizer-use efficiency. Improved fertility improves WP and can therefore stabilize production and enable crops to exploit favourable rainfall in good years. Given the inherent low fertility of many dry-area soils, judicious use of fertilizer is particularly important. Under rain-fed conditions, the application rate of N fertilizer is not high. In northern Syria, 50 kg N ha^{-1} is sufficient under rain-fed conditions. However, with water applied by SI, the crop responds to nitrogen up to 100 kg N ha^{-1}, after which no benefit is obtained. This rate of N greatly improves WP. It is also important that there is adequate available phosphorus in the soil so that the response to N and applied irrigation is not constrained (Ryan, 2000). Cereal–fallow and continuous cereal cropping are the predominant crop rotations in WANA. The poor productivity and deterioration of the natural-resources base of such cropping systems are obvious. Including legumes (for human food and/or animal feed) in the rotation has proved to be beneficial for sustainable crop production. The major beneficial effect of legumes is generally attributed to their addition of fixed N to cropping systems. However, other effects, such as increased cereal yield, improved WP and soil conditions and interruption of disease and pest cycles, are also important.

Among the major soil factors affecting WP are depth, texture, structure and crusting, salinity and fertility. Tillage (form, depth, frequency and timing) and soil-surface management play important roles in enhancing WP, particularly in dry areas. Calcareous soils, formed from limestone residuum, predominate in the WANA region, with variable textures, depths and slopes. Organic-matter content is generally low.

Most documented research on WP is on single crops, in which the performance of each crop is studied separately. To obtain the optimum output of crop production per unit input of water, the monocrop WP should be extended to a multicrop WP in which more than one crop is sharing the use of the unit of input. WP of a multicrop system is usually expressed in economic terms, such as profit or revenue. While economic considerations are important, they are not adequate as indicators of sustainability, environmental degradation and natural-resources conservation. What may appear to be economically viable in the short run may be disastrous in the long run.

Good soil- and crop-management practices can considerably increase the efficiency with which water available from precipitation and irrigation can be used. Improved WP can be achieved if the crops are well established and adequately fertilized, weeds are controlled and appropriate crop rotations are used (Pala and Studer, 1999). These activities should also be considered together with the proper management of the soil if productivity is to be sustained and resources are to be conserved in the long term. As mentioned before, soils of the WANA region are predominantly calcareous, frequently deficient in phosphates, with variable depths and textures governing the maximum amount of water that can be stored and, hence, the effective length of the growing seasons. Maximizing the use of water available for crop growth is mainly done through increasing the water supply to crops, increasing their transpiration and decreasing evaporation from the soil surface (Gregory, 1991). The suggested technology packages vary with agroecological conditions and farmers' objectives.

Many dry-area soils are inherently low in fertility, as was pointed out before, and the correct application of fertilizers is therefore essential. Extensive work in Syria (Pala *et al.*, 1996) demonstrated the benefits of appropriate fertilization for WP and therefore for production and yield stability, especially of wheat and barley, in WANA. In deficient soils, seedbed phosphate (usually together with a small dose of nitrogen) enhances the rate of leaf expansion, tailoring, root growth and phenological development, ensuring faster ground cover and canopy closure, and earlier completion of the growth cycle before rising temperatures increase the atmospheric demand (Gregory, 1991).

An example of the interaction between fertilizer application and WP is presented in Fig. 11.3 (Oweis, 1997). The data show that an additional 50 kg N ha^{-1} may double the WP of SI. However, the optimum level of N is site-specific and dependent on the irrigation depth.

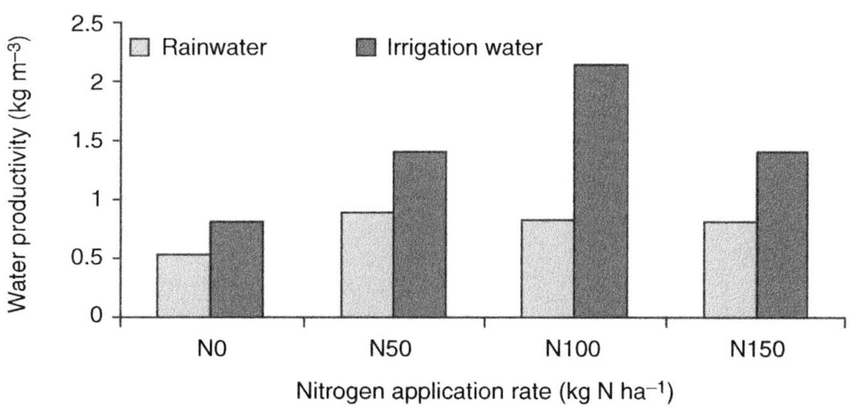

Fig. 11.3. Water productivity of wheat as affected by the rate of nitrogen application under rain-fed and supplemental irrigation in northern Syria (from Oweis, 1997).

More efficient crop varieties

Among ICARDA's research outputs are technologies designed to increase productivity while conserving and enhancing the natural-resources base. Germplasm-improvement programmes focused on increasing the productivity of barley, durum and bread wheat and food and forage legumes, along with integrated pest management in cereal- and legume-based cropping system. ICARDA-improved varieties cover about 90% of spring bread wheat in WANA, with a remarkable increase in yield for the benefit of resource-poor farm households. This programme operates in close partnership with national agricultural research systems (NARS) and other sister centres and advanced research institutions. The application of decentralized breeding and farmer participatory methods has increased the efficiency of variety development by enabling researchers to work directly with farmers in assessing varieties for specific adaptation.

Exploitation of the interaction of genotype and management

The identification of crops and cultivars with the optimum physiology, morphology and phenology for local environmental conditions and especially for the pattern of water availability is an important area of research. For example, the selection for improved response to irrigation has been conducted in lentil and chickpea (Hamdi *et al.*, 1992). Breeding and selection for improved WP and the use of genotypes best adapted to specific conditions can improve soil water use and increase WP (Studer and Erskine, 1999).

As was mentioned before, combining more appropriate cultivars with improved management practices results in major improvements in crop yield and WP. The following two case histories illustrate this simultaneous change in both genotype and management, with the first involving early sowing in the food legume chickpea and the second describing the use of SI in wheat production.

Early sowing of chickpea

In the Mediterranean region, rain falls predominantly in the cool winter months of November to March. Traditionally, chickpea is sown in late February and early March. From March onwards, the crop experiences increasingly strong radiation and a rapid rise in temperature, which cause an increase in the rate of leaf-area development, with consequent high evapotranspiration. This period of high evaporative demand occurs at the end of the rainfall period, when the residual soil moisture is inadequate to meet the evaporative demand. Therefore, the crop experiences drought stress during the late vegetative growth and reproductive growth and produces a low yield. Changing from the traditional spring sowing to winter sowing is possible but only with cultivars possessing cold tolerance and resistance to key fungal diseases (Singh *et al.*, 1997; Studer and Erskine, 1999).

The average gains in seed yield from early-sowing chickpea over three sites and ten seasons is 70% or 690 kg ha^{-1}, which translates into an increase in WP of 70% (Fig. 11.4; Erskine and Malhotra, 1997). In 30 on-farm trials comparing winter with spring chickpea in northern Syria, the mean benefit of winter sowing in seed yield and WP was 31% (Pala and Mazid, 1992). Currently, an estimated 150,000 ha of chickpea is winter-sown in the WANA region (Singh and Saxena, 1996).

Improved wheat cultivars under supplemental irrigation

The use of SI is another example of a concurrent change in both management practice and water-responsive cultivars to increase WP. This example demonstrates the need to combine changes in management with the use of adapted varieties in SI of wheat. SI requires varieties that are adapted to or suitable for varying amounts of water application. Appropriate varieties need first to manifest a strong response to limited water applications, which means that they should have a relatively high

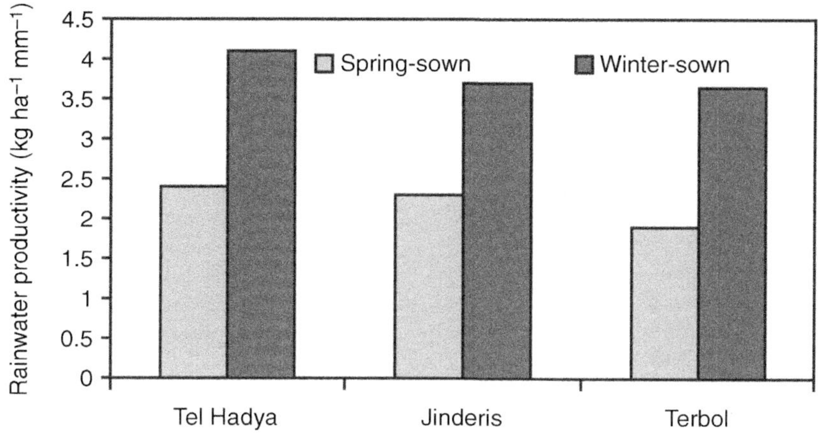

Fig. 11.4. Rainwater productivity of winter- and spring-sown chickpea grains in north Syria (from Erskine and Malhotra, 1997).

yield potential. At the same time, they should maintain some degree of drought resistance and hence express a good plasticity. In addition, the varieties should respond to the higher fertilization rates that are generally required under SI (Oweis, 1997; Oweis *et al.*, 1999) and should resist lodging, which can occur in traditional varieties under irrigation and fertilization. Figure 11.5 shows the variations in the response of two durum- and two bread-wheat varieties to various water-management options.

Water productivity versus land productivity

The case of wheat

In WANA, where water is more limiting than land, the objective of irrigated agriculture should be to maximize the return per unit of water and not per unit of land. This should yield higher overall production, since the saved water can be used to irrigate new land with higher production. Higher WP is linked with higher yields. This parallel increase in yields and WP, however, does not continue all the way. At some high level of yield (pro-

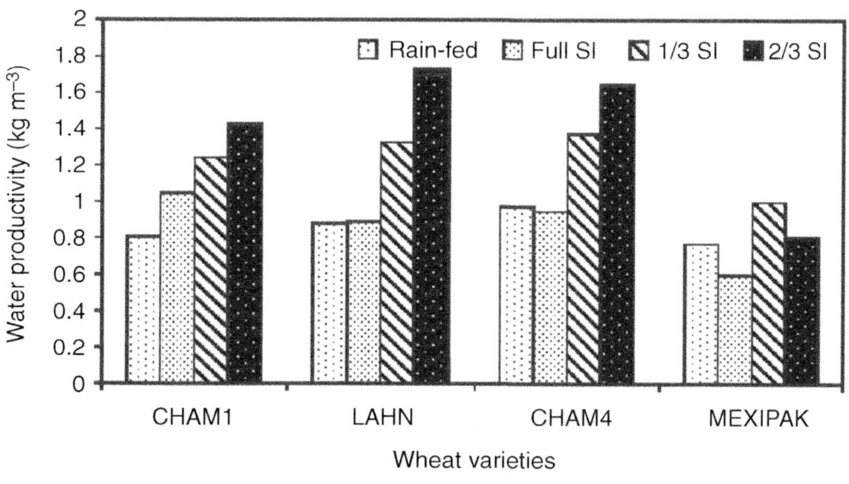

Fig. 11.5. Four-year averages of rain and supplemental irrigation water productivity for wheat varieties grown in northern Syria (from T. Oweis 2001, unpublished data). CHAM1 and LAHN are durum.

duction per unit of land), incremental yield increase requires higher amounts of water. This means that WP starts to decline as yield per unit land increases above certain levels. Figure 11.6 shows the relation between yield increase and WP increase for durum wheat under SI in Syria.

It is clear that the amount of water required to produce the same amount of wheat at yield levels beyond 5 t ha^{-1} is much higher than the water requirement at lower levels. It would be more economical to produce only 5 t ha^{-1} and then use the saved water to irrigate new land than to produce maximum yield with excessive amounts of water at low WP. This, of course, applies only when water, and not land, is limiting and without sufficient water to irrigate all the available land. When the curvilinear relationship of Fig. 11.6 applies, which is not always the case, maximum WP occurs at less than the maximum yield level per unit area.

The association of high WP values with high yields has important implications for crop management for achieving efficient use of water resources in water-scarce areas (Oweis *et al.*, 1998a). For example, attaining higher yields with increased WP is only economical when the increased gains in crop yield are not offset by increased costs of other inputs. The curvilinear WP–yield relationship makes clear the importance of attaining relatively high yields for efficient use of water. Policies for maximizing yield and/or net profit should be considered carefully before they are applied under water-scarce conditions. For example, guidelines for recommending irrigation schedules under normal water availability (Allen *et al.*, 1998) may need to be revised when applied in water-scarce areas.

The Syrian case-study

As earlier reported, research has shown that applying only 50% of full SI requirements (over that of rainfall) causes a yield reduction of only 10–15%. This finding, in light of the increasing water scarcity in Syria, encouraged ICARDA and the Extension Department of the Ministry of Agriculture to further test deficit SI strategies at farmers' fields. The hypothesis was that applying 50% of SI requirements to the whole field, while maximizing WP, will be more beneficial to the farmer than applying 100% of wheat irrigation requirements to half of the field, while

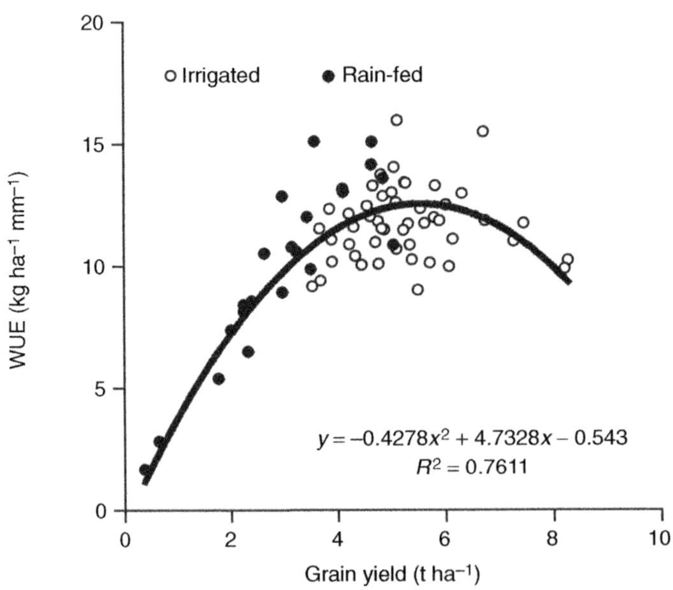

Fig. 11.6. Relationship between crop water productivity and crop grain yield for durum wheat under supplemental irrigation in Syria (from Zhang and Oweis, 1999). WUE, water-use efficiency.

leaving the other half rain-fed. The demonstrations were conducted on farmers' fields and managed collectively by the farmers, the researchers and the extension officials.

The farmer-managed demonstration plots were established over 6 years in the rain-fed areas with annual rainfall ranging from 250 to 450 mm. Rain-fed wheat yields in this area are generally low (less than 2 t ha^{-1}) and variable from one year to the next. Supplemental irrigation is practised in the area and has shown good potential to increase and stabilize production. However, it was observed that farmers tended to over-irrigate and the groundwater in the region had been continuously depleted.

Each farmer's land was divided into four 1 ha parts: the first was left rain-fed, the second was irrigated in the usual manner by the farmer, but water amounts were measured, the third part was irrigated such that no moisture stress occurred and the fourth part was irrigated with 50% of the full irrigation requirements. Water requirements were determined using evaporation from class A pans installed in the field, using appropriate pan and crop coefficients. Rain was also measured at the farm. Irrigation water was given from wells or public canals and measured by calibrating the flow rate and determining the time needed to apply the required amount. At the end of the season, the crop yields were measured and other data were collected. The farmers used improved wheat cultivars and recommended inputs and cultural practices at each site.

When there is not enough water to provide full irrigation for the whole farm, the farmer has two options: to irrigate part of the farm with full irrigation, leaving the other part rain-fed, or to apply deficit SI to the whole farm. Assuming that, under a limited water resource, only 50% of the full irrigation required by the farm would be available, the option of deficit irrigation was compared with other options. The results are summarized in Table 11.5. They show that, under the rainfall conditions prevailing in Syria during the years 1994–2000, a farmer with a 4 ha farm would, on average, produce 33% more grain from his/her farm if he/she adopted deficit irrigation than if full irrigation was applied. The advantage of applying deficit irrigation increased the benefit by over 50% compared with that of the farmer's usual practice of over-irrigation. Thus, the application of deficit SI, when water resources are limited, could potentially double the land area under irrigation. The results of this programme point to the possibility of producing more food with less water.

Present Needs and Future Directions

In rain-fed areas, water-conservation measures are of primary importance. As discussed above, they include such practices as fallow management, control of runoff and water harvesting. Integrated with these practices are the selection or development of

Table 11.5. Wheat-grain production scenarios for 4 ha farms with various strategies of supplemental irrigation in Syria.

Management strategy	Rain-fed (342 mm)	Farmer's practice	Full SI	Deficit SI
Total water applied (m^3)		2980	2220	1110
Grain yield (t ha^{-1})	1.8	4.18	4.46	4.15
Water productivity (kg m^{-3})	0.53	0.70	1.06	1.85
Possible 4 ha farm production (t) if water is not limiting	7.2	16.7	17.8	16.6
Possible 4-ha farm production (t) under limited water (50% of full irrigation requirement is assumed to be available)	7.2	10.8	12.5	16.6

high-yielding, drought-tolerant crop varieties, efficient use of fertilizers, combating pests and diseases, crop rotation and optimal planting dates to maximize the probability of rainfall use during critical periods of crop growth. The collective effects of such practices are complex when integrated with rain-fed farming systems and yet they are even more pronounced under irrigated agriculture.

Until recently, large irrigation projects have been given high priority, while small-scale water development for agriculture has received inadequate attention. It becomes evident now that small-scale irrigation, including SI for rain-fed agriculture, and a variety of water-harvesting techniques have considerable potential to meet agricultural and domestic water needs and to improve WP. Small-scale water-development programmes can fulfil many local water needs and have considerable global potential for the achievement of sustainable agricultural development. In a small-scale water-development scheme, individual farmers or communities develop and operate most project activities. However, technical assistance is often necessary during survey, design, construction and maintenance. Such undertakings can often contribute to both development and conservation, while enhancing local involvement in environmental management, promoting equity, improving the standard of living and thus helping to slow or prevent migration to urban areas.

Modern irrigation technologies in developed countries are very sophisticated and expensive. They are automated and computerized, equipped with such components as sensing devices, pressure regulators, filters and sensors. All this is helpful because it saves labour, which is usually expensive in industrial countries, but irrigation technologies do not need to be so complicated and expensive in developing countries. It is possible to simplify these technologies and adapt them to the needs of the resource-poor farmers of developing countries. Irrigation can be made a small-scale operation for poor farmers or communities, who have a need for the most efficient irrigation system to produce enough food for themselves and others.

The problem is how to use water more efficiently, while preventing environmental damage, in order to get a better return for the cost involved in making water available and in applying it. Applying too much water to the land causes a host of adverse effects, such as salinity build-up if drainage is poor, decline in crop yield due to aeration problems and loss of nutrients and energy and water wastage. We should remember that salinization was a major factor in the failure of past civilizations in many parts of the world.

Major Research Issues in Water-scarce Areas

There is no doubt that improving the productivity of water in dry areas will continue to be a priority. Efforts to direct new research and the transfer of available technologies to overcome water shortages are very much needed. Coordination of these efforts within an agreed-upon framework may enhance their impact. Elements of the research and technology framework would include:

1. The development of alternative land-use systems and cropping patterns for improved water use that are economically competitive and that respond to changing markets and demands in various agroecologies and socio-economic situations.
2. The development and transfer of alternative irrigation technologies with high water productivity and suitable for irrigation in these alternative land-use systems.
3. Developing new guidelines for irrigation scheduling under water-scarce conditions. Conventional guidelines are suitable only under normal water supply.
4. Developing methodologies for the assessment of water use at basin level of representative areas for evaluating the amounts of depleted and recoverable water and the economic returns.
5. Improving crop materials (germplasm) for higher WP in addition to the conventional target of high yield.
6. Evaluating the environmental consequences of conservative management of scarce water and ways to mitigate adverse effects.

7. Maintaining a balance between water allocation for food and for the environment under dry conditions.

8. Providing socio-economic incentives for improved water management at the farm level and development of appropriate policies.

References

Ahmad, M. (1996) Sustainable water policies in the Arab region. Paper presented at the Symposium 'Water and Arab Gulf development: problems and policies', University of Exeter, UK.

Allen, R.G., Pereira, L.S., Raes, D. and Smith, M. (1998) *Crop Evapotranspiration. Guidelines for Computing Crop Water Requirements*. FAO Irrigation Drainage Paper 56, FAO, Rome, 300 pp.

Ayars, J.E., Phene, C.J., Hutmacher, R.B., Davis, K.R., Schoneman, R.A., Vail, S.S. and Mead, R.M. (1999) Subsurface drip irrigation of row crops: a review of 15 years' research at the Water Management Research Laboratory. *Agricultural Water Management* 42, 1–27.

Ceccarelli, S. and Grando, S. (1991) Environment of selection and type of germplasm in barley breeding for low-yielding conditions. *Euphytica* 57, 207–219.

Ceccarelli S., Grando, S. and Impiglia, A. (1998) Choice of selection strategy in breeding barley for stress environments. *Euphytica* 103, 307–318.

Duivenbooden, N.V., Pala, M., Studer, C. and Bielders, C.L. (eds) (1999) *Efficient Soil Water Use: the Key to Sustainable Crop Production in the Dry Areas of West Asia and North and Sub-Saharan Africa. Proceedings of the Workshops Organized by the Optimizing Soil Water Use Consortium; Niamey, Niger, 26–30 April 1998 and Amman, Jordan, 9–13 May 1999*. International Center for Agricultural Research in the Dry Areas (ICARDA), Aleppo, Syria, 490 pp.

El Beltagy, A. (2000) Strategic options for alleviating conflicts over water in dry areas. A paper presented at the Crawford Fund Sixth Conference on Food, Water and Wars: Security in a World of Conflict, 15 August 2000, Canberra, Australia.

English, M., Musick, J.T. and Murty, V.V.N. (1990) Deficit irrigation. In: Hoffman, G.J., Howell, T.A. and Solomon, K.H. (eds) *Management of Farm Irrigation Systems*. ASAE, St Joseph, Michigan, pp. 631–663.

Erskine, W. and Malhotra, R.S. (1997) Progress in breeding, selecting and delivering production packages for winter sowing chickpea and lentil. In: *Problems and Prospects of Winter Sowing of Grain Legumes in Europe, AEP Workshop, December 1996, Dijon*. AEP, France, pp. 43–50.

Gregory, P.J. (1991) Concepts of water use efficiency. In: Harris, H.C., Cooper, P.J.M. and Pala, M. (eds) *Soil and Crop Management for Improved Water Use Efficiency in Rain-fed Areas. Proceedings of an International Workshop, Ankara, Turkey 1989*. ICARDA, Aleppo, Syria, pp. 539–547.

Hamdi, A., Erskine, W. and Gates, P. (1992) Adaptation of lentil seed yield to varying moisture supply. *Crop Science* 34, 987–991.

Harris, H.C. (1991) Implications of climatic variability. In: Harris, H.C., Cooper, P.J.M. and Pala, M. (eds) *Soil and Crop Management for Improved Water Use Efficiency in Rain-Fed Areas. Proceedings of an International Workshop 1989, Ankara, Turkey*. ICARDA, Aleppo, Syria, p. 352.

Margat, J. and Vallae, D. (1999) *Water Resources and Uses in the Mediterranean Countries: Figures and Facts*. Blue Plan, Regional Activity Center, UNEP, Aleppo, Syria.

Nordblom, T. and Shomo, F. (1995) *Food and Feed Prospects to 2020 in West Asia/ North Africa Region*. Social Studies Paper No. 2, NRMP, ICARDA, Aleppo, Syria.

Oweis, T. (1997) *Supplemental Irrigation: a Highly Efficient Water-use Practice*. ICARDA, Aleppo, Syria, 16 pp.

Oweis, T. and Hachum, A. (2001) Reducing peak supplemental irrigation demand by extending sowing dates. *Agricultural Water Management* 50, 109–123.

Oweis, T. and Taimeh, A. (1996) Evaluation of small basin water harvesting systems in the arid region of Jordan. *Water Resources Management* 10, 21–34.

Oweis, T., Pala, M. and Ryan, J. (1998a) Stabilizing rain-fed wheat yields with supplemental irrigation and nitrogen in a Mediterranean-type climate. *Agronomy Journal* 90, 672–681.

Oweis, T., Oberle, A. and Prinz, D. (1998b) Determination of potential sites and methods for water harvesting in central Syria. *Advances in GeoEcology* 31, 83–88.

Oweis, T., Hachum, A. and Kijne, J. (1999) *Water Harvesting and Supplemental Irrigation for Improved Water Use Efficiency in the Dry Areas*. SWIM Paper 7, International Water Management Institute, Colombo, Sri Lanka.

Oweis, T., Zhang, H. and Pala, M. (2000) Water use efficiency of rain-fed and irrigated bread wheat in a Mediterranean environment. *Agronomy Journal* 92, 231–238.

Oweis, T., Prinz, D. and Hachum, A. (2001) *Water Harvesting: Indigenous Knowledge for the Future of the Drier Environments.* ICARDA, Aleppo, Syria.

Pala, M. and Mazid, A. (1992) On-farm assessment of improved crop production practices in northwest Syria: chickpea. *Experimental Agriculture* 28, 175–184.

Pala, M. and Studer, C. (1999) Cropping systems management for improved water use efficiency in dryland agriculture. Paper presented at the International Conference on Water Resources Conservation and Management in Dry Areas, 3–6 December 1999, Amman, Jordan.

Pala, M., Matar, A. and Mazid, A. (1996) Assessment of the effects of environmental factors on the response of wheat to fertilizer in on-farm trials in a Mediterranean type environment. *Experimental Agriculture* 32, 339–349.

Rodriguez, A. and Thomas, N. (1999) *Mapping Rural Poverty and Natural Resource Constraints in Dry Areas.* Social Sciences Paper No 6, Natural Resources Management Programme, ICARDA, Syria.

Ryan, J. (ed.) (2000) *Plant Nutrient Management under Pressurized Irrigation Systems in the Mediterranean region. Proceedings of the IMPHOS International Workshop Organized by the World Phosphate Institute, 25–27 April 1999, Amman, Jordan.* ICARDA, Aleppo, Syria, 440 pp.

Seckler, D., Molden, D. and Barker, R. (1999) *Water Scarcity in the Twenty First Century.* Water Brief 1, IWMI, Colombo, Sri Lanka.

Singh, K.B. and Saxena, M.C. (1996) *Winter Chickpea in Mediterranean-type Environments. A Technical Bulletin.* ICARDA, Aleppo, Syria.

Singh, K.B., Malhotra, R.S., Saxena, M.C. and Bejiga, G. (1997) Winter sowing of chickpea in the Mediterranean region. *Agronomy Journal* 89(1), 112–118.

Studer, C. and Erskine, W. (1999) Integrating germplasm improvement and agricultural management to achieve more efficient water use in dry area crop production. Paper presented at the International Conference on Water Resources Conservation and Management in Dry Areas, 3–6 December 1999, Amman, Jordan.

Zhang, H. and Oweis, T. (1999) Water–yield relations and optimal irrigation scheduling of wheat in the Mediterranean region. *Agricultural Water Management* 38, 195–211.

Zhang, H., Oweis, T., Garabet, T.S. and Pala, M. (1998) Water use efficiency and transpiration efficiency of wheat under rain-fed conditions and supplemental irrigation in a Mediterranean-type environment. *Plant and Soil* 201, 295–305.

12 Efficient Management of Rainwater for Increased Crop Productivity and Groundwater Recharge in Asia

S.P. Wani,[1] P. Pathak,[1] T.K. Sreedevi,[2] H.P. Singh[3] and P. Singh[1]
[1]*International Crops Research Institute for the Semi-Arid Tropics, Andhra Pradesh, India;* [2]*Andhra Pradesh Rural Livelihoods Project, Andhra Pradesh, India;* [3]*Central Research Institute for Dryland Agriculture, Andhra Pradesh, India*

Abstract

Rainwater is the main source of water for agriculture but its current use efficiency for crop production ranges between only 30 and 45%. Annually, 300–800 mm of seasonal rainfall are not used productively, as the rainfall becomes surface runoff or deep drainage. The International Crops Research Institute for the Semi-Arid Tropics (ICRISAT)'s long experience, in partnership with national agricultural research systems, in integrated watershed management has clearly demonstrated that areas with good soils in the semi-arid tropics (SAT) in Asia can support double-cropping, while surplus rainwater could recharge the groundwater. In the integrated watershed approach, the emphasis is on *in situ* conservation of rainwater at farm level, with the excess water being taken out of the fields safely through community drainage channels and stored in suitable low-cost structures. The stored water is used as surface irrigation or for recharging groundwater. Following conservation of the rainwater, its efficient use is achieved through choosing appropriate crops, improved varieties, cropping systems and nutrient and pest-management options for increasing productivity and conserving natural resources. Long-term, on-station watershed experiments have demonstrated that Vertisols with a rainfall of 800 mm have the capacity to feed 18 persons ha^{-1} (4.7 t of food grains ha^{-1}) compared with their current productivity of 0.9 t ha^{-1} supporting four persons ha^{-1}. This increased productivity can be achieved if the productivity of rainwater is doubled (from 30% to 67%) and the soil loss is reduced by 75% compared with the loss under traditional methods of cultivation. By adopting such a holistic approach to the management of rainwater in partnership with the communities, crop productivity in the watersheds is substantially increased (up to 250%), groundwater levels improved and soil loss minimized. Results from such on-farm integrated watersheds are discussed. Conditions for success in the improved management of rainwater are: community participation, capacity building at local level through appropriate technical guidance and the use of new scientific tools to manage the watersheds efficiently. To sustain agricultural productivity in the SAT, this holistic approach of watershed management needs to be scaled up through appropriate policy and institutional support and its on-site and off-site impacts need to be studied.

 Water Productivity in Agriculture: Limits and Opportunities for Improvement (eds J.W. Kijne, R. Barker and D. Molden)

Introduction

Water is the primary constraint in the semi-arid tropics (SAT) and its scarcity confounds the sustainability of agriculture in the SAT. If not managed properly, water adversely affects crop productivity and causes land degradation through runoff and associated soil loss. The SAT cover parts of 55 developing countries; they are the home of over 1.4 billion people, of whom 550 million are below the poverty line. Seventy per cent of all the poor people live in rural areas, where the key occupation is agriculture. The SAT are characterized by high water demand, with a mean annual temperature greater than 18°C. Rainfall exceeds evapotranspiration for only 2–4.5 months in the dry SAT and for 4.5–7 months in the wet–dry SAT (Troll, 1965). The coefficient of variation of annual rainfall ranges between 20 and 30% in these dry regions.

The rising demand for water for non-agricultural uses is proportionally reducing the water availability for agriculture. Thus efficient management of rainwater through water harvesting and improved water-use technologies helps increase productivity, reduces poverty and maintains the natural-resources base in the SAT.

Watershed as a Unit for Efficient Management

The watershed is a logical unit for the efficient management of rainwater in the dry regions. Along with water, other natural resources, such as soil, vegetation and biota, can also be managed efficiently by adopting an integrated watershed-management approach.

Based on impressive successes, with on-station watersheds using new technologies for double-cropping on Vertisols, researchers expected that this approach could be 'transferred' to farmers' fields, thereby enhancing the productivity of rain-fed systems. The whole process evolved around the 'demonstration' of the technology package and of its possible benefits under farmers' conditions. The two basic assumptions were that:

- All Vertisols faced the same degree of waterlogging, which could be alleviated by the adoption of broad bed and furrow (BBF).
- Farmers would adopt the technology once its benefits were demonstrated to the farmers under their specific conditions.

The Tadannapally village, Medak district in Andhra Pradesh, India, served as a test area for on-farm watershed trials by scientists of the International Crops Research Institute for the Semi-Arid Tropics (ICRISAT) in collaboration with Andhra Pradesh Agricultural Department officials. The Vertisol technology package was demonstrated in the village watershed. It included land smoothing, drain construction, the introduction of the BBF system, use of a bullock-drawn Tropicultor, summer cultivations, dry seeding and the use of appropriate nutrient and pest-management options along with improved high-yielding crop varieties. Yields in the improved watershed were compared with those in the traditional farmers' system. The trials performed during 1981/82 confirmed that on-farm yields could be similar to those from operational research watersheds. Of the latter, the improved productivity system with a sorghum + pigeonpea intercrop produced higher grain yields (1.9 t ha^{-1}) and net returns of Rs 3838 ha^{-1} $year^{-1}$ compared with those from the traditional farmers' fields, which recorded 0.55 t ha^{-1} of grain yield and net returns of Rs 1234 ha^{-1} $year^{-1}$. Similar on-farm evaluations were done at several locations in Maharashtra, Gujarat, Madhya Pradesh, Karnataka and Andhra Pradesh.

However, subsequent evaluation of these watersheds after 15 years revealed that, in most of them, the farmers went back to their normal practices and that only selected practices were continued. As part of the watershed evaluation exercise, hundreds of farmers were interviewed and a multidisciplinary team of scientists analysed the process, farmers' interviews and possible reasons for the low adoption of the technology package.

Lessons Learned

Many lessons were learned from these studies, which need to be carefully applied in order to sustain the existing agricultural production systems in the SAT. Joshi *et al.* (1999) list the following:

- Components of Vertisol watershed technology, such as placement of seeds and fertilizers, improved varieties, use of fertilizers and summer cultivation, were already known and widely adopted by the farmers. However, their adoption increased after demonstration on the farmers' fields.
- The technology was found to be biased towards large farmers.

The whole technology package was not adopted by the farmers, but different components were. Several constraints affected the adoption of technology and higher adoption rates were observed in assured high-rainfall Vertisol areas.

ICRISAT's scientists have articulated the following additional lessons learned from years of working with watershed technologies (Wani *et al.*, 2001):

- Efficient technical options are needed to manage natural resources for sustaining systems.
- Mere on-farm demonstration of technologies by the scientists does not guarantee their adoption by the farmers.
- The contractual mode of farmers' participation adopted during Vertisol technology evaluation did not achieve the expected results. There is a need to have a higher degree of farmers' participation through a consultative to cooperative mode, from the planning stage up to the evaluation stage.
- Appropriate technology applications to address region-specific constraints need to be identified and simple broad recommendations do not help, e.g. Vertisols and BBF.
- Developmental projects lacked technical support so technical guidance is essential. No single organization can provide answers to all the problems in a watershed; thus, a consortium of organizations is needed for technical guidance.
- The process of partnership selection for each watershed has to be undertaken carefully and a generalized formula-based selection does not guarantee success.
- Technical change is intimately bound up with the broader institutional context of the watershed and the role of institutions and different players varies from location to location.
- Individual farmers should first realize tangible economic profits from the watersheds; it is only then that they come forward to participate in community-based activities in the watershed.
- A holistic-systems approach through the convergence of different activities is needed and it should improve farmers' livelihoods and not merely conserve soil and water in the watershed.
- Technological packages as such are not adopted and farmers adopted specific components that they found beneficial.
- There is no beginning or end to watershed inventions, and capacity building is critical for all the stakeholders. It is a continuous learning process.
- Women and youth groups play an important role in decision-making in the families.

New Integrated Watershed-management Model for Efficient Management of Natural Resources

A new model for efficient management of natural resources in the SAT has emerged from the lessons learned from extensive watershed-based research. The important components of the new integrated watershed-management model are as follows:

- The farmers' participatory approach through the cooperation model and not through the contractual model.
- The use of new science tools for management and monitoring of watersheds.
- Linking of on-station and on-farm watersheds.

- A holistic system's approach to improve livelihoods of people and not merely conservation of soil and water.
- A consortium of institutions for technical guidance on the on-farm watersheds.
- A microwatershed within the watershed, where farmers conduct strategic research with technical guidance from the scientists. Minimize free supply of inputs for undertaking the evaluation of technologies.
- Low-cost soil- and water-conservation measures and structures.
- The amalgamation of traditional knowledge and new knowledge for efficient management of natural resources.
- Emphasis on individual farmer-based conservation measures for increasing productivity of individual farms along with community-based soil- and water-conservation measures.
- Continuous monitoring and evaluation by the stakeholders.
- Empowerment of the community of individuals and strengthening of village institutions for managing natural watersheds.

Since 1999, using the new integrated water-management model, we have initiated new on-farm benchmark watersheds in India, Thailand and Vietnam. Five on-farm and three on-station watersheds in different agroecological, socio-economic and technological situations have been selected and work is ongoing in India, Thailand and Vietnam. As a case study, one on-farm watershed, the Adarsha watershed at Kothapally, Ranga Reddy district, in Andhra Pradesh, India, is described here. In addition, as illustrations of specific components of the new model, examples from other benchmark watersheds are also presented.

Use of New Science Tools for Managing and Monitoring Watersheds

Water budgeting using simulation models

For prioritization and selection of target regions for watershed development, first-order water budgeting using a geographic information system (GIS)-linked water-balance model is employed. Such a simulation model, used with monthly rainfall and soil data, generates output that can be used effectively to prioritize the regions and strategies for improved management of rainwater (Fig. 12.1). Once the target region is selected, then, for selection of appropriate benchmark sites, second-order water-budgeting studies using simulation models are applied For selected sites in the SAT of India, the WATBAL model (Keig and McAlpine, 1974) and weekly rainfall data of the past 30 years allowed the analysis of various soil-water availability and runoff (water surplus) scenarios. This is shown in Fig. 12.2 for four sites. High-rainfall locations selected were Bhopal, Nagpur, Indore and Adilabad, with annual rainfall ranging from 1000 to 1200 mm. The soils have a high water-holding capacity (≈ 200 mm). For these locations, the mean water surplus ranged from 270 to 508 mm during the season. Water surplus in 70% of the years (at the 30th percentile) ranged from > 130 to > 270 mm across locations. In 50% of the years it was > 230 to > 475 mm, indicating a tremendous opportunity to harvest rainfall in surface ponds or to recharge the groundwater.

At the medium rainfall (> 700 mm) locations, such as Hyderabad, Solapur, Aurangabad and Bangalore, the mean water surplus ranged from 66 to 187 mm annually. The soils in this region are Alfisols, Vertic Inceptisols and Vertisols, ranging in water-holding capacity from 100 to 200 mm in the root zone. Considering the depth of the soils at Hyderabad and Solapur, the opportunity for water harvesting exists for 50% of the years or less. However, on low water-holding capacity soils, such as Alfisols, it will be possible to harvest water in at least 70% of the years. At Aurangabad and Bangalore, the opportunities for water harvesting are greater, as the soils are shallower and of lower water-holding capacity. This analysis of the water balance indicates the opportunities for water harvesting and improved water management in different regions of the SAT, India, which would raise crop production from the existing low levels. It also provides information for selecting appropriate technologies, such as water harvesting or *in situ* water-conservation methods, which would be cost-effective and

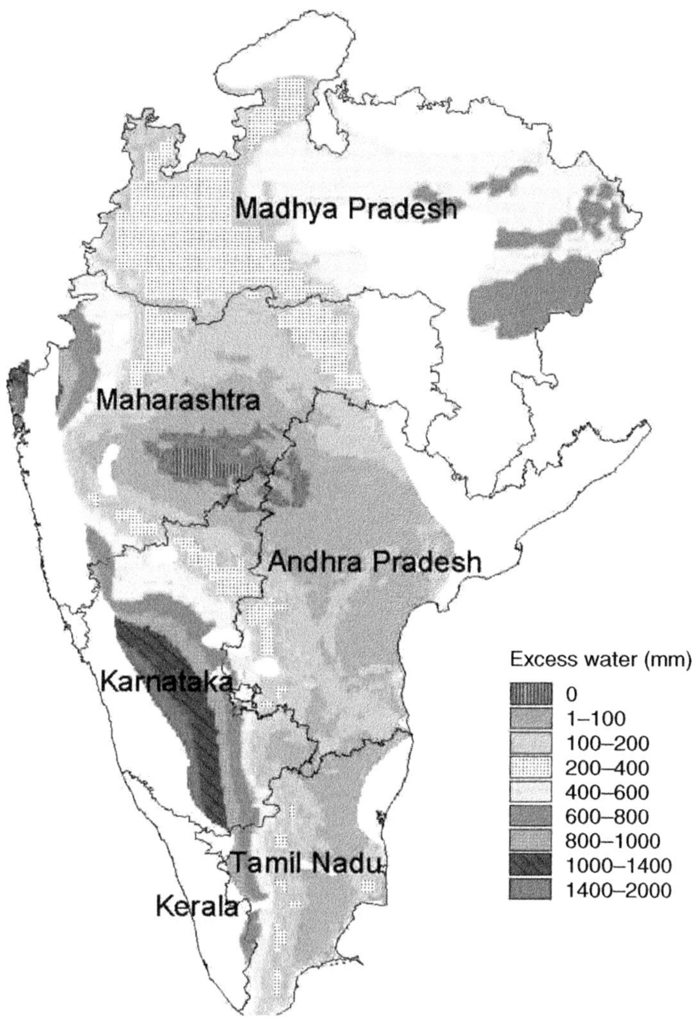

Fig. 12.1. Excess water available for harvesting as runoff in the states of the semi-arid tropics, India (June–October).

more impact-oriented about representative benchmark sites in the target ecoregion.

The CERES family of models has proved to be effective in simulating the water balance of soils with vertical drainage, which is often an unrealistic assumption. Runoff produced by such models is only from a point in space and no account is taken of water accumulation over space and time. In partnership with the Michigan State University (MSU), USA, through a US linkage grant to ICRISAT and with funding support from the Asian Development Bank, we have attempted to integrate the topographic features of the watershed in the hydrological models. The automation of terrain analysis and the use of digital elevation models (DEMs) have made it possible to quantify the topographic attributes of the landscape for hydrological models. These topographic models, commonly called digital terrain models (DTMs), partition the landscape into a series of interconnected elements, based on the topographic characteristics of the landscape, and are usually coupled to a mechanistic soil-water-balance model. The partitioning between

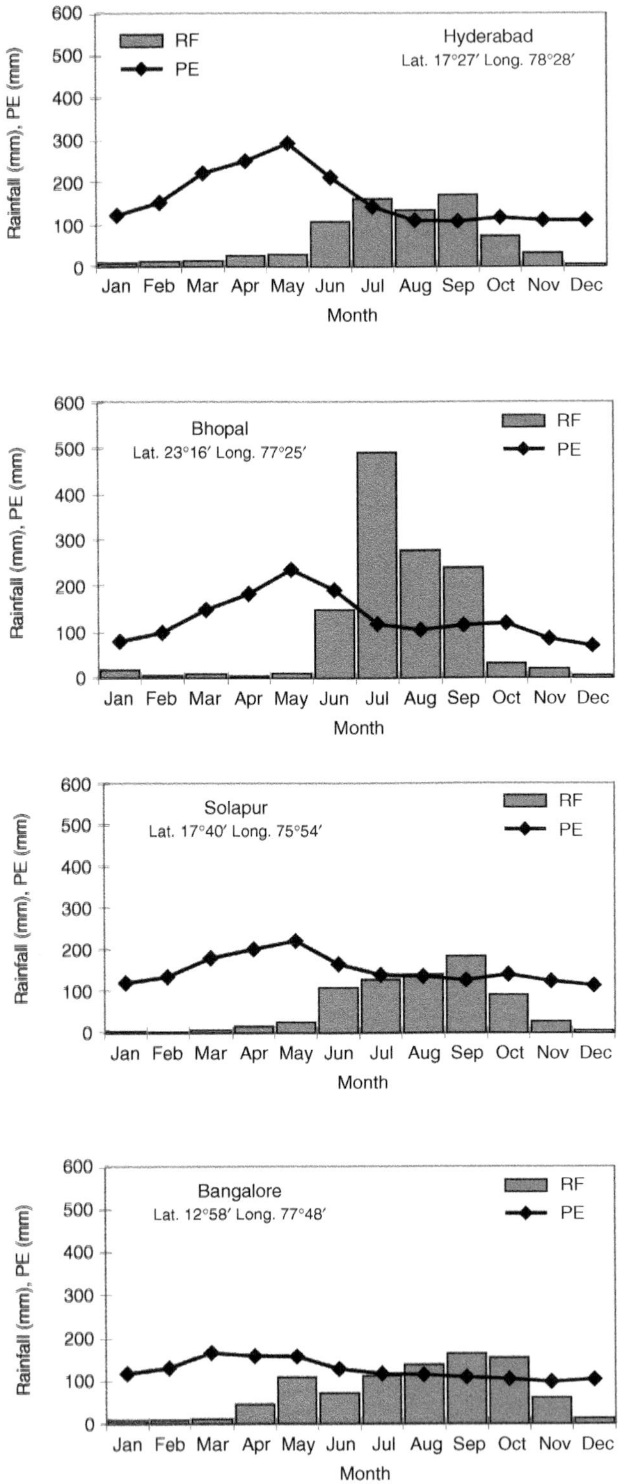

Fig. 12.2. Rainfall (RF) and pan-evaporation (PE) analysis for selected sites in the SAT, India.

vertical and lateral movement at a field-scale level helps to predict the complete soil-water balance and consequently the available water for the plants over space and time.

The data generated in the Black Watershed (BW) 7 on-station watershed at ICRISAT was used for validating the model developed at MSU. This partnership research led to the development of SALUS-TERRAE, a DTM for predicting the spatial and temporal variability of soil-water-balances. A regular grid DEM provided the elevation data for SALUS-TERRAE. We have successfully applied the SALUS-TERRAE, which has a functional spatial soil-water balance model, at a field scale to simulate the spatial soil-water-balance and identify how the terrain affects the water routing across the landscape. The model provided excellent results when compared with the field-measured soil-water content.

Feasibility studies for providing harvested water for crop production

For the Akola region, the simulated probabilities of getting 40, 60, 80 and 100 mm of water for supplemental irrigation from the runoff-harvesting structure are shown in Fig. 12.3. The probabilities of getting water for irrigation from the tank are high for most of the growing season. However, the high probability of getting 100 mm of irrigation water was limited to only 3 months, namely September, October and November. High runoff and low seepage loss are the main reasons for adequate availability of water in a harvesting structure. The 10 years of mean cumulative water-outflow data from the runoff-harvesting structure indicate that the structure could be enlarged, since approximately 2200 m^3 runoff water overflows from the structure every year. Overall, the analysis indicates a good prospect of runoff-water harvesting in the Akola region.

Crop simulation models for identifying the constraints and yield-gap analysis

We have validated the Decision Support System for Agricultural Technology (DSSAT) model for CROPGRO soybean and CROPGRO chickpea using the data sets generated from an on-station watershed at Patancheru. The validated models were used for estimating the potential soybean–chickpea system's yields in the target ecoregion, using the historical weather data for estimating the yield gaps. The soybean model and weather records of the past 22 years from Patancheru

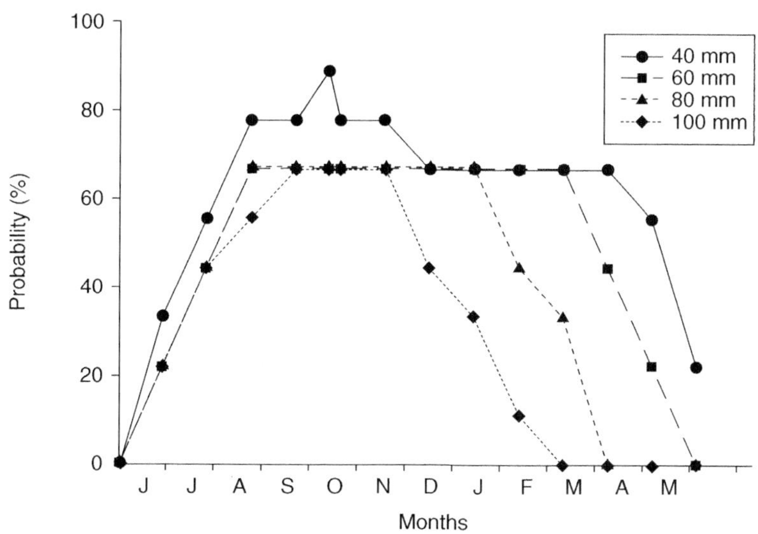

Fig. 12.3. Probabilities of obtaining 40, 60, 80 and 100 mm of water for irrigation from a tank at Akola (based on 10 years' simulated data).

were successfully used to evaluate the effect of soil depth on soybean yields. From the non-linear yield–soil-depth relationship obtained, it was observed that at Patancheru – even during a normal rainfall year – soybean cannot be grown in a soil with a depth of less than 37.5 cm. The analysis also revealed that, in 70% of the years, the soybean–chickpea system's yield at Patancheru could be 3.5 t ha^{-1} on medium-depth soil and 3.0 t ha^{-1} on shallow-depth soil (< 50 cm).

Crop simulation models, using a scenario analysis for yield-gap and constraint identification, simulate the crop yields in a given climate and soil environment. ICRISAT researchers have adopted DSSAT version 3.0, a soybean crop-growth model, to simulate the potential soybean yield in Vertisols at different benchmark locations (Tsuji *et al.*, 1994). The mean simulated yield was compared with the mean observed yield of the last 5 years to calculate the yield gap. The results (shown in Table 12.1) indicate that there is a considerable potential to bridge the yield gap between the actual and potential yield through the adoption of improved resource-management technologies. Such a scenario analysis helps the researchers to identify the high-potential areas where large yield gaps exist and considerable gains in productivity can be achieved.

Economic evaluation of tank irrigation systems

The economic evaluation of tank irrigation for high-rainfall Vertisol areas has been carried out using a simulation model (Pandey, 1986). The model consisted of several component modules for rainfall, runoff, soil moisture and yield response to irrigation and tank-water balance. Simulations were run for three different seepage rates, namely, 0, 10, 20 mm day^{-1}, for a test site on a Vertisol in central India (Madhya Pradesh). Results obtained from the simulation indicate that, as the seepage rate increases, the optimal tank size also increases, while the optimal size of the command area and other factors, such as runoff volume and availability of irrigable land, become constraints. It was found that tanks are quite attractive for the

Table 12.1. Simulated soybean yields and yield gap for selected locations in India.

Location	Mean sowing date	Mean harvest date	Simulated yields (t ha^{-1})		Mean observed yield[a] (t ha^{-1})	Yield gap (t ha^{-1})
			Mean	SD		
Primary zone						
Raisen	22 June	11 Oct.	3.05	1.28	–	–
Betul	19 June	8 Oct.	2.37	0.64	0.86	1.51
Guna	30 June	14 Oct.	1.69	1.96	0.84	0.85
Bhopal	16 June	8 Oct.	2.31	0.61	1.00	1.31
Indore	22 June	10 Oct.	2.30	0.98	1.12	1.18
Kota	3 July	16 Oct.	1.24	0.98	1.01	0.23
Wardha	17 June	6 Oct.	3.00	0.65	1.04	1.95
Secondary zone						
Jabalpur	23 June	11 Oct.	2.24	0.48	0.90	1.35
Amaravathi	18 June	8 Oct.	1.62	0.74	0.94	0.68
Belgaum	17 June	30 Sept.	1.99	0.66	0.57	1.42
Tertiary zone						
Hyderabad (shallow soil)	20 June	5 Oct.	2.70	0.69	–	–
Hyderabad (medium-deep soil)	20 June	5 Oct.	2.66	0.70	–	–

[a]Mean of reported yields of last 5 years.
SD, standard deviation.

soybean–wheat cropping pattern, the most common in the region, even at seepage rates as high as 20 mm day^{-1}. With the soybean + pigeonpea intercrop, the tank is profitable at seepage rates of less than 10 mm day^{-1}.

Linking On-station Strategic Research with On-farm Watersheds

The operational-scale watersheds at ICRISAT, used since 1976 and aimed at increasing productivity and improving soil quality through an integrated watershed approach, were a logical choice to study rainwater harvesting for increased productivity and groundwater recharge. The technology package developed by ICRISAT for enhancing productivity on Vertisols consists of summer cultivation, BBF for draining excess rainwater safely out of the field, dry planting, grassed waterways, use of an improved bullock-drawn Tropicultor for field operations, improved stress-tolerant crop varieties and appropriate nutrient and pest-management options. This package has shown promising results.

Improved vs. conventional systems – Vertisol watershed

In an improved system with all the options mentioned above, the average productivity was 4.7 t ha^{-1}, which indicates a carrying capacity of 18 persons ha^{-1} year^{-1}, whereas the traditional system with farmer-adopted practices yielded only about 0.9 t ha^{-1} and had a carrying capacity of only four persons ha^{-1} year^{-1} (Fig. 12.4). Along with this higher productivity, the improved system could also sequester more carbon (0.335 t ha^{-1} year^{-1}) and improve soil quality (Wani *et al.*, 2000). Most importantly, in the improved system, 67% of the rainfall was used by the crops, while 14% of the rainfall was lost as runoff and 19% as evaporation and deep percolation. In the traditional system, only 30% of the total rainfall was used by the crops, while 25% was lost as runoff and 45% as soil evaporation and deep percolation. The soil loss in the improved system was only 1.5 t ha^{-1}, compared with the traditional system, where the soil loss was 6.4 t ha^{-1}.

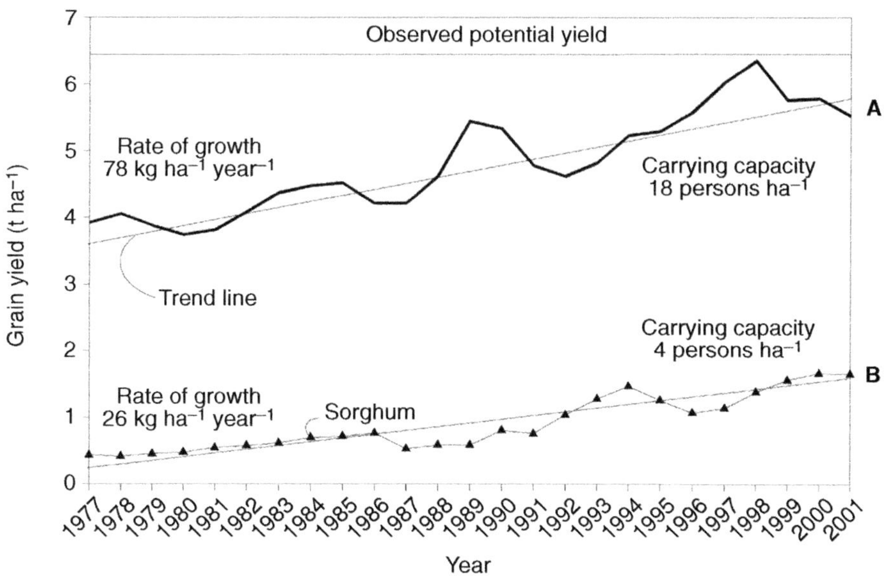

Fig. 12.4. Three-year moving average of grain yield under improved (A) and traditional (B) technologies on a Vertisol watershed at ICRISAT (1977–2001).

Increased productivity – Vertic Inceptisol watershed

At ICRISAT, Patancheru, crop productivity and resource use were studied for a soybean–chickpea sequential and soybean + pigeonpea intercrop systems on two land-forms (BBF and flat) and with two soil depths (shallow and medium-deep) at a watershed scale on a Vertic Inceptisol. The results show that, during 1995–2000, the improved BBF system recorded on average 0.1 t ha^{-1} more grain yield than the flat land-form. During 2000/01, when recorded rainfall was 958 mm (31% above normal rainfall), the BBF system yielded 500 kg more grains in the soybean–chickpea sequential system than in the flat land-form treatment. Similarly, an increased crop yield of 2.9 t ha^{-1} of soybean intercropped with pigeonpea on BBF was recorded compared with 2.63 t ha^{-1} in the flat land-form treatment. The total runoff was higher in the flat land system (23% of the seasonal rainfall) than on the improved system (15% of the seasonal rainfall). The BBF had more deep drainage than the flat land system, especially for the shallow soil. The runoff figure in the flat land system (190 mm), with a peak runoff rate of 0.096 $m^3\ s^{-1}\ ha^{-1}$, compared unfavourably with the BBF system, which had a lower runoff (150 mm) and a lower peak runoff rate (0.086 $m^3\ s^{-1}\ ha^{-1}$). Hence, the BBF system was useful in decreasing runoff and increasing rainfall infiltration. The soil loss in the flat land system was 2.2 t ha^{-1} versus 1.2 t ha^{-1} in the BBF system.

These studies clearly demonstrate the potential of Vertisols and Vertic Inceptisols with 800 mm of annual average rainfall at the watershed level. They also show that similar high yields could probably be achieved at the field scale if the same approach is followed.

Response of crops to supplemental irrigation

Once the rainwater has been harvested, it needs to be used efficiently to increase the system's productivity. The option to use the harvested rainwater for supplemental irrigation during a stress period was evaluated at ICRISAT and other research stations in India.

Benefits of supplemental irrigation in terms of increasing and stabilizing crop production have been impressive even in dependable rainfall areas of both Alfisols and Vertisols (El-Swaify *et al.*, 1985; Vijayalakshmi, 1987; Pathak and Laryea, 1991; Oswal, 1994; Singh *et al.*, 1998). As shown in Table 12.2, good yield responses to supplemental irrigation were obtained on Alfisols in both rainy and post-rainy seasons. The average irrigation water productivity (WP) (ratio of increase in yield to depth of irrigation water applied) varied with the crop, e.g. for sorghum it was 14.9 kg $ha^{-1}\ mm^{-1}$ and for pearl millet it ranged from 8.8 to 10.2 kg $ha^{-1}\ mm^{-1}$. Tomatoes responded very well to supplemental water application, with an average WP of 186.3 kg $ha^{-1}\ mm^{-1}$. In the sorghum + pigeonpea intercrop, two irrigation turns of 40 mm each gave an additional gross return of Rs 3950 ha^{-1}. The largest additional gross return from the supplemental irrigation was obtained by growing tomato (Rs 13,870 ha^{-1}).

On Vertisols, the average additional gross returns due to supplemental irrigation were about Rs 830 ha^{-1} for safflower, Rs 2400 ha^{-1} for chickpea and Rs 3720 ha^{-1} for chilli. The average WP was largest for chickpea, with 5.6 kg $ha^{-1}\ mm^{-1}$, followed by chilli, with 5.3 kg $ha^{-1}\ mm^{-1}$, and safflower, with 2.1 kg $ha^{-1}\ mm^{-1}$.

Farmers' Participatory Approach: Selection of Watershed, Prioritization and Execution of Works

The adoption of integrated watershed management on farm is possible through community initiatives and strength of local participation. People's participation in planning, developing and executing the watershed activities is indispensable.

ICRISAT, Drought Prone Area Project (DPAP) officials, non-governmental organizations (NGOs) and farmers formed a consortium and visited three priority villages in the targeted Ranga Reddy district in Andhra Pradesh. The consortium partners jointly selected the Kothapally watershed as the participatory on-farm watershed, as the village did not have a single tank for community use.

Table 12.2. Grain-yield response (t ha^{-1}) of cropping systems to supplemental irrigation on an Alfisol watershed, ICRISAT Centre.

One irrigation turn of 40 mm	Increase due to irrigation	WAE[a] (kg ha^{-1} mm^{-1})	Two irrigation turns of 40 mm each	Increase due to irrigation	WAE[a] (kg ha^{-1} mm^{-1})	Combined WAE (kg ha^{-1} mm^{-1})
Intercropping system						
	Pearl millet			Pigeonpea		
2.353	0.403	10	1.197	0.423	5.3	6.8
	Sorghum			Pigeonpea		
3.155	0.595	14.9	1.22	0.535	6.7	9.4
Sequential cropping system						
	Pearl millet			Cowpea		
2.577	0.407	10.2	0.735	0.425	5.3	6.9
	Pearl millet			Tomato		
2.215	0.35	8.8	26.25	14.9	186.3	127.1

[a]Water application efficiency (WAE) = $\frac{\text{Increase in yield due to water application.}}{\text{Depth of irrigation}}$

The maximum area was cultivated with rainfed crops and the yields were low (1–1.5 t ha^{-1}). Moreover, during the initial visit and subsequent reconnaissance surveys, farmers showed a keen interest in participation in the watershed programme. The *Gram Sabha* (a general meeting of all the villagers) ratified the decision to select the watershed and agreed to take an active part in the watershed programmes. Subsequently, villagers' committees, self-help groups and user groups did all the planning and execution of the various watershed works.

Microwatershed Development as an Island for Testing Technology, Evaluation and Monitoring

Within a watershed of 470 ha, a microwatershed of 30 ha was delineated and developed, and subsequently the impact of watershed development on runoff, soil loss and nutrient losses was monitored. Both developed and undeveloped microwatersheds were fully instrumented with automatic runoff-recording and sediment-loss-gauging stations. In addition, rain gauges were fixed across the watershed to measure the rainfall variation in the watershed. In the microwatershed, farmers conducted simple trials to compare improved crop varieties, land-form treatments, balanced-nutrient schedules, integrated pest-management (IPM) and integrated nutrient-management (INM) options, etc. Farmers were given technical support but no inputs were provided free of cost for evaluating the technologies. The type of tests farmers conducted included comparing improved land-form treatments, such as BBF and contour planting, using an improved bullock-drawn Tropicultor versus the normal practice of sowing crops with the traditional wooden plough. Other trials involved fertilization and the various improved crop-management options mentioned above. Field experimentation by the farmers did not remain confined to the microwatershed, as a large number of farmers conducted trials throughout the watershed.

Increased productivities with improved management practices at Adarsha watershed, Kothapally

At Kothapally, farmers evaluated improved management practices, such as sowing on a

BBF land-form, flat sowing on contour, fertilizer application, nutrient-management treatment along with *Rhizobium* or *Azospirillum* sp. inoculations and using an improved bullock-drawn Tropicultor for sowing and intercultural operations. Farmers obtained a twofold increase in yield in 1999 (3.3 t ha^{-1}) and a threefold increase in 2000 (4.2 t ha^{-1}), as compared with the yields of sole maize (1.5 t ha^{-1}) in 1998 (Table 12.3). Intercropped maize with improved practice in pigeonpea gave a fourfold maize yield (2.7 t ha^{-1}) compared with yields on traditional farmers' fields of 0.7 t ha^{-1}. In the case of sole sorghum, the improved practices increased yields threefold within 1 year. In 1999/2000, farmers achieved the highest systems productivity, total income and profit from improved maize–pigeonpea and improved sorghum–pigeonpea intercropping systems (Table 12.4). Moreover, the cost–benefit ratio of the improved systems was more (3.5 times) than the traditional cotton-based systems (Wani, 2000). In 2000/01, several farmers evaluated BBF and flat land-form treatments for shallow and medium-depth black soils using different crop combinations. On average, farmers harvested 250 kg more pigeonpea and 50 kg more maize per hectare using BBF on medium-depth soils than with the flat land-form treatment. Furthermore, even with the flat land-form treatment, farmers harvested 3.6 t of maize and pigeonpea using the improved management options compared with 1.72 t of maize and pigeonpea grains using the normal cultivation practices (Table 12.5). Farmers with shallow soils and with other cropping systems reported similar benefits from the improved BBF land-form and

Table 12.3. Average crop yields from on-farm evaluation of improved technologies in Adarsha watershed, Kothapally, 1998, 1999 and 2000.

		Yield (t ha^{-1})	
Crop	1998 baseline	1999	2000
Sole maize	1.50	3.25	3.75
Intercropped maize	–	2.70	2.79
(farmers' practice)		0.70	1.60
Intercropped pigeonpea	0.19	0.64	0.94
(farmers' practice)		0.20	0.18
Sole sorghum	1.07	3.05	3.17
Intercropped sorghum	–	1.77	1.94

Table 12.4. Total productivity, cost of cultivation for different crops at Kothapally watershed during crop season 1999/2000.

Cropping systems	Total productivity (t ha^{-1})	Cost of cultivation (Rs ha^{-1})	Total income (Rs ha^{-1})	Profit (Rs ha^{-1})	Cost: benefit ratio
Maize/pigeonpea (improved)	3.3	5,900	20,500	14,600	1:3
Sorghum/pigeonpea (improved)	1.57	6,000	15,100	9,100	1:2
Cotton (traditional)	0.9	13,250	20,000	6,750	1:1
Sorghum/pigeonpea (traditional)	0.9	4,900	10,700	5,800	1:2
Green gram (traditional)	0.6	4,700	9,000	4,300	1:2

Table 12.5. Productivities in different on-farm trails at Kothapally during 2000/01.

System	Soils	Land-form	Yield (t ha^{-1}) (1)	(2)	Total systems productivity (1 + 2)
Maize/PP	Shallow	BBF	1.75	0.38	2.13
Maize/PP	Shallow	Flat	1.68	0.29	1.97
Maize/PP	Medium	BBF	2.83	1.07	3.90
Maize/PP	Medium	Flat	2.78	0.82	3.60
Sorghum	Medium	BBF	3.00	–	3.00
Maize/PP	(Local farmers' practice)		1.49	0.22	1.71
Sorghum/PP	(Local farmers' practice)		0.47	0.11	0.59
Sorghum	(Local farmers' practice)		1.01	–	1.01

1. Main crop (maize or sorghum).
2. Component crop (pigeonpea (PP)).

other management improvements. In this area, rainfall during 1999 was 559 mm, which was 30% below normal rainfall, and in 2000 the rainfall was 958 mm, 31% above normal. In spite of this variation in rainfall (Tables 12.3 –12.5), productivity of the crops continued to show a marked increase during these years.

Nutrient-budgeting approach – boron and sulphur amendments

At the Lalatora watershed, a detailed characterization of soils revealed that they are deficient in boron (B) and sulphur (S), while both these nutrients are critical for optimizing productivity of soybean-based systems. Farmers were made aware of the results and some farmers came forward to evaluate the response of B and S application in their fields along with the improved management options. Farmers applied 10 kg of borax (1 kg B) and 200 kg ha^{-1} of gypsum (30 kg S). The treatments studied were: best-bet (control) treatment, B application, S application and B + S application. In 2000, all the farmers reported significant differences in soybean plant growth with B, S and B + S treatments over the control treatment. Soybean yields increased by 19–25% percent over the best-bet control treatment (Table 12.6). In 2000, soybean yields in the control were 1.52 t ha^{-1} – that is, 18% more than the 1999 best-bet treatment yields of 1.28 t ha^{-1}. The results indicate that B and S amendments not only increase soybean yields over the best-bet treatment but also benefited the subsequent wheat crop without further application of B and S. This residual benefit of B and S amendments for the subsequent wheat crop were to the tune of 31 to 40.6% over the best-bet treatment. The system's productivity when soybean was followed by wheat increased by 27–34% over the best-bet treatment. The farmers were so much impressed with their experimentation that for the 2001 season they indented B and S for their use well in advance on cost basis through the NGO the Bharatiya Agro

Table 12.6. Soybean yields with boron, sulphur and boron + sulphur treatments.

Treatment	Grain yield (t ha^{-1}) Soybean	Wheat	Soybean–wheat system
Boron	1.87 (23.2)[a]	3.74 (40.6)	5.61 (34.2)
Sulphur	1.81 (19.1)	3.5 (31.9)	5.31 (27.0)
Boron + sulphur	1.91 (25.6)	3.57 (34.2)	5.48 (31.1)
Control (best-bet treatment)	1.52	2.66	4.18

[a]Values in parentheses are percentage increases over control (best-bet treatment).

Industries Foundation (BAIF). Noting the results of these farmers' experiments in the Lalatora subwatershed, farmers in other subwatersheds of the Milli watershed also volunteered to conduct these experiments in their fields during the 2001 rainy season.

Consortium Approach for Technical Guidance

A consortium of various institutes and organizations, as shown in Fig. 12.5, provides technical support for each on-farm benchmark watershed.

Empowering the Stakeholders through Training

Farmers were exposed to new methods and technologies for managing natural resources through training and field visits to on-station and on-farm watersheds. Farmers and landless families were trained and encouraged to undertake income-generating activities in the watershed, which can be of help in sustaining its productivity. The training sessions for farmers included training in on-farm operating implements and IPM and INM options. Other key agents of change, such as watershed committee members and agricultural and extension officials, were also trained at ICRISAT on different aspects of integrated watershed management. Special efforts were made to educate and increase the awareness of women farmers regarding new management options, as women play a key role in the adoption of a new technology. Many women were trained in vermicomposting technology at Kothapally. Educated youth were trained in skilled activities such as nuclear polyhedrosis virus (NPV) production and vermicomposting, which provided them with a source of income.

Continuous Monitoring and Evaluation

To know the impact of watershed management, continuous monitoring and impact assessment were done in respect of various determinants. Where relevant, examples of initial results of the monitoring exercise are inserted between square brackets.

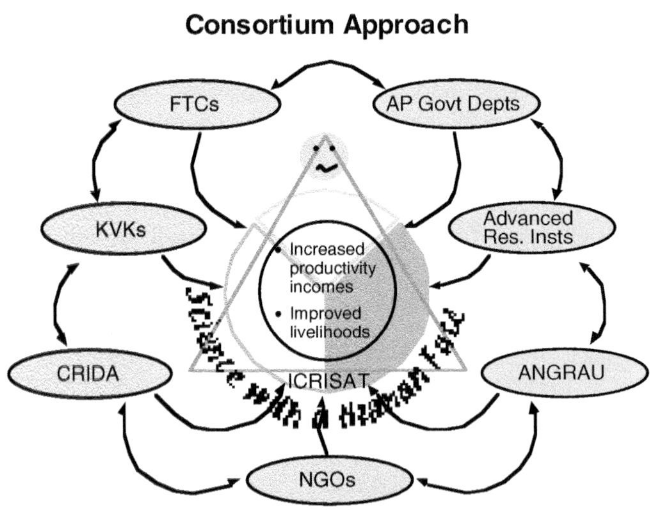

Fig. 12.5. A consortium of various institutions and organisations that provide technical support to each on-farm benchmark watershed. FTCs, farmers' training centres; KVKs, Krishi Vigyan Kendras (farm science centres); CRIDA, Central Research Institute for Dryland Agriculture; NGOs, non-governmental organizations; ANGRAU, Acharya NG Ranga Agricultural University; AP, Andhra Pradesh.

- Weather: an automatic weather station is installed to continuously monitor the weather parameters.
- Groundwater: open wells in the watershed are georeferenced and regular monitoring of water levels is carried out. [Hydrological investigations of the existing wells in the watershed indicated a rise in groundwater levels (5–6 m) at Kothapally (Fig. 12.6).]
- Runoff, soil and nutrient loss: these are monitored using automatic water-level recorders and sediment samplers. [Runoff as a ratio of the seasonal rainfall was observed to be 7% in the undeveloped watershed and 0.6% in a developed watershed, where soil- and water-conservation measures, such as gully plugging and bunding, had been adopted.]
- Pest monitoring: pheromone traps were installed to monitor *Helicoverpa* populations and, where appropriate, pest-control measures through IPM options have been started.
- Crop productivity: yields are recorded for each crop every year. [Data were analysed in terms of net income and the results from 1999–2001 were described in the previous section.]
- Nutrient budgeting: soil-nutrient levels are monitored and studies are being conducted to determine the optimum doses of fertilizers to maintain the soil-nutrient balance. Biological nitrogen fixation in farmers' fields is quantified using the N difference method and ^{15}N isotope-dilution method.
- Satellite monitoring: changes in cropping intensity, greenery, water bodies and groundwater levels are monitored. [GIS maps indicating soil types, soil depths and crops grown during the rainy and post-rainy season have been prepared.]

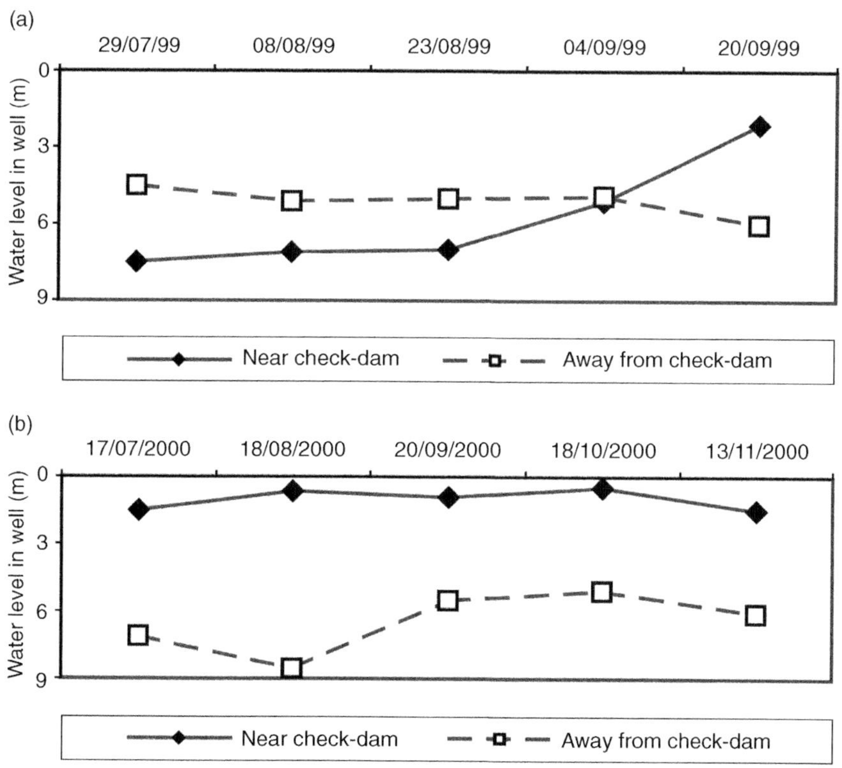

Fig. 12.6. (a) Groundwater levels before construction of check-dam in Adarsha watershed at Kothapally during 1999; (b) Effect of check-dam on groundwater levels in Adarsha watershed at Kothapally during 2000.

Emerging Issues

Despite the adoption of the integrated watershed approach for water harvesting and efficient use of natural resources, certain important issues need to be addressed as they have a bearing on sustainable production in the SAT. One of them is the need to better understand the motives behind collective action and the role of gender in the adoption of new technologies in the watershed framework. These issues are equally as important as the technical and economic factors. In the past, little attention was given in on-farm watershed work to farmers' participation, community action, group formation and empowerment of farmers. This has undoubtedly contributed to very low adoption rates, as well as to the unsustainability of many watershed technologies.

Some emerging issues are as follows:

- An integrated watershed is a continuous process, and the issue is how to plan and finance the activities involved. What training and incentives are most successful?
- How to institutionalize technical guidance for the watersheds.
- How to harmonize existing village institutions with committees especially set up for managing the watershed, and with other self-help groups. How to increase the efficiency of all these efforts through collective action.
- How to develop and enforce policies for rainwater harvesting. Who is entitled to its use? Who is responsible for the maintenance of rainwater-harvesting structures, wells and groundwater recharge? How to sustain the management of the watershed, i.e. how to make the community aware of the continuous efforts required for sustaining the productivity of the watershed, and how to ensure the ongoing participation of all stakeholders.
- How to include in the monitoring and assessment studies an evaluation of all on-site and off-site impacts of the watershed-development programmes.
- How to plan an exit strategy from watersheds and ensure sustainability through development of institutional and policy options.

Conclusion

An in-depth analysis of the possible scenarios of SAT farming systems reveals that the key elements of efficient management of rainwater are community participation, capacity building at local level through technical guidance by a consortium of organizations and use of high-science tools to manage the watershed efficiently. To sustain the productivity in the SAT, a holistic approach of integrated watershed management needs to be scaled up through appropriate policy and institutional support and its on-site and off-site impacts need to be studied.

Acknowledgements

The financial assistance provided by the Asian Development Bank through RETA 5812 is gratefully acknowledged. The efforts of staff of DPAP, BAIF, Central Research Institute for Dryland Agriculture and the M Venkatarangaiah Foundation are greatly appreciated. At all the benchmark sites, teams of scientists from national agricultural research systems (NARS) and ICRISAT were involved in undertaking research, and their efforts are also acknowledged. The contribution from farmers, who are the important partners, is gratefully acknowledged.

References

El-Swaify, S.A., Pathak, P., Rego, T.J. and Singh, S. (1985) Soil management for optimized productivity under rainfed conditions in the semi-arid tropics. *Advances in Soil Science* 1, 1–64.

Joshi, P.K., Bantilan, C. and Pathak, P. (1999) Watershed research at ICRISAT: lessons for priority setting. Paper presented in: *National Workshop on Watershed Approach for Managing Degraded Lands in India: Challenges for the 21st Century, 27–29 April 1998.* Vigyan Bhavan, New Delhi.

Keig, G. and McAlpine, J.R. (1974) *WATBAL: a Computer System for the Estimation and Analysis of Soil Moisture Regimes from Simple Climatic Data*, 2nd edn. Technical Memorandum 74/7, Australian Division of Land Research, CSIRO, Canberra.

Oswal, M.C. (1994) Water conservation and dryland crop production in arid and semi-arid regions. *Annals of Arid Zone* 33(2), 95–104.

Pandey, S. (1986) Economics of water harvesting and supplementary irrigation in the semi-arid tropics of India: a systems approach. PhD dissertation submitted to the University of New England, Australia.

Pathak, P. and Laryea, K.B. (1991) Prospects of water harvesting and its utilization for agriculture in the semi-arid tropics. In: *Proceedings of the Symposium of the SADCC Land and Water Management Research Program Scientific Conference, 8–10 October 1990*. Gaborone, Botswana, pp. 253–268.

Singh, A.K., Kumar, A. and Singh, K.D. (1998) Impact of rainwater harvesting and recycling on crop productivity in semi-arid areas – a review. *Agriculture Review* 19(1), 6–10.

Troll, C. (1965) Seasonal climate of the earth. In: Rodenwaldt, E. and Jusatz, H. (eds) *World Maps of Climatology*. Springer Verlag, Berlin, pp. 19–25.

Tsuji, G.Y., Jones, J.W. and Balas, S. (eds) (1994) *DSSAT v. 3*. University of Hawaii, Honolulu, Hawaii, USA.

Vijayalakshmi, K. (1987) Soil management for increasing productivity in semi arid red soils: physical aspect. Alfisols in the semi-arid tropics. In: *Proceedings of the Consultants Workshop on the State of the Art and Management Alternatives for Optimizing the Productivity of SAT Alfisols and Related Soils, 1–3 December 1983*. ICRISAT, Patancheru, India, pp. 109–111.

Wani, S.P. (2000) Integrated watershed management for sustaining crop productivity and reduced soil erosion in Asia. Paper presented at MSEC Meeting, 5–11 November 2000, Semarang, Indonesia.

Wani, S.P., Singh, P., Pathak, P. and Rego, T.J. (2000) Sustainable management of land resources for achieving food security in the semi-arid tropics. In: Gawande, S.P., Bali, J.S., Das, D.C., Sarkar, T.K., Das, D.K. and Narayanamurthy, G. (eds) *Advances in Land Resources Management for the 21st Century*. Angrakor Publishers, New Delhi, India.

Wani, S.P., Sreedevi, T.K., Pathak, P., Singh, P. and Singh, H.P. (2001) Integrated watershed management through a consortium approach for sustaining productivity of rainfed areas: Adarsha watershed, Kothapally, AP – a case study. Paper presented at Brainstorming Workshop on Policy and Institutional Options for Sustainable Management of Watersheds, 1–2 November 2001, ICRISAT, Patancheru, Andhra Pradesh, India.

13 Water Productivity in Forestry and Agroforestry

C.K. Ong and B.M. Swallow
International Centre for Research in Agroforestry, Nairobi, Kenya

Abstract

Forests are the biggest users of water worldwide and extensive forested areas have been lost or are undergoing conversion to agriculture, creating concerns about loss of hydrological functions and increasing the competition for scarce water between agriculture, urban centres, industries and wildlife. The challenge is to improve the sustainability and productivity of land and water use, especially for the growing populations of many developing countries. In this chapter we review recent findings on the hydrology of forests and agroforestry systems and indicate how modifications in tree-based systems might increase water productivity.

In forestry, the focus of research has moved from the hydrological functions of upland forest reserves that are close to settlements to a greater recognition of the roles played by upland communities in the management of water resources. A major source of conflict over water resources is the contrasting perceptions of 'watershed functions' between forest managers and local people, which are often based more on myths of forest functions than on science – for example, the idea that forests increase rainfall. These myths continue to dominate the views of policy makers and institutions and should be revised. The challenge is to gain a better insight into how farmer-developed land-use mosaics have modified watershed-protection functions. Priority must be given to the perceptions, experiences and strategies of local communities.

Trees on farms have the potential for improving productivity in two ways. Trees can increase the amount of water that is used on farm as tree or crop transpiration. Trees can also increase the productivity of the water that is used by increasing biomass of trees or crops produced per unit of water used. Plot-level evidence shows that improvements in water productivity as a consequence of modifications to the microclimate of the crop are likely to be limited. Instead, evidence from semi-arid India and Kenya showed that the greater productivity of agroforestry systems is primarily due to the higher amount of water used. Almost half of the total water use occurred during the dry season, when cropping was impossible, and the rest was extracted from soil reserves. This implies a high temporal complementarity between the crop and tree components of the landscape mosaic. Research is needed to examine the impact of the increased water use on the drainage and base flow at the landscape level. This chapter also describes some of the technical approaches that can be used to improve land and water management, the role of trees and its relation to hydrology and the challenges for rational land-use decision-making.

 Water Productivity in Agriculture: Limits and Opportunities for Improvement (eds J.W. Kijne, R. Barker and D. Molden)

Introduction

Forests are the biggest users of water worldwide. The tropical forests in Brazil, the Congo Democratic Republic, Indonesia, Peru and Venezuela form a large proportion of the closed forests, which are vital for the well-being of the planet and, therefore, clearing of such forests is strongly opposed. Thus, much of the future increase in food and wood production in the humid tropics and elsewhere will have to be achieved from land and water resources already in use. Therefore, the central challenge is to improve the productivity with which existing land and water resources are used.

Over the past half-century, great progress has been achieved in land- and water-use productivity. In agriculture, the advances are generally referred to as the Green Revolution. In forestry, advances have been brought about through a variety of improvements in forest-management systems, including fast-growing high-yielding plantations, and through genetic improvement. In the early stages of the green revolution, research was directed mainly at plant breeding, fertilizer use and plant protection. However, the pace of advances by these means is slowing. The annual increase in cereal yields in developing countries, which from 1967 to 1982 was 2.9%, has fallen to nearly 1%. As a consequence, more attention has recently been directed at greater productivity in the use of land and water resources – for example, through nutrient recycling and soil and water conservation. A further powerful incentive in this direction has come from considerations of sustainability. Applied to land and water, sustainability means meeting the production needs of present land users while conserving, for future generations, the resources on which that production depends.

Forest Hydrology: Myths and Perceptions

Forest hydrology deals with the hydrological cycle of water from precipitation, interception by the vegetation, infiltration into the soil, drainage to groundwater and runoff. The conventional hydrological approach is to seek at least a 30-year record of stream flows and then it is a straightforward statistical exercise to predict future flows, on the assumption that the data provided are samples from a continuous distribution, unaffected by perturbations. In practice, this assumption is not valid in many developing countries, as entire upland catchment areas are cleared and converted to agriculture within a few years. Nevertheless, such long-term studies of catchments in Europe, America and East Africa gave rise to the widely accepted concept of the benefits of forest protection and rehabilitation in mountainous areas (McCulloch and Robinson, 1993), which still shapes the forest policy of many developing countries.

A major difficulty with the conventional approach is that the findings are rarely appropriate for extrapolation to other areas in similar environments or for situations of rapid changes in land use. For this reason, a physical-process approach with micro-meteorological measurements, which is more complex and expensive, was adopted in a few sites to fully understand the sensitivity to climatic variability and vegetation change (Calder, 1998).

After more than a century of forest hydrology there are still a few controversial issues, or so-called 'myths', which hamper rational land-use decision-making. Calder (1998) summarized these issues as follows:

1. Forests increase rainfall. This is mostly myth because the effects of forests are likely to be small, except for cloud forests.
2. Forests increase runoff. Evidence shows that there is less runoff from forests compared with shorter vegetations, because of higher evaporation losses from trees.
3. Forests regulate flows and increase dry-season flows. This depends more on the water-infiltration properties of soils than the forests *per se*. Many studies show less flow with trees, except for cloud forests.
4. Forests reduce erosion. This depends largely on the management methods employed. Some species of trees, such as teak, may actually cause more erosion than shorter vegetations!

5. Forests reduce floods. There is little scientific evidence to prove a direct relationship between forests and floods. Management activities, such as cultivation, road construction and compaction, are more important.
6. Forests improve water quality. This is mostly true, although bad land management is even more significant.

Unfortunately, these myths persist and continue to dominate the views of policy makers and institutions, creating unnecessary conflicts between the government and local communities. For example, development authorities attributed the disastrous floods in Bangladesh and northern India to the deforestation of the Himalayas, even though the frequency or magnitude of flooding has not increased over the last 120 years! Research by Hofer and Messerli (1998) has shown that precipitation and runoff from the Himalayas do not seem to be important causes of floods in Bangladesh. Instead, the main cause appears to be the rainfall patterns in the Meghalaya hills, followed by the Brahmaputra catchment. Curiously, while politicians and engineers perceive floods as the major hazard in Bangladesh, local farmers consider river erosion a much bigger problem than monsoonal floods, which deposit rich organic soil on their fields and increase crop yield! Based on these findings, traditional thinking regarding flood processes, common practices on flood management and even the prioritization of different hazards in Bangladesh must be revised and differentiated. An important lesson from the Bangladesh case is that priority must be given to the perceptions, experiences and strategies of local communities. Furthermore, the underlying causes of conflict probably hold for many other watersheds and are related to the lack of insight into how landscape mosaics influence watershed functions.

Agroforestry

Agroforestry offers one promising option for efficient and sustainable use of land and water. In simplified terms, agroforestry means combining the management of trees with productive agricultural activities. Agroforestry provides opportunities for forest conversion in the true sense of the term – that is, replacement of natural forests with other tree-based land-use systems. There are also opportunities to use agroforestry for the prevention or reversal of land degradation in the humid tropics (Cooper *et al.*, 1996). There are numerous potential benefits that agroforestry systems can achieve, ranging from diversification of production to improved natural-resources utilization. The key benefits in terms of natural-resources use are as follows:

1. Soil conservation in terms of protection against erosion.
2. Improvement or maintenance of soil fertility.
3. Water conservation and more productive use of water.
4. Providing environmental functions required for sustainability.

A recent review by Wallace *et al.* (2003) has described the above benefits of agroforestry, while this chapter will focus on the water utilization of agroforestry systems.

Can Trees Increase the Productive Use of Rainfall?

Successful plant mixtures appear to be those that make 'better' use of resources by using more of the resource, using it more efficiently or both. In terms of the water use of an agroforestry system, a central question is, therefore, does intercropping woody and non-woody plants increase total harvestable produce by making more effective use of rainfall? It is possible, at least theoretically, that a mixture of trees and crops may improve the overall rainfall-use efficiency – either directly, by more rain being used as transpiration, or indirectly, by increasing water productivity (WP), i.e. the ratio of biomass or yield over volume of water depleted (Seckler *et al.*, Chapter 3, this volume). Analysis of these two effects requires a systematic study of the water balance of agroforestry systems, such as that carried out by

Ong *et al.* (2000) for an agroforestry system in a subhumid part of Kenya. Wallace *et al.* (2003) describe the complexity of the water balance of an agroforestry system on sloping land at Machakos, Kenya. The interception process in agroforestry systems differs from that of forests in two main ways. First, many agroforestry systems tend to have relatively sparse tree densities and, secondly, additional complexity is introduced by the crop component of the system with its rapidly varying canopy cover. The sparse nature of the tree component of agroforests affects two key factors that influence the interception, i.e. the amount of water stored on the tree canopy and the rate of evaporation from the tree canopy.

In semi-arid agroforestry systems, such as those found in Machakos, Kenya (i.e. 10–50% cover), annual interception loss is between 3 and 10% of rainfall. Higher interception losses have been reported in the much denser multistorey agroforestry systems in Costa Rica, where the rainfall is higher and more intense. High interception losses have also been reported for montane forests in humid tropical regions, e.g. as much as 50% by Schellekens *et al.* (1999). The main reason put forward for these high forest interception losses in humid regions is the advection of energy from nearby oceans.

Significant quantities of water can be lost as evaporation from the soil surface, particularly in tropical regions with frequent rainfall, high radiation and sparse ground cover. In agroforestry systems, the presence of a tree canopy decreases the radiation intensity at the ground, thereby reducing soil evaporation compared with cropping systems. This is because total soil evaporation is determined (at least in part) by the radiant energy reaching the soil surface. Direct measurements of soil evaporation made using minilysimeters show reductions in soil evaporation of up to 30% due to the presence of the tree canopy. The reduction in soil evaporation is smaller with sparser tree canopies, 15% of rainfall when cover is ~0.5% and 6% of rainfall when cover is ~0.2% (Wallace *et al.*, 1999).

Clearly, the reductions in soil evaporation produced by tree-canopy shade can help offset the losses of water associated with the tree-canopy interception. The analysis by Wallace *et al.* (1999) indicates that, when annual rainfall is low, the saving in soil evaporation due to canopy shade may be greater than the interception loss. However, once rainfall exceeds ~700 mm per annum, interception losses generally exceed saving in soil evaporation. The exact point at which the two effects completely offset each other will depend mainly on rainfall intensity and soil type.

When rainfall reaches the soil surface, some of it will normally infiltrate into the soil. If the rainfall rate is greater than the infiltration rate, the excess water starts to collect at the surface and, when the surface storage is exceeded, runoff will occur. Infiltration is, therefore, a dynamic process that changes during the course of a rainstorm depending on the soil characteristics, the slope of the land and the rainfall intensity. Where the intercropping of woody and non-woody plants alters any of these factors, then infiltration and runoff may be affected (Kiepe, 1995a). Soil characteristics that affect infiltration are surface crusting, surface storage, saturated hydraulic conductivity and the presence or absence of plant residues. Vegetation cover generally increases infiltration and reduces runoff by altering one or more of these factors. It is well known that conversion from forestry to agriculture can dramatically reduce infiltration within 2–3 years, but restoration of infiltration by reforestation might take several years on severely degraded watersheds. This hysteresis effect is rarely acknowledged and more research is needed to determine how to speed up the restoration of infiltration in conjunction with water-harvesting structures.

There are a number of agroforestry practices that are designed to conserve water and reduce runoff by their direct effect on soil slope. Planting of trees or hedgerows on the contours of sloping land can have the effect of forming natural terraces, as water and soil are collected on the up-slope side of the hedgerow. The barrier effect of the hedgerow not only reduces soil loss but also runoff, commonly to about one-third of its value without hedges. Measurements by drip infiltrometer at Machakos, Kenya, showed that, on a Lixisol (Alfisol) soil with a 14% slope, rates of infiltration were 69 mm h^{-1} under hedgerows and 11 mm h^{-1} under the cropped alleys (Kiepe, 1995a,b). This increased infiltration rate also reduced runoff in these contour-hedgerow systems. Drainage is the component of the water balance that is most difficult to measure directly. At Machakos, it was concluded that drainage from the tree/crop mixture was much less than from the sole crop.

Another way in which trees can affect soil moisture is via the possibility of 'hydraulic lift', in which water taken up by plant roots from moist zones of soil is transported through the root system and released into drier soil (Dawson, 1993). Rainfall captured through stem flow, especially by a woody canopy, can be stored deep in the soil for later use when it is returned to the topsoil beneath the canopy by hydraulic lift. Recently, the opposite of hydraulic lift has been reported in Machakos and elsewhere, i.e. water is taken from the topsoil and transported by roots into the subsoil (Burgess *et al.*, 1998; Smith *et al.*, 1999). This mechanism, termed 'downward siphoning' by Smith *et al.* (1999), would lead to the opposite effect of hydraulic lift and would enhance the competitiveness of deep-rooted trees and shrubs.

The likely effect on each of the water-balance components of the combination of trees with a crop compared with growing the crop alone is summarized in Table 13.1.

Interception losses are around 10% in semi-arid areas but they can be between 10 and 50% in humid tropical climates, depending on whether the location is continental, montane or coastal. This loss will be completely compensated for by a decrease in soil evaporation in a semi-arid climate, but only partially in a humid tropical climate. Runoff, soil moisture and drainage are all likely to decrease in an agroforest in either climatic regime, with the amount varying according to soil type, slope and species. The extra canopy and the ability of tree roots to exploit water at depth in the soil will lead to a general increase in transpiration in the agroforestry system.

Water Productivity in Tree/Crop Mixtures

The WP, or water-use ratio, of any crop or tree/crop mixture is inversely proportional to the mean saturation deficit (expressed in kPa) of the atmosphere, d (Monteith, 1986):

$$\mathrm{WP} = k/d \qquad (13.1)$$

where k is a physiological characteristic specific to a given species. WP can be expressed as kg m^{-3}. The total dry-matter production or grain yield is simply the product of WP and the amount of water used by the vegetation. Theoretical considerations and experimental studies have shown that (at least, under fairly idealized conditions) the product of WP and d is quite conservative among species groups (Ong *et al.*, 1996). Therefore,

Table 13.1. Differences in water-balance components between an agroforestry system with 50% tree cover and a monocrop (from Wallace *et al.*, 2003).

Water-balance component	Difference between agroforestry and a monocrop (% of rainfall), semi-arid climate	Difference between agroforestry and a monocrop (% of rainfall), humid tropical climate
Interception loss	+10%	+10–50%
Runoff	Decrease	Decrease
Soil moisture	Decrease	Decrease
Soil evaporation	−10%	−5%
Transpiration	Increase	Increase
Drainage	Decrease	Decrease

the net effect of atmospheric humidity on any given species is one of the most important factors affecting productivity, since dry-matter production per unit of water transpired decreases by a factor of two as saturation deficit increases from ~2 kPa in moist temperate climates to ~4 kPa in semi-arid areas. For example, experiments in India under similar mean saturation deficits (2.0–2.5 kPa) provided season-long values of 3.9 and 4.6 kg m^{-3} for millet, compared with 1.5–2.0 kg m^{-3} for groundnut. WP for grain yield is usually about half of the values indicated above. However, WP is not always higher in C_4 species, since similar values have been reported for drought-tolerant C_3 species, such as cowpea and cotton, and relatively drought-sensitive cultivars of the C_4 species, sorghum and maize.

Equation 13.1 shows that there are two ways that overall production could be increased. The first is by increasing *k*, the physiological characteristic, which depends on the biochemistry controlling the photosynthetic processes in plant cells. This may be achieved by plant selection (e.g. C_3 or C_4 species) or by breeding or genetically engineering crops with a higher value of *k*. The second way to increase WP is to reduce *d*, either by manipulating the microclimate or by growing plants in a more suitable macroclimate. This means that agroforests growing in humid tropical regions, where the air is more humid (i.e. low *d*), will have higher WP.

In theory, the potential of agroforestry to improve WP is limited compared with intercropping, as the understorey crops are usually C_4 species and the overstorey trees are invariably C_3 species. Improvement in WP is most likely if the understorey crop is a C_3 species, which is usually light-saturated in the open so partial shade may have little effect on its assimilation. However, the shade will reduce transpiration, with the result that WP increases. Evidence from both semi-arid India and subhumid Kenya indicates that WP is about 10% higher in agroforestry systems with a C_3 understorey compared with those with a C_4 understorey (Ong *et al.*, 1996). This may explain why cotton yield in the Sahel is not reduced by the heavy shading of karite (*Vitellaria paradoxa*) and nere (*Parkia biglobosa*) in parklands, while yields of millet and sorghum are reduced by 60% under the same trees (Kater *et al.*, 1992). The same process may explain the observation that in the South and Central American savannahs C_3 grasses are found only under trees and never grow in open grasslands dominated by C_4 grasses.

There is also the potential for microclimate modification in agroforestry systems, due to the presence of an elevated tree canopy. This may alter not only the radiation, but also the humidity and temperature around an understorey crop. Some evidence for this has been found where crops have been grown using trees as shelter-belts, and decreases in *d* have been reported for several crops (Brenner, 1996). Data from an agroforestry trial in Kenya also show that the air around a maize crop growing beneath a *Grevillea robusta* stand is more humid than the free atmosphere above the trees (Ong *et al.*, 2000).

Evidence from a series of shade-cloth trials on maize and bean at Machakos shows small but beneficial effects of shading on crop temperature and crop production when rainfall is inadequate for crop production (Ong *et al.*, 2000), but, unlike the savannah situations, the crops failed because below-ground competition consistently outweighed the benefits of shade. In contrast, Rhoades (1997) reported increased soil water (4–53% greater than in the open) in the crop root zone beneath *Faidherbia albida* canopies in Malawi. In theory, trees can increase soil water content underneath their canopies if the water 'saved' by reduced soil evaporation and funnelling of intercepted rainfall as stem flow exceeds that removed by the root systems beneath tree canopies (Ong and Leakey, 1999). At high tree densities, the proportion of rainfall 'lost' as interception by tree canopies and used for tree transpiration would exceed that 'saved' by shading and stem flow, resulting in drier soil below the tree canopy. Van Noordwijk and Ong (1999) expressed this as the amount of water used per unit of shade. This may be one of the most important factors for the observed difference between savannah and alley-cropping systems and between cloud-forest vegetation and fast-growing tree plantations. Below is a list of the situations in which agroforestry can increase water productivity.

1. Understorey vegetation comprises C_3 plants, e.g. cotton and C_3 grasses.
2. Tree shade increases humidity of understorey vegetation in semi-arid climates, e.g. parkland systems and wind-breaks.
3. Planting of trees as contour hedgerows on hill slopes increases infiltration and reduces runoff.
4. Presence of deep water beyond the reach of crop rooting systems.
5. Trees can use rains that fall outside the cropping season.
6. Trees have canopy architecture that intercepts high amounts of water per unit shade.

Can Agroforestry Mimic the Ecological Functions of Natural Ecosystems?

It is often assumed that appropriate agroforestry systems can provide the environmental functions needed to ensure sustainability and maintain microclimatic and other favourable influences, and that such benefits may outweigh the disadvantages of a more complicated management (Sanchez, 1995). Secondly, it is also assumed that agroforestry might be a practical way to mimic the structure and function of natural ecosystems, since components of the latter result from natural selection towards sustainability and the ability to adjust to perturbations (Van Noordwijk and Ong, 1999). Recent reviews of agroforestry findings, however, have highlighted several unexpected but substantial differences between intensive agroforestry systems and their natural counterparts that would limit their adoption for solving some of the critical land-use problems in the tropics (Rhoades, 1997; Ong and Leakey, 1999; Van Noordwijk and Ong, 1999). The most intractable problems for agroforestry appear to be in the semi-arid tropics. In this section, we describe recent insights into the physiological mechanisms between trees and crops in agroforestry systems and how they might be employed to reduce the trade-offs between environmental functions and crop productivity, i.e. retain the positive effects of trees observed in natural ecosystems.

Resource Capture: Complementarity or Competition?

The principles of resource capture have been used to examine the influence of agroforestry on ecosystem function, i.e. the capture of light, water and nutrients (Ong and Black, 1994), and to better understand the ecological basis of sustainability of tropical forests. For example, Cannell *et al.* (1996) proposed that successful agroforestry systems depend on trees capturing resources that crops cannot. The capture of growth resources by trees and crops can be grouped into three broad categories to show competitive, neutral or complementary interactions. In the neutral or trade-off category, trees and crops exploit the same pool of resources, so that increases in capture by one species result in proportional decreases in capture by the associated species. If trees were able to tap resources unavailable to crops, then the overall capture would be increased, as shown by the convex curve, i.e. complementary use of resources. In the third category, negative interactions between the associated species could result in serious reduction in the ability of one or both species to capture growth resources. It is important to bear in mind that tree–crop interactions may change from one category to another depending on the age, size and population of the dominant species, as well as the supply and accessibility of the limiting growth resources.

Such ideas on capture of deep water and nutrients, coupled with recent innovations in instrumentation (minirhizotrons, sap-flow gauges), have stimulated a new interest in root research (Van Noordwijk and Purnomosidhi, 1995; Khan and Ong, 1996) and increased attention on spatial complementarity in rooting distribution and the potential beneficial effects of deep rooting. Agroforestry is also considered as critical for maintaining ecosystem functioning in parts of Australia where deep-rooted perennial vegetation has been removed and replaced by annual crops and pastures, leading to a profound change in the pattern of energy capture by vegetation, rising water tables and associated salinity (Lefroy and Stirzaker, 1999). The Australian example showed that, compared with the natural ecosystem it replaced, the agricultural system is 'leaky' in terms of resource capture, which gives rise to salinity because of the salts accumulated over millions of years in the Australian continent. Recent investigations in West Africa suggest that a similar magnitude of 'leakiness' is possible when native bush vegetation or woodland, which provides little runoff or groundwater recharge, is converted into millet fields. In West Africa, there is no likelihood of salinity associated with the greater recharge but nutrients are leached to lower depths. The expectation is that agroforestry systems will be able to improve nutrient cycling because of their extensive tree-root systems. Earlier research on South African savannahs has shown that tree roots extend into the open grassland, providing a 'safety net' for recycling water and nutrients and accounting for 60% of the total below-ground biomass (Huntley and Walker, 1982).

Manipulation of Water Use and Root Function

Early studies of spatial complementarity in agroforestry began by examining the rooting architecture of trees and crops grown as pure stands. For example, Jonsson *et al.* (1988) described the vertical distribution of five tree species at Morogoro, Tanzania, and concluded that the root distribution of trees and maize were similar except for *Eucalyptus camaldulensis*, which had a uniform distribution of 1 m. Thus, they concluded that there is little prospect of spatial complementarity if these trees and crops were grown in combination. Recent reviews of the rooting systems of agroforestry systems by Gregory (1996) and Ong *et al.* (1999) essentially supported the earlier conclusion of Jonsson *et al.* (1988).

What is the extent of spatial complementarity in water use when there is such a considerable overlap of the two rooting systems? Results at Machakos, Kenya, consistently showed that there was no advantage in water uptake when there was little water recharge below the crop root zone (Jackson *et al.*, 2000). However, when recharge occurred following heavy rainfall, tree roots were still able to exploit more moisture below the rooting zone of the crops, even when there was a complete overlap of the root systems of trees and crops.

Where groundwater is accessible to tree roots, there is clear evidence for spatial complementarity. For instance, measurements of stable isotopes of oxygen in plant sap, groundwater and water in the soil profile of wind-breaks in the Majjia valley in Niger showed that neem trees, *Azadirachta indica*, obtained a large portion of their water from the surface layers of the soil when rain was abundant, but during the dry season tree roots extracted groundwater (6 m depth) or deep reserves of soil water. In contrast, at a site near Niamey, West Africa, where groundwater was at a depth of 35 m, they found that both the trees and millet obtained water from the same 2–3 m of the soil throughout the year (Smith *et al.*, 1997).

Recently, it has been shown that it is worthwhile to manage below-ground competition by root pruning. For example, Singh *et al.* (1989) demonstrated that root barriers to 50 cm depth are extremely effective in reducing competition between 4-year-old *Leucaena leucocephala* hedgerows and associated crops in semi-arid India. However, the beneficial effects lasted only one season because tree roots reinvaded the crop rooting zone from beneath the root barriers. In contrast, studies in Bangladesh (Hocking, 1998; Hocking and Islam, 1998) revealed that below-ground competition from a wide range of tree species (mainly fruit trees) was virtually eliminated by pruning the lateral roots off the trees. Likewise, studies in Uganda show that competition by *Maesopsis emini*, the fastest growing of 12 tree species compared, was completely eliminated by root pruning (Ong *et al.*, 2002). Results with all species showed that overall tree transpiration was not reduced after root pruning because unsevered roots that were located deeper increased their rates of sap flow to satisfy transpirational demand from the atmosphere. More importantly, root pruning dramatically improved crop growth.

Long-term studies of the effects of root pruning are needed because such information is crucial for promotion of the technology to farmers. While many farmers appreciate the benefits of reduced tree–crop competition following crown pruning, ideas of below-ground competition are completely new to most of them. The experience in Bangladesh indicated that root pruning is feasible when land is relatively scarce (average of 0.8 ha per household), crop yields and earnings are quite low and there is a need to grow more trees for household needs and income generation (Hocking, 1998). While crown pruning yields immediate products (firewood and fodder) and offers longer-term gains in crop yield, benefits from root pruning are delayed and thus farmers need convincing that the effort is worthwhile. In Africa, many farmers consider root pruning too difficult and impractical to execute. Fortunately, the techniques themselves can be quickly and easily demonstrated in the field and experience has shown that farmers can readily change their minds regarding the practicality of incorporating root pruning into their cultivation cycles. However, long-term studies of root-pruned trees are needed to address the following questions:

1. Does forcing tree roots to extract most of their water from beneath the crop-rooting zone influence soil-water recharge at depth, and what are the implications for the long-term water balance?
2. Is the growth of the tree and its stability in the wind significantly influenced by root pruning?
3. Does the loss of fine roots and mycorrhizas diminish the capacity of the tree roots to intercept and recycle plant nutrients that leach from near the soil surface?
4. What are the implications of severing surface roots on N_2 fixation and mycorrhizal activity?

Progress and Challenges Ahead

This review has shown that considerable progress has been made in terms of the hydrology of protected forest catchments and agroforestry plots. Much of this process information has been incorporated into various models in order to extrapolate the findings to other environments (Lawson *et al.*, 1995; Van Noordwijk and Lusiana, 1999). A major challenge is how to look beyond the plot and farm level in order to deal with interactions between the plots that comprise a land-use mosaic at the landscape, watershed and regional scales. The conventional approach is to sum across areas of similar hydroecological conditions, assuming that the factors involved in scaling up are proportional to the area occupied by each zone. However, this approach might overstate the beneficial effects of water saved at the plot level, since water that is used in one plot is not available to down-slope plots. This approach also misses a potentially more important effect: the effect of land use on the quality of water available to down-slope users. Swallow *et al.* (2002) discuss how filters and channels affect lateral flows. A contour hedgerow, for example, may occupy a very small part of the landscape but have disproportionately large effects on reducing surface runoff. A boundary planting of trees running down the slope, on the other hand, will have very little beneficial effect on surface runoff.

These lateral-flow effects need to be taken into account in an assessment of water productivity at the catchment or river-basin scale. Computer-based simulation models can be useful tools for predicting the effects of different land- and water-use regimes on catchment hydrology. Catchment experiments in different sizes and shapes of catchment are needed to fully appreciate the cross-temporal and cross-spatial effects of different configurations of agroforestry, forests, agriculture and other land uses on catchment hydrology. Catchment experiments need to be fully participatory throughout the planning and implementation stages, with research, development and monitoring activities very well integrated (Johnson *et al.*, 2002).

The importance of obtaining more information using a catchment-wide approach is underlined by pointing out that current understanding of resource capture by agroforestry systems is based on well-managed small plots, often in research stations, in which about 30–45% of the rainfall is used for transpiration. Such a level of rainfall utilization is rarely achieved in subsistence agriculture or on a watershed scale and there are still ample opportunities for increasing water use by incorporating trees in the landscape. For example, Rockström (1997) reported that only 6–16% of the total rainfall in a watershed in Niger was utilized by pearl millet for transpiration and the remainder was lost by soil evaporation (40%) or deep drainage (33–40%). Thus, future opportunities for simultaneous agroforestry systems should be explored within the landscape as well as on underutilized niches within and around the farms, such as boundary plantings. Increases in productivity may also be achieved by combining agroforestry with small water-harvesting structures (Rockström *et al.*, Chapter 9, this volume).

Another important challenge is resolving the contrasting perceptions of 'forest functions' by various stakeholders. Existing institutions and policies are largely based on a forest–agricultural land-use dichotomy and this may lead to an unnecessary sense of conflict. For example, Verbist *et al.* (2003) proposed that some farmer-developed agroforestry mosaics in Sumatra are as effective in watershed protection functions as the original forest cover! If this is true, then conflicts between state officials and local communities can be resolved to mutual benefit. Experience from the floods of Bangladesh illustrates the importance of understanding the perceptions of local people and the impacts of land-use mosaics and climate.

In tropical countries where forested catchments are located on submontane and montane elevations, there is a growing concern that deforestation is associated with the decline in river flows, although there is no hard evidence to show that link between deforestation, rainfall and river flow. Nevertheless, evidence from elsewhere showed that montane or 'cloud' forests play a vital role in intercepting moist air and maintaining low flow, which cannot be reproduced by planting fast-growing trees, such as pines and eucalptyus (Schellekens *et al.*, 1999). More research is clearly needed to determine ways to restore the hydrological functions of such vital catchments.

Although there appears to be limited scope for spatial differentiation in rooting between trees and crops, i.e. spatial complementarity in water-limited environments, it is worthwhile to manage below-ground competition by shoot and root pruning. Pruning of lateral roots could redirect root function and be a powerful tool for improving spatial complementarity, provided that there are adequate resources at depth (Ong *et al.*, 2002). Research is needed to examine how the downward displacement of functional tree roots following root pruning affect their role in intercepting nutrients leaching from the zone of crop rooting and the long-term hydrological implications.

In the humid tropics, agroforestry systems offer opportunities for conversion of forested land to productive use, while retaining many of the beneficial effects of watershed functions. Multistrata systems (forest gardens, agroforests) and perennial-crop combinations appear to be the most appropriate agroforestry systems for sustainable land use in the humid tropics, including on sloping land; these systems are commonly found acceptable by farmers. Research is needed to examine their impacts on the quantity and quality of water of the stream flow.

In semi-arid environments, it may be more worthwhile to focus attention on the selection of trees that provide more direct and immediate benefits to farmers (rather than the selection for soil enrichment), with minimum loss of crop productivity. It is perhaps not surprising that farmers are already beginning to experiment with such systems. For example, in the drylands of eastern Kenya, farmers have recently developed an intensive parkland system, using a fast-growing indigenous species, *Melia volkensii* (*Meliaceae*), which provides high-value timber in 5–8 years and fodder during the dry season without apparent loss in crop productivity (Ong *et al.*, 2002).

Finally, although there is clearly great potential for agroforestry systems to conserve and improve resource use, it is by no means suggested that agroforestry automatically brings about all of the above benefits. In order to do so, an agroforestry system must be appropriate for the environment (climate, soil, etc.), practicable (within the local and on-farm constraints), economically viable and acceptable to the farmer. Finally, as with any system of agriculture or forestry, to achieve the potential benefits an agroforestry system needs to be well managed. If these conditions are fulfilled, there is considerable potential for agroforestry to combine production with sustainable land use.

Acknowledgements

This chapter is largely drawn from a review presented by J.S. Wallace, A. Young and C.K. Ong at the United Nations Educational, Scientific and Cultural Organization (UNESCO) Meeting on Water, People and Forest, Kuala Lumpur, August 2001, Malaysia.

References

Brenner, A.J. (1996) Microclimate modifications in agroforestry. In: Ong, C.K. and Huxley, P.A. (eds) *Tree–Crop Interactions. A Physiological Approach.* CAB International, Wallingford, pp. 156–186.

Burgess, S.S.O., Adams, M.A., Turner, N.C. and Ong, C.K. (1998) Redistribution of water within plant root systems. *Oceologia* 115, 306–311.

Calder, I.R. (1998) *Water-resource and Land-use Issues.* SWIM Paper 2, International Water Management Institute, Colombo, Sri Lanka.

Cannell, M.G.R., Van Noordwijk, M. and Ong, C.K. (1996) The central agroforestry hypothesis: the trees must acquire resources that the crop would not otherwise acquire. *Agroforestry Systems* 34, 27–31.

Cooper, P.J.M., Leakey, R.R.B., Rao, M.R. and Reynolds, L. (1996) Agroforestry and the mitigation of land degradation in the humid and sub-humid tropics of Africa. *Experimental Agriculture* 32, 235–290.

Dawson, T.E. (1993) Hydraulic lift and water use by plants: implications for water balance, performance and plant–plant interactions. *Oceologia* 95, 565–574.

Gregory, P.J. (1996) Approaches to modelling the uptake of water and nutrients in agroforestry systems. *Agroforestry Systems* 34, 51–65.

Hocking, D. (1998) Trees in wetland rice fields: a successful tree management technology developed through participatory action–research in Bangladesh. *Agroforestry Today* 10, 4–6.

Hocking, D. and Islam, K. (1998) Trees on farms in Bangladesh. 5. Growth of top and root-pruned trees in wetland rice fields and yield of understorey crops. *Agroforestry Systems* 39, 101–115.

Hofer, T. and Messerli, B. (1998) *Floods in Bangladesh. Process Understanding and Development Strategies.* United Nations University Press, 32 pp.

Huntley, B.J. and Walker, B.H. (eds) (1982) *Ecology of Tropical Savannas.* Springer Verlag, Berlin, 669 pp.

Jackson, N.A., Wallace. J.S. and Ong, C.K. (2000) Tree pruning as a means of controlling water use in an agroforestry system in Kenya. *Forest Ecology and Management* 126(2), 133–148.

Johnson, N., Ravnborg, H.M., Westermann, O. and Probst, K. (2002) User participation in watershed management and research. *Water Policy* 6(3), 507–520.

Jonsson, K., Fidjeland, L., Maghembe, J.A. and Hogberg, P. (1988) The vertical distribution of fine roots of five tree species and maize in Morogoro, Tanzania. *Agroforestry Systems* 6, 63–69.

Kater, L.J.M., Kante, S. and Budelman, A. (1992) Karite (*Vitellaria paradoxa*) and nere (*Parkia biglobosa*) associated with crops in South Mali. *Agroforestry Systems* 18, 89–118.

Khan, A.A.H. and Ong, C.K. (1996) A low cost heat pulse method for measuring tree root water uptake. *Agroforestry Today* 7(2), 19–22.

Kiepe, P. (1995a) *No Runoff, No Soil Loss: Soil and Water Conservation in Hedgerow Barrier Systems.* Tropical Resource Management Papers 10, Agricultural University, Wageningen, The Netherlands.

Kiepe, P. (1995b) Effect of *Cassia siamea* hedgerow barriers on soil physical properties. *Geoderma* 66, 113–120.

Lawson, G.J., Crout, N.M.J., Levy, P.E., Mobbs, D.C., Wallace, J.S., Cannell, M.G.R. and Bradley, R.G. (1995) The tree crop interface: representation by coupling forest and crop process-models. *Agroforestry Systems* 30, 199–221.

Lefroy, E.C. and Stirzaker, R.J. (1999) Agroforestry for water management in southern Australia. *Agroforestry Systems* 45, 277–302.

McCulloch, J.S.G. and Robinson, M. (1993) History of forest hydrology. *Journal of Hydrology* 150, 189–216.

Monteith, J.L. (1986) How do crops manipulate water supply and demand? *Philosophical Transactions of the Royal Society, London, B* 316, 245–259.

Ong, C.K. and Black, C.R. (1994) Complementarity in resource use in intercropping and agroforestry systems. In: Monteith, J.L., Scott, R.K. and Unsworth, M.H. (eds) *Resource Capture by Crops*. Nottingham University Press, Loughborough, UK, pp. 255–278.

Ong, C.K. and Leakey, R.R.B. (1999) Why tree–crop interactions in agroforestry appear at odds with tree–grass interactions in tropical savannahs. *Agroforestry Systems* 45, 109–129.

Ong, C.K., Black, C.R., Marshall, F.M. and Corlett, J.E. (1996) Principles of resources capture and utilisation of light and water. In: Ong, C.K. and Huxley, P.A. (eds) *Tree–Crop Interactions – a Physiological Approach*. CAB International, Wallingford, UK, pp. 73–158.

Ong, C.K., Deans, J.D., Wilson, J., Mutua, J., Khan, A.A.H. and Lawson, E.M. (1999) Exploring below ground complementarity in agroforestry using sap flow and root fractal techniques. *Agroforestry Systems* 44, 87–103.

Ong, C.K., Black, C.R., Wallace, J.S., Khan, A.A.H., Lott, J.E., Jackson, N.A., Howard, S.B. and Smith, D.M. (2000) Productivity, microclimate and water use in *Grevillea robusta* based agroforestry systems on hillslopes in semi-arid Kenya. *Agriculture, Ecosystems and Environment* 80, 121–141.

Ong, C.K., Wilson, J., Deans, J.D., Mulatya, J., Raussen, T. and Wajja-Musukwe, N. (2002) Tree–crop interactions: manipulation of water use and root function. *Agricultural Water Management* 53, 171–186.

Rhoades, C.C. (1997) Single-tree influences on soil properties in agroforestry: lessons from natural forest and savanna ecosystems. *Agroforestry Systems* 35, 71–94.

Rockström, J. (1997) On-farm agrohydrological analysis of the Sahelian yield crisis: rainfall partitioning, soil nutrients and water use efficiency of pearl millet. PhD dissertation, University of Stockholm.

Sanchez, P. (1995) Science in agroforestry. *Agroforestry Systems* 30, 5–55.

Schellekens, J., Scatena, F.N., Bruijnzeel, L.A. and Wickel, A.J. (1999) Modelling rainfall interception by a lowland tropical rain forest in northeastern Puerto Rico. *Journal of Hydrology* 225, 168–184.

Singh, R.P., Ong, C.K. and Saharan, N. (1989) Above- and below-ground interactions in alley cropping in semiarid India. *Agroforestry Systems* 9, 259–274.

Smith, D.M., Jarvis, P.G. and Odongo, J.C.W. (1997) Sources of water used by trees and millet in Sahelian windbreak systems. *Journal of Hydrology* 198, 140–153.

Smith, D.M., Jackson, N.A., Roberts, J.M. and Ong, C.K. (1999) Reverse flow in tree roots and downward siphoning of water by *Grevillea robusta*. *Functional Ecology* 13(2), 256–264.

Swallow, B.M., Van Noordwijk, M. and Garrity, D.P. (2002) The effects of scales, flows and filters on property rights and collective action in watershed management. *Water Policy* 6(3), 457–474.

Van Noordwijk, M. and Lusiana, B. (1999) Wanulcas, a model of water, nutrient and light capture in agroforestry systems. *Agroforestry Systems* 43, 217–242.

Van Noordwijk, M. and Ong, C.K. (1999) Can the ecosystem mimic hypotheses be applied to farms in African savannahs? *Agroforestry Systems* 45, 131–158.

Van Noordwijk, M. and Purnomosidhi, P. (1995) Root architecture in relation to tree–soil–crop interactions and shoot pruning in agroforestry. *Agroforestry Systems* 30, 161–173.

Verbist, B.J.P., Van Noordwijk, M., Tameling, A.C., Schmitz, K.C.L. and Ranier, S.B.L. (2003) A negotiation support tool for assessment of land use change impacts on erosion in a previously forested watershed in Lampung, Sumatra, Indonesia (in press).

Wallace, J.S., Jackson, N.A. and Ong, C.K. (1999) Modelling soil evaporation in an agroforestry system in Kenya. *Agricultural and Forest Meteorology* 94, 189–202.

Wallace, J.S., Young, A. and Ong, C.K. (2003) The potential of agroforestry for sustainable land and water management. *Journal of Hydrology* (in press).

14 Water Productivity and Potato Cultivation

Walter T. Bowen
International Fertilizer Development Center, Muscle Shoals, Alabama, USA and International Potato Center (CIP), Quito, Ecuador

Abstract

This chapter provides a review of work done at the International Potato Center (CIP) on improving water productivity in potato. Generally, potato is shallow-rooted and sensitive to even mild water deficits. Most of CIP's work related to water productivity was done in the 1980s as part of a research programme to develop improved germplasm and agronomic practices for potato production in warm tropical environments. Heat-tolerant as well as drought-tolerant materials were selected and tested under a range of warm climates, with studies conducted to quantify evapotranspiration, stomatal conductance, leaf water potential, soil water dynamics and root growth. These same parameters were also determined in agronomic field experiments designed to quantify the effects of mulching, intercropping and close plant spacing on yield and water-use efficiency. Although needed, similar detailed studies on water-productivity components have yet to be done for potato grown more commonly in cooler environments at high altitudes in the tropics.

Introduction

In terms of global production, potato (*Solanum tuberosum* L.) is the fourth most important food crop after maize, rice and wheat (Table 14.1). The current production of 306 million t represents a modest increase worldwide of 15.5% since the early 1960s. Such global statistics, however, mask the much greater expansion of potato production that has taken place in developing countries versus developed countries during the past 40 years. Hence, a more revealing story is told through the statistics shown in Fig. 14.1. While global production has increased from 265 to 306 million t, the proportion of that production coming from developed countries has decreased from 89% to 58%, which translates into an actual decrease in production in developed countries. Meanwhile, the proportion of global production coming from developing countries has increased from 11% to 42%, representing a remarkable increase in production of 100 million t, i.e. from 29 to 129 million t (340.9% increase from 1961–1963 to 1998–2000).

Much of the increase in potato production in developing countries has occurred in Asia, most notably in China and India. Although yields have improved in both countries, the increase in production can be attributed mainly to a continuous expansion of area planted to potato (Fig. 14.2). From 1985–1987 to 1995–1997, the average annual rate of

Table 14.1. Global area, yield and production for the four most important food crops averaged over the years 1998–2000 (FAOSTAT, October 2001).

Crop	Area (million ha)	Yield (t ha^{-1})	Production (million t)
Maize	139	4.4	604
Rice	154	3.9	595
Wheat	215	2.7	588
Potato	19	15.9	306

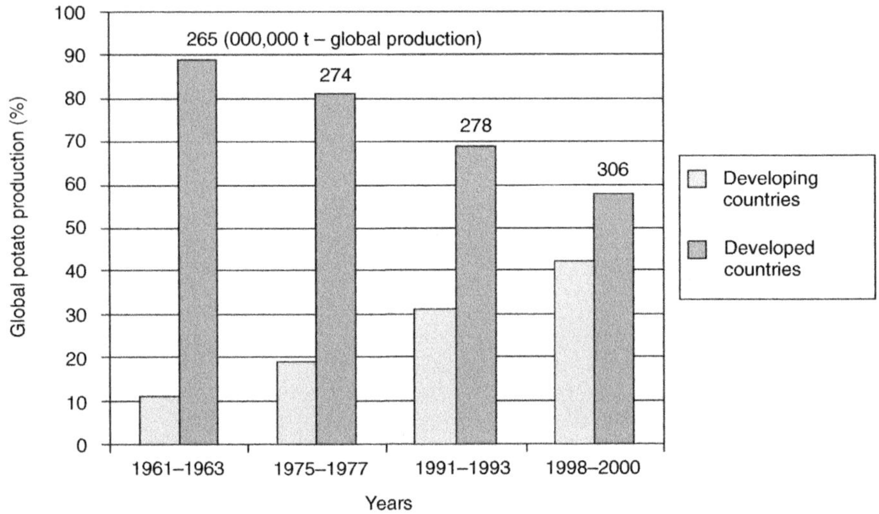

Fig. 14.1. Percentage of global potato production coming from developed and developing countries from 1961–1963 to 1998–2000 (FAOSTAT, October 2001).

growth in potato production was 6.2% in China, with a growth rate in area planted of 3.3% $year^{-1}$ (CIP, 1998). Most of the growth in potato production in Asia has occurred in the interior highlands of China and on the Indo-Gangetic plains of India. Potato has emerged as an important food crop on the Indo-Gangetic plains following an expansion in irrigation infrastructure and the construction of large cold-storage facilities for storing potato before sale and as a seed crop during summer (Bardhan Roy *et al.*, 1999). Whereas potato is grown as a cool, dry-season (winter) irrigated crop on the Indo-Gangetic plains, in China it is grown mostly under rain-fed conditions during summer.

Potato production is expected to continue to increase in developing countries, providing an important source of food, nutrition and income. Recent projections for developing countries show an expected annual growth rate in potato production of 2.7% during the period 1993–2020 (Scott *et al.*, 2000). For global potato production, Scott *et al.* (2000) estimate that 80% of the projected increase will come from developing countries, with 64% coming from Asia alone. They also go on to project that, in Asia, most of the increase in production will have to come from improvements in yield, since area expansion is expected to decrease substantially.

The growing demand for potato – as both a fresh and processed food – and a decreasing availability of land for area expansion mean that yields will have to be improved through some combination of germplasm enhancement, better crop protection and more efficient and productive management

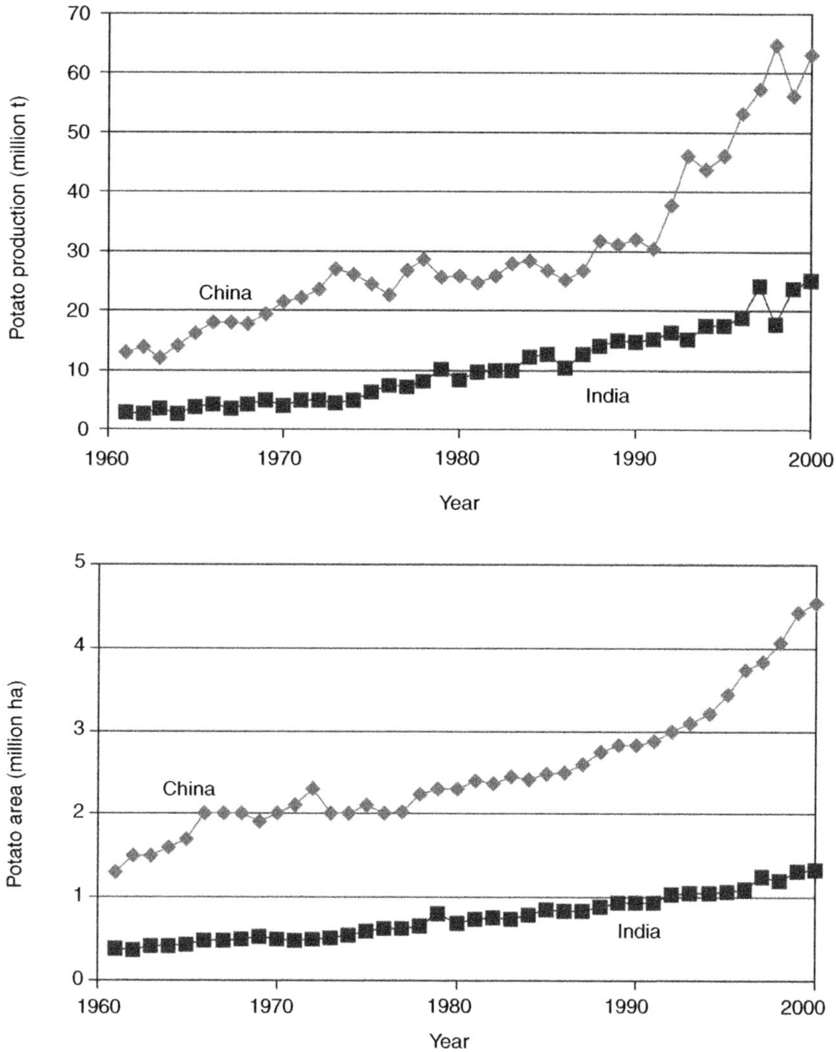

Fig. 14.2. Potato production and area harvested in China and India from 1961 to 2000 (FAOSTAT, October 2001).

systems. The average yield of 15.9 t ha^{-1} currently estimated at the global level (Table 14.1) is much below the yields of 30–50 t ha^{-1} commonly obtained across a range of environments and management systems, so it would seem that there is considerable scope for improvement (Allen and Scott, 1992). Critical to achieving improved yields will be access to an adequate water supply, including more efficient use of all available water in both irrigated and rain-fed systems.

The purpose of this chapter is to review water-use issues in potato production, at least to the extent that the research programme at the International Potato Center (CIP) has been able to address them. More specifically, the review will focus on what has been done at CIP that could be useful for improving the efficiency or productivity of water in both rain-fed and irrigated potato systems in developing countries. Topics to be addressed include an analysis of the relationship between yield and water supply, genotypic differences in response to water supply, the impact of agronomic practices on water use and the potential role of simulation models as both research and application tools. Since research on these topics at CIP has been limited, the reader may wish to refer to recent reviews of more extensive work by the wider research community in Wright and Stark (1990), Gregory and Simmonds (1992), Vayda (1994) and Jefferies (1995).

Yield and Water Supply

Most of the work done at CIP on water use and potato production was conducted in the 1980s as part of a research programme to develop improved germplasm and agronomic practices for potato production in warm tropical environments (Midmore and Rhoades, 1987; Midmore *et al.*, 1991; Midmore, 1992). Warm tropical environments were generally defined as those having a day length of 10–14 h, minimum night-time temperatures of 18–20°C, mean maximum temperatures greater than 25°C and mean annual soil temperatures at 50 cm depth of 22°C or more. Since high temperature was considered to be a primary cause of low yields in warm climates, the programme led to an active breeding effort to develop heat-tolerant clones.

The clones that were found to possess heat tolerance were those that were strongly induced to initiate tubers under high temperatures, were more efficient in the conversion of intercepted radiation to dry matter under high temperatures and matured earlier than non-tolerant clones. Later, field studies showed that selection for heat tolerance had also improved water-use efficiency, as calculated from the per unit increase in fresh tuber yield per unit of applied water, but only in warm climates (Trebejo and Midmore, 1990). In this section, we shall examine more closely the water-use efficiencies defined by the Trebejo and Midmore (1990) studies. Further reference to water-use efficiency will stress instead the term water productivity, which will be taken here to mean the ratio of fresh tuber yield to applied water expressed as kg ha^{-1} mm^{-1}.

Using a single line-source-sprinkler irrigation system, Trebejo and Midmore (1990) studied the growth and yield response of three potato clones to variable rates of water supply in contrasting hot and cool seasons in the coastal desert environment of Lima, Peru (12°05′S at 240 m above sea level). Like most of coastal Peru, there is no effective rainfall in Lima and plants only grow with irrigation. The hot (summer) and cool (winter) seasons in this environment are unique not only because of differences in temperature regimes, but also because the winter season has much less solar radiation and higher humidity due to the impact of the cold Humbolt current on local weather. Basically, the winter remains somewhat cloudy, with frequent periods of fog. During El Niño years, when warmer waters displace the current, the winter season is distinctly cooler, with greater solar radiation and less humidity. The Trebejo and Midmore (1990) study was conducted in 1985, which was not an El Niño year, so the winter and summer seasons were typical of most years.

The three potato clones chosen by Trebejo and Midmore (1990) included two heat-tolerant clones (DTO33 and LT1) and one Peruvian cultivar (Revolución) well adapted to normal coastal winter conditions. All three were planted in rows parallel to the single-line source and managed equally, except for the amount of water during the summer and winter seasons of 1985. The single-line source was set up to apply water in a decreasing gradient away from the line, with growth and yield analyses conducted on plants in rows 2, 4 and 6 extending out from the line. In both seasons, irrigation water was applied so that row 2 received an amount needed to replace water evaporated from a class A evaporation tank. The other rows therefore received progressively less water, and there was no effective rainfall during either season. No runoff from the surface or drainage below the rooting zone was assumed to occur, so all applied water was available for plant uptake.

The response of fresh tuber yield to applied water, averaged across the three clones, is shown in Fig. 14.3. Clearly, water productivity was much greater during the

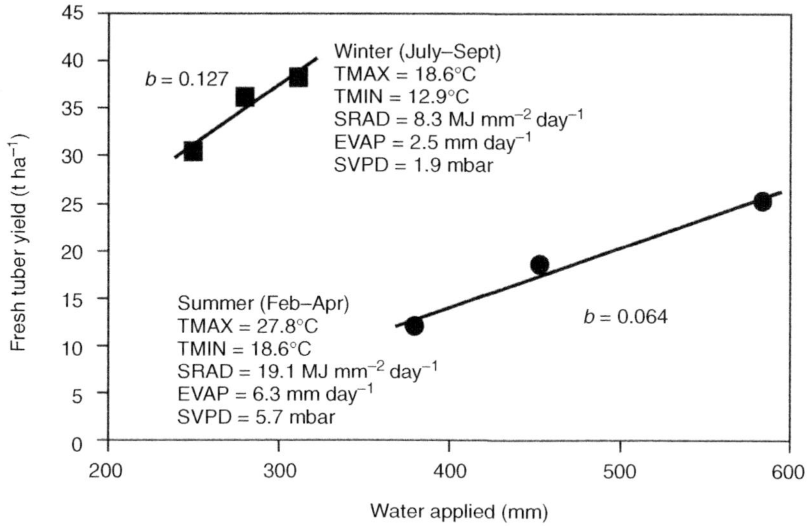

Fig. 14.3. Relationship between fresh tuber yields averaged across three potato clones and total water applied during the 1985 summer and winter seasons in Lima, Peru. The abbreviations represent weather-related variables averaged daily over respective growing seasons: TMAX is maximum temperature, TMIN is minimum temperature, SRAD is global solar radiation, EVAP is evaporation and SVPD is saturation vapour-pressure deficit of the air (data from Trebejo and Midmore, 1990).

winter season compared with the summer season. The relationship between fresh tuber yield and water applied was linear for both seasons, which is generally expected if soil evaporative losses and vapour-pressure deficits are equal across treatments within the same season (Sinclair *et al.*, 1984). The slopes of the linear relationships indicated that water productivity during the winter was 127 kg ha^{-1} mm^{-1}, but only 64 kg ha^{-1} mm^{-1} during the summer. These values are close to those of 54–120 kg ha^{-1} mm^{-1} reported by Wright and Stark (1990) for several studies with potato.

The amount of water applied to the potato crop grown during the more humid winter ranged from 250 mm to 312 mm, with the latter representing what was needed to replace class A pan-evaporation. A reduction of only 62 mm in water applied resulted in a decrease in fresh tuber yield from 38.2 t ha^{-1} to 30.3 t ha^{-1}. For the summer crop, applied water ranged from 380 mm to 584 mm and the associated yields ranged from 12.1 t ha^{-1} to 25.4 t ha^{-1}. Therefore, about half as much water was used during the winter to produce 150% more yield than that obtained in the summer. The same studies referred to earlier by Wright and Stark (1990) showed that seasonal water use averaged 607 mm (range 450–800 mm) and yield levels averaged 56 t ha^{-1} (range 33–72 t ha^{-1}).

For both seasons, irrigation based on class A pan-evaporation did not appear to result in excessive water application, since maximum yields were obtained using this method. Nevertheless, seasonal evapotranspiration (ET) estimates that were obtained from gravimetric soil samples showed that cumulative water applied by irrigation was in excess of cumulative ET by about 90 mm (Trebejo and Midmore, 1990). Concerns about excessive irrigation or poor drainage are always valid in potato cultivation, since yield is as sensitive to reduced aeration as it is to drought stress, although yield loss due to the latter is more common (Wright and Stark, 1990). When less water was applied than that needed to replace class A pan-evaporation, water deficits obviously occurred and yield levels were depressed (Fig. 14.3).

Needless to say, potato farming in coastal Peru occurs during the winter, when the cool humid conditions are favourable for growth and more efficient use of irrigation water. The cooler temperatures result in delayed maturity, which provides more time for the interception of solar radiation and the conversion of intercepted radiation to dry matter. For the Trebejo and Midmore (1990) study, the mean harvest date of all three clones was 91 days after planting for the summer and 110 days after planting for the winter. During winter, less soil evaporation and the smaller vapour-pressure deficit of the air also combine to enhance water productivity when compared with the summer. Generally, more humid environments provide greater water productivity because of lower vapour-pressure deficit (Sinclair *et al.*, 1984).

Genotypic Differences in Response to Water Supply

Data from the Trebejo and Midmore (1990) study can also be used to illustrate genotypic differences in response to water supply. While the previous section compared water-productivity values averaged across all three clones, this section will discuss results for each clone averaged across all water-application levels.

Figure 14.4 shows the fresh tuber yield and the water productivity calculated for each clone during the winter and summer seasons. The two heat-tolerant clones, LT1 and DTO33, yielded more in the summer than Revolución, the cultivar better adapted to coastal winter conditions. Selection for heat tolerance has evidently improved adaptability to warmer climates, which is also reflected in greater water productivity when the heat-tolerant clones are compared with Revolución in the summer. None the less, better yields and water productivity were realized for all clones when grown during the winter compared with the summer. The highest yield and water productivity were obtained with Revolución grown during the winter.

In a comparison of the two heat-tolerant clones, LT1 probably performed better than DTO33 because of its later maturity. In the summer, LT1 was harvested 92 days after planting (DAP) versus 81 DAP for DTO33. In the winter, LT1 was harvested 112 DAP versus 103 DAP for DTO33.

Other evidence for genotypic differences in water productivity comes from a study of leaf resistances for 14 potato cultivars, conducted by Ekanayake and de Jong (1992). Greater water productivity could conceivably be obtained through manipulation of stomatal behaviour so that midday water stress triggers midday stomatal closure to prevent high transpiration rates and minimize damage to the crop (Sinclair *et al.*, 1984). Ekanayake and de Jong (1992) did find significant genotypic differences in leaf resistance or stomatal behaviour, indicating the possibility of improving water productivity and drought resistance through improved germplasm.

Another avenue for improving the water productivity of potato could be through deeper rooting to extract soil moisture from deeper in the profile. Recent investigation of a large number of clones in the desert conditions of Lima has shown genotypic differences in depth to rooting (N. Pallais, International Potato Center, Lima, Peru, 2000, personal communication).

Impact of Agronomic Practices on Water Use

Agronomic practices that reduce soil evaporation should tend to increase water productivity. In a study of the benefits of surface mulches on yield, Midmore *et al.* (1986b) showed that mulch increased tuber yield during the summer in Lima by 20%. Although it was not directly determined how much water productivity might have been affected, Midmore *et al.* (1986a) did conclude that mulch always increased soil-moisture retention. Thus enhanced yields obviously mean increased water productivity for the same amount of applied water, at least for the summer season in Lima. These same studies also showed that mulch resulted in earlier tuber initiation and greater tuber bulking rates.

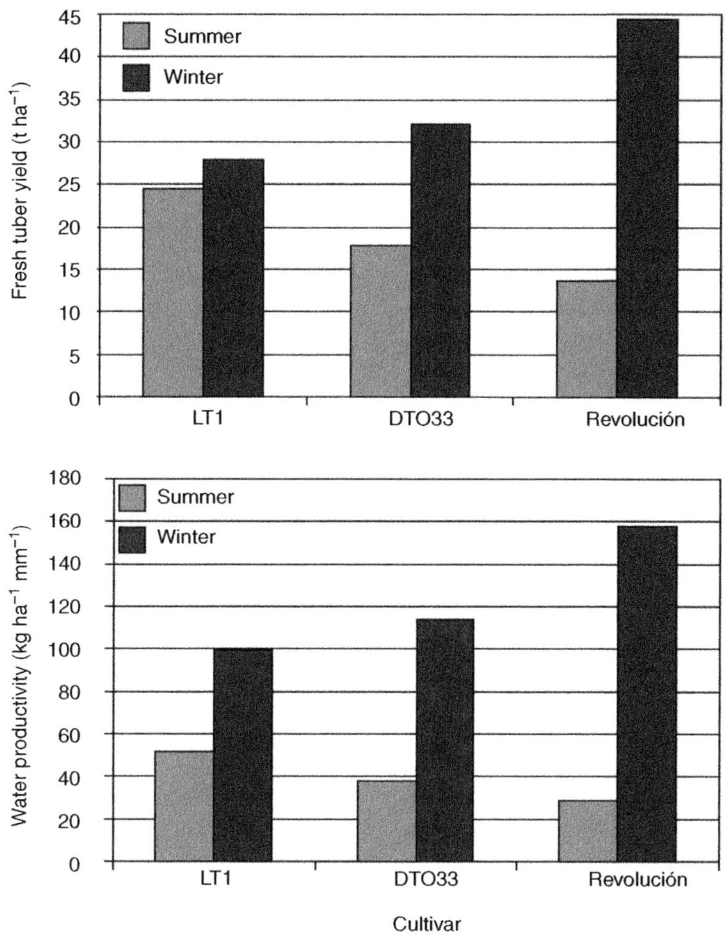

Fig. 14.4. Fresh tuber yields and estimated water productivity (expressed as kg fresh tubers ha^{-1} mm^{-1} of applied water) for three cultivars grown during the 1985 summer and winter seasons in Lima, Peru (data from Trebejo and Midmore, 1990).

Manrique and Meyer (1984) also studied the impact of mulches on potato yield during winter and summer seasons in Lima. They saw no effect on yields during the winter, but summer yields were increased by 58% with surface mulch, which improved soil-moisture retention.

Potential Role of Simulation Models

Owing to advances in computer technology and accessibility, models of soil and plant systems have become increasingly valuable instruments for assimilating knowledge gained from experimentation. Their use within a research programme has the potential to increase efficiency by emphasizing process-based research, rather than the study of merely site-specific net effects. Consequently, a modelling approach lends structure to a research programme, helping to focus on the quantitative description of soil and plant processes. This information can then be used to predict how the system might respond to different environmental and management factors. A modelling approach also provides a dynamic, quantitative framework for multidisciplinary input.

Although there are several potato models available, at CIP we have chosen to initially work with the SUBSTOR-Potato model (Ritchie *et al.*, 1995). This model simulates, on a daily basis, the accumulation and partitioning of biomass and the phenological development of a potato crop as influenced by temperature, photoperiod, intercepted radiation and soil water and nitrogen (N) supply. To illustrate the comprehensive nature of the model, a listing of the processes that are simulated when accounting for N demand and supply are provided in Table 14.2. The environmental factors that the model uses to simulate each process are also shown. The model is thought to be especially valuable for studies of the interaction of water and N supply. SUBSTOR has been tested in various environments and has generally performed well when simulated data have been compared with measured data (Griffin *et al.*, 1995; Mahdian and Gallichand, 1995; Ritchie *et al.*, 1995; Travasso *et al.*, 1997; Bowen *et al.*, 1999).

We have begun to calibrate and test the SUBSTOR-Potato model across a wide range of environments and management systems in the Andes and more recently in the Indo-Gangetic plains of India. A comparison of simulated and observed tuber yields obtained from field studies in the Andes is shown in Fig. 14.5. This limited testing shows that the model realistically

Table 14.2. Major processes that are simulated and environmental factors that affect those processes in the N submodel of SUBSTOR-Potato (version 3).

Process simulated	Main factors influencing process
Crop N demand	
Growth	Solar radiation, temperature
Development	Photoperiod, temperature
Soil N supply	
Mineralization/immobilization	Soil temperature, soil water, C/N ratio
Nitrification	Soil temperature, soil water, soil pH, NH_4^+
Denitrification	Soil temperature, soil water, soil pH, soil C
NO_3^- leaching	Drainage
Urea hydrolysis	Soil temperature, soil water, soil pH, soil C
Uptake	Soil water, NO_3^-, NH_4^+, crop demand, root length density

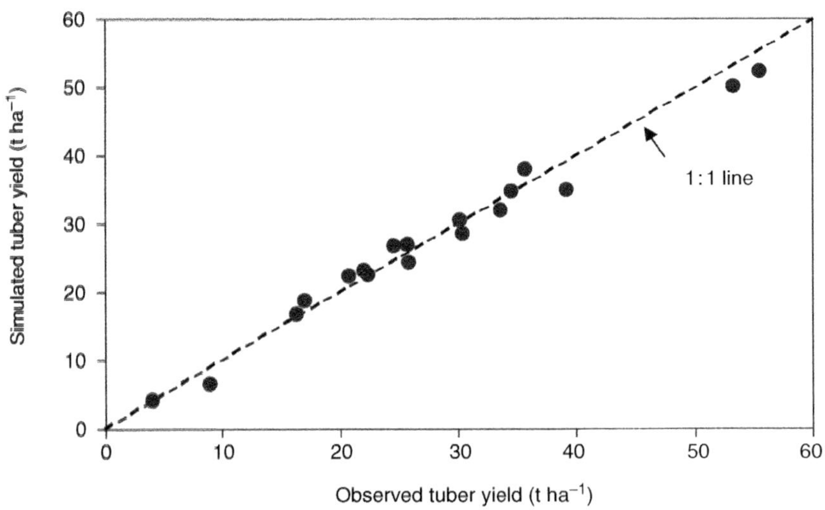

Fig. 14.5. Relationship between simulated and observed fresh weight of potato tubers for sites in Bolivia, Ecuador, Peru and Venezuela.

simulated fresh tuber yields that ranged from 4 t ha^{-1} to 56 t ha^{-1} due to differences in weather, soils, cultivars and management. Further testing is under way as we continue to critically evaluate the performance of SUBSTOR and search for ways to improve it and other models as research, education and management tools.

We hope that ongoing evaluation of the SUBSTOR-Potato model will provide us with a powerful tool for analysing the impact of weather variability on potato yields and water productivity under different management practices. The impact of weather variability on a given management option can be quantified by running the model with many different years of weather data. The weather data may be the daily values actually recorded at the site, or the model may statistically generate weather using historical data.

Conclusions

To date, the only major research done at CIP on addressing water-productivity issues has come about indirectly from its research programme in the 1980s to develop potato adapted to warm climates. Although this programme was successful in identifying heat-tolerant clones that can produce tubers under warm-temperature regimes and that may also have higher water productivity when grown in such environments, potato continues to be best adapted to cool-temperature regimes for both tuber-yield potential and greater water productivity. Water productivity will inevitably continue to be higher in humid conditions with low vapour-pressure deficit gradients. Nevertheless, it is clear that there exists a useful range of genetic variability that could be taken advantage of for more drought-tolerant and water-use-efficient genotypes, which should prove beneficial for both rainfed and irrigated potato systems. Other potentially fruitful lines of research for increasing water productivity at the field level include the investigation of more efficient irrigation systems, e.g. drip irrigation, and soil-management practices that emphasize less tillage and the maintenance of more residue on the soil surface.

References

Allen, E.J. and Scott, R.K. (1992) Principles of agronomy and their application in the potato industry. In: Harris, P. (ed.) *The Potato Crop: the Scientific Basis for Improvement,* 2nd edn. Chapman & Hall, London, pp. 816–881.

Bardhan Roy, S.K., Walker, T.S., Khatana, V.S., Saha, N.K., Verman, V.S., Kadian, M.S., Haverkort, A.J. and Bowen, W. (1999) Intensification of potato production in rice-based cropping systems: a rapid rural appraisal in West Bengal. In: *CIP Program Report 1997–1998.* International Potato Center, Lima, Peru, pp. 205–212.

Bowen, W., Cabrera, H., Barrera, V. and Baigorria, G. (1999) Simulating the response of potato to applied nitrogen. In: *CIP Program Report 1997–1998.* International Potato Center, Lima, Peru, pp. 381–386.

CIP (International Potato Center) (1998) *CIP Potato Facts: a Compendium of Key Figures and Analysis for 32 Important Potato-producing Countries.* International Potato Center, Lima, Peru.

Ekanayake, I.J. and de Jong, J.P. (1992) Stomatal response of some cultivated and wild tuber-bearing potatoes in warm tropics as influenced by water deficits. *Annals of Botany* 70, 53–60.

Gregory, P.J. and Simmonds, L.P. (1992) Water relations and growth of potatoes. In: Harris, P. (ed.) *The Potato Crop: the Scientific Basis for Improvement,* 2nd edn. Chapman & Hall, London, pp. 214–246.

Griffin, T.S., Johnson, B.S. and Ritchie, J.T. (1995) *A Simulation Model for Potato Growth and Development: SUBSTOR-Potato Version 2.0.* IBSNAT Research Support Series 02, Department of Agronomy and Soil Science, University of Hawaii, Honolulu, Hawaii.

Jefferies, R.A. (1995) Physiology of crop response to drought. In: Haverkort, A.J. and MacKerron, D.K.L. (eds) *Potato Ecology and Modelling of Crops under Conditions of Limiting Growth.* Kluwer Academic Publishers, Dordrecht, The Netherlands, pp. 61–74.

Mahdian, M.H. and Gallichand, J. (1995) Validation of the SUBSTOR model for simulating soil water content. *Transactions of the ASAE* 38, 513–520.

Manrique, L.A. and Meyer, R. (1984) Effects of soil mulches on soil temperature, plant growth and potato yields in an aridic isothermic environment in Peru. *Turrialba* 34, 413–420.

Midmore, D.J. (1988) Potato (*Solanum* spp.) in the hot tropics. VI. Plant population effects on soil temperature, plant development and tuber yield. *Field Crops Research* 19, 183–200.

Midmore, D.J. (1992) Potato production in the tropics. In: Harris, P. (ed.) *The Potato Crop: the Scientific Basis for Improvement*, 2nd edn. Chapman & Hall, London.

Midmore, D.J. and Rhoades, R.E. (1987) Applications of agrometeorology to the production of potato in the warm tropics. *Acta Horticulturae* 215, 103–136.

Midmore, D.J., Berrios, D. and Roca, J. (1986a) Potato (*Solanum* spp.) in the hot tropics. II. Soil temperature and moisture modification by mulch in contrasting environments. *Field Crops Research* 15, 97–108.

Midmore, D.J., Roca, J. and Berrios, D. (1986b) Potato (*Solanum* spp.) in the hot tropics III. Influence of mulch on weed growth, crop development, and yield in contrasting environments. *Field Crops Research* 15, 109–124.

Midmore, D.J., Berrios, D. and Roca, J. (1988) Potato (*Solanum* spp.) in the hot tropics V. Intercropping with maize and the influence of shade on tuber yields. *Field Crops Research* 18, 159–176.

Midmore, D.J., Ekanayake, I.J., Benz, J.S. and Payton, F.V. (1991) Contribution of recent physiological studies to the improvement of potato production in warm climates. In: Govinden, N., Julien, M.H.R., Hunt, G.L.T. and Autrey, L.J.C. (eds) *Production, Post-harvest Technology and Utilization of Potato in the Warm Tropics. Proceedings of a Workshop, 23–27 July 1990, Réduit, Mauritius.* Mauritius Sugar Industry Research Institute, Reduit, Mauritius, pp. 45–51.

Ritchie, J.T., Griffin, T.S. and Johnson, B.S. (1995) SUBSTOR: functional model of potato growth, development and yield. In: Kabat, P. *et al.* (eds) *Modelling and Parameterization of the Soil–Plant–Atmosphere System: a Comparison of Potato Growth Models.* Wageningen Pers., Wageningen, pp. 401–435.

Scott, G.J., Rosegrant, M.W. and Ringler, C. (2000) Global projections for root and tuber crops to the year 2000. *Food Policy* 25, 561–597.

Sinclair, T.R., Tanner, C.B. and Bennett, J.M. (1984) Water-use efficiency in crop production. *Bioscience* 34, 36–40.

Travasso, M.I., Caldiz, D.O. and Saluzzo, J.A. (1997) Yield prediction using the SUBSTOR-Potato model under Argentinian conditions. *Potato Research* 39, 305–312.

Trebejo, I. and Midmore, D.J. (1990) Effect of water stress on potato growth, yield and water use in a hot and a cool tropical climate. *Journal of Agricultural Science, Cambridge* 114, 321–334.

Vayda, M.E. (1994) Environmental stress and its impact on potato yield. In: Bradshaw, J.E. and Mackay, G.R. (eds) *Potato Genetics.* CAB International, Wallingford, UK, pp. 239–255.

Wright, J.L. and Stark, J.C. (1990) Potato. In: *Irrigation of Agricultural Crops.* Agronomy Monograph 30, pp. 859–888.

15 Rice–Wheat Cropping Systems in the Indo-Gangetic Plains: Issues of Water Productivity in Relation to New Resource-conserving Technologies

Peter R. Hobbs[1] **and Raj K. Gupta**[2]
[1]*Centro Internacional de Mejoramiento de Maize y Trigo, Mexico City, Mexico;*
[2]*Rice–Wheat Consortium for the Indo-Gangetic Plains, New Delhi, India*

Abstract

The rice–wheat cropping system is found on 13.5 million ha in south Asia and is one of the most important cropping patterns for food self-security in the region. This system is found in the fertile, hot semi-arid to hot subhumid regions of the Indus and Gangetic alluvial plains of Bangladesh, India, Nepal and Pakistan. Irrigation is commonly used to stabilize the productivity of this system, using canal and tube-well water. Area and yield growth have been responsible for continued production growth for these cereals over the past 30 years and have matched the population growth and demand for food. This growth over the past 30 years was based on key inputs, such as variety, fertilizer and irrigation, with most of the investment coming from the public sector. Future growth required to meet population growth will be close to 2.5% $year^{-1}$ and must come from yield rather than from area growth, since the latter will decline as urbanization and industries spread to prime agricultural land. Competition for water will be a major challenge for agriculture and it is imperative that this scarce resource is used efficiently. This chapter describes various resource-conserving technologies that are being promoted by the rice–wheat consortium (one of seven Consultative Group on International Agricultural Research (CGIAR) ecoregional initiatives) to attain the goal of raising productivity in the region and meeting food-security needs while, at the same time, efficiently using natural resources, including water, providing environmental benefits, improving the rural livelihoods of farmers and helping to alleviate poverty. This technology of the post-green revolution will depend on farmer adoption and investment. Increasing and improving stakeholder participation in experimentation and fine-tuning of the technology will be a key to success.

Introduction

The rice–wheat (RW) cropping system is found on 13.5 million ha in south Asia and is one of the most important cropping patterns for food self-security in the region. Another 10 million ha are found in China, mostly in the central areas of the Yangtze River valley. This system is found in the fertile, alluvial Indo-Gangetic plains (IGP) of Bangladesh, India, Nepal and Pakistan of south Asia. In this system, rice is grown in the warm, sub-humid, monsoonal, summer months and wheat in the cooler, drier, winter season.

 Water Productivity in Agriculture: Limits and Opportunities for Improvement (eds J.W. Kijne, R. Barker and D. Molden)

Both crops are grown in one calendar year. Other crops, including pulses, oilseeds, potatoes, vegetables, fodder and sugarcane, are also grown, particularly in the winter. Irrigation is a common feature of this system, either from extensive surface-canal systems or from shallow wells and tube wells (shallow or deep). Rain-fed RW also exists, but most of the farmers apply at least one irrigation for wheat and many apply a full irrigation schedule.

The population growth rate in the IGP is still about 2% $year^{-1}$ and will take a few decades to stabilize. It is estimated that about 2.5% growth in cereal production will be required to meet food demands in the next decade (Hobbs and Morris, 1996). During the past 30 years, agricultural production has been able to keep pace with population demand for food. This came about through significant area and yield growth. Area growth was a result of new lands being farmed and through increases in cropping intensity, from a single crop to double or even triple crops per calendar year. Area growth will be less important in its contribution to production growth in the future, as more land is used for urban areas and industry. Yield growth will have to be the mainstay for providing the means for meeting future food demands unless food imports start to play a major role in south Asia. Evidence from some long-term experiments, however, show that problems of stagnating yields at levels far below the potential productivity and even yield declines are occurring in some areas in the RW systems of south Asia (Hobbs and Morris, 1996; Dawe *et al.*, 2000; Duxbury *et al.*, 2000; Regmi *et al.*, 2002a,b). The total factor productivity is declining and farmers have to apply more fertilizer to obtain the same yields. Soil organic matter is declining; new weeds, pests and diseases are creating more problems; and paucity of irrigation water in the north-west is resulting in excessive pumping of groundwater. Farmers are complaining about high input costs and low prices for their produce. Marketing of excess production is a burden for farmers and storage is a problem for governments. There is, therefore, a huge challenge ahead in the region to sustainably meet future food demands without damaging the natural-resources base on which agriculture depends, producing food at a cost that is affordable by the poor, and with incentives to farmers that allow them to improve their livelihoods and ultimately alleviate poverty.

Water will become a major limiting factor for sustained production in the next decade in the IGP. Rapidly growing urban areas and industry will compete with agriculture for good-quality water. There are already reports of declining water tables in some areas (Harrington *et al.*, 1993), leading to more costly pumping of groundwater and increased costs of production. In several other canal command areas of the IGP, water tables are rising, leading to secondary soil salinization. The interbasin transfer of irrigation water to meet evapotranspiration demands of the RW system is a key feature of intensively cultivated irrigated agriculture in the IGP. The demands for water from the RW system exceed those available from rain and canal supplies. Farmers often rely on groundwater, which in places is low in quality, either due to excessive salt content or due to the presence of residual alkalinity, with detrimental effects to soil health. Long-term, regional hydrological salt and water balances, as influenced by existing and alternative management practices and as driven by policies (e.g. pricing, common-property management), are items of crucial information if we are to achieve sustainable agriculture in the region. Scientists are using crop-growth simulation and risk-analysis models to evaluate risk-efficient water-use strategies at the district level. Initial results suggest that improved water and energy pricing policies could reduce water use by 25%.

This chapter describes various resource-conserving technologies (RCTs) that are being promoted by the RW Consortium (RWC) to attain the goal of raising productivity in the region and meeting food-security needs while, at the same time, efficiently using natural resources, including water, providing environmental benefits and improving the rural livelihoods of farmers.

RWC and the IGP

The RWC was established in 1994 as a CGIAR ecoregional initiative. A CGIAR ecoregional programme is a combination of natural-resources management and production (extension) in a defined geographical area with site-specific socio-economic and policy environments (Fig. 15.1). The RWC for the IGP is a successful partnership between national programmes (Bangladesh, India, Nepal and Pakistan) in south Asia, several international centres of the CGIAR (the International Maize and Wheat Improvement Center (CIMMYT), the International Rice Research Institute (IRRI), the International Crops Research Institute for the Semi-Arid Tropics (ICRISAT), the International Potato Center (CIP) and the International Water Management Institute (IWMI)) and various advanced international institutions (Cornell University; International Agricultural Centre (IAC), Wageningen; Institute of Arable Crops Research (IACR), Rothamsted; CAB International (UK) and Melbourne University). The RWC focuses on issues of raising the productivity and sustainability of RW systems of south Asia in an efficient way and by conserving natural resources, leading to improved livelihoods and reduction of poverty.

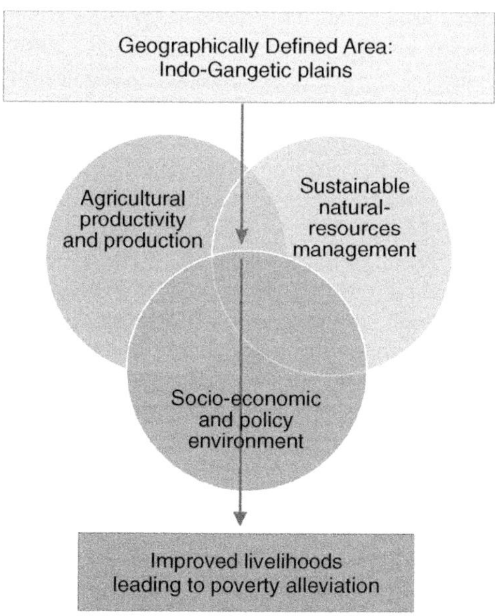

Fig. 15.1. Conceptual diagram of a CGIAR ecoregional project.

The major success of the RWC in the last few years has been the development and deployment of RCTs with farmers in the RW systems of south Asia (Hobbs, 2001; Hobbs *et al.*, 2002). RCTs range from simple surface seeding, where wheat seed is broadcast on the non-ploughed soil, and zero tillage with a special opener for placing seed in the soil, also without ploughing, to reduced tillage and bed planting. In 2000, 100,000 ha of zero-tillage wheat were planted in Pakistan and India in the RW areas. This has benefited farmers through less cost, more yield and more income.

One major hurdle to acceptance was changing the mind-sets of all partners concerned since the phrase 'the more you till, the more the yield' was stubbornly adhered to and difficult to overcome. In this chapter, RCTs are defined as any practice that will result in improvement of the efficiency of natural resources. Water is the major natural resource described in this chapter. Another term for RCTs – used by the Food and Agriculture Organization (FAO) – is Conservation Agriculture. Globally, these technologies are rapidly gaining popularity among farmers, as they result in higher production at less cost, with significant benefits to the environment and more efficient use of natural resources. Ultimately, these result in higher profits, cheaper food and improved farmer livelihoods. Crop diversification is also easier as less land is needed to produce staple cereals, freeing land for other crops. Interestingly, farmers are bearing the capital cost of this new technology, as opposed to the public sector.

Another major challenge in the region is how the knowledge-intensive technologies are transferred to the farmers, especially in areas where extension services are weak. This has been accomplished by the RWC through the promotion of participatory approaches and expanded partnerships with stakeholders.

Various RCTs

A basket of RCT options are being developed and made available to farmers for experimentation and adoption. Some are based on reduced tillage for wheat, including zero tillage. Bed-planting systems are being promoted to increase water productivity and, when combined with reduced tillage in a permanent bed system, they provide even more savings. Laser levelling, combined with these tillage systems, provides additional benefits. Many of the benefits of the tillage options for wheat are lost when rice soils are traditionally puddled (ploughed while wet). System-based technologies are now being promoted that do away with puddling so that total system productivity is raised. The use of groundwater to obtain early rice planting and efficient use of rainwater is another technology. The various technologies are briefly described in the following paragraphs.

Surface seeding

Surface seeding is the simplest zero-tillage system being promoted. In this tillage option, wheat seed is placed on to a saturated soil surface without any land preparation. This is a traditional farmer practice for wheat, legume and other crops in parts of eastern India and Bangladesh. Wheat seed is broadcast either before (relay planting) or after the rice crop is harvested.

Promotion of surface seeding for planting wheat has been done for several years in areas where the soils are fine-textured and drain poorly and where land preparation is difficult and often results in a cloddy tilth. The key to success with this system is having the correct soil moisture at seeding. Too little moisture results in poor germination and too much moisture can cause seed to rot. A saturated soil is best. The roots corkscrew into the moist soil and follow the saturation fringe as it drains down the soil profile. The high soil moisture reduces soil strength and thus eliminates the need for tillage. Additional irrigation may not be needed if the roots can penetrate the surface layers while it is still moist. In this case the roots penetrate the soil before the surface soil dries and soil strength increases so the roots can then follow the water table down the profile. In drier soils, the roots cannot penetrate the soil surface because of the higher soil strength and miss the opportunity to follow the water table as it drains down the profile.

In China, where surface seeding is also practised, farmers apply cut straw to mulch the soil, to reduce evaporative losses of moisture and to control weeds. The standing stubble also protects the young seedlings from birds. However, relay planting can be done only if the soil moisture is correct for planting.

One of the major pluses of surface seeding is that no costly equipment is needed and any farmer can easily adopt this practice. The use of a drum or simple seeder for line sowing is found to be advantageous. This system is being monitored in farmer fields to determine if continuous surface seeding is possible or whether a rotation of tillage systems may be needed to control future weed problems.

Zero tillage with inverted-T openers

This is another RCT, where the seed is placed into the soil by a seed drill without prior land preparation. This technology, which has been tested in Pakistan (Aslam *et al.*, 1993; Sheikh *et al.*, 1993), is currently being tested in other areas of the Indo-Gangetic floodplains, including India and Nepal. This technology is more relevant in the higher-yielding, more mechanized areas of north-western India and Pakistan, where most land preparation is now done with four-wheel tractors. However, to extend the technology in eastern parts of the IGP in Bangladesh, equipment is being modified for two-wheel hand tractors and bullocks.

The basis for this technology is the inverted-T openers (Fig. 15.2), which were developed in and imported from New Zealand. This coulter and seeding system places the seed into a narrow slot made by the inverted-T opener as it is drawn through the soil by the four-wheel tractor. The coulters can be rigid or spring-loaded, depending on the design and cost of the machine. This type of seed drill works very well in sit-

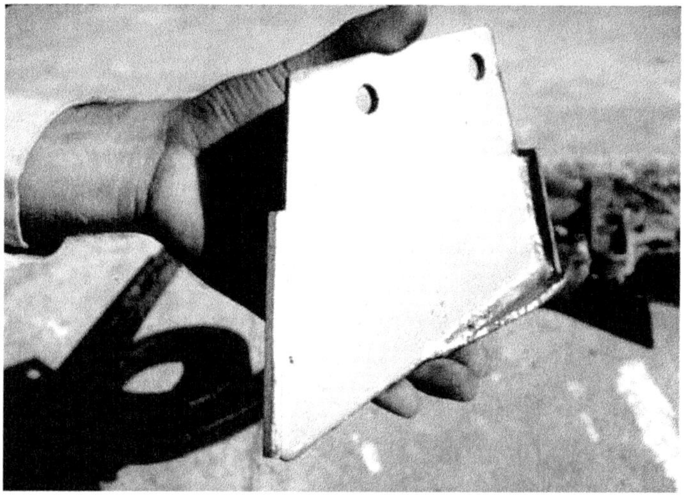

Fig. 15.2. A photograph of the inverted-T coulter used for zero tillage in the IGP.

uations where there is little surface residue after harvesting of rice. This usually occurs after manual harvesting.

Where combine harvesting is becoming popular, loose straw and residue create a problem for the inverted-T opener. Farmers currently burn residues to overcome this problem of loose stubble, whether they use zero tillage or the traditional system. Since the RWC wants to discourage this practice, which has major environmental and air-pollution implications, future strategies will look at alternative machinery and techniques to overcome this problem. Leaving the straw as mulch on the soil surface has not been given much thought in Asian agriculture. However, results from rain-fed systems and some preliminary results in Asia suggest that this may be very beneficial to the early establishment and vigour of crops planted this way (Sayre, 2000) and for soil moisture conservation, water infiltration and erosion. Studies are needed to explore the regional-scale benefits and longer-term consequences of this practice, already being adopted in some way in the zero-tillage wheat system.

Interestingly, significantly fewer weeds were found under zero tillage compared with conventional tillage (Verma and Srivastava, 1989, 1994; Singh, 1995), which is the opposite of the experience with zero-tillage systems in developed countries (Christian and Bacon, 1990; Kuipers, 1991). This observation has been confirmed in many other locations. Results from 336 on-farm trials in Haryana are shown in Fig. 15.3. Fields with both zero tillage and normal tillage were sprayed with weedicide, but significantly lower weed counts were found in fields with zero tillage both before and after herbicide application. This difference can be explained by the nature of the weeds found in the RW cropping system. Most of the weeds affecting the wheat crop germinate during the crop season and, since the soil is disturbed less under zero tillage, fewer weeds are exposed and fewer germinate. Also, before the weeds are able to grow and compete, the main crop is able to cover up the surface and significantly reduce the weed biomass. Weed problems are typically more severe under conventional tillage than under zero tillage, at least in the short term. Longer-term research is needed to anticipate future consequences of tillage changes for weed populations (e.g. to quantify likely shifts in weed species and the effects of those shifts on yield stability).

Earlier planting is the main reason for the additional yields obtained under zero tillage (Table 15.1). Zero-tilled plots were planted as close as possible to 14–20 November, the optimum date for planting wheat in India and Pakistan. The results of many trials suggest that the longer the farmer delays planting, the lower the yield. This finding has

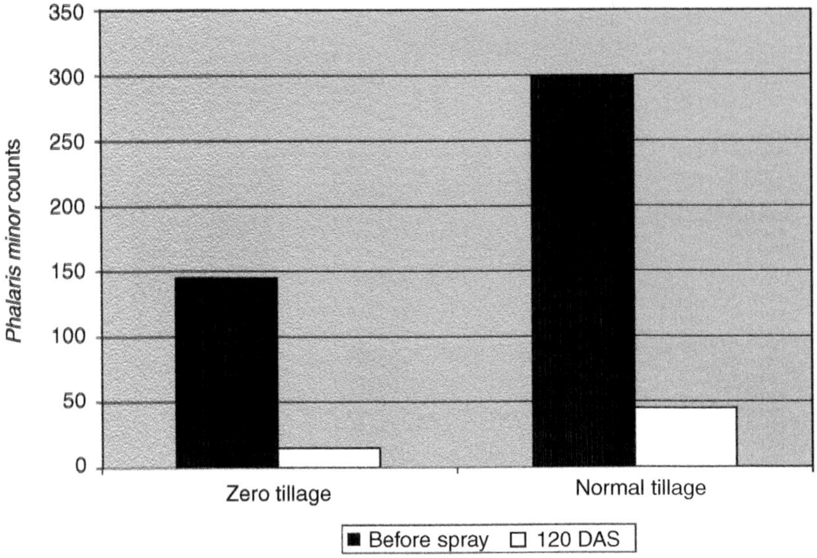

Fig. 15.3. *Phalaris minor* population before herbicide spray and 120 days after seeding (DAS) in zero-till and normal-tilled plots in 1998/99, Haryana Agricultural University, Hisar, India.

Table 15.1. Comparison of zero tillage and farmers' practice for establishing wheat after rice in Punjab, Pakistan, at locations where planting dates for the two methods differ (from Aslam *et al.*, 1993).

	Wheat yield (kg ha^{-1})		
Location	Zero tillage	Farmers' practice	Difference in the no. of days
Daska, site 2	3143	3209	10
Daska, site 1	3842	2735	13
Ahmed Nagar	4308	3526	20
Maujianwala	2689	2198	22
Mundir Sharif	4245	2660	33
Daska, site 3	3838	3420	44
Mean	3677a	2598b	24

Means followed by the same letter do not differ significantly at the 5% level using Duncan's multiple range test.

been confirmed in trials throughout the IGP in the past few years. In Haryana, surveys and crop cuts have shown that zero tillage produces 400–500 kg ha^{-1} more grain than traditional systems. This is attributed to earlier, timely planting, fewer weeds, better plant stands and improved fertilizer efficiency because of placement with the seed drill. Some farmers are now in their fourth year of adoption of zero tillage and find no deleterious effects that would make them revert to the traditional system.

Reduced tillage

The Chinese have developed a seeder for their 12 hp, two-wheel diesel tractor, which prepares the soil and plants the seed in one operation. This system consists of a shallow rotovator, followed by a six-row seeding system and a roller for compaction of the soil. Several tractors and implements were imported from Nanjing, China, into Nepal, Pakistan and India, where they have been tested over the past few years with positive results. In Bangladesh, farmers are using more than 200,000 hand tractors imported from China. Soil moisture was found to be critical in this reduced-tillage system. The rotovator fluffs up the soil, which then dries out faster than with normal land preparation. The seeding coulter does not place the seed very deep, so soil moisture must be high during seeding to ensure germination and root extension before the soil dries appreciably. Modification of the seed coulter to place the seed a little deeper would help correct this problem.

The main drawback of this technology is that the tractor and the various implements are not easily available and spare parts and maintenance are major issues. It would help if the private or public sector in south Asian countries could import this machinery or develop a local manufacturing capability. As it becomes more costly to keep and feed a pair of bullocks for a year, more farmers in the region are turning to significantly cheaper mechanized options of land preparation. One of the benefits of this tractor is that it comes with many options for other farm operations; it includes a reaper, rotary tiller and a mouldboard plough, and it can also drive a mechanical thresher, winnowing fan or power source for pumping water. However, most farmers are attracted to the tractor because it can be hitched to a trailer and used for transportation. For smaller-scale farmers who cannot afford their own tractors, custom hiring is a common alternative.

The Chinese tractor can also be used with a rotovator (Bangladesh) to quickly prepare the soil and incorporate the seed after a second pass. This speeds up the planting and results in better stands with less cost than traditional methods. However, the Chinese seeder attachment does a better job, because the seeds are placed at a uniform depth in the single pass. Engineers are experimenting with removing some of the blades that rototill the soil. In this way, a strip of soil is cultivated rather than the whole area. This reduces the power needs and costs and makes it easier for farmers to manage the tractor. In India, a four-wheel tractor version of this 'strip-tillage' machinery is available.

Bed-planting systems

In bed-planting systems, wheat or other crops are planted on raised beds. This practice has increased dramatically in the last decade or so in the high-yielding, irrigated, wheat-growing area of north-western Mexico (Meisner *et al.*, 1992; Sayre and Moreno Ramos, 1997). Bed planting in Mexico rose from 6% of farmers surveyed in 1981 to 75% in 1994, and farmers have given the following reasons for adopting the new system:

- Management of irrigation water is improved.
- Bed planting facilitates irrigation before seeding and thus provides an opportunity for weed control prior to planting.
- Plant stands are better.
- Weeds can be controlled mechanically, between the beds, early in the crop cycle.
- Wheat seed rates are lower.
- After wheat is harvested and straw is burnt, the beds are reshaped for planting the succeeding soybean crop. Burning can also be eliminated.
- Herbicide dependence is reduced, and hand-weeding and roguing are easier.
- Less lodging occurs.

This system is now being assessed for suitability in the Asian subcontinent. At the Punjab Agricultural University, two bed widths and two or three rows of wheat planted per bed were compared with conventional flat-bed planting. Two rows on 70 cm beds were best. Two of the major constraints on higher yields in north-western India and Pakistan are weeds and lodging. Both can be reduced in bed-planting systems. The major weed species affecting wheat, *Phalaris minor*, is normally controlled using the herbicide Isoproturon, which is not always effective. Farmers do not always apply Isoproturon either well or on time; in addition, recent reports have confirmed that *P. minor* has developed Isoproturon resistance (Malik and Singh, 1995; Malik, 1996; Malik *et al.*, 1998). Alternative integrated weed strategies must be developed to overcome this problem. Preliminary observations indicate that *P. minor* is less prolific on dry tops of raised beds than on the wetter soil found in conventional flat-bed planting. Cultivating between the beds can also reduce weeds. Thus bed planting provides farmers with additional options for controlling weeds.

Lodging is also less of a problem on raised beds. Additional light enters the canopy and strengthens the straw, and the soil around the base of the plant is drier. Reduced lodging can have a significant effect on yield, since many farmers in the Punjab do not irrigate after heading precisely because they want to avoid lodging. As a result, water can become limiting during grain filling, resulting in lower yields. On raised beds, this irrigation need not be avoided, for the reasons stated.

Results show that there is no significant difference in yield between flat- and bed-planted systems, which means that yield was not sacrificed by moving to a bed system (Table 15.2). The data in Table 15.2 were obtained in a 1 year on-station experiment. Further studies are needed to investigate possible significant interactions between bed planting and variety, with more spreading varieties yielding better in this system than upright ones. Figure 15.4, although based on

Table 15.2. Effects of bed size and configuration on wheat yield, Punjab Agricultural University, Ludhiana, India, 1994/95 (from unpublished data from S.S. Dhillon, wheat agronomist, Punjab Agricultural University).

	Sowing Method					
		75 cm beds		90 cm beds		
Variety	On the flat 25 cm row	2 rows per bed	2 + 1 rows	3 rows per bed	3 + 1 rows	Mean
	Wheat yield (kg ha^{-1})					
PBW 226	5740	6170	6390	6160	6320	6160a
WH 542	6290	5830	6360	6000	6040	6110a
CPAN 3004	6020	5530	6140	5630	5600	5780b
PBW 154	5460	5110	6000	5930	5880	5680b
HD 2329	5770	4660	6190	5580	5810	5600b
PBW 34	5650	5610	5800	5580	5630	5650b
Mean	5820	5490	6150	5810	5880	

Means followed by the same letter do not differ significantly at the 5% level using Duncan's multiple range test.

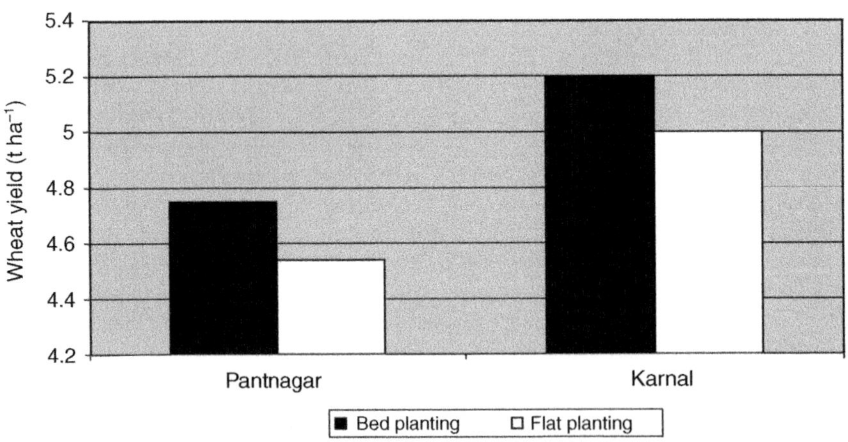

Fig. 15.4. Comparison of bed versus flat planting of wheat in farmers' fields in Pantnagar ($n = 3$) and Karnal ($n = 8$), India, in 1998/99.

trials in only a few farmers' fields in Haryana in 1998/99, shows a significant difference in wheat yield resulting from the planting method. Similar results have been found since then and the planting of wheat on beds is gaining popularity in Haryana and the Punjab of India and in the Punjab of Pakistan. Farmers are particularly pleased with the water savings they obtain from bed systems.

An additional advantage of bed planting becomes apparent when beds are 'permanent' – that is, when they are maintained over the medium term and not broken down and re-formed for every crop. In this system, wheat is harvested and straw is left or burnt. Passing a shovel down the furrows reshapes the beds. The next crop (soybean, maize, sunflower, cotton, etc.) can then be planted into the stubble in the same bed. Research in farmers' fields has also shown that rice can be grown on beds, making this system feasible in the RW pattern. Rice can be grown on beds by either transplanting seedlings or direct seeding. At the moment, transplanting on beds is best since normal herbicides used for transplanted rice can be used to control weeds. As dry-seeded herbicides become available and weeds can be managed, dry-seeded rice on beds will become more attractive. One farmer in the Punjab obtained 9 t ha^{-1} of rice transplanted on beds and saved more than 50% of the water he normally applied on flat, transplanted rice. This was confirmed through monitoring water use in bed-planted plots in 2002.

The use of beds also provides a way for improving fertilizer-use efficiency. This is achieved by placing a band of fertilizer in the bed at planting or as a top-dressing. Using slow-release formulations and experimenting with urea supergranules can make further improvements. Both can be applied in the bed at the time of planting with the seed-cum-fertilizer drill.

Multiuse of Low-quality Water

Low-quality water is often used in a cyclic mode in the IGP. At times, it is blended with canal water in watercourses to improve the total water supply and also the flow rates. Blending of low- and good-quality water has previously been discouraged because of its adverse effect on crop productivity. However, if the resultant electrical conductivity of the blended water supplies is less than the threshold conductivity, they can be used safely. In combination with new RCTs, blending of multiquality water supplies in on-farm water-storage reservoirs improves the quality of water having residual sodium carbonates and overcomes problems associated with this water. It can also improve the use of rainwater and the water productivity and yields of a bed-planted wheat crop. Preliminary results of a trial conducted in Pakistan have been very encouraging. The bed-planting system also offers scope for use of even saline water. When saline water is applied in a raised-bed–furrow land configuration, it permits salt movement to the top of the raised beds, keeping the root zone relatively free of salts below the furrow. This improves the ability of the plants to avoid early salt injury at seedling stage and subsequently improves the salt tolerance of the crop due to crop ontogeny. When combined with mulching or residue retention, bed planting has the potential to reduce evaporation losses from the soil surface and salinization and to further improve crop productivity in saline environments.

Non-puddling for rice

The benefits of the new resource-conserving tillage options listed above are lost when rice soils are puddled (ploughed when wet). Therefore, the RWC is encouraging research on station and with farmers to find ways to eliminate this soil-degrading process. Most rice farmers in south Asia traditionally puddle their soils to help pond water, reduce percolation losses and control weeds. Initial data indicate that rice-fields do not need to be flooded after the first few weeks and that puddled soils have more cracking and need more water once the fields dry. Initial flooding, though, is important to promote tillering and to more effectively control weeds. Studies are also being initiated to determine the exact water balance for the puddled and non-puddled conditions at the field, water-course and command level. This is being done for fields where bed planting is practised and in fields with flat planting, with and without tillage. As mentioned above, farmers feel that bed-planted rice saves water over the traditional system. Quantitative data will be available soon to confirm this.

Data presented in Fig. 15.5 show that wheat yields are significantly better when wheat is planted with zero tillage after non-puddled rice than after puddled rice. The data also show that rice yields are similar between puddled and non-puddled situations if weeds can be controlled. This shows that RCTs need to be assessed on a systems basis and not on a single commodity.

Laser levelling

All the above technologies can benefit from levelled fields. This is being promoted in Pakistan as a means of improving water efficiency. However, when this is combined with zero tillage, bed planting and non-puddled rice culture, plant stands are better, growth is more uniform and yields are higher. The use of permanent-bed systems and zero tillage results in less soil disturbance and reduces the need for future levelling. India is also starting work on this and promoting levelling in farmers' fields in Haryana and western Uttar Pradesh (UP).

Supplemental water use in eastern India

The winter season following the long rainy season is short in eastern India. Long-term analysis of the rainfall data clearly indicates that there are three distinct periods of moisture availability. The early moist period (evaporation exceeds rainfall) extends over 12–18 days, followed by 93–139 days of a humid moist period, wherein precipitation exceeds potential evapotranspiration. This is followed by a moist period of 17–22 days, where once again rainfall is less than evapotranspiration. If the rice seedlings and crop can be established early in the first moist period, before the humid period, the rice crop can benefit from the monsoonal rain and grow without the need for irrigation. Timely transplanting of rice also results in earlier harvests and allows timely planting of the next wheat crop. The results of farmer-participatory field trials showed that the strategy of timely transplanting of rice improves wheat yields. The productivity of the RW system was nearly 12–13 t ha^{-1} when

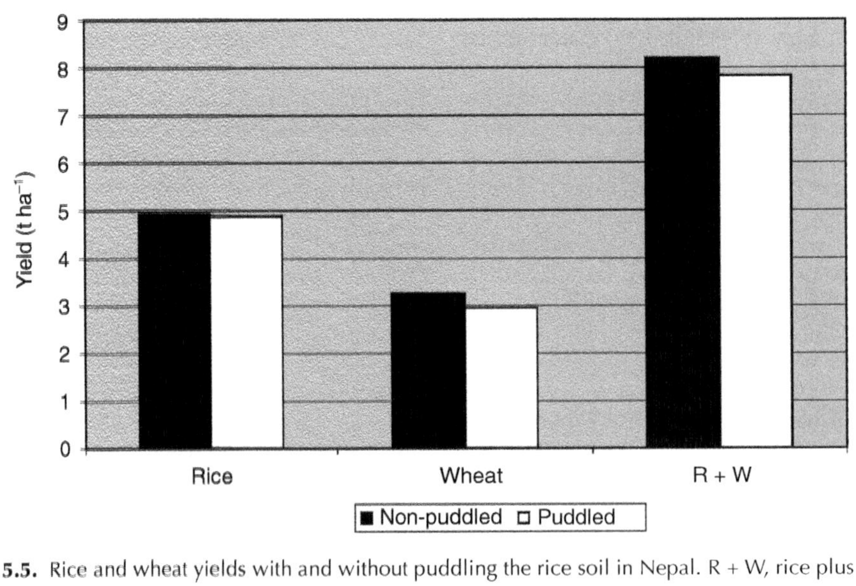

Fig. 15.5. Rice and wheat yields with and without puddling the rice soil in Nepal. R + W, rice plus wheat.

rice was transplanted before 28 June. This was reduced by more than 40% when fields were planted after 15 August (to 6–7 t ha^{-1}).

It was also reported that peripheral bunds with a height of 18–20 cm around fields could store nearly 90% of the total rainwater *in situ* for improved growth and production of rice.

Benefits of the RCTs in Terms of Water Use

Farmers are adopting the new RCTs quickly. Figure 15.6 shows the rapid adoption of zero tillage in the region. More than 100,000 ha of wheat were grown that way last year and this is expected to increase to a million in the next few years. There is no evidence from farmers' interviews and other surveys that farmers are going back to the old system. On the contrary, adoption of RCTs could be even faster if it were possible to have sufficient machinery available from small-scale manufacturers. Farmer feedback on water savings with these new technologies essentially says that they save water. For zero tillage, farmers report about 25–30% savings. This comes in several ways. First, zero tillage is possible just after rice harvest and any residual moisture is available for wheat germination. In many instances, where wheat planting is delayed after rice harvest, farmers have to pre-irrigate their fields before planting. Zero tillage saves this irrigation. Savings in water also come from the fact that an untilled soil has less infiltration than a tilled soil and so water flows faster over the field. That means farmers can apply irrigation much faster. Because zero tillage takes immediate advantage of residual moisture from the previous rice crop, as well as cutting down on subsequent irrigation, water use is reduced by about 10 cm-ha, or approximately 1 million l ha^{-1}. One additional benefit is less waterlogging and yellowing of the wheat plants after the first irrigation, which are common occurrences on normal ploughed land. In zero tillage, less water is applied in the first irrigation and this yellowing is not seen.

Farmers also report water savings in bed planting. Farmers commonly mention 30–50% savings in this system. Farmers also indicate that it is easier to irrigate with bed planting. Obviously, half the space is used for water and so less water is used. The question is whether farmers need to apply more frequent turns of irrigation with this system. This is being studied in a newly started RWC Asian Development Bank (ADB)

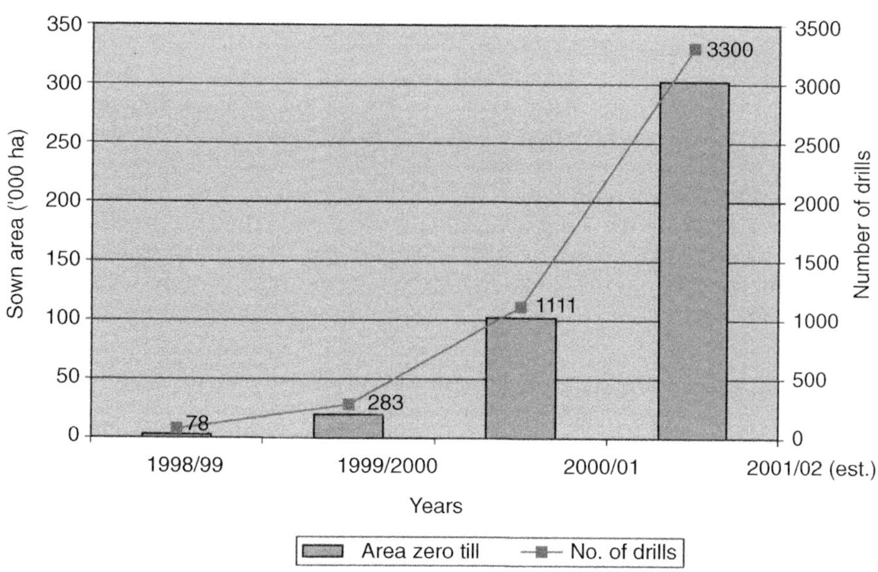

Fig. 15.6. Trends in wheat zero-tillage area in rice–wheat systems in India and Pakistan.

project in Pakistan and India. In the initial year of planting rice on beds, farmers estimated that they used 50–65% less water than on the flat. They kept the beds flooded for the first week, but were then able to cut down on irrigation frequency later. This also needs to be confirmed with good quantitative data.

Members of the On-Farm Water-Management staff of Pakistan's Punjab comparing RCT farms have collected data on water productivity (Gill *et al.*, 2000). This is presented in Table 15.3 for one location. All the systems provided better water productivity compared with traditional systems. The average water saved with laser levelling, zero tillage and bed planting over the traditional was 715, 689 and 1329 m^3 ha^{-1}, with a value of Rs 522, 503 and 970 ha^{-1} based on a water rate of Rs 900 per acre-foot for private tube wells for the year 1999/2000.

We still need to collect data for water use under puddled and non-puddled rice cultivation. Definitely, water percolation will be higher in non-puddled situations, but the total water use may be less, since no water is needed for seedling raising or puddling the main field. Also, when puddled soils dry – and many farmers cannot keep their fields continuously flooded – soils crack and so the field needs more water to fill the profile when water is next added. Less cracking occurs in non-puddled soils. Balasubramanian and Hill (2002) reported from Bhagat *et al.* (1999) that maintaining saturated soil conditions throughout the crop growth period (of puddled, transplanted rice) saved more than 40% of irrigation water compared with continuous shallow flooding when weeds were controlled with herbicides in the Philippines. This suggests that standing water is needed early to help tillering and control weeds but is not essential after that.

Importance of Participatory Technology Development

Adoption of RCTs in south Asia has been rapid over the past few years, especially for zero tillage with the inverted-T planter. Figure 15.6 shows that more than 100,000 ha were grown in 2001. This success was possible because of the application of participatory approaches for accelerating adoption. The traditional extension system, which was so effective in the early years of the Green Revolution, was based on the development of recommendations and packages and then having the extension service demonstrate the technology to farmers. Seed and fertilizer were easily packaged and it was possible to lay out many trials at low cost. When this traditional extension system was used for extending RCTs, problems arose. The first problem was the availability of the machinery to conduct the demonstrations. However, the main constraint was convincing farmers, extension workers and, at times, scientists that this technology had any benefit. Success came once partners were allowed to work together and experiment with the technology. Local manufacturers had to be involved in the development and manufacture of the equipment. Machinery had to be of high quality and yet at a cost within the budgets of farmers. Farmers had to be shown how the drill worked and then allowed to experiment with the equipment before they could be persuaded to accept this radical technology. Stories abound of how the farmers who first tried the technology were ridiculed by their neighbours for trying something so alien. But, once the seed germinated, farmers begged the innovators to help them sow their fields. It is now felt that zero-tillage wheat is an acceptable technology and will

Table 15.3. Effect of sowing on irrigation water productivity (WP) in wheat production in Mona Project, Pakistan (from Gill *et al.*, 2000).

	Laser levelling	Zero tillage	Bed planting	Normal planting
Water applied (m^3 ha^{-1})	2849	2933	2281	3610
Yield (t ha^{-1})	4764	4188	4134	3968
WP (kg m^{-3})	1.67	1.43	1.81	1.10

be part of the recommendations for planting wheat. Similarly, other RCTs, such as bed planting, will become accepted practices as machinery is made available and more farmers experiment with the system.

One question that is often asked is 'who can benefit from this technology and is it just for the large, commercial farmer?' The answer appears to indicate that this technology is scale-neutral and that farmers from all social classes can benefit from the many advantages that this system brings to wheat cultivation. A survey, conducted in 2000 in Haryana, of 20 villages and 91 farmers showed that 24% of the zero-tillage adopters owned a tractor, while the rest used service providers. The average farm size of adopters ranged from 0.8 to 20.2 ha. Twenty-two per cent had less than 2 ha each and 37% had farms of 2–4 ha each. All but one farmer agreed that zero tillage was highly profitable and data show that zero tillage resulted in US\$75 ha^{-1} more than with conventional practices. Yields from zero tillage were 5.4 t ha^{-1} while for conventional tillage they were 5.1 t ha^{-1}. Resource-poor farmers and farmers without tractors are using contract services to plough their fields at the moment. It is becoming too expensive to keep a pair of bullocks just for land preparation and so using a service provider is more economical. When this is applied to zero tillage or bed planting, the benefits are even more pronounced. In this case, the farmer has only to rent the service once and his/her fields are planted. This saves him/her money and provides time to do other activities. Data from socio-economic and impact-assessment surveys in India and Pakistan show this to be true. The first innovators are larger, better-endowed tractor owners. Later less well-endowed farmers adopt the technology as they see the benefits and obtain the services for this technology.

Farmer responses to the RCTs, especially zero tillage, provide valuable feedback for scientists in the RWC for improving the technology. At the same time, scientists have been monitoring the fields where these technologies are being adopted and collecting data on soil, biotic and resource use.

Conclusions

This chapter has described various RCTs available for testing the RW systems of the IGP. It describes the benefits to the farmers of the region adopting various RCTs in terms of improved production at lower cost, improving the efficiency of natural resources, benefits to the environment and improved livelihoods of farmers, all of which ultimately help in alleviating poverty. Water is particularly highlighted, since farmers indicate that all the technologies result in water savings. There is a research need to more accurately measure these savings and a recent ADB project will do just that. It is also important to look at water balances at different scales to determine if water savings at the field level will also give savings at the watercourse, command and basin levels.

The success of resource-conserving technologies is dependent on rapid adoption by farmers. Accelerated adoption was mainly the result of the change in the paradigm for extending the technology. Instead of the linear approach to extension commonly found in the Green Revolution era, the importance of partnerships, expanded stakeholders and participatory approaches, where farmers could experiment and feed back information, soon became apparent. RCTs are a key to ensuring sustainable food production in south Asia in the next decade. Overcoming mind-sets that hold traditional beliefs about excessive tillage and providing the enabling factors that allow exposure of the technology to all those involved in agriculture will be key factors for future success. This technological revolution is seen as one way to sustainably increase food production to meet future demands while conserving natural resources, improving farmer livelihoods and reducing the negative effects on the environment. Water is placed high on the list of natural resources and its use and productivity can definitely be improved with these new technologies. Of course, all of these benefits will be of little use unless nations in south Asia control their population growth.

References

Aslam, M., Majid, A., Hashmi, N.I. and Hobbs, P.R. (1993) Improving wheat yield in the rice–wheat cropping system of the Punjab through zero tillage. *Pakistan Journal of Agricultural Research* 14, 8–11.

Balasubramanian, V. and Hill, J.E. (2002) Direct seeding of rice in Asia: emerging issues and strategic research needs for the 21st century. In: Pandey, S., Mortimer, M., Wade, L., Tuong, T.P., Lopez, K. and Hardy, B. (eds) *Direct Seeding: Research Issues and Opportunities*. Proceedings of the International Workshop on Direct Seeding in Asian Rice Systems: Strategic Research Issues and Opportunities, 25–28 January, 2000, Bangkok, Thailand. International Rice Research Institute, Los Banos (Philippines), pp. 15–39.

Bhagat, R.M., Bhuiyan, S.I. and Moody, K. (1999) Water, tillage and weed management options for wet seeded rice in the Philippines. *Soil Tillage Research* 52, 51–58.

Christian, D.G. and Bacon, E.T.G. (1990) A long-term comparison of ploughing, tine cultivation and direct drilling on the growth and yield of winter cereals and oilseed rape on clayey and silty soils. *Soil and Tillage Research* 18, 311–331.

Dawe, D., Dobermann, A., Moya, P., Abdulrachman, S., Bijay Singh, Lal, P., Li, S.Y., Lin, B., Panaullah, G., Sariam, O., Singh, Y., Swarup, A., Tan, P.S. and Zhen, Q.X. (2000) How widespread are yield declines in long-term rice experiments in Asia? *Field Crops Research* 66, 175–193.

Duxbury, J.M., Abrol, I.P., Gupta, R.K. and Bronson, K.F. (2000) Analysis of long-term soil fertility experiments with rice–wheat rotations in South Asia. In: Abrol, I.P., Bronson, K.F., Duxbury, J.M. and Gupta, R.K. (eds) *Long-Term Soil Fertility Experiments with Rice–Wheat Rotations in South Asia.* Rice–Wheat Consortium Paper Series No. 6, Rice–Wheat Consortium for the Indo-Gangetic Plains, New Delhi, pp. 7–22.

Gill, M.A., Kahlown, M.A., Choudhary, M.A. and Hobbs, P.R. (2000) *Evaluation of Resource Conservation Technologies in Rice–Wheat System of Pakistan.* Report of the OFWM Division, Punjab Government, Lahore, Pakistan.

Harrington, L.W., Fujisaka, S., Morris, M.L., Hobbs, P.R., Sharma, H.C., Singh, R.P., Chaudhary, M.K. and Dhiman, S.D. (1993) *Wheat and Rice in Karnal and Kurukshetra Districts, Haryana, India: Farmers' Practices, Problems and an Agenda for Action.* Haryana Agricultural University (HAU), Indian Council for Agricultural Research (ICAR), CIMMYT, and the International Rice Research Institute (IRRI), Mexico, DF.

Hobbs, P.R. (2001) Tillage and crop establishment in South Asian rice–wheat systems: present and future options. *Journal of Crop Production* 4(1), 1–23.

Hobbs, P.R. and Morris, M.L. (1996) *Meeting South Asia's Future Food Requirements from Rice–Wheat Cropping Systems: Priority Issues Facing Researchers in the Post-Green Revolution Era.* NRG Paper 96–01, CIMMYT, Mexico, DF.

Hobbs, P.R., Singh, Y., Giri, G.S., Lauren, J.G. and Duxbury, J.M. (2002) Direct seeding and reduced tillage options in the rice–wheat systems of the Indo-Gangetic Plains of South Asia. In: Pandey, S., Mortimer, M., Wade, L., Tuong, T.P. and Hardy, B. (eds) *Direct Seeding in Asian Rice Systems: Strategic Research Issues and Opportunities.* International Rice Research Institute, Philippines, pp. 201–205.

Kuipers, H. (1991) Agronomic aspects of ploughing and non-ploughing. *Soil and Tillage Research* 21, 167–176.

Malik, R.K. (1996) Herbicide resistant weed problems in the developing world and methods to overcome them. In: *Proceedings of the 2nd International Weed Science Congress, Copenhagen, Denmark.* FAO, Rome, pp. 665–673.

Malik, R.K. and Singh, S. (1995) Littleseed canarygrass resistance to isoproturon in India. *Weed Technology* 9, 419–425.

Malik, R.K., Gill, G. and Hobbs, P.R. (1998) *Herbicide Resistance – a Major Issue for Sustaining Wheat Productivity in Rice–Wheat Cropping System in the Indo-Gangetic Plains.* Rice–Wheat Consortium Research Paper Series No. 3, Rice–Wheat Consortium for the Indo-Gangetic Plains, New Delhi, India, 36 pp.

Meisner, C., Acevedo, E., Flores, D., Sayre, K., Ortiz-Monasterio, I. and Byerlee, D. (1992) *Wheat Production and Grower Practices in the Yaqui Valley, Sonora, Mexico.* Wheat Special Report No. 6, CIMMYT, Mexico, DF.

Regmi, A.P., Ladha, J.K., Pathak, H., Pasuquin, E., Bueno, C., Dawe, D., Hobbs, P.R., Joshy, D., Maskey, S.L. and Pandey, S.P. (2002a) Analyses of yield and soil fertility trends in a 20-year rice–rice–wheat experiment in Nepal. *Soil Science Society of America Journal* 66, 857–867.

Regmi, A.P., Ladha, J.K., Pasuquin, E.M., Pathak, H., Hobbs, P.R., Shrestha, M.L., Gharti, D.B. and Duveiller, E. (2002b) The role of potassium in sustaining yields in a long-term rice–wheat experiment in the Indo-Gangetic Plains of Nepal. *Biology and Fertility of Soils* 36, 240–247.

Sayre, K.D. (2000) Effects of tillage, crop residue retention and nitrogen management on the performance of bed-planted, furrow irrigated spring wheat in northwest Mexico. Presented at the 15th Conference of the International Soil Tillage Research Organization, 2–7 July 2000, Fort Worth, Texas.

Sayre, K.D. and Moreno Ramos, O.H. (1997) *Applications of Raised-bed-planting Systems to Wheat.* Wheat Program Special Report No. 31, CIMMYT, Mexico, DF.

Sheikh, A.D., Khan, M.A., Aslam, M., Ahmad, Z. and Ahmad, M. (1993) *Zero Tillage Wheat Cultivation Technology in the Rice–Wheat Areas of the Punjab.* National Agricultural Research Council (NARC), Islamabad, Pakistan.

Singh, Y. (1995) *Rice–Wheat Cropping System NARP Basic Research Sub-project. Progress Report 1994–95.* G.B. Pant University of Agriculture and Technology, Pantnagar, Uttar Pradesh, India.

Verma, U.N. and Srivastava, V.C. (1989) Weed management in wheat under zero and optimum tillage conditions. *Indian Journal of Agronomy* 34, 176–179.

Verma, U.N. and Srivastava, V.C. (1994) Production technology of non-tilled wheat after puddled transplanted rice. *Indian Journal of Agricultural Science* 64, 277–284.

16 Land and Water Productivity of Wheat in the Western Indo-Gangetic Plains of India and Pakistan: a Comparative Analysis

Intizar Hussain,[1] R. Sakthivadivel[2] and Upali Amarasinghe[1]
International Water Management Institute, [1]Colombo, Sri Lanka and [2]Tamil Nadu, India

Abstract

In much of the recent discussions on wheat yields for India and Pakistan, attention has been drawn to irrigated-wheat-yield differences in Bhakra (India) and Punjab (Pakistan). The average wheat yields generally reported in Bhakra (about 4 t ha^{-1}) are almost double the yield reported for Punjab (about 2 t ha^{-1}). These discussions have raised an important research question on why wheat yields vary so much under fairly similar agroclimatic, socio-economic and management conditions.

The purpose of this study is to analyse variations in wheat yields and to assess the range of factors affecting wheat yields and the profitability of wheat production in selected irrigation systems in India and Pakistan. The study attempts to identify constraints and opportunities for closing the existing yield gap. It is hypothesized that substantial gains in aggregate yields can be obtained by improved water-management practices at the farm and irrigation-system levels.

The study was conducted in the Bhakra canal system (BCS) in the Kaithal irrigation circle in India and the Lower Jehlum canal system (LJCS) in the Chaj sub-basin in Pakistan. Six watercourses, one each on the head, middle and tail reaches of one distributary in each country, were selected for detailed field-level data collection.

Results show that the average wheat yield in the selected irrigation system in India is somewhat higher (4.48 t ha^{-1}) than that in the selected system in Pakistan (4.11 t ha^{-1}), but not by as much as is generally perceived. However, the overall yield gap across farms is much wider in the study area in LJCS-Pakistan than that in BCS-India. Wheat-yield differences are much higher across watercourses (i.e. at the distributary level) than across distributaries.

There is a significant inequity in distribution of canal water in the study areas in both BCS-India and LJCS-Pakistan, with tail reaches receiving less canal water than head and middle reaches. Groundwater use, as expected, is higher in reaches receiving less canal water and vice versa. The average productivity of consumed water is similar for the selected systems in both countries, i.e. 1.36 kg m^{-3} in India and 1.37 kg m^{-3} in Pakistan. However, average productivity of diverted water is higher for BCS-India (1.47 kg m^{-3}) than for LJCS-Pakistan (1.11 kg m^{-3}).

In the study areas of both countries, average land productivity is lower in locations where groundwater is of relatively poorer quality. The groundwater quality within a distributary deteriorates towards the middle and tail reaches (except for Khadir in LJCS-Pakistan, where groundwater is less saline in the tail ends), and these reaches currently receive less canal water. Thus, intradistributary canal-water allocation is an important issue in reducing the yield gap.

Using farm-level data, yield functions were estimated to analyse the effects of a range of production factors. Results show that, in addition to improved farm-management practices, such as adopting new varieties, avoiding sowing delays and improved input applications, the improvements in water-manage-

 Water Productivity in Agriculture: Limits and Opportunities for Improvement (eds J.W. Kijne, R. Barker and D. Molden)

ment practices at the system level will also contribute to increased wheat yields and overall profitability. Improving timings of canal-water deliveries and adopting an effective canal-water reallocating strategy will result in overall socio-economic gains.

In both BCS-India and LJCS-Pakistan, the profitability of wheat production decreases with the overall quality of the water used. The study presents alternative scenarios for the impacts of changes in the allocation of canal and groundwater on the socio-economics of wheat production. It is concluded that overall gains from wheat production can be increased by adopting effective reallocation of canal water at the distributary level. Many of the gains under the scenario will be in locations where groundwater is of poorer quality. The policy implication of this is that, under conditions of canal-water scarcity and variations in the quality of groundwater, joint management of canal water and groundwater is essential to increase overall gains from crop production. The study presents an example of 'institutional water scarcity' that could be addressed through effective institutions, leading to improved management of available surface-water and groundwater resources.

Introduction

Wheat production in both India and Pakistan has increased significantly over the past three decades, due to expansion in area sown to wheat, as well as yield improvements. Average wheat yields have increased at 3.21% annually in India (0.84 t ha^{-1} in 1961–1963 to 2.55 t ha^{-1} in 1996–1998) and 2.72% annually in Pakistan (0.82 t ha^{-1} in 1961–1963 to 2.10 t ha^{-1} in 1996–1998). The wheat area has increased by 1.9% annually in India and 1.53% annually in Pakistan over the same period. However, in the 1990s, the average yield growth in both countries has been slower than in the past (1.7% in India and 1.6% in Pakistan) with only slight year-to-year fluctuations. Deceleration in yield growth rate has caused concerns among policy makers and planners in both countries. Also, in much of the recent discussions on wheat yields for India and Pakistan, attention has been drawn to irrigated-wheat-yield differences in Bhakra (India) and Punjab (Pakistan), with average wheat yields in Bhakra (around 4 t ha^{-1}) almost double those in Punjab (around 2 t ha^{-1}). These discussions have raised an important research question on why wheat yields vary widely under fairly similar agroclimatic, socio-economic and management conditions.

The primary objective of this study is to understand farm-level wheat-yield variations and to identify constraints and opportunities for increasing yields and overall profitability of wheat production. The specific objectives are to:

- Analyse intercountry and intracountry variations in wheat yields in the selected irrigated agricultural systems in India and Pakistan.
- Analyse factors contributing to such variations.
- Identify constraints and opportunities and possible methods to reduce existing yield gaps and to increase production.

The key hypothesis to be tested is that substantial gains in aggregate yields and overall profitability of wheat production can be obtained by improved water-management practices at the farm and irrigation-system levels.

There is a plethora of literature on analysing determinants of wheat yields in India and Pakistan. Tyagi and Sharma (2001) and Mudasser *et al.* (2001) give comprehensive reviews of the literature on determinants of wheat productivity in the India and Pakistan. Ahmed and Chaudhry (1996) discuss productivity differentials of the Indian and Pakistani Punjabs. Some of the specific studies at farm level in India include the degree of deficit irrigation and perceived reliability of canal-water supply (Narayanamoorthy and Perry, 1997), effects of irregularity and inadequacy of water supplies on wheat yields (Mishra and Tyagi, 1988), effects of delay in sowing on wheat productivity (Chaudhary and Bhatnagar, 1980; Rehman, 1986; Altaf, 1994; Nagarajan, 1998), decisions on the number of irrigations to be applied on the field and its relationship to groundwater availability (Pintus,

1997), number of irrigations and wheat yields (Aslam, 1998), effects of soil type and quality on wheat yields (Doorenbos *et al.*, 1979; Siddiq, 1994) and the effects of mixing of fresh water and saline/sodic waters on wheat yields (Minhas *et al.*, 1998).

Most past studies analysing determinants of wheat productivity have focused on soil, agronomic factors and water individually, with only a few attempting to analyse water-related factors at the system and farm levels in a more rigorous manner. This study takes a holistic approach by rigorously analysing a fairly comprehensive set of factors, including soil, agronomic and water-related factors (quantity, quality and timing of applications) and their influence on wheat yields in the selected irrigation systems in India and Pakistan. The analysis of factors is undertaken at both farm and irrigation system/subsystem levels. The study adds to the previous literature by developing a set of scenarios for improved water management and its socio-economic implications for farmers.

Study Locations

The study was conducted in two irrigation systems, namely the Bhakra canal system (BCS) of Haryana and Punjab in India and the Lower Jehlum canal system (LJCS) of Punjab in Pakistan. Specific study sites were chosen from two distributaries in each location selected from each of these systems. The key characteristics of these systems and of specific study sites are given below.

India

The Bhakra system was planned to serve the arid tracts of Punjab, Haryana and parts of Rajasthan. In Haryana, the Bhakra canal service is divided into five irrigation circles. One of them is the Kaithal circle, in which the study site is located. For the present study, two minors of the Kaithal irrigation circle, Batta minor (Sirsa branch) and Rohera minor (Habri branch), were selected.

The climate of the study area is semi-arid. The normal annual rainfall varies from 500 to 600 mm year^{-1}. The rainy season starts from 15 June and continues up to September and contributes about 70–80% of the total annual rainfall. The winter season starts from November and extends up to February. During this season, the temperature varies from 5°C to 20°C. Soils of the study area are light- to medium-textured, varying from sandy loam to clay loam, and are low in organic matter. The phosphorus content is medium but the potassium contents vary from medium to high. The soil pH ranges from 7.8 to 9.5. The fields in the tail end of the selected minors are generally saline in nature.

Pakistan

In Pakistan, the study was conducted in the Chaj Doab sub-basin of the Upper Indus basin. Chaj Doab is irrigated by both the Lower Jhelum canal and the Upper Jhelum canal. Two distributaries, namely, Lalian and Khadir, located in Laluwali and Khadir irrigation subdivisions, respectively, were selected for this study.

The climate of the study area is hot summers and cold winters. Summers start in late March, and May–July are the hottest months. The mean minimum and maximum temperatures are 25°C and 39°C, respectively. During summer, maximum rainfall occurs during July (136 mm) and August (76 mm). Winters start from late October/early November and extend up to February. During this season the temperature varies from 6°C to 21°C. The winter season also receives part of the annual rainfall, in December (27 mm) and January (33 mm). The soils of Chaj Doab are mostly calcareous loamy soils.

Study Design, Methodology and Data

Two distributaries were selected in the BCS in India and in the LJCS in Pakistani Punjab, representing a relatively inadequate canal-water environment, practising conjunctive use of canal water and groundwater of differing quality and having large variations in farm-level wheat yields. For comparison purposes, a consistent study design and methodology were adopted for both locations.

Three watercourses, one each at the head, middle and tail ends, were selected from each selected distributary. The selection of watercourses along the distributary was based on the total length, command area and number of watercourses of the distributary. In Pakistan, one more watercourse was selected from the middle part of the Lalian distributary, where the Food and Agriculture Organization (FAO) is implementing demonstrative interventions on the effects of laser levelling and raised-bed–furrow cultivation practices on crop yields.

After considering the requirement for reliable statistical and econometric analyses and research manageability and logistics, a sample of 36 farms along each watercourse was selected. This includes 12 farms each on head, middle and tail ends of each watercourse. The total sample size was 216 farms in BCS-India and 218 in LJCS-Pakistan. A selected farm may have several field parcels. Yields between parcels may vary, due to possible differences in dates of planting and input. Considering these intrafarm yield differences, only one parcel on each farm was selected randomly for in-depth data collection, including water measurements at the plot level. Data were also collected for the remaining plots on each selected farm, but these data represent averages across the remaining plots on each farm.

All primary data for this study were collected during Rabi 2000/01, i.e. from October 2000 through to May 2001. Two types of questionnaires were used to collect primary farm/plot-level data:

1. General questionnaire – to collect basic information, including farm location, size, tenurial status, crop areas and production activities during the season (Rabi 2000/01).
2. Process questionnaire – to record daily observations from the beginning of the crop season till crop harvesting, on farmers' production activities on each of the selected plots, including water measurements at the plot level (water from both surface-water and groundwater sources).

In addition, data on farmers' warabandi schedule, water measurements at the watercourse level, fluctuations water-table depth (at head, middle and tail ends of each watercourse), salinity of both surface water and groundwater, soil salinity and rainfall were also collected on a regular basis.

Characteristics of selected watercourses

Table 16.1 provides key characteristics of the selected watercourses in both locations.

From the point of view of comparability in size, both Lalian and Khadir in LJCS-Pakistan have a large gross command area (GCA) (20,000–26,000 ha) compared with Batta and Rohera in BCS-India (roughly 4000 ha). On the other hand, the GCAs of the selected watercourses in BCS-India and LJCS-Pakistan are of comparable sizes. The GCA of the selected watercourses varies from around 81 ha to 457 ha, with relatively higher GCAs of tail-end watercourses.

The water allowance per hectare at the watercourse level in the Indian system is more or less uniform (0.0017 $m^3 s^{-1}$). On the other hand, water allowance per hectare in the LJCS-Pakistan system varies over a range from 0.0001 $m^3 s^{-1}$ to 0.0002 $m^3 s^{-1}$. Also, there is a distinct difference in water allowance between the two distributaries in LJCS-Pakistan. For Lalian, it is nearly twice that of Khadir.

Wheat-yield variations in the study area

Average wheat yields are higher in the study area in BCS-India (4.48 t ha^{-1}) than in the study area in LJCS-Pakistan (4.11 t ha^{-1}) (Fig. 16.1). However, these yield differences are not as high as is generally perceived (as discussed earlier). Yet the variations of wheat yield in the distributaries in Pakistan are higher (coefficient of variation (CV) = 33%) than the variations of wheat yields in India (CV = 12%).

In BCS-India, the CV of yields is the same across the two distributaries and the intra-watercourse CV is generally less than that at the distributary level. In LJCS-Pakistan, the CV of the yields differs between the two distributaries (37% for Lalian and 27% for

Table 16.1. General characteristics of selected watercourses.

Outlet/distributary	Gross command area (ha)	Design capacity ($m^3 s^{-1}$)	Average discharge[a] ($m^3 s^{-1}$)	Water allowance per ha ($m^3 s^{-1}$)	Ground-water EC ($dS m^{-1}$)
Batta – head	167	0.027	0.028	0.00016	1.37
Batta – middle	226	0.037	0.039	0.00016	4.22
Batta – tail	254	0.042	0.047	0.00017	5.76
Batta – all	3,669		–	–	3.81
Rohera – head	146	0.023	0.021	0.00016	1.41
Rohera – middle	81	0.013	0.020	0.00016	2.41
Rohera – tail	204	0.034	0.036	0.00017	5.04
Rohera – all	4,131		–	–	2.95
Batta and Rohera	–		–	–	3.39
Lalian – head	179	0.036	0.039	0.00020	1.07
Lalian – middle	130	0.026	0.040	0.00020	0.66
Lalian – middle (FAO)	189	0.038	0.062	0.00020	1.56
Lalian – tail	248	0.049	0.033	0.00020	1.71
Lalian – all	19,785	–	–	–	1.31
Khadir – head	180	0.018	0.018	0.00010	1.05
Khadir – middle	178	0.027	0.023	0.00015	1.02
Khadir – tail	457	0.049	0.018	0.00011	0.79
Khadir – all	25,859	–	–	–	0.95
All	–	–	–	–	1.13

[a]Measured in the field.

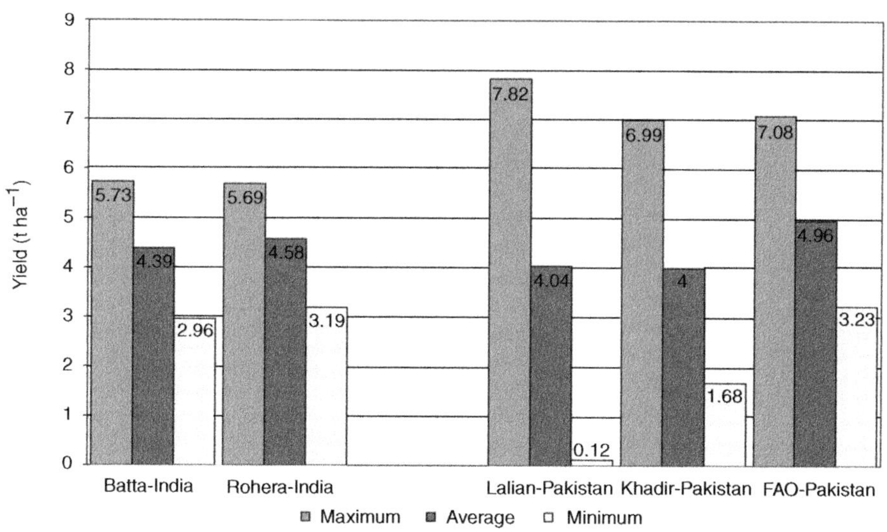

Fig. 16.1. Wheat-yield variations in the study areas of the Bhakra canal system in India and the Lower Jehlum canal system in Pakistan.

Khadir) and it varies significantly within and across watercourses. This finding has an important research and policy implication as to what should be the unit of analysis and what type of efforts should be directed where.

There are also significant differences in wheat yields across head, middle and tail reaches within and across watercourses along the two distributaries. In general, wheat yields are higher in head-reach watercourses and decrease towards tail-reach watercourses in both locations, except for the Khadir distributary in LJCS-Pakistan. In the Khadir distributary, the yields in tail ends are higher than those in the head and middle reaches, basically reflecting the availability and use of good-quality groundwater (Table 16.2).

Factors Affecting Wheat Productivity

The above results indicate that there are significant yield differences within and across watercourses. What are the key factors that influence the crop-productivity differences between different locations?

The productivity of wheat depends on a range of factors, including land- and water-related factors (location of farms, quality of land, source of water, quality and quantity of water, timing of water application, etc.), climatic factors (rainfall, temperature, sunshine, wind, frost), agronomic factors (quality, quantity and timing of applications of inputs such as fertilizers, weedicides, labour, etc.), socio-economic factors (educational level, experience in farming, farm size, tenancy terms, land fragmentation, availability of credit, etc.) and farm-management factors (adoption of production technology, farm planning and management practices), etc. Some of these factors may be interrelated. The effects of some of these may be much smaller than those of others. We focus here on major factors influencing wheat productivity.

Soil quality

Soils of the study areas in both countries are loamy soils. In BCS-India, average soil electrical conductivity (EC) across six watercourses

Table 16.2. Average wheat yield (t ha^{-1}) of different watercourses in India and Pakistan, 2000/01 (based on crop-cutting experiment, 2000/01).

Location of watercourse	Location of watercourse			Location of watercourse		
	Head	Middle	Tail	Head	Middle	Tail
BCS-India	Batta			Rohera		
Head	4.81	4.73	4.42	4.92	4.83	4.28
Middle	4.56	4.42	4.22	4.89	4.79	3.98
Tail	4.35	4.31	3.72	4.91	4.67	3.55
Average	4.57	4.49	4.12	4.91	4.76	4.04
(CV)	(0.11)	(0.07)	(0.13)	(0.09)	(0.08)	(0.10)
Dist. average	4.39			4.58		
(CV)	(0.12)			(0.12)		
LJCS-Pakistan	Lalian			Khadir		
Head	5.18	4.02	2.96	4.56	3.00	4.51
Middle	4.92	3.31	3.01	3.32	3.51	4.57
Tail	4.79	4.5	3.59	4.22	3.62	4.69
Average	4.95	3.92	3.19	4.03	3.37	4.59
(CV)	(0.20)	(0.44)	(0.47)	(0.25)	(0.29)	(0.20)
Dist. average	4.04			4.00		
(CV)	(0.37)			(0.27)		

CV, coefficient of variation; Dist., distributary.

varies from 1.85 to 5.63 dS m^{-1}. The average soil quality of the watercourse command areas in Batta (average EC of 3.83 dS m^{-1}) is lower than that in Rohera (average EC of 2.86 dS m^{-1}). There are significant locational variations within the distributaries in both BCS-India and LJCS-Pakistan, with soil quality generally deteriorating towards tail-end locations. The average EC and pH for the tail ends of Batta and Lalian (two areas of relatively poorer-quality soils) are 5.63 dS m^{-1} and 8.25 and 3.15 dS m^{-1} and 8.34, respectively.

Water use

Both canal (surface) water and groundwater are used in most parts of the study areas of both countries. In general, groundwater use is high where canal water is in short supply. Overall, the proportion of groundwater use per hectare is higher in BCS-India than in LJCS-Pakistan (Table 16.3). Groundwater use is much higher in the Rohera distributary in BCS-India and the Khadir distributary in LJCS-Pakistan, contributing on average around 90% of total water use at the farm level, compared with Batta in BCS-India (73%) and Lalian in LJCS-Pakistan (55%). However, there are significant variations in water use from the two sources across various reaches of the canal systems. In both the study areas, groundwater use is much higher in the tail-end reaches than in the head and middle reaches, where canal-water supply is relatively higher.

The location of farms/watercourses is directly related with the use of both surface water and groundwater. The head and middle reaches receive more canal water than the tail ends in both BCS-India and LJCS-Pakistan. This is indicated by measurements of the canal-water flow at the outlet level for each of the selected watercourses and the amount of canal water applied at the field level. The average canal water applied for wheat in BCS-India is 550 m^3 ha^{-1} compared with 980 m^3 ha^{-1} in LJCS-Pakistan. Canal-water use is higher in Batta (BCS-India) and Lalian (LJCS-Pakistan), averaging 816 m^3 ha^{-1} and 1458 m^3 ha^{-1}, respectively, compared with Rohera (285 m^3 ha^{-1}) and Khadir (465 m^3 ha^{-1}). Data on outlet-level discharges and farm/field-level water supplies suggest that there are wide locational variations in canal-water supplies and hence unequal distribution of water to farmers across reaches of distributaries in both BCS-India and LJCS-Pakistan.

Inequity in canal-water distribution

Overall inequity in canal-water distribution is higher in the study area in LJCS-Pakistan than in the study area in BCS-India. The estimated Gini coefficients[1] for BCS-India and LJC-Pakistan are 0.29 and 0.42, respectively. Gini coefficients are higher for distributaries where canal-water supply per hectare is relatively less (Rohera – BCS-India; Khadir – LJCS-Pakistan). Except for the head end of the Batta watercourse in BCS-India, Gini coefficients for tail-end watercourses are higher than their respective head-end watercourses. In general, inequity in canal-water distribution prevails both within watercourses and across watercourses along distributaries.[2]

Water quality

The quality of canal water is generally good for irrigation in both BCS-India and LJCS-Pakistan, with EC levels of 0.22, 0.24, 0.25

[1] Gini coefficient is based on the Lorenz curve and is a commonly used measure of inequity. The value of the Gini coefficient ranges between 0 and 1. A zero value shows a completely equal distribution. The greater the value of Gini, the greater the degree of inequity in distribution.

[2] The head–tail equity ratio is another measure of inequity. The results indicate that the head–tail equity ratios for average per hectare canal-water use in selected distributaries are 1.72:1 and 3.90:1 in BCS-India and LJCS-Pakistan, respectively. These results further suggest that head–tail inequities in LJCS-Pakistan are much greater than in BCS-India.

Table 16.3. Water and other input use for wheat production in BCS-India and LJCS-Pakistan.

Outlet/dist./ minor	No. of canal irrigations	Total no. of irrigations	Amount of tube-well water applied (m^3)	Amount of canal water applied (m^3)	Amount of total water applied (m^3)	Tube-well water as % of total water applied	Seed (kg ha^{-1})	NPK (kg ha^{-1})	No. of ploughings
BCS-India									
Batta – head	1.1	4.3	1829	849	2678	68	100	248	3.6
Batta – middle	1.0	4.1	2219	897	3116	71	108	229	3.7
Batta – tail	1.1	4.5	2545	700	3245	78	99	193	3.9
Rohera – head	0.7	4.3	2194	584	2778	79	109	199	4.0
Rohera – middle	0.4	4.4	3011	148	3159	95	106	202	3.6
Rohera – tail	0.2	5.0	3225	109	3334	96	124	244	4.4
Batta – all	1.1	4.3	2197	816	3013	73	102	223	3.7
Rohera – all	0.4	4.6	2806	282	3088	91	113	215	4.0
All	0.8	4.4	2500	550	3050	82	108	219	3.9
LJCS-Pakistan									
Lalian – head	2.1	4.4	1845	1500	3345	55	116	169.4	3.5
Lalian – middle	2.5	3.8	1304	2745	4049	32	126	195.6	3.8
Lalian – tail	1.5	4.6	2146	345	2491	86	124	88.8	2.9
Khadir – head	1.5	4.4	2704	606	3311	82	130	154.7	5.7
Khadir – middle	1.3	4.7	3591	600	4191	86	131	139.2	3.2
Khadir – tail	1.3	5.4	5088	187	5275	96	127	153.2	4.6
Lalian – all	2.0	4.2	1758	1458	3185	55	123	144.5	3.5
Khadir – all	1.9	4.8	3794	465	4259	89	130	148.9	4.5
All	1.9	4.5	2748	980	3702	74	126	146.7	3.9

and 0.27 dS m^{-1} for Lalian, Batta, Rohera and Khadir, respectively. However, the quality of the groundwater is generally low in the study areas of both countries. The average EC level of groundwater for BCS-India (3.39 dS m^{-1}) is much higher than that for LJCS-Pakistan (1.13 dS m^{-1}). Therefore, overall groundwater salinity levels are relatively higher in the study area in BCS-India than that in LJCS-Pakistan.

The groundwater quality varies significantly across distributaries. Groundwater is more saline in Batta in BCS-India and Lalian in LJCS-Pakistan (the two distributaries currently receiving relatively more canal water) than that in Rohera in India and Khadir in Pakistan. Groundwater quality varies significantly across head, middle and tail reaches of the distributaries. In general, the groundwater quality deteriorates towards the middle and tail reaches, except for Khadir in LJCS-Pakistan, where the groundwater salinity levels decrease towards the middle and tail reaches. (The reason for the good quality of groundwater in Khadir tail ends is that it is closer to the Chenab River.) Highly saline groundwater reaches are the ones receiving less canal water. Thus, the present strategy of canal-water allocation at the distributary level, i.e. more canal water to areas of highly saline groundwater – Batta and Lalian – compared with areas of relatively less saline groundwater areas – Rohera and Khadir – makes sense. However, the main problem lies within a distributary where saline groundwater reaches are receiving less canal water. Tail reaches of Batta and Lalian are the worst-affected areas (Tables 16.1 and 16.3).

Fertilizer and other inputs

Table 16.3 provides data on average quantities of key non-water inputs used for wheat production.

Number of ploughings

The number of ploughings is the same across irrigation systems in both countries. On average, there are four ploughings in the study areas of both countries.

Seeds

Overall use of seed per hectare in LJCS-Pakistan (126 kg ha^{-1}) is higher than that in BCS-India (108 kg ha^{-1}). This may be because most farmers in LJCS-Pakistan use older seed varieties (mostly from the home storage) as compared with those in BCS-India.

Fertilizer

There is a significant difference in the use of NPK per hectare across the two countries. Average NPK use per hectare in BCS-India is substantially higher (222 kg ha^{-1}) than that in LJCS-Pakistan (146 kg ha^{-1}). Most farmers in BCS-India have applied NPK in line with recommended amounts, and there is not much variation across and within distributaries. On the other hand, NPK use in LJCS-Pakistan is lower on most farms than the recommended levels (for medium soil-fertility levels, the recommended amount of NPK is 253 kg ha^{-1}). In the study area of Pakistan, there are significant differences in quantities of NPK used across farms and watercourses. NPK application rate is higher in Khadir (148 kg ha^{-1}) compared with Lalian (145 kg ha^{-1}).

For LJCS-Pakistan, NPK and yield show a strong positive relationship, yields increasing with increasing amounts of NPK applications. Given the complementary relationships between NPK and water, average NPK use is higher on farms and watercourses where water supplies are also higher and vice versa. Also, NPK use is directly related to reliability of water supplies. Farmers using a higher percentage of good-quality groundwater also use higher amounts of fertilizers and vice versa. The least amount of NPK use is found on farms in Lalian tail ends (89 kg ha^{-1}), where groundwater is of poorer quality, canal water supplies are the least and, consequently, yields are low. Other factors that may influence yields include quantity of weedicides, wheat seed variety and sowing time.

Water Use Versus Wheat Yield

Generally, with adequate, reliable/timely and good-quality groundwater, yields can be expected to be higher than those with canal water. This is true in Khadir in LJC-Pakistan, where the quality of groundwater is good. Increasing the proportion of good groundwater in total water applied resulted in improved wheat yields in this distributary. On the other hand, in the remaining three distributaries, the use of saline groundwater had a negative impact on wheat yields. The overall significance of impacts of groundwater use and its quality are quantified in the yield function developed below.

Yield-function analysis

The yield function estimates the effects of various factors of production on wheat yields. Separate analysis was undertaken for the samples of BCS-India and LJCS-Pakistan. The yield function was specified using a range of variables, including those discussed earlier, and estimated with a set of functional forms including linear, log-linear, log–log (Cobb–Douglas) and quadratic. The popular econometric and statistical criteria, such as predictive power of the equation, consistency and plausibility of estimated coefficients, algebraic signs and numerical magnitudes and their statistical significance, were used to select the functional form that had the best fit for the given data set. The following yield functions for BCS-India and LJCS-Pakistan were finally estimated with a set of independent variables, as given below:

BCS-India:

$$Y_i = \alpha_0 + \alpha_1 D_{mi} + \alpha_2 D_{ti} + \alpha_3 V_i + \alpha_4 S_i + \alpha_5 F_i + \alpha_6 W_i + \alpha_7 WD_i + \alpha_8 NW_i + \alpha_9 T_i + \alpha_{10} ECTW_i + U_i \ldots \quad (16.1)$$

LJCS-Pakistan

$$Y_i = \alpha_0 + \alpha_1 D_{mi} + \alpha_2 D_{ti} + \alpha_3 V_i + \alpha_4 S_i + \alpha_5 F_i + \alpha_6 W_i + \alpha_7 W^2_i + \alpha_8 NW_i + \alpha_9 T_i + \alpha_{10} ECTW_i + U_i \ldots \quad (16.2)$$

where:

Y = wheat output/yield in tons per hectare;

D_m = dummy for middle location of farmers on the distributary (D_m = 1 if the location is middle, D_m = 0 otherwise);

D_t = dummy for tail location of farmers on the distributary (D_t = 1 if the location is tail, D_t = 0 otherwise);

V = dummy for variety (for LJCS-Pakistan V = 1 if variety is MH97, V = 0 otherwise; and for BCS-India V = 1 if variety is WH-542 and PBW-343, V = 0 otherwise; these are relatively newer varieties);

S = sowing week (for LJCS-Pakistan first actual sowing week is 16–22 October 2000; for BCS-India first actual sowing week is 1–7 November); delay in sowing is hypothesized to negatively affect yields.

F = quantity of fertilizers – NPK – in kg ha^{-1};

W = quantity of total irrigation water applied (m^3);

WD = weedicides use as a fraction of recommended dosage;

NW = number of irrigations or waterings to wheat during the entire growing season;

T = for LJCS-Pakistan, time gap between pre-sowing and first post-sowing; for BCS-India, time gap between second and third irrigation/watering;[3]

$ECTW$ = percentage of groundwater in total water applied measured at field outlet (%), times electrical conductivity (EC) of groundwater (dS m^{-1});

α_s = coefficients to be estimated;

i = denotes farm;

U = error term.

Estimated coefficients (α) measure absolute change in wheat yield per unit change in one factor, holding the others constant. Location dummies capture the influence of location-specific factors other than those included in the yield function (particularly, soil salinity, land quality and rainfall). The coefficient of the dummy variable for seed, α_3, measures the net contribution of improved seed vari-

[3] In the estimation process, we also tried time gaps between irrigations other than these.

eties relative to all other seed varieties. The results of the estimated equations are presented in Table 16.4.

Among the wide range of factors that could possibly affect wheat yields, the location, seed variety, quantity of irrigation water and fertilizers for LJCS-Pakistan, quantity of weedicides (for BCS-India), number and timing of irrigation/waterings and quality of groundwater are found to be significant in influencing wheat yields. While the coefficients of determination of the estimated equations are low for both equations, it is acceptable given the type of data being used in estimations (cross-sectional).

The coefficients of location dummies indicate that wheat yields on middle and tail locations are lower than those at the head ends by 0.11 t ha^{-1} and 0.44 t ha^{-1}, respectively, for BCS-India and 0.70 t ha^{-1} and 0.53 t ha^{-1}, respectively, for LJCS-Pakistan. For LJCS-Pakistan, the lower coefficient for the tail end indicates the dominant effect of relatively good-quality groundwater on yields at Khadir. However, the magnitude of the effect of other factors on yields varies significantly across locations – as indicated by marginal productivities calculated based on the above coefficients using appropriate units (Table 16.5).

In BCS-India new seed varieties (WH-542, PBW-343) contribute an additional 97 kg ha^{-1} to average wheat yields, while in LJCS-Pakistan a new variety (MH 97) contributes an additional 995 kg ha^{-1} to average wheat yields (after accounting for locational differences). A delay of 1 week in sowing reduces wheat yield by 105 kg ha^{-1} in BCS-India and by 121 kg ha^{-1} in LJCS in Pakistan. An additional 10 kg of NPK use would increase yield by 29 kg ha^{-1} in LJCS-Pakistan and decrease yield by 9 kg ha^{-1} in BCS-India, indicating the scope for reducing fertilizer use in average yields. A volume of 100 m^3 of more water contributes 24 kg ha^{-1} to yields in Pakistan, while only a marginal increase of 0.13 kg ha^{-1} is seen in BCS-India, indicating that average yields obtained are closer to the highest point on the yield–water curve (and that farmers are applying water fairly well in line with crop water requirements). Therefore, there is not much scope to increase yields by further increasing the quantity of irrigation water per hectare.

The total quantity of water per hectare now supplied over one season when given in more frequent waterings positively influences yields, with each additional watering increasing yield by 48 kg ha^{-1} and 183 kg

Table 16.4. Estimated coefficients and their significance.

	BCS-India		LJCS-Pakistan	
Variable	Coefficient	*t* value	Coefficient	*t* value
Constant	4.456	11.08	3.583	5.74
D_m	−0.1058	−1.08	−0.701	−3.66
D_t	−0.4384	−3.65	−0.526	−2.58
V	0.2028	2.71	1.696	5.01
S	−0.0146	−2.87	−0.121	−3.31
F	−0.000745	−1.43	0.00292	2.25
W	1.3 E-6	0.02	0.000538	2.32
W^2	–	–	−0.0000000445	−1.87
WD	0.183	2.07	–	–
NW	0.047938	0.59	0.183	2.35
T	0.004385	1.01	−0.0777	−3.37
ECTW	−0.058609	−2.93	−0.364	−2.28
R^2	0.44		0.40	
N	216		218	

E-6 = 0.0000013.

Table 16.5. Marginal productivities of factors of production.

Factor	BCS-India: marginal productivity (kg ha^{-1})	LJCS-Pakistan: marginal productivity (kg ha^{-1})
Wheat seed variety (MH 97 for Pakistan, and WH-542 and PBW-343 for India)[a]	97	995
Sowing delay by week	−105	−121
NPK quantity in kg per 10 kg unit	−7	29
Irrigation water (m^3) per 100 m^3	0.13	24
Number of irrigations/waterings	48	183
Time gap between irrigations/waterings (for Pakistan, time gap between pre-sowing and first post-sowing; for India time gap between second and third irrigation/watering) (week)	4	−78
Per cent of groundwater in total water applied times present level of average EC of groundwater, at 100% groundwater-use level	−199	−411

[a]Considering the locational factors, marginal productivity of MH 97 (LJCS-Pakistan) would be 1696 kg ha^{-1} at the head, 995 kg ha^{-1} at the middle and 1521 kg ha^{-1} at the tail reaches. In BCS-India, marginal productivity of WH-542 and PBW-343 would be 202 kg ha^{-1} at the head, 97 kg ha^{-1} at the middle and −236 kg ha^{-1} at the tail (because the negative locational effect is greater than the positive effect of variety).

ha^{-1} for sample farms in BCS-India and LJCS-Pakistan,[4] respectively. The period after crop emergence is critical for crop growth, and prolonged delays in watering influence crop yields negatively[5] in the case of LJCS-Pakistan. A delay of 1 week in the first post-sowing watering from the appropriate period reduces wheat yields by 78 kg ha^{-1}. In the case of weedicides, with the application of recommended doses, there is the considerable increase in yields of 183 kg ha^{-1} in BCS-India. Only 11%, 20% and 2.5% of sample farmers have applied 60%, 80% and 110% of the recommended dosage of weedicides in their fields, respectively, while 7% of the sample farmers have not applied any weedicides.

In addition to timeliness, quality of water is also an important factor influencing yields. At the present level of groundwater EC (dS m^{-1}), the use of only groundwater (i.e. 100% groundwater with no canal water) reduces wheat yields on average by 199 kg ha^{-1} and 411 kg ha^{-1} in BCS-India and LJCS-Pakistan, respectively. Overall, yield response to groundwater use and its quality varies across locations in the distributaries. It is clear from the above discussion that seed variety (in LJCS-Pakistan), correct dosage of weedicide application (in BCS-India) and the quality of groundwater (in both BCS-India and LJCS-Pakistan) are the three most important factors influencing wheat yields.

As noted above, the marginal productivity of irrigation water is much lower in BCS-India than in LJCS-Pakistan. However, while the average productivity of consumed water is much the same, the average productivity of diverted water is much higher in BCS-India than in LJCS-Pakistan (Table 16.6).

Improving Wheat Productivity

This study identifies several factors influencing land and water productivity of wheat and indicates scope for improving land and water productivity and the profitability of

[4] In the Pakistani Punjab, the general recommendation for wheat is three to five waterings, depending on climatic conditions and groundwater-table depths (Government of Punjab, 2000).

[5] The Pakistani Punjab Agriculture Department recommends that, for wheat, watering after sowing be done within 20–25 days if it is sown after cotton, maize or sugar cane and within 30–40 days if it is sown after rice.

Table 16.6. Average productivity of water.

Outlet/ distributary/ minor	Average land productivity/ yield (kg ha^{-1})	Average productivity of consumed water (kg m^{-3})	Average productivity of total water applied (kg m^{-3})
BCS-India			
Batta – head	4569	1.38	1.71
Batta – middle	4485	1.36	1.44
Batta – tail	4119	1.25	1.27
Rohera – head	4908	1.49	1.77
Rohera – middle	4761	1.44	1.51
Rohera – tail	4043	1.23	1.21
Batta – all	4391	1.33	1.46
Rohera – all	4576	1.39	1.48
All	4483	1.36	1.47
LJCS-Pakistan			
Lalian – head	4946	1.60	1.48
Lalian – middle	3917	1.29	0.97
Lalian – tail	3188	1.08	1.28
Khadir – head	4033	1.31	1.22
Khadir – middle	3372	1.10	0.80
Khadir – tail	4590	1.62	0.87
Lalian – all	4206	1.39	1.32
Khadir – all	3998	1.34	0.94
All	4106	1.37	1.11

wheat production in general. From a policy point of view this could be:

- Improving agronomic/farm-management practices through: promoting new seed varieties (such as MH 97 in Pakistan and WH-542 and PBW-343 in India), providing/enhancing the role of extension services to farmers for dissemination of up-to-date knowledge on appropriate sowing dates and quantities and timing of application of inputs, particularly irrigation water.
- Improving water-management practices, including improving timeliness of water delivery, increasing canal-water supply, reallocating canal water within and across distributaries and encouraging the use of relatively good-quality groundwater wherever possible.

However, the reallocation option would be justified only if overall economic and social gains from such an exercise are higher than from the present situation. We have studied the socio-economic impacts of canal-water reallocation and present scenarios and strategies for canal-water reallocation.

Impact of Canal-water Reallocation

We analyse the impacts of the use of groundwater and canal water on wheat productivity and profitability in BCS-India and LJCS-Pakistan by using the yield functions estimated by Equations 16.1 and 16.2, respectively. In this analysis, we assume that all other factors, including total quantity of water applied and the price of wheat, remain at current levels across various canal reaches; only the source of water or a combination of proportions of water from the two sources changes. In order to generate various scenarios, we estimate the gross margin of wheat production with the following equation:

$$\hat{GM}_L = (P_L \times \hat{Y}_L) - COP_L \qquad (16.3)$$

where, $\hat{GM}_L$ is the estimated gross margins (US\$ ha^{-1}); $\hat{Y}_L$ is predicted wheat yield in t ha^{-1}; P_L is the price of wheat; COP_L is the cost of production (US$ ha^{-1}); and L (L = 1, 2 and 3) is for farm location (head, middle and tail). The predicted wheat yields are obtained using average values of independent variables in Equations 16.1 and 16.2.

We have generated several scenarios of canal-water reallocation and the results are presented in detail in Hussain *et al.* (2003). Only two scenarios are presented here (Tables 16.7 and 16.8). These would be the most likely scenarios out of all that were tried because of the inherent limitations in canal-water supplies:

- Scenario 0. Base level (at present levels of groundwater and canal-water use in all reaches).
- Scenario 1 (BCS-India) – 10% canal water with 90% groundwater in head reaches, 20% canal water with 80% groundwater in middle reaches and 30% canal water with 70% groundwater in tail reaches.
- Scenario 1 (LJCS-Pakistan) – 10% of canal-water use each in head, middle and tail reaches of Khadir (with 90% groundwater); 25% canal-water (and 75% groundwater) use each in the Lalian head and middle reaches; and 50% groundwater (and 50% canal water) in the Lalian tail reach.

The reallocation strategy (scenario 1) results in overall gains in both BCS-India and LJCS-Pakistan. In BCS-India, average yields and production increase from the base level by 0.12 t ha^{-1} and 26 t, respectively (Table 16.7). Gross margins and total value of production increase from the base level by US$15 ha^{-1} and US$3250. Highest gains are achieved on both the Batta and the Rohera tail reaches, with the Batta minor gaining US$12 ha^{-1} and Rohera US$17 ha^{-1}. There is only a marginal decrease in gross margins on the Batta head (US$2 ha^{-1}).

In LJCS-Pakistan, average yields and production increase from the base level by 0.21 t ha^{-1} and 77 t, respectively. Gross margins and total value of production increase from the base level by US$10 ha^{-1} and US$3569 (Table 16.7). Average yields and total production increase across all reaches of both distributaries. However, gross margins and total value of production decrease marginally on the Lalian head and middle reaches, and the largest gains are achieved on the Lalian tail reaches (US$35 ha^{-1}). Marginal losses on the head reaches can be avoided through influencing other factors, such as timeliness and reliability of irrigation supplies. Overall, the reallocation strategy presents a win–win situation. Murray-Rust and Vander Velde (1992) present a related discussion on the conjunctive water use and canal-water reallocation, and discuss the complexity in implementing reallocation of canal water and pumped groundwater between head- and tail-end farmers (in terms of farmers' acceptability of this option, particularly of those at the head ends of the systems).

Conclusions and Policy Implications

The study was conducted to understand the causes of differences in land and water productivity in wheat production across farms and reaches of selected irrigation systems in BCS-India and LJCS-Pakistan. These sites were selected because of similar agroclimatic characteristics and yet they were reported to have large variations in wheat yields. The study analysed the impacts of both land–water and non-land–water factors on productivity. Key findings of the study are summarized below.

- The hypothesis that significant gains in aggregate wheat yields can be obtained by improved water-management practices at the farm and irrigation-system levels was accepted for the irrigation systems selected for this study.
- The difference of average wheat yields in the studied irrigation systems in India (4.48 t ha^{-1}) and in Pakistan (4.11 t ha^{-1}) is not as high as generally perceived.
- There are significant differences in yields across farms and locations in selected irrigation systems in both countries, with much greater yield variations in LJCS-Pakistan than in BCS-India.
- The differences in wheat yield between watercourses are greater than between farms within a watercourse command area.

Table 16.7. Impact of canal-water reallocation on each of the selected watercourses – BCS-India.

Item/scenario	Batta – head	Batta – middle	Batta – tail	Batta	Rohera – head	Rohera – middle	Rohera – tail	Rohera	All
Wheat yield									
Scenario 0	4.63	4.53	4.40	4.52	4.83	4.81	4.38	4.67	4.60
Scenario 1	4.68	4.66	4.56	4.63	4.89	4.91	4.59	4.80	4.72
Total production (t)									
Scenario 0	167	163	158	488	174	173	158	505	993
Scenario 1	168	168	164	500	176	177	165	518	1,019
Total value (US$)									
Scenario 0	11,938	13,685	11,230	36,815	12,310	10,594	10,139	33,205	69,910
Scenario 1	11,862	14,219	12,092	38,146	12,425	10,987	11,503	34,959	73,163
Gross margins (US$ ha^{-1})									
Scenario 0	332	380	312	341	342	294	282	307	324
Scenario 1	330	395	336	353	345	305	320	324	339

Table 16.8. Impact of canal-water reallocation on each of the selected watercourses – LJCS-Pakistan.

Item/scenario	Lalian – head	Lalian – middle	Lalian – tail	Lalian	Khadir – head	Khadir – middle	Khadir – tail	Khadir	All
Wheat yield									
Scenario 0	4.82	4.43	3.47	4.24	4.29	3.73	4.42	4.15	4.19
Scenario 1	4.93	4.50	3.70	4.38	4.56	4.02	4.67	4.41	4.40
Total production (t)									
Scenario 0	176	198	225	932	235	103	331	652	1,582
Scenario 1	181	202	240	963	250	111	349	694	1,659
Total value (US$)									
Scenario 0	11,791	12,774	15,638	62,833	12,495	5,798	17,776	35,414	96,374
Scenario 1	11,550	11,712	17,894	63,304	13,167	6,632	19,489	40,674	99,943
Gross margins (US$ ha^{-1})									
Scenario 0	322	285	241	225	228	210	237	225	255
Scenario 1	315	261	276	259	241	240	260	259	265

- The total water applied varies significantly: an average of 3050 m^3 compared with crop water requirements of 3300 m^3 in BCS-India, and an average of 3702 m^3 compared with crop water requirements of 3009 m^3 in LJCS-Pakistan.
- There is significant inequity in distribution of canal water between tail reach versus head and middle reaches in both BCS-India and LJCS-Pakistan, with inequities much higher in LJCS-Pakistan than in BCS-India.
- The average productivity of consumed water is similar for both countries, i.e. 1.36 kg m^{-3} for BCS-India and 1.37 kg m^{-3} for LJCS-Pakistan. However, the average productivity of diverted water is higher for BCS-India (1.47 kg m^{-3}) than for LJCS-Pakistan (1.11 kg m^{-3}).
- The quality of groundwater is relatively poor in both locations and more so in BCS-India, while the average productivity per hectare is lower where groundwater is of poorer quality.
- In both countries, more canal water is supplied to distributaries where groundwater is more saline, and this is a rational strategy. However, groundwater quality varies significantly across reaches within a distributary. In general, groundwater quality deteriorates towards the middle and tail reaches. The reaches with saline groundwater currently receive less canal water, and the productivity of wheat is low in these reaches. Thus, intradistributary canal-water allocation is an important issue in relation to the profitability of wheat production.
- The locational differences in distribution of canal water, quality of groundwater and level of input use lead to significant variations in profitability of wheat production.

The results of the estimated yield functions suggest that, in addition to location-specific factors, such as soil salinity, land quality and rainfall, factors such as seed variety, application of recommended doses of weedicides, planting dates, irrigation-application dates and timing of water supplies and groundwater quality are important contributing factors in yield differences. Promoting on-farm agronomic practices, such as newer seed varieties, and dissemination of knowledge on planting dates and timings and application rates of inputs, especially water and fertilizers and proper dosage of weedicides, are important for reducing yield gaps.

In addition, improvements in water-management practices at the system level will also contribute to increased yields and the overall profitability of wheat production. Improving on timings of canal-water deliveries and adopting an effective strategy for allocation of canal water will result in overall significant socio-economic gains in wheat production. The results of the study suggest that poor groundwater quality, leading to accumulation of salts, is one of the key factors influencing wheat yields, and that groundwater quality varies significantly across reaches in command areas of the systems. Wheat production is highly profitable with only canal-water use and least profitable with the sole use of poor-quality groundwater. As the results show, a reliable supply of good-quality water could indeed significantly increase wheat yields and enhance the profitability of wheat production in irrigated lands of the Indian subcontinent.

The chapter presents two scenarios on the impacts of water use from two sources on the socio-economics of wheat production. The results indicate that overall gains from wheat production will increase if canal water is re-allocated such that more canal water is supplied to canal reaches where groundwater is of poorer quality. The reallocation strategy in scenario 1 would increase average gross margins by US$15 and US$10 in BCS-India and LJCS-Pakistan, respectively. Much of the gain from this reallocation will be achieved in tail reaches. The policy implication of these findings is that, under conditions of canal-water scarcity and locational variations in the quality of groundwater, joint management of canal water and groundwater is essential to increase overall gains from crop production. The study presents an example of 'institutional water scarcity' that could be addressed through enhancing existing institutions or developing institutional arrangements explicitly designed to effectively manage the available surface-water and groundwater resources holistically.

References

Ahmed, B. and Chaudhry, M.A. (1996) *Productivity Differential between Pakistan and Indian Punjabs: an Analysis.* Agricultural and Social Sciences Research Centre, Faculty of Agricultural Economics and Rural Sociology, University of Agriculture, Faisalabad, Pakistan.

Altaf, Z. (1994) Rice–wheat systems in Pakistan – The sustainability questions. In: *Expert Consultation on 'Sustainability of the Rice–Wheat Production Systems in Different Agro-Ecological Settings in Asia', Organized by RAPA in Bangkok.* RAPA Publication 1994/11, Bangkok, pp. 197–203.

Aslam, M. (1998) *Improved Water Management Practices for the Rice–Wheat Cropping Systems in Sindh Province, Pakistan.* Report No. R-70–1a, International Irrigation Management Institute, Lahore, Pakistan, 96 pp.

Chaudhary, T.N. and Bhatnagar, B.K. (1980) Wheat root distribution, water extraction pattern and grain yield as influenced by time and rate of irrigation. *Agricultural Water Management* 3, 115–124.

Doorenbos, J., Kassam, A.H., Benvelsen, C.I.M., Bransheid, V., Plusje, J.M.G.A., Smith, M., Uittenbogaard, G.O. and Wal, H.K.V.D. (1979) Yield response to water. In: *FAO Irrigation and Drainage Paper.* Food and Agriculture Organization of the United Nations, Rome, pp. 164–170.

Government of Punjab (2000) Agriculture Department, Lahore, Pakistan.

Hussain, I., Sakthivadivel, R., Amarasinghe, U., Mudasser, M. and Molden, D. (2003) *Land and Water Productivity of Wheat in the Western Indo-Gangetic Plains of India and Pakistan: a Comparative Analysis.* Research Report, International Water Management Institute, Colombo, Sri Lanka.

Minhas, P.S., Sharma, O.P. and Patil, S.G. (eds) (1998) *25 years of Research on Management of Salt-affected Soils and Use of Saline Water in Agriculture.* CSSRI, Karnal, pp. 150–170.

Mishra, A. and Tyagi, N.K. (1988) Improving canal water deliveries with auxiliary storage. *Journal of Irrigation and Drainage Engineering* 114, 535–545.

Mudasser, M., Intizar, H. and Muhammad, A. (2001) Constraints to wheat productivity in Pakistan: a review of literature. International Water Management Institute (IWMI), Colombo, Sri Lanka. Duplicated.

Murray-Rust, H. and Vander Velde, E. (1992) Conjunctive use of canal water and groundwater in Punjab Pakistan: management and policy options. In: *Advancements in IIMI's Research (1992).* International Irrigation Management Institute, Colombo, Sri Lanka pp. 161–188.

Nagarajan, S. (1998) Perspective on wheat demand and research. In: Nagarajan, S., Singh, G. and Tyagi, B.S. (eds) *Wheat Research Needs Beyond 2000.* Directorate of Wheat Research, Karnal, pp. 13–28.

Narayanamoorthy, S.G. and Perry, C.J. (1997) *Farmer Response to Rationed and Uncertain Irrigation Supplies.* Research Report 24, International Irrigation Management Institute (IIMI), Colombo, Sri Lanka.

Pintus, F. (1997) *Impact of Irrigation, Salinity and Cultural Practices on Wheat Yields in Southeastern Punjab, Pakistan.* Report No. R-30, IIMI, Lahore, Pakistan, 70 pp.

Rehman, M.S.U. (1986) Contribution of main factors of production on the yield of wheat. MSc (Hons) Agricultural Economics thesis, submitted to Department of Agricultural Economics, University of Agriculture, Faisalabad, Pakistan, 42 pp.

Siddiq, A. (1994) Sustainability of Indus basin: impact of tertiary salinity/sodicity on wheat productivity, damage assessment and future public policy. PhD dissertation, presented to Graduate College, University of Illinois at Urbana-Champaign, USA.

Tyagi, N.K. and Sharma, D.K. (2001) Improving wheat productivity in Indo-Gangetic plains: constraints and technological options: a review of literature. International Water Management Institute (IWMI), Colombo, Sri Lanka. Duplicated.

17 Reform of the Thai Irrigation Sector: is there Scope for Increasing Water Productivity?

François Molle[1]
International Water Management Institute, Colombo, Sri Lanka

Abstract

Most major water basins in Thailand, especially the Chao Phraya River basin, are now nearing closure. An increasing amount of water is being diverted out of agriculture, and intrabasin allocation generates tensions. Water productivity can be potentially raised by two economic measures with three possible effects. The measures are water pricing and reallocating water away from agriculture. Water pricing may: (i) elicit water saving and the adoption of water-saving technologies; and (ii) encourage shifts toward non-rice crops with a higher economic return per unit of water consumed. Reallocating water out of agriculture to other uses, possibly through market mechanisms, may also be conducive to overall economic gains.

This chapter shows that, in the case of Thailand, the benefits of such reforms are much fewer than expected and that transaction costs and political risks probably outweigh the possible gains. The case of the Chao Phraya River basin suggests that the closure of a basin is accompanied by several endogenous adjustments to water scarcity and that the scope for significant productivity gains is reduced. It is stressed that the current physical, institutional and legal settings do not allow the implementation of such economics-based regulations. While emphasis is placed on the gap between the rhetoric of economic tools and the conditions of the real world, the chapter also sketches guidelines for reform of the water sector.

Introduction

> A water tax could be levied, in a manner similar to the paddy land tax, over the whole area at present cultivated and the future extension of this area, as far as the fields are benefited by the [irrigation] system . . . water rates could in general be assessed in some proportion to the quantity of water utilised, and would most probably be a suitable taxation for dry season crops and garden cultivation.
>
> (van der Heide, 1903)

The hindsight provided by history, though often neglected, is sometimes the best short cut to understanding that what may appear as desirable is not always feasible or even logical when seen from a different perspective. The above statement is issued not from a recent consultant report, as one might easily believe, but from the *General Report on Irrigation and Drainage in the Lower Menam (Chao Phraya) Valley* submitted in 1903 to the government of Siam by H. van der Heide, a

[1] The author thanks Randy Barker, Bryan Bruns and Madhusudan Bhattarai for their useful comments on an earlier version of this chapter.

Dutch engineer. Clearly, all the calls for pricing water issued during the 20th century were, until recent years, mostly motivated by a concern for cost recovery.[2] Early legislation on water also included some provisions on pricing. The Royal Irrigation Act of 1942 was the first to allow for the collection of a fee that was to remain under ceilings of 5 baht rai^{-1} (1 rai = 0.16 ha)[3] and 0.50 baht m^{-3} for factories, but these rates have not been revised hitherto (Wongbandit, 1997). At present, only a few non-agricultural users using canal water are paying a fee.

It was only recently that water pricing was proposed as a way to regulate water use, in terms of volume or allocation. Such a proposal was the consequence of growing water scarcity in the country, as well as the interest of donors and some academics in the water sector to initiate measures of 'demand management'. In fact, despite being a tropical country with a monsoonal season, Thailand has joined a host of countries currently facing water shortages. With the exception of the southern region and some forest areas along the border, hydrological data show that the annual average rainfall in Thailand varies between 1100 mm and 1600 mm. During the 6 driest months of the year, from December to May, the country relies chiefly on the water available in 28 main storage dams. However, only 15% of the 200 billion m^3 (Bm^3) annual runoff remains trapped in the dams (ESCAP, 1991).

Gradually, due to the concomitant development of irrigated and urban areas, constraints on water resources started to be felt, particularly in the Chao Phraya River basin, where irrigated areas have been developed beyond the potential expressed by the available water resources (a situation qualified by the World Bank as 'overbuilt'). The expansion of the Bangkok metropolitan area (BMA) led to the gradual extraction of a significant share of the basin resources for urban and industrial water uses. Increasing competition for water materialized through recurrent water shortages, occurring principally in the dry season and mostly affecting rice cultivation and prompting restrictions in the water supply of the BMA (in 1994 and 1999). Solutions proposed to solve the current water-shortage situation span a wide ideological range, from those supporting the development of more water resources (new dams, diversion from the Mekong River or Salaween River) or the reform of the concerned administrations to those advocating a gradual privatization and commoditization of water. This issue recently entered the limelight following an announcement that the granting of Asian Development Bank (ADB) funds to the country (presented as being crucial to the country's economic recovery following the 1997 crisis) would be conditional on its subscribing to and applying the overall principle of water pricing. The public debate has been significantly obfuscated by the conflicting and often confusing views on water pricing, as reflected in newspaper declarations, interviews, consultants' reports and non-governmental organization (NGO) literature.[4]

---

[2] See, for example, De Young (1966):

> 'The light taxation affects any large scale government programme to improve conditions for the peasants. It is evident that not until the government has assurance of steady and increased income from local taxes can it expect to support large scale farm improvement projects … As yet the government has not come to the conclusion that at least a partial support of such a project should come from equitable taxation of the peasants. Any program designed to aid the farmer, such as large scale irrigation, is recognised now only as a national investment and a responsibility of the government. That this policy sooner or later must change is self-evident, for without local taxation the peasants' demands for agricultural, educational, health, and transportation improvements cannot be met.'

[3] US$1.00 = 40 baht in 2000.

[4] An examination of official declarations reported in national newspapers gives a measure of the fluctuating argumentation, reflecting the unsettled nature of the negotiations, the general nature of the arguments and the lack of consensus even within a given administrative body (see Molle *et al.*, 2001a).

Increasing water productivity covers several meanings. First, it means that the output (say, in t ha^{-1}) of a given crop per m^3 of water applied is raised. This is tantamount to achieving water savings (while maintaining yields), which can occur at the plot level and/or at the irrigation-system level, with or without adopting new technologies. Secondly, it means that the economic productivity of irrigated agriculture can be increased by shifting to crops with a higher benefit (in baht ha^{-1}) per unit of water used (m^3). This implies the selection of cash crops with higher returns and less water demand than rice. Thirdly, it means considering all alternative uses of water, including those outside the agriculture sector, and allocating water preferentially to those that yield a higher economic value (baht m^{-3}). This chapter reviews whether, in the Thai case, these three objectives are sound and whether they can be achieved through economic tools, such as water pricing[5] or water markets. It is necessary to distinguish here between small- and medium/large-scale irrigation projects. The former are often epitomized by the traditional *muang fay* (river diversion) systems of northern Thailand, while the latter are best represented by the Chao Phraya delta. Unless otherwise mentioned, what follows refers to medium/large-scale projects, which make up two-thirds of the country's irrigated area. The discussion also centres on the dry season, when water scarcity is an issue, rather than on the rainy season.

Water Pricing and Water Savings

The Director General of the Royal Irrigation Department (RID) was recently seen on a Thai national TV channel explaining, somewhat contritely, that water efficiency was very low in Thailand (around 30%) and that this had to be remedied in the face of the water shortage experienced by the country. This short sequence epitomizes better than anything else the extent to which such a statement has become conventional wisdom. A thorough probe into the literature, however, provides little evidence that such a value and the general validity of the statement are established.[6] Rather, it suggests that such a view is derived from general analyses, such as those of the Thailand Development Research Institute (TDRI, 1990) or Postel (1992a) (which may have a positive role in raising the general awareness of the problems lying ahead but may be totally misleading when applied to a particular case), and is further disseminated by repetition.

International agencies (and sometimes, in their footsteps, local officials) commonly report that Thai farmers are 'guzzling' water or are showing 'water greed' (*The Nation*, n.d.), furthering the general idea that efficiency in large state-run irrigated schemes is often as low as 30% and sticking to this overall vision without questioning it any further. Yet research conducted in recent years has shown that water basins tend to 'close' when demand builds up and that little water is eventually 'lost' out of the system. There has been widespread recognition that focusing on relatively low irrigation efficiency at the on-farm or secondary levels could be totally misleading (Keller *et al.*, 1996; Molden and Sakthivadivel, 1999; Perry, 1999; Seckler *et al.*, 2002, Chapter 3, this volume). When analysed at the macro-level and the basin level, many systems – river deltas accounting for the most significant of them – are eventually found to operate with extremely high overall efficiency. Thus the scale of analysis of water-use efficiency is crucial.

[5] Therefore, the chapter does not address the relevance of pricing for cost recovery or other purposes not directly related to crop or economic productivity.

[6] The values encountered in reports and theses are by no means straightforward. They mix values at the plot or scheme level and never consider the macro-level of the basin. Most of the drainage of small run-of-river or pumping schemes usually returns to the river. Regarding large-scale schemes, recent reports, such as JICA (1992), take 65% for the west bank (conservation area), while Binnie and Partners (1997) consider values of 45% (but give no clue as to why such values are adopted). In all instances, the focus is always on 'classical efficiency' and not on how it relates to the basin-level water flows and water balance.

In the dry season, the Chao Phraya delta provides an illustrative example of such a closed system. Most of the return flow from fields is reused downstream and most of the drains have been gated to capture or retain superficial and subsuperficial flows in the dry season. Several tens of thousands of tube wells have been dug to tap shallow aquifers wherever suitable. Water releases at Bhumipol and Sirikit dams (see Box 17.1), as well as at the Chai Nat diversion dam, are nowadays better attuned to user requirements and this results in little waste. If we consider the efficiency of irrigation at the macro-level, we see that the only 'waste water' (i.e. not depleted for production purposes) is water that evaporates in waterways or fallow lands or that eventually flows out of the delta system into the sea in excess of what is needed to control pollution and intrusion of salinity in the mouth of the river (in the dry season). As this water is now extremely limited, it follows that very little water is lost.[7] The second component of water 'loss' is that of infiltration, either to shallow aquifers or to deep aquifers. In the first case, water is tapped again through shallow tube wells (forming secondary water sources) or soon flows to the drainage system, where it is reused. In the second case, the infiltrated water reaches deep aquifers, which are notoriously overexploited in the Bangkok area, resulting in land subsidence and horrendous costs in the upgrading of flood protection and in flood damages.[8] Therefore, we may state that infiltration losses in the delta are not sufficient to offset the depletion of the aquifers. The water balance in the basin (Molle *et al.*, 2001a) shows that in the dry season the overall efficiency of controlled[9] water use is around 88%.

Box 17.1. Water allocation in the Chao Phraya basin.

The Chao Phraya basin can be conveniently divided in three parts. The upper part (upstream of the two main storage dams: Bhumipol and Sirikit dams), the middle part (from the dams to Chai Nat) and the lower part, or the delta proper (see figure). The dams are operated by the Energy Generation Authority of Thailand (EGAT). In the dry season, according to the year, between 2 and 8 Bm^3 are released to be distributed by the RID among 25 subunits called 'Irrigation Projects'. Priority of water goes first to Bangkok, then to the control of saline intrusion, next to the supply of orchards and shrimp ponds and last to inland transportation and rice cultivation. The irrigation sector, despite receiving the largest share on average, has to cope with a high interannual fluctuation of the amount of water apportioned to it. Allocation is a top-down process, where the shares of the Projects are centrally defined. Water abstraction in the middle basin cannot be fully controlled by RID and has been increasing dramatically (to 35% of dams' releases). In the dry season, pumping from waterways is the most common way to access water.

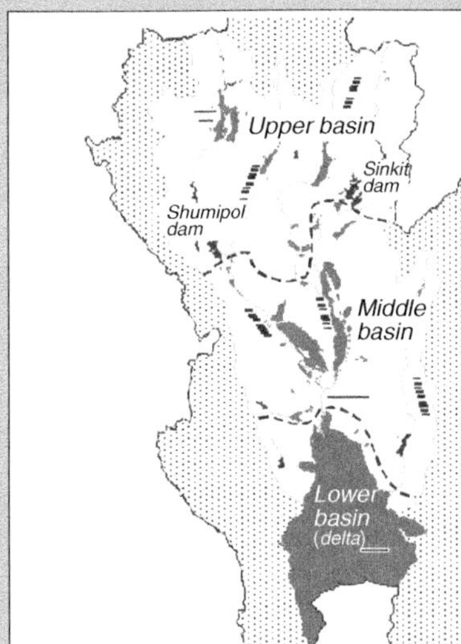

[7] In past years, the Energy Generation Authority of Thailand (EGAT) may have released water only for the purpose of energy generation, thus resulting in fresh water being lost to the sea. However, this has been extremely rare during the dry seasons of the last 10 years. Whether this should still be permitted by EGAT, even in the wet season, is discussed in Molle *et al.* (2001a). In all cases, such losses are controlled and deliberate and, therefore, cannot be considered as decreasing the efficiency.

[8] It is estimated that damages from the 1995 flood amounted to 50 billion baht (i.e. US$2 billion).

[9] Includes water released from the dams, diverted from the Mae Klong basin and extracted from shallow and deep wells.

Even when we carefully examine plot irrigation, it is hard to find the criticized pattern of wasteful practices. The main reason is that most farmers access water through pumping. This is true for all the farmers located in the lower delta (in this so-called flat conservation area, water is integrally and individually pumped from a dense network of waterways) and for approximately 60% of the farmers in the upper delta. It follows that, altogether, about 80% of farmers resort to pumping, the great majority using low-lift axial pumps powered by two-wheel tractors. Although the Chao Phraya and Mae Klong schemes were designed to supply water by gravity, RID experienced difficulties in managing reduced flows in the dry season. To offset this constraint, farmers have developed an impressive individual pumping capacity allowing them to tap whatever little flow might appear in the canal. Because of the costs incurred by these water-lifting operations, there is little likelihood that farmers may be squandering water.[10] This is consistent with recent estimates of water use in the delta, which show that scheme efficiency (evapotranspiration (ET)/net diverted water) is remarkably high (60%), with only 10,000 m^3 diverted ha^{-1} and per crop, including 15% of rainfall (Molle *et al.*, 2001a).

The consequence of all these elements is that few overall water savings can be expected from a hypothetical change in the behaviour of water users, because the efficiency in the Chao Phraya delta, and probably in other closed basins of Thailand, is already high. Molle *et al.* (2001a) have investigated the different paths that may lead to improved efficiency and equity (dam management, shifts in cropping calendars, etc.) but have shown that emphasis on irrigation-use efficiency would be misplaced. In addition, it is both self-evident and widely recognized that the individual volumetric pricing of water is not feasible in the context of small-scale rice farming in large gravity-irrigation schemes. Thus all incentives to save water embedded in volumetric pricing are lost when we are forced to shift to a water fee per unit of land or other proxy.[11]

This drastic constraint is generally dealt with by turning to the alternative of 'water wholesaling', in which water is attributed to groups of users ('water management blocks', for TDRI (2001)), for example, to those farmers who are served by the same lateral canal, on whom would fall the burden and the responsibility to allocate and manage water, solve conflicts and collect a water charge. This alternative also has the advantage of 'forcing' farmers to act collectively to achieve greater efficiency/equity within the command area of their canal and to constitute a form of bargaining power to discuss issues of water allocation with RID. In such a case, the incentive is passed on to the group, which is expected to derive its own internal arrangements aimed at saving water and hence reducing the water charge of the group as a whole and of each of its members in particular.

Such volumetric pricing could theoretically even elicit investment in water-saving technologies, if the investments compare favourably with the corresponding financial savings. The Iran case described by Perry (2001) suggests that technological change is too expensive for farmers, irrespective of the cost of water, and that the net value of water consumed (in \$ m^{-3}) is comparable to the costs of reducing consumption through improved technologies. In addition, such investments are to be made (collectively) by upstream farmers to the benefit of the downstream farmers, a scenario that is difficult to

[10] In some cases, the costs of pumping may even discourage farmers from growing a second or third crop. These costs, combined with poor levelling, also explain the low use of water in sugarcane cultivation.

[11] In addition, the introduction of such a fee per area is doomed to encounter severe difficulties in situations where access to water is highly heterogeneous. This is the case, for example, in the upper delta, where some farmers may access water throughout the year, while elsewhere others receive a very uncertain supply. In addition, this access can be partly provided by gravity and partly by pumping, and their respective shares can vary greatly from one year to the next. Therefore, quantifying the real benefit of irrigation water for hundreds of thousands of farmers, when this benefit is, spatially and temporally, highly heterogeneous, is deemed impractical.

envision without public intervention. In the Thai case, there is no available technology (hardware) that could bring about drastic water savings in rice cultivation, but such a mechanism might encourage technical innovation regarding water management at the plot level.[12]

This appealing solution of water wholesaling features nicely in paper proposals of consultants and academics, and is credited with some success in Mexico or Andhra Pradesh. However, it implies a series of prerequisites that are often not given due attention (Molle, 2001; Molle *et al.*, 2001a). A detailed review of these conditions is beyond the scope of this chapter but it can be mentioned that the main difficulties lie in the definition of a 'service' to which the fee would correspond. This includes the question of both allocation (the process to define each year how the fluctuating water stock in the dams is to be apportioned) and distribution (ensure the timing and the discharges of deliveries as agreed upon). The degree of technical and institutional control over the whole water basin is at present insufficient to ensure this. On the other hand, it is debatable whether there is enough social capital within a rather heterogeneous farming population to carry out all the tasks that the groups are expected to perform.

In sum, water pricing on an individual basis is possible only if based on the plot area and is, therefore, tantamount to an additional tax with, at best, no impact on water productivity.[13] The 'wholesaling' of water is an option that requires far-reaching improvements to be brought about at the technical and institutional levels prior to implementation. Even in such a case, there is no strong empirical evidence that the turnover of management to water-user groups has any significant impact on water productivity (Samad, 2001). In addition, a careful analysis of field water use, as well as water accounting for the basin, does not point to significant water losses (but some improvements in dam management and scheduling are nevertheless desirable and possible). This suggests that the heavy transaction costs incurred by the establishment of some form of water pricing would far outweigh the meagre potential gains in productivity.

A corollary from this conclusion is that the refrain 'water is consistently undervalued, and as a result is chronically overused' (Postel, 1992a) may well have little validity in closed basins. In Thailand, many observers, such as Christensen and Boon-Long (1994), who believe that 'since water is not appropriately priced, it is used inefficiently, and consumers have no incentive to economize', have also considered this postulate as self-evident for irrigation.[14] Ironically, despite severely lacking consistency, it is presented as the main justification for water pricing and gains apparent consistency only under the effect of repetition.[15]

[12] Experiences from China and Madagascar suggest that yields can be maintained with innovative water-management techniques conducive to water savings. At the moment, there is no clear picture as to whether this is allowed by particular socio-economic and cultural factors or whether there is scope for the dissemination of these innovations. For the Madagascar case, see Moser and Barett (2001) for a pessimistic view on such a hope.

[13] It is often noted (Moore, 1989; Meinzen-Dick and Rosegrant, 1997) that the impact is more likely to be negative, as farmers paying for water feel that they have acquired a right to more 'comfort' in use and are less concerned with how much water they consume.

[14] This is an extrapolation of the experience with urban water, which differs markedly from irrigation.

[15] See the declarations of an official of the Ministry of Agriculture: 'Water should be priced in order to increase the efficiency of its use in the farm sector' (*The Nation*, 2000a); 'Agricultural experts agree that water-pricing measures would help improve efficiency in water use among farmers' (*The Nation*, 1999); the Director of the National Water Resources Committee: 'In reality water is scarce, and the only mechanism to save water and encourage efficient use is to give it a price' (*The Nation*, 2000b); the resident adviser for the ADB in Thailand: 'International best practices suggest that efficiency in water management can be improved considerably through imposition of nominal water user fees' (*Bangkok Post*, 2000); 'Currently, most farmers don't have to pay for irrigation water and, thus, have little incentive to conserve water or to use it efficiently on high-value crops. As a result, irrigation efficiency is under 30 percent' (TDRI, 1990), etc.

Water Productivity and Crop Choice

Conventional wisdom admittedly considers rice as a water-consuming crop.[16] The possibility of achieving water conservation by inducing a shift away from rice to field crops, such as mung bean, groundnut, maize, or chilli, which consume (ET) approximately 60% of the amount of water needed for rice, has long been underlined by policy makers and has formed the cornerstone of public projects aimed at fostering agricultural diversification (Siriluck and Kammeier, 2000). This was already a recommendation of the Food and Agriculture Organization (FAO) as early as the 1960s, as well as the alternative that 'received the most attention' from Small (1972) in his study of the delta. Australia and Japan were jointly engaged in agronomic tests in the late 1960s and 1970s in order to propose field crops for irrigated areas. 'In recent years, low export prices for rice, and the difficulties encountered by Thailand in maintaining her export markets have further intensified the interest in stimulating the production of upland crops', noted Small in 1972. Such a concern has been constantly expressed for at least four decades. Even nowadays, it is not rare to hear officials complaining off the record that 'farmers are stubborn', that 'they lack knowledge and only know how to grow rice', and that 'they oppose any change' described by outsiders as beneficial.

Planting crops with lower water requirements would, ideally, allow more farmers to benefit from a second crop in the dry season. If the economic benefit of such crops compares favourably with rice, then there is an overall gain in such a shift. This reasoning is implicitly based on average values of farmers' income, despite the fact that, in peasant agriculture, risk is a much more relevant concern. Scott (1976) has shown that the sustainability of peasant economies was more closely governed by vagaries in yields than by average values, and it has also been shown that people resented smaller, fixed taxes much more than larger ones indexed on real yields. It can be argued that yields in the irrigated areas discussed in this chapter are made stable by the use of irrigation. It must not be overlooked, however, that risk in production, in any case not negligible (diseases, grasshoppers, etc.), has been replaced by risk in marketing, further compounded by the higher requirements of cash input demanded by commercial crops. As a general rule, the potential return of capital investments is strongly correlated to the level of risk attached to the undertaking (Molle *et al.*, 2001b). This is clearly exemplified by Szuster *et al.* (2003) in their comparative study of rice and shrimp farming in the Chao Phraya delta. In other words, on average, cash crops may fetch higher prices but they are also subject to more uncertainty, either in terms of yields or of farm-gate prices. Thus, only those farmers with enough capital reserve to weather the losses experienced in some years can benefit from the overall mid-term higher returns; others go bankrupt and remain indebted. Shrimp farming, again, provides a good example of such a situation.

This situation differs significantly from that of Western agriculture, where bottom prices or 'intervention schemes' are generally established to compensate for economic losses when they occur (more on this later). In addition, Western farmers generally benefit from insurance (against exceptional yield losses), which comes with stronger cooperative and professional structures.

It could be argued, however, that the price of rice is also highly uncertain and that rice production suffers from uncertainty as much as other crops. If the price of rice does fluctuate, its crucial importance for the rural economy brings it under more scrutiny. Despite recurring complaints, echoed in newspapers, that rice farmers lose money when producing rice, the political impact of possible low prices, in reality, largely shields them from dropping

[16] This is derived from the vision of the large amount of water that must be diverted, in particular to meet land-preparation requirements and seepage/infiltration losses, but much less so on a purely agronomic basis (water depleted by ET).

under the reproduction threshold. Ad hoc public interventions are always implemented when such a risk arises (even though their impact generally falls short of expectations and benefits tend to be captured by millers and other actors in the rice industry). This does not hold, however, for secondary or marginal crops (which invariably include the desirable 'cash crops'), and complaints of scattered producers have little chance of being heard in the case of depressed prices. A typical example of such a cash crop is chilli, a rather capital- and labour-intensive crop, which can fetch 25 baht kg^{-1} in one year (providing a high return) and 2–3 baht kg^{-1} in the following year (with a net loss for farmers).

Theoretically, a shift to non-rice crops could be elicited by differential taxes for crop type or water use (when individual or group volumetric pricing is possible). However, such a measure will only be significant if the tax differential represents a significant share of the income, say 10% or more. Perry (1996) found that volumetric charges in Egypt were an unrealistic means of encouraging significant reductions in demand because, in order to have an influence on demand, charges would have to be very high.[17] Raising (fixed) taxation to such levels would only increase the risk attached to non-rice crops, thus producing an effect opposite to that desired.

Evidence of the dynamics of diversification in the delta (Kasetsart University and IRD, 1996) points to the fact that farmers display great responsiveness to market changes and opportunities (a point definitely confirmed by the recent spectacular development of inland shrimp farming (Szuster and Flaherty, 2000)). Good transportation and communication networks allow marketing channels to perform rather efficiently. The main weak point remains the risk attached to the frequent fluctuation of the prices of field crops, which discourages farmers from shifting significantly to non-rice crops. As long as the economic environment of field-crop production remains unattractive and uncertain, there is little incentive for farmers to adopt such crops and a limited basis to sustain criticism of their growing rice, as many have incurred losses by growing field crops (either of their own accord or at the suggestion of extension services).

In addition, there are several other constraints (agroecology – heavy soil with little drainage, not favourable to growing field crops, labour[18] and capital requirements, skill learning, development of proper marketing channels, etc.) that have an impact on the process of diversification and it is doubtful whether 'pushing' for it would be eventually beneficial. Siriluck and Kammeier's (2000) study of a large-scale public programme aimed at encouraging crop diversification shows that such interventions meet with mixed success and are not flexible enough to adapt to different physical and socio-economic environments. Contrary to common rhetoric, farmers do not need to have their water priced to shift to other productions. They will increasingly do so if the uncertainty about water and commodity prices is reduced. They have time and again shown dramatic responsiveness to constraints on other production factors, such as land and labour (Molle and Srijantr, 1999), and have already sufficiently experienced the scarcity of water to adapt their cropping patterns, should conditions be favourable.

Water Productivity and Sectorial Allocations

The last form of achieving economic gains in productivity is to reallocate water used in agriculture to other sectors, which invariably display a higher return per m^3 used. There is a conspicuous and widespread argument that (public) centralized water allocation in Thailand has reached its limits and that water rights and water markets would provide a flexible mechanism to allow the reallocation of scarce resources towards the most economically profitable uses. This is strongly reminis-

[17] The price required to induce a 15% fall in demand for water would have reduced farm incomes by 25%.
[18] For example, the harvest of mung bean, a typical supplementary crop with no additional water requirements, is often a problem because of labour shortage.

cent of the deadlock experienced in the western USA, where water rights[19] are locked in uses of low productivity and where market mechanisms constitute one of the ways out of the stalemate (see Huffaker *et al.*, 2000). The claim that central agencies have failed in properly allocating water has become a refrain supporting the idea of markets as an alternative.

In the Thai context, commentators do not hesitate to incorporate this concern into their rationale, asserting that the state has proved inefficient in allocating water to the most beneficial uses.[20] It is intriguing to see the ubiquity of this argument, even outside its 'original' context, and how it permeates debates even in settings where this problem has been handled relatively successfully. Contrary to the alleged government failure in allocating water resources, sectorial allocation in Thailand (as in most countries) has been driven by a clear priority in use, which mirrors the economic return of all activities. Cases of non-agricultural activities, in particular industrial ones, that would have been constrained or impeded by the lack of water are unheard of and it is hard to see how criticism of central allocation can fly in the face of such evidence. The deadlock experienced in the western USA is unknown here and establishing a water market might create exactly the kind of problems it is assumed to solve, should, as is apparent in the USA, the rural sector be reluctant to relinquish its established rights.

It seems that the argument is loosely based on the implicit (but fallacious) assumption that, if the agriculture sector uses a share of Thai (controlled) waters as high as 80% then it is likely to enjoy a sort of privilege, to the detriment of other activities. It is also often (rightly) stressed that saving 5% of water in agriculture would represent a huge amount for other activities, but not that the latter are not directly claiming it, as they are effectively served first. To present the agriculture sector as the spoilt, unrepentant and ungrateful child of the nation does little justice to the fact that farmers are, in fact, served with the (fluctuating) left-over water in the system. This share happens to be the largest one only because other uses have not yet developed to a wider magnitude (and also because the government (not the farmers) has invested in infrastructure allowing the use of this water for irrigation). The argument glosses over the facts that: (i) this share will decline in the future (as agriculture is usually deprived of its water when other sectors grow);[21] and (ii) the unwritten 'rights' of farmers being limited to the left-over water, the farm sector has to cope with a very fluctuating supply, which also generates severe

[19] There is some irony in the evidence that, if the Thai legal system had been based on prior appropriation rights, as in the western USA, the delta would have been granted senior rights on water since the 1960s or earlier and Bangkok would now be trying to buy these rights from farmers. In such a case, farmers would at present not be asked to pay but, on the contrary, courted to accept money as compensation!

[20] A typical example is provided by Christensen and Boon-Long (1994): 'a concern which could raise problems in the area of basin management involves the authority of the basin [administration] to impose allocation priorities ... The burden of proof for such an initiative is to show that command and control could result in better allocations and less market failure.' Israngkura (2000), for his part, considers that 'the returns on the irrigation dam investment have been low due to the lack of effective water demand management that could prevent less productive water utilisation'. This suggests that irrigation and its assumed low return have deprived other potentially more productive uses, whereas irrigation is, in fact, allocated the leftovers in the system (after the prioritization of water to BMA and energy production). TDRI (2001) posits that 'the current command and control system are unable to meet structural and cyclical changes in the demand and supply of natural resources, including water', while Kraisoraphong (1995) states: 'Past experience has shown the government's role to be ineffective and thus an alternative proposed by economists and the academic circles has been to use economic instruments such as water pricing'.

[21] Experiences from Israel, the USA, India or China indicate (Postel, 1992a) that, in all cases, the share of agriculture was decreased to the benefit of cities.

difficulties for management and for ensuring equity in allocation (see Molle, *et al.*, 2001a).

In addition, there are practical considerations that relegate water transactions to the category of fancy mind games. Reallocation of water is difficult to achieve because it requires not only an accurate definition of individual rights but also a very high degree of control of water and transportation facilities to transfer water from one user to the other. The assertion that 'if the price of rice is low, [Thai] farmers would be happy to cede their right to industrialists' (Wongbandit, 1997), runs counter to the most basic evidence. Industrialists or cities are served first and would do nothing with more water allotted to them when the price of rice is low, let alone the fact that the physical constraints of the distribution network make such a reallocation impossible. How would the 'rights' of a group of farmers in, say, Kamphaeng Phet (middle basin) be transferred to a given golf-course or factory in the suburbs of Bangkok?

Central allocation may appear as a problem to farmers, who are, effectively, gradually dispossessed of their unwritten 'rights' as other uses grow, but this is not a problem to other economic sectors, which are served at low or no cost[22] and on a priority basis. The definition of entitlements and their transfer within a 'bank' or a market mechanism would, indeed, have the positive consequence of providing a mechanism through which the ineluctable dispossession of farmers would be accompanied by financial compensation. In any case, we are very far from a situation in which individuals rights could be defined. The transfer of group-based entitlements would lead to extremely high transaction costs and to internal conflicts, so that such an option is both illusory and unattractive under present conditions.

Lastly, the very notion of economic productivity as a macro-level aggregate must also be scrutinized through the lens of its social and equity implications. The idea is basically that 'if an irrigator can earn more by selling water to a nearby city than by spreading it on alfalfa, cotton or wheat, transferring that water from farm to city use is economically beneficial' (Postel, 1992b), this reallocation being either occasional or permanent. The theory works as long as the reallocation of factors occurs between activities that constitute alternatives for investments and between users who also have a range of opportunities and compete in a perfect market. In other words, this holds for the logic of capitalistic investment, which constitutes the underpinning and driving force of the proposed economic mechanisms. The small peasant, however, often distinguishes him/herself by a lack of choice or, rather, by an alternative which is, willingly or not, quitting the farm sector.[23] If farmers who are unduly exposed to the competition of sectors with a much higher profitability were eventually led to leave their lands fallow (or to sell them to big farmers), they could ultimately swell the ranks of the unemployed (and even the slum population in the capital if there is a strong push process at work). It is hard to see how the overall benefit of the society would be maximized by such a scenario, despite the fact that macro-indicators would (deceivingly) suggest an overall gain. The impact of the diversion of water out of agriculture is a complex issue (Rosegrant and Ringler, 1998), but in developing countries with large agriculture sectors and percentages of rural poor there is often little

[22] Non-agricultural users pay for (part of) the cost of production (abstraction, treatment, transfer) but not for water itself.

[23] Similarly, it is often inferred from observations that some farmers, in particular contexts (such as Pakistan), are led to pay high amounts of money for secure water and that 'farmers are *willing* to pay' (Postel, 1992b; World Bank, 1993). A less optimistic reading would be to assume that many of these farmers do so because they have no choice and because survival, indeed, entails a high 'willingness to pay'. This would be consistent with observations that these informal markets are sometimes not competitive, and the prices charged are higher than theoretically expected.

room to manoeuvre.[24] This concern is also echoed by the World Bank economist W. Price (1994):

> In time, markets in water may expand, but only in locations with extreme scarcity of resources and where municipal or industrial users can afford to pay a large amount per unit of water to an agricultural user – enough for a farmer to invest in another business or to become economically independent. The conditions in South Asia are a long way from this.

Advocates for free markets may place excessive emphasis on aggregated economic values and tend to ignore differences among actors. Schiller and Fowler (1999), for example, stress that 'Ag-urban transfers allow California *as a whole* to use water more efficiently. Because they are *voluntary*, such transfers constitute positive-sum, or 'win–win' situations in which both parties come out ahead' (emphasis added). The point is that 'as a whole' and 'voluntary' might in fact not always be realized and could conceal situations of 'no choice' or 'win–lose' situations with no alternative for one party in the transaction.[25] The seductive perspective to reach an automatic and optimal 'match of supply and demand' is, again, a macro-level aggregated vision that ignores how the demand is characterized and what happens to those who cannot even formulate their demand because they cannot compete with bigger players.

Constraints and Opportunities for Water Reform

The meagre benefits that can be expected on the productivity side, in all senses of the term, do not imply that the status quo is the best option. Although this takes us beyond the limited scope of this chapter, a few comments are given here regarding the reform of the water sector.

Current disruptions in the Thai water systems relate to difficulties in both allocation and distribution (Molle *et al.*, 2001a). In small basins of the north, water diversion needs sometimes exceed the available flow and there is a lack of technical and legal criteria to referee the disputes that arise. In the Chao Phraya basin, the supply to irrigated areas, notably the delta, is made chaotic because of the lack of control over users in the middle basin: over a span of 15 years, the percentage of dam releases diverted (often 'hijacked'!) by these users in the dry season moved up from 5% to 35%. Unscheduled planting of rice, often done by using residual surface water or groundwater, also contributes to creating local mismatches between effective supply and demand, triggering political interventions and raising the uncertainty in supply. Achieving equity in allocation is also made difficult by the fact that available water stocks (from storage dams) vary, for each dry season, between 2 and 8 Bm^3. As a result, it has proved unsustainable to stick to the rotational allocation policy established in the early 1980s, in which half of each project was to receive water in 1 out of 2 years, because this 'right' could not be ensured.

There is a wide (rhetorical) consensus that 'water rights' must be defined, that the administrative management of the water sector must be simplified and that a water law and basin organizations are needed. This fits a vague picture of modernization along the lines of what is presented as international 'best practices' or standards, and meets little opposition. Some wishful thinking helps one assume that such reforms will take place

[24] This is, in reality, not peculiar to developing countries. In the western USA, Frederik (1998) reports that 'when farmers want to sell water to cities, irrigation districts resist, fearing the loss of agricultural jobs that accompany rural water use', while Wahl (1993) acknowledges that 'most agricultural water districts have viewed the potential for water transfers only very tentatively out of concern over the security of their water rights and potentially adverse effects on the districts and local communities'.

[25] Similarly 'users' is a neutral word that tells us little about their heterogeneity in terms of strategies and factor endowment. See, for example, World Bank (1994): 'Reliance on the price mechanism is in the interest of *users* because it directs provision towards preferences determined by users rather than by bureaucrats.'

by their own momentum, but there is limited debate on the substance of such reforms, and heavy doubt over whether provisions would be eventually enforced. Legal provisions are obviously useless without a basic capacity for law enforcement and penalties, an aspect in which Thailand admittedly has an unimpressive record (Christensen and Boon-Long, 1994; Wongbandit, 1995; Flaherty *et al.*, 1999). Countries like Sri Lanka and certain states of India have been debating water laws for 30 years without effectively enacting a law (Shah *et al.*, 2000) and, when they did, the most critical aspects either were removed from the final version or remained a dead letter (see also the example of Vietnam (Malano *et al.*, 2000)).

If such reforms are well intentioned and probably sound as a general guideline for long-term changes, it needs to be recognized that their implementation must be phased and conceived as a long-term process. For example, before considering establishing rights, participatory water-allocation processes at different relevant levels of the basin should be geared towards designing ways to define seasonal entitlements, which also implies regaining control over scheduling, over the expansion of irrigated areas and over unofficial water abstraction. This, in turn, has far-reaching administrative, technical and political implications, which are not subject to full control: in other words, reforms or laws are like water off a duck's back if they are not strongly backed by politicians and officials. What is known about the resilience of the Thai 'bureaucratic polity' (see, for example, Nelson, 1998; Arghiros, 1999) should preclude any optimism on the extent of the decentralization process,[26] as well as on the propensity of the administration to hand over its power swiftly and willingly. It is often implicitly assumed that the state bureaucracy is a neutral monolithic agency, sensitive to rational arguments about cost-effectiveness or public welfare. Pinstrup-Andersen (1993)[27] has shown that this was unrealistic and that the failure to incorporate knowledge of goals and behaviours of agencies and politicians was the most common feature of poor policies. A positive way of looking at the ongoing processes is to view these initiatives as part of a learning process. However, there is a risk that a partial failure would also make the participation of farmers increasingly difficult in the future.

Conclusions

The justification for the current proposals for a reform of the Thai water sector rests heavily on assumptions of low irrigation efficiency and poor economic productivity, despite the wide irrelevance of these arguments in the Thai context. There is a risk that well-intentioned reforms will draw upon blueprints based more on some ideological drive[28] than on in-depth and site-specific analyses of the situation, and will end up being superimposed on the Thai context. The ubiquitous caveat found in many conclusions of papers dealing with the economics-based regulation of water use is found to be often widely disregarded in practice: it cautions against applying general principles without due consideration being given to the historical, geographical, cultural, socio-economic and political contexts. Policies that are believed to have proved successful are often replicated blindly and lead to resounding failure. This applies to various aspects of the water sector, including irrigation-system design, water institutions (Shah *et al.*, 2000; Molle, 2003b) or water legislation (e.g. the replica versions of the Chilean *Código de Aguas* (see Dourojeanni and Jouravlev, 1999)).

---

26 However unsatisfying in the short term, the decentralization process is nevertheless a far-reaching political process that will in the long run bring more democratization. But this time frame, again, is in opposition to that of the proposed reforms.

27 His focus was on food and nutrition policies but his conclusions can be applied to water policy as well.

28 On how ideology shapes public interventions and policy in the Thai water sector, see Molle (2003b).

If most of the irrelevance of the arguments based on efficiency is linked to the closed nature of the Chao Phraya basin, then we must recognize the importance of devising reforms that distinguish between different types of basins, and even between different hydronomic zones (Molden *et al.*, 2001). The Mae Klong basin, which also ends up in the Chao Phraya delta, presents a different picture. The average annual inflow into the main two upstream storage dams is approximately 30% above the average requirements in the basin. This means that the possible low efficiency of irrigation is hardly an issue. At the other end of the spectrum, water-short basins, such as the Chao Phraya basin, have gradually developed means of raising efficiency in use (gating of drains, conjunctive use of groundwater, pumping water from ponds and other low-lying areas, improving the management of dams, etc.) and may not lend themselves to significant improvements in that respect (Molle, 2003a). At present, only 12% of dam water is wasted by evaporation or going to the sea in the dry season (Molle *et al.*, 2001a).

It has also been shown that the centralized water-allocation system has handled the issue of allocating water to activities with higher economic return relatively well, and that the alleged 'lion's share' of water for agriculture is actually the (fluctuating) leftover water in the system after allocation to higher-prioritized uses has been met. With reduced scope for achieving water savings or economic reallocation, the prospects for achieving significant gains in productivity are slim, and the concepts of a water charge or water markets lose most of their appeal. However, the 'virtuous' linkage existing between structural, managerial, institutional and financial approaches is also recognized (Small, 1996). The strongest argument about water pricing is the 'glue factor', where pricing is considered as a mere reinforcing factor of a contractual binding between the RID and groups of users. The 'wholesaling' of water to groups is an option that comes with several prerequisites, and emphasis is placed on the existing gap between these conditions and the current situation. However, if joint management and farmers' financial participation are desirable, there is still little empirical evidence of the impact of turnover on productivity (Samad, 2001); the gains are unlikely to be large, especially when no volumetric pricing is possible.

In contrast to the more appealing justifications based on the idea of 'saving water', which readily relates to the concrete experience of water shortage, it appears that the major changes to be brought about by reforms relate to water allocation within the agriculture sector (with full participation of users), to the control of new diversions, to equity and to the control of environmental impacts.

References

Arghiros, D. (1999) Political reform and civil society at the local level: the potential and limits of Thailand's local government reform. Paper presented to the 7th International Conference on Thai Studies, Amsterdam.

Bangkok Post (2000) Farmers say no to water burden. 11 June.

Binnie and Partners (1997) *Chao Phraya Basin Water Management Strategy (Main Report and Annexes).* Binnie and Partners, Bangkok.

Christensen, S.R. and Boon-Long, A. (1994) *Institutional Problems in Thai Water Management.* Thailand Development Research Institute, Bangkok.

De Young, J.E. (1966) Village life in modern Thailand. University of California Press, Berkeley and Los Angeles, 224 pp.

Dourojeanni, A. and Jouravlev, A. (1999) *El codigo de aguas de Chile: entre la ideologia y la realidad.* Division de Recursos Naturales e Infrastrutura, CEPAL, Santiago de Chile, 83 pp.

ESCAP (Economic and Social Commission for Asia-Pacific) (1991) *Assessment of Water Resources and Water Demand by User Sectors in Thailand.* ESCAP, Bangkok, 99 pp.

Flaherty, M., Vandergeest, P. and Miller, P. (1999) Rice paddy or shrimp pond: tough decisions in rural Thailand. *World Development* 27(12), 2045–2060.

Frederik, K.D. (1998) *Marketing Water: the Obstacles and the Impetus*. Resources for the Future, Issue 132. http://www.rff.org/resources_archive/pdf_files/132_water.pdf

Huffaker, R., Whittlesey, N. and Hamilton, J.R. (2000) The role of prior appropriation in allocating water resources into the 21st century. *Water Resource Development* 16(2), 265–273.

Israngkura, A. (2000) Why can't Thailand afford more irrigation dams? *TDRI Quaterly Review* 15(3), 3–7.

JICA (Japan International Cooperation Agency) (1992) *Study on the West Bank of the Lower Chao Phraya Delta*. JICA, Bangkok.

Kasetsart University and Institut de Récherche pour la Développement (IRD) (ex-ORSTOM) (1996) *Identification of Agricultural and Irrigation Patterns in the Central Plain of Thailand: Prospects for Agricultural Research and Development*. DORAS Project, Bangkok, 220 pp.

Keller, A., Keller, J. and Seckler, D. (1996) *Integrated Water Resources Systems: Theory and Policy Implications*. Research Report 3, International Irrigation Management Institute, Colombo, Sri Lanka.

Kraisoraphong, K. (1995) Evolving water policy in the Bangkok metropolitan region. PhD thesis, University of British Columbia, Canada, 335 pp.

Malano, H.M., Bryant, M.J. and Turral, H.N. (2000) Management of water resources: can Australian experiences be transferred to Vietnam? *Water Resources Journal* 24(4), 307–315

Meinzen-Dick, R. and Rosegrant, M.W. (1997) Water as an economic good: incentives, institutions and infrastructure. In: Kay, M., Franks, T. and Smith, L. (eds) *Water: Economic Management and Demand*. E. & F.N. Spon, London, UK.

Molden, D. and Sakthivadivel, R. (1999) Water accounting to assess use and productivity of water. *Water Resources Development* 15(1/2), 55–71.

Molden, D., Sakthivadivel, R. and Keller, J. (2001) *Hydronomic Zones for Developing Basin Water Conservation Strategies*. Research report No. 56, International Water Management Institute, Colombo, Sri Lanka, 30 pp.

Molle, F. (2001) *Water Pricing in Thailand: Theory and Practice*. Research Report No. 7, DORAS Project, Kasetsart University, Bangkok, 78 pp. http://std.cpc.ku.ac.th/delta/conf/Acrobat/Papers_Eng/pricing.pdf

Molle, F. (2003a) The closure of river basins: a perspective on technical and institutional adjustments in the Chao Phraya basin, Thailand. *Water International* (in press).

Molle, F. (2003b) Allocating and accessing water resources: practise and ideology in the Chao Phraya delta. In: Molle, F. and Srijantr, T. (eds) *Thailand's Rice Bowl: Perspectives on Social and Agricultural Change in the Chao Phraya Delta*. White Lotus, Bangkok.

Molle, F. and Srijantr, T. (1999) *Agrarian Change and the Land System in the Chao Phraya Delta*. Research Report No. 6, DORAS Project, Kasetsart University, Bangkok, 191 pp.

Molle, F., Chompadist, C., Srijantr, T. and Keawkulaya, J. (2001a) *Dry-season Water Allocation and Management in the Chao Phraya Delta*. Research Report No. 8, DORAS Project, Kasetsart University, Bangkok, 250 pp.

Molle, F., Srijantr, T., Latham, L. and Thepstitsilp, P. (2001b) *The Impact of the Access to Irrigation Water on the Evolution of Farming Systems: a Case Study of 3 Villages in the Chao Phraya Delta*. Research Report No. 11, DORAS Centre, Kasetsart University, Bangkok, 75 pp.

Moore, M. (1989) The fruits and fallacies of neoliberalism: the case of irrigation policy. *World Development* 17(11), 1733–1750.

Moser, C.M. and Barett, C. (2001) The disappointing adoption dynamics of a yield-increasing, low external input technology: the case of SRI in Madagascar. Draft, Cornell University.

Nelson, M. (1998) *Central Authority and Local Democratisation in Thailand*. Studies in Contemporary Thailand No. 6, White Lotus, Bangkok, 325 pp.

Perry, C.J. (1996) *Alternative to Cost Sharing for Water Service to Agriculture in Egypt*. IIMI Research No. 2, International Irrigation Management Institute (IIMI), Colombo, Sri Lanka.

Perry, C.J. (1999) The IWMI water resources paradigm: definitions and implications. *Agricultural Water Management* 40(1), 45–50.

Perry, C.J. (2001) *Charging for Irrigation Water: the Issues and Options, with a Case Study from Iran*. Research Report No. 52, International Water Management Institute, Colombo, Sri Lanka, 17 pp.

Pinstrup-Andersen, P. (ed.) (1993) *The Political Economy of Food and Nutrition Policies*. International Food Policy Research Institute, Washington, DC, USA, 278 pp.

Postel, S. (1992a) *The Last Oasis: Facing Water Scarcity*. Norton, New York.

Postel, S (1992b) http://www.unesco.org.uy/phi/libros/efficient_water/wcap5.html

Price, W. (1994) Water markets in South India. In: *Water Policy and Water Markets: Selected papers and pro-*

ceedings from the World Bank's Ninth Annual Irrigation and Drainage Seminar, Annapolis, Maryland, 8–10 December 1992. Le Moigne, G., Easter, K.W., Ochs, W.J. and Giltner, S. (eds) Technical Paper No. 249, World Bank, pp. 107–111.

Rosegrant, M.W. and Ringler, C. (1998) Impact on food security and rural development of transferring water out of agriculture. *Water Policy* 1(6), 567–586.

Samad, M. (2001) *Impact of Irrigation Management Transfer on the Performance of Irrigation Systems: a Review of Selected Asian Experiences*. ACIAR Water Policy Workshop, Bangkok, 15 pp.

Schiller, E. and Fowler, E. (1999) *Ending California's Water Crisis: a Market Solution to the Politics of Water*. Pacific Research Institute, San Francisco, 38 pp.

Scott, J.C. (1976) *The Moral Economy of the Peasant*. Yale University Press, New Haven, Connecticut, 246 pp.

Shah, T., Makin, I. and Sakthivadivel, R. (2000) Limits to leapfrogging: issues in transposing successful river management institutions in the developing world. In: Abernethy, C. (ed.) *Intersectoral Management of River Basins*. International Water Management Institute, Colombo, Sri Lanka, pp. 89–114.

Siriluck, S. and Kammeier, H.D. (2000) Government policy and farmers' decision making: the agricultural diversification programme for the Chao Phraya river basin, 1993–2000. In: *Proceedings of the International Conference 'The Chao Phraya Delta: Historical Development, Dynamics and Challenges of Thailand's Rice Bowl,' December 2000*, Vol. 2. Kasetsart University, Bangkok, pp. 63–96. http://std.cpc.ku.ac.th/delta/conf/prog_list.htm

Small, L.E. (1972) An economic evaluation of water control in the northern region of the Greater Chao Phraya Project of Thailand. PhD dissertation, Cornell University, 400 pp.

Small, L.E. (1996) Financial tools for improving irrigation performance. In: Sampath, R.K. and Young, R.A. (eds) *Social, Economic, and Institutional Issues in Third World Irrigation Management*. Westview Press, Boulder, Colorado, pp. 147–268.

Szuster, B.W. and Flaherty, M.S. (2000) Inland low-salinity shrimp farming in the central plains region of Thailand. In: *Proceedings of the International Conference 'The Chao Phraya Delta: Historical Development, Dynamics and Challenges of Thailand's Rice Bowl,' December 2000*, Vol. 1. Kasetsart University, Bangkok, pp. 159–170. http://std.cpc.ku.ac.th/ delta/conf/prog_list.htm

Szuster, B.W., Molle, F., Flaherty, M.S. and Srijantr, T. (2003) Socio-economic and environmental implications of inland shrimp farming in the Chao Phraya delta. In: Molle, F. and Srijantr, T. (eds) *Thailand's Rice Bowl: Perspectives on Social and Agricultural Change in the Chao Phraya Delta*. White Lotus, Bangkok.

TDRI (Thailand Development Research Institute) (1990) *Water Shortages: Managing Demand to Expand Supply*. Thailand Development Research Institute, Bangkok, 101 pp.

TDRI (2001) *Water Resources Management: Policy Guidelines for Thailand*. Thailand Development Research Institute, Bangkok.

The Nation (n.d.) Water greed threatens Asian farmers.

The Nation (1999) Government to consider ADB terms. 17 February.

The Nation (2000a) Groups against farmers paying to use water. 21 April.

The Nation (2000b) Water-pricing test project to start soon. 23 April.

van der Heide, H. (1903) *General Report on Irrigation and Drainage in the Lower Menam (Chao Phraya) Valley*. Ministry of Agriculture, Bangkok, 149 pp.

Wahl, R.W. (1993) *Water Marketing in California: Past Experience, Future Prospects*. Reason Public Policy Institute, Los Angeles, 28 pp.

Wongbandit, A. (1995) Water law in Thailand: constraint or facilitation for sustainable development? In: *Proceedings of the Third Chulabhorn Science Congress, Water and Development: Water is Life*. Consultant Report, Bangkok.

Wongbandit, A. (1997) *Legal Aspects, Annexe G of the Report 'Chao Phraya Basin Water Resources Management Strategy'*. Binnie & Partners, Bangkok, 74 pp.

World Bank (1993) *Water Resources Management: a World Bank Policy Paper*. World Bank, Washington, DC.

World Bank (1994) *Infrastructure for Development*. Oxford University Press, Oxford.

18 Upscaling Water Productivity in Irrigated Agriculture Using Remote-sensing and GIS Technologies

Wim Bastiaanssen,[1] Mobin-ud-Din Ahmad[2] and Zubair Tahir[3]
[1] *International Institute for Aerospace Survey and Earth Sciences, Enschede, The Netherlands;* [2] *International Water Management Institute, Colombo, Sri Lanka;* [3] *International Water Management Institute, Lahore, Pakistan*

Abstract

Reliable information on water depletion for agricultural production is much needed when freshwater resources are getting scarcer. This is the case in the irrigated Indus basin. Despite their importance, data required to monitor the productivity of the land and water resources over vast areas are usually not available or accessible. Satellite measurements from the National Oceanic and Atmospheric Adminstration (NOAA) weather satellite are combined in this study with ancillary *in situ* data into a geographic information system (GIS). Remote-sensing measurements are converted to crop yield, to actual evapotranspiration and, indirectly, to net groundwater use. The GIS data consist of canal-water deliveries and rainfall records. For each of the canal commands, the productivity of water is calculated. Large variability in the data is found from the different canal commands in the Indus basin. It is concluded that water productivity is controlled more by crop yields than by the water input. The spatial variability of productivity per unit water diverted is greater than per unit depleted. This can be ascribed to wide variations in the relationship between canal-water supply and actual evapotranspiration. This is an issue covered by classical irrigation efficiencies. Upscaling of water productivity for the Indus basin was achieved by aggregating the various canal command areas from the upstream end of the system downwards. The results show that the productivity of water tends to a constant value at a spatial scale of 6 million ha and higher. At that scale, water diversion and water depletion are equal, which implies that groundwater systems, to a large extent, regulate losses and reuse of water resources. The Indus basin is an example of substantial groundwater recycling and this needs to be taken into account in analytical frameworks of water productivity.

Introduction

When freshwater resources are getting scarcer, such as in the irrigated Indus basin, it is necessary to have an accurate description of the depletion of the water resource as a result of agricultural production. Frameworks for the formulation and assessment of water productivity have been developed by Molden *et al.* (1998) and Seckler *et al.* (Chapter 3, this volume), and have been used in water-management studies (e.g. Droogers and Kite, 1999).

Water-management techniques often focus on 'saving' water at field level, but, in water-scarce conditions, water is diverted at one

 Water Productivity in Agriculture: Limits and Opportunities for Improvement (eds J.W. Kijne, R. Barker and D. Molden)

place and used at another. It is, therefore, of extreme importance to gain an insight into the efficiencies and productivities at larger scales. Traditional field surveys and field-scale water-balance measurements cannot give a comprehensive description of the water flows at the regional scale. Processes, such as recharge, capillary rise and groundwater extractions, are difficult to measure or estimate for subsystems. These water terms are mentioned in particular, as they are reflected in the processes of water recycling. However, information on crop acreage, yields and canal-water deliveries is also difficult to obtain, as actual canal operation may differ substantially from the planning and design discharges.

Lack of data required for monitoring the productivity of the land and water resources, especially over vast irrigation schemes and river basins, can often hamper the application and understanding of the water-productivity framework. The aim of this chapter is to demonstrate how remote-sensing and geographic information system (GIS) tools can help in assessing water productivity and how productivity varies with spatial scale.

Hydrological Approach

The soil-water balance and crop-production values form the basis for the water-productivity analysis. The soil-water balance relates total supply to total consumption and has a storage term for cases when inflow and outflow are not balanced (see also Fig. 18.1).

$$\Delta S = (P + I_{cw} + I_{tw} + q^{\uparrow}) - (ET_a + q^{\downarrow}) \quad \text{(mm)} \qquad (18.1)$$

where ΔS is the storage change, P is precipitation, I_{cw} is canal-water supply, I_{tw} is groundwater supply through tube wells, $q^{\uparrow}$ is capillary rise, ET_a is the actual evapotranspiration and $q^{\downarrow}$ is the recharge. Since several terms of Equation 18.1 are difficult to quantify, the three groundwater terms are taken together:

$$NGW = I_{tw} + q^{\uparrow} - q^{\downarrow} \quad \text{(mm)} \qquad (18.2)$$

where NGW is the net groundwater use, i.e. the extractions of groundwater minus the net recharge. NGW represents the net withdrawal of groundwater, which is important for the sustainability analysis. The recharge $q^{\downarrow}$ comprises the return flow from tube-well irrigation, I_{tw}, but can also arise from precipitation, P, and canal-water irrigation, I_{cw}. After combining Equations 18.1 and 18.2, the simplified water-balance equation is:

$$\Delta S = P + I_{cw} + NGW - ET_a \quad \text{(mm)} \qquad (18.3)$$

For the current case study in the Indus basin, P is taken from rain gauges, I_{cw} from flow records and ET_a from remote sensing. The two unknowns are then ΔS and NGW. If, in addition, the storage changes are ignored, which is not correct in all canal command areas, NGW remains as the residual term of the water balance. The storage changes depend on groundwater-table fluctuations, which in some cases can be as much as 100 mm year^{-1} (see Ahmad and Bastiaanssen, 2003). A sufficient number of piezometric readings was not available to estimate ΔS in a systematic manner across the entire Indus basin. Because of this limitation, ΔS was disregarded, as is usually done in hydrological studies for longer time periods.

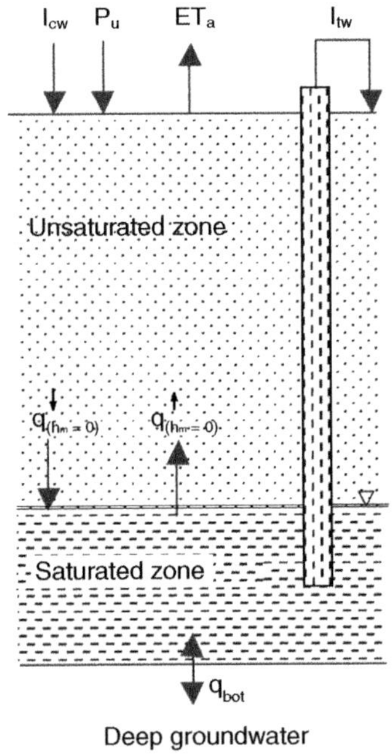

Fig. 18.1. Schematic presentation of the soil-water balance.

Hydrological Results

Data on precipitation and canal-water supply were taken from Habib *et al.* (1999) and from Tahir and Habib (2000). The Indus basin comprises 44 canal command areas, some of which have incomplete data and were, therefore, not further considered in this study. The data on canal-water supply per unit culturable command area show a variation of 40 to 830 mm during the rabi (dry winter) season (Fig. 18.2). This suggests a very non-uniform distribution of irrigation water across the Indus basin during the dry winter season. Similar heterogeneity in canal-water supply was found for kharif (wet summer). These large deviations may result, in part, from measurement and interpretation errors in the main canals, as flows through these huge irrigation canals are not easy to measure accurately. The numbers used in this study for canal command areas, with complete data sets for all water-balance terms, are presented in the Appendix.

Actual evapotranspiration data are taken from Bastiaanssen *et al.* (2002), who based their analysis on remotely sensed data. Raw data from the National Oceanic and Atmospheric Administration – Advanced Very High Resolution Radiometer satellite (NOAA-AVHRR) were used. The surface energy-balance algorithm for land (SEBAL) has been applied to convert the raw satellite data into broadband surface albedo, vegetation index and surface temperature. The major objective of SEBAL is to explore the range of: (i) surface albedo values for describing net radiation; (ii) vegetation indices to assess the variability of soil heat flux; and (iii) surface temperatures for estimating sensible heat flux. The energy-balance equation is used to compute actual evapotranspiration from the energy left for the latent heat flux:

$$LE_{24} = R_{n24} - H_{24} \qquad (Wm^{-2}) \qquad (18.4)$$

where LE_{24} is the 24 h latent heat flux associated with evapotranspiration, R_{n24} is the 24 h net radiation and H_{24} is the 24 h sensible heat flux. The soil heat flux on a 24 h basis is usually small and can be ignored. LE_{24} can be converted into actual evapotranspiration (mm day^{-1}) from the energy required to vaporize 1 kg of water at a given temperature. Equation

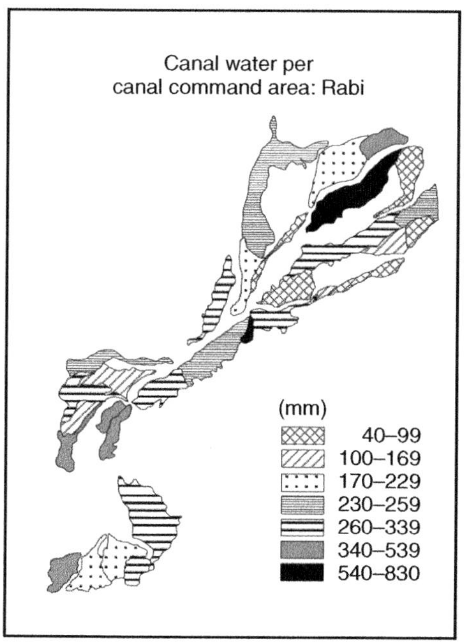

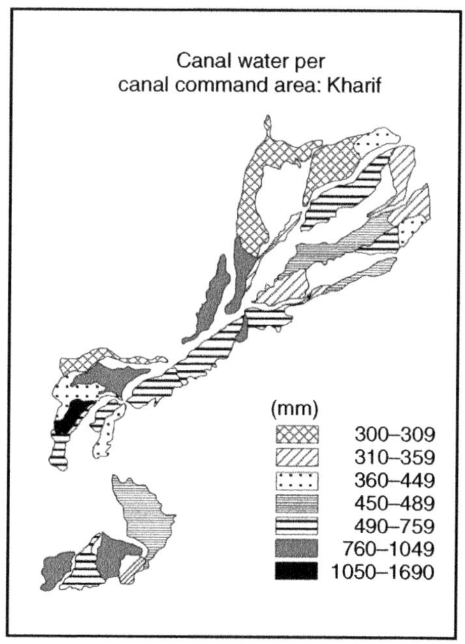

Fig. 18.2. Canal-water use in rabi (1993/94) and kharif (1994) for the Indus basin based on secondary data.

18.4 was used to compute the actual evapotranspiration for cloud-free NOAA images acquired during 20 different days throughout an annual cycle. Individual day results were temporally integrated by preserving the evaporative fraction between two successive satellite acquisition days. The evaporative fraction on a daily time basis is equal to LE_{24}/R_{n24}. This energy partitioning was fixed until the next available AVHRR image. Since net radiation changes considerably due to cloud cover that may arise during satellite flyover days, day-to-day variations of R_{n24} have been taken into account to compute LE_{24} from the temporally preserved LE_{24}/R_{n24} fraction. The Indus basin was divided for this purpose into five climatic zones, and daily global radiation (short-wave radiation reaching the land surface) was computed for every climatic zone.

Figure 18.3 shows the map of annual actual evapotranspiration. Validation in the Indus basin was realized through the application of the well-calibrated field-scale transient moisture-flow model SWAP (Sarwar *et al.*, 2000), *in situ* Bowen ratio measurements (Ahmad *et al.*, 2002) and water-balance residual analyses for an area of 3 million ha. The accuracy of assessing time-integrated evaporative depletion was found to vary from 0.3% at field scale to 4.5% at the regional scale of 3 million ha.

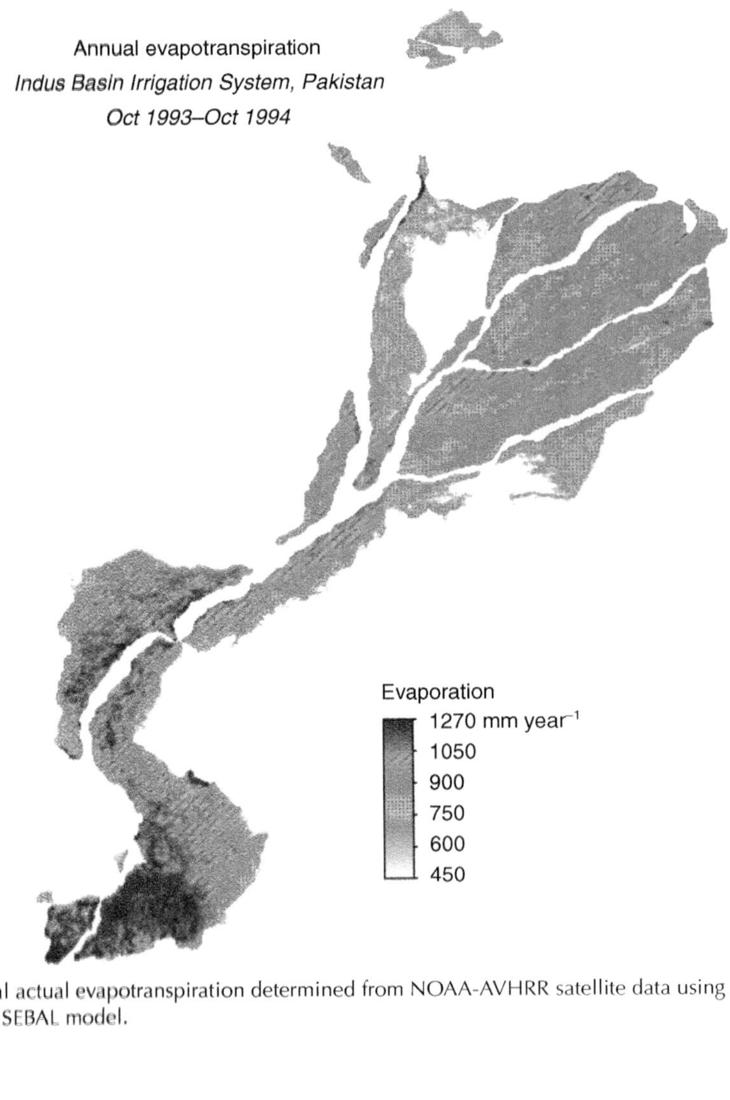

Fig. 18.3. Annual actual evapotranspiration determined from NOAA-AVHRR satellite data using the physically based SEBAL model.

Actual evapotranspiration during the rabi was, on average, 350 mm, while kharif had a total consumption of 620 mm. The spatial variation is again – as for canal-water supply – very high, with annual evapotranspiration values ranging from 450 to 1270 mm $year^{-1}$!

Water supply to the cropped area can come from three difference sources, i.e. canal irrigation, I_{cw}, groundwater irrigation, I_{tw}, and net precipitation, P_n (gross precipitation P minus interception losses P_i and surface runoff). For the sustainability of irrigation systems, it is important to estimate the extent to which irrigated agriculture depends on groundwater resources. The groundwater–resource ratio, ξ , is defined as the ratio of groundwater irrigation, I_{tw}, to the total inflow from all sources:

$$\xi = I_{tw}/(I_{cw} + I_{tw} + P_n) \qquad (-) \qquad (18.5)$$

A map of the groundwater–resource ratio ξ is shown in Fig. 18.4. It demonstrates that there is little contribution of groundwater during kharif and a relatively significant amount of groundwater use during the dry rabi season (the rainfall in many areas varies from 25 to 50 mm). During rabi, some areas rely for 80% of their water resources on groundwater.

Crop Yield

Crop yield is a major input in water-productivity frameworks. Crop-yield information is classically collected through field surveys. This is a laborious activity, especially when one has to deal with vast areas. To aid the ground sampling and to swiftly obtain an overall picture of the crop development, a remote-sensing model for crop-yield prediction was developed and applied (Bastiaanssen and Ali, 2003). This model is based on Monteith's equation for biomass production, which reads in its simplest form as:

$$B_{io} = APAR\ \varepsilon \qquad (kg\ m^{-2}\ day^{-1}) \qquad (18.6)$$

where B_{io} (kg m^{-2} day^{-1}) is the biomass production, APAR (MJ m^{-2}) is the absorbed photosynthetic active radiation and ε (kg MJ^{-1}) is the light-use efficiency. Incoming solar radiation and light interception by leaves control APAR. Solar radiation was computed from the actual hours of sunshine, and the leaf presence from the normalized difference vegetation index (NDVI), being derived from the NOAA-AVHRR sensor. The light-use efficiency ε depends not only on the type of crop (C_3 or C_4), but also

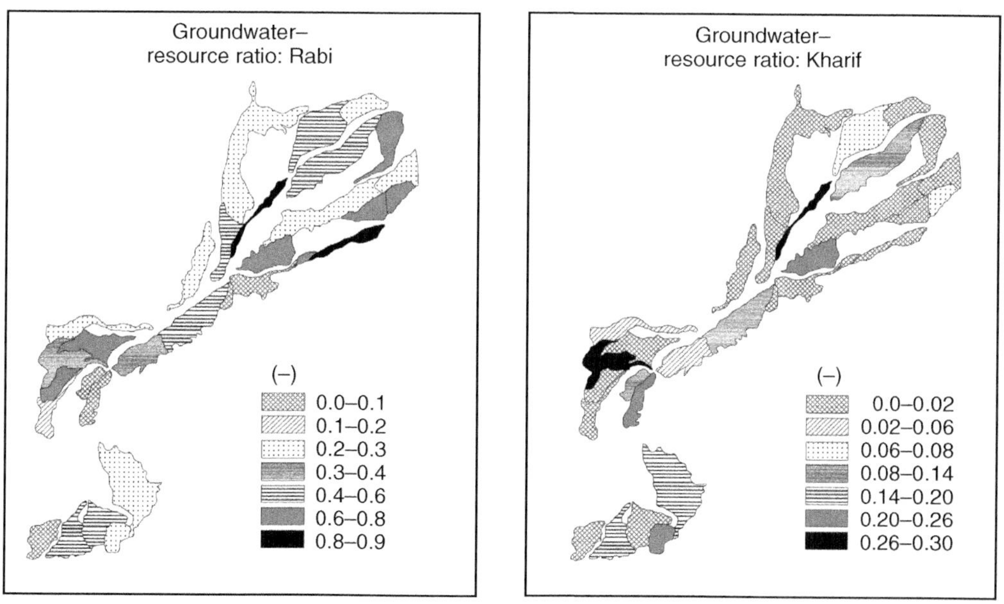

Fig. 18.4. Groundwater–resource ratio (fraction net groundwater use/total water resources available) for canal command areas in rabi (1993/94) and kharif (1994).

on the soil-moisture availability, which affects leaf water potential. Moisture stress reduces the light-use efficiency, and this feedback was taken into account by incorporating the evaporative fraction (LE_{24}/R_{n24}) into the light-use efficiency. The biomass production rates for single NOAA acquisition days were further integrated in time by considering day-to-day variation of cloud cover, which affects APAR because clouds reflect and scatter solar radiation. The light-use efficiency ε was made quasi-variable by adjusting the value between consecutive NOAA images.

Remote-sensing estimates of crop yield have been validated against secondary data collected by the Agriculture Department of Pakistan. The validation revealed a root mean square error of 525, 616, 551 and 13,484 kg ha^{-1} for wheat, rice, cotton and sugarcane yield, respectively. The deviation between secondary data and remote-sensing data shows that the yield of wheat, rice and sugarcane can be mapped for approximately 80% of the cases within the 95% confidence levels of the secondary field data. On average, crop yields in Pakistan are on the lower side. The yields are 2276, 1756, 1293 and 47,929 kg ha^{-1} for wheat, rice, cotton and sugarcane, respectively. A comparison with the study of Hussain *et al.* (2000), who collected crop-cutting experimental data in Sindh, confirmed the wheat yields to be low in Sindh. In Fig. 18.5, yields of wheat, cotton and rice are presented for 26 out of the 44 canal commands. The canal command areas are numbered from the upstream to the downstream end. It is evident that, except for wheat, location in the basin does not significantly affect yields.

Water Productivity

One of the first issues in water-productivity data is to identify which 'crop' and which 'drop' are referred to. To differentiate between water productivity per unit depleted and per unit canal-water supply makes sense, as the former describes how productively water that leaves the basin is used, whereas the latter illustrates the return from canal-management efforts and irrigation-sector investments.

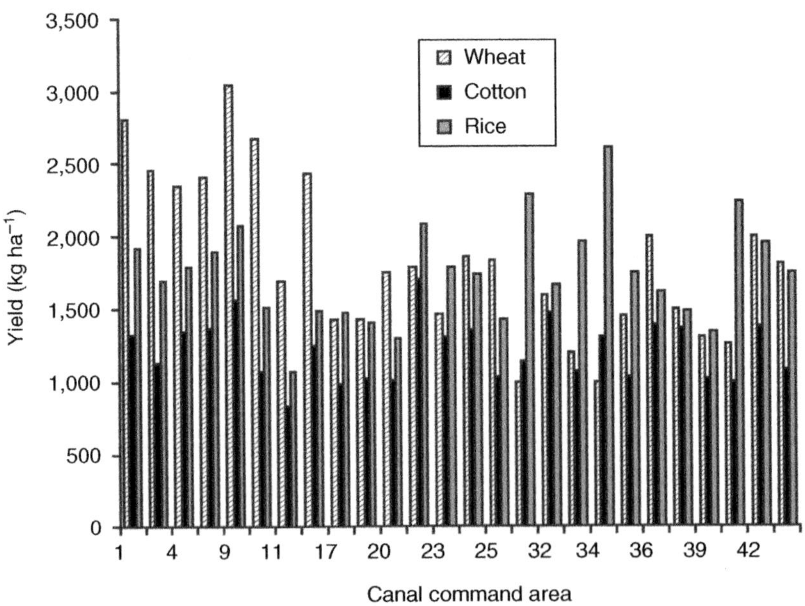

Fig. 18.5. Canal command area yield data for wheat, cotton and rice. The yield data have been computed from NOAA-AVHRR data.

Molden *et al.* (1998) suggested defining water productivity per unit diverted irrigation supply, the latter meaning *'surface irrigation water diverted to the command area plus net removals from groundwater'* (writers' italics). There are reasons why we believe that the use of 'unit canal-water supply' in the denominator may have some advantages for the Indus basin:

- The irrigation manager is responsible for canal-water supply and prefers to understand the impact of the expensive irrigation infrastructure. Groundwater management is usually delegated to institutions other than the irrigation departments.
- Adding together water flows originating from different sources (P, I_{cw}, I_{tw}) prevents the study of the impact of separate sources. In particular, the important role of groundwater in respect of water productivity gets hidden if it is included in the diverted water.

We propose using the following set of definitions:

$$WP_{ETa} = Y_a/ET_a \quad (kg\ m^{-3}) \quad (18.7)$$

$$WP_{Icw} = Y_a/I_{cw} \quad (kg\ m^{-3}) \quad (18.8)$$

$$WP_{\$} = GVP/I_{cw} \quad (US\$\ m^{-3}) \quad (18.9)$$

where Y_a (kg ha^{-1}) is the actual crop yield and GVP (US\$ kg^{-1}) is the gross value of production. GVP is computed from the crop production of every crop, its market price and its acreage. The indicator $WP_{\$}$ is especially suitable as it comprises the total production of different crops. Also, it can be used in the comparison with water-productivity values of other users, such as fish production, ecosystems, etc.

Y_a and ET_a raster data from satellites can be easily combined to make crop-specific evaluations of WP_{ETa}; this is not straightforward for Y_a/I_{cw}, as crop-specific I_{cw} data are seldom available. Hence, WP_{ETa} has the advantage that it can be used to make crop-specific evaluations. Table 18.1 contains an overview of the basin-wide crop-specific productivity values. It shows that sugarcane and cotton have higher water consumption than rice because of their longer growing period. Cotton has the lowest WP_{ETa} values and sugarcane the highest. However, the use of world market prices of agricultural products for 1994 shows that cotton is more economically productive than rice and wheat. This, by itself, shows that evaluating agricultural production and water-resources depletion is not straightforward.

With data on crop yield, actual evapotranspiration and canal-water flow available in the GIS database, it became feasible to compute WP_{ETa} and WP_{Icw}. The case of wheat is given as an example (Fig. 18.6). WP_{ETa} varies from 0.2 to 0.8 kg m^{-3}, which is really a low value. A literature search on WP_{ETa} for wheat showed an average value of approximately 1.0 kg m^{-3}; hence, Pakistan is performing poorly in terms of WP_{ETa}, as the whole range is less than the worldwide average. The WP_{ETa} trend in Fig. 18.6 shows that the response of wheat yield to evaporation is not constant; the value for WP_{ETa} increases with higher yields ($R^2 = 0.83$). The obvious conclusion, then, is that wheat with a higher yield is more efficient in terms of water depletion. This is worth exploring further in future studies.

Table 18.1. Average output in terms of physical and economical production per unit water depleted in the Indus basin during rabi 1993/94 and kharif 1994.

Crop	Evapotranspiration (mm)	Crop yield (kg ha^{-1})	Productivity per unit consumed (kg m^{-3})	GVP per unit consumed (US\$ m^{-3})
Cotton	579	1,293	0.22	0.43
Rice	414	1,756	0.42	0.13
Wheat	357	2,276	0.64	0.10
Sugarcane	965	47,929	4.97	–

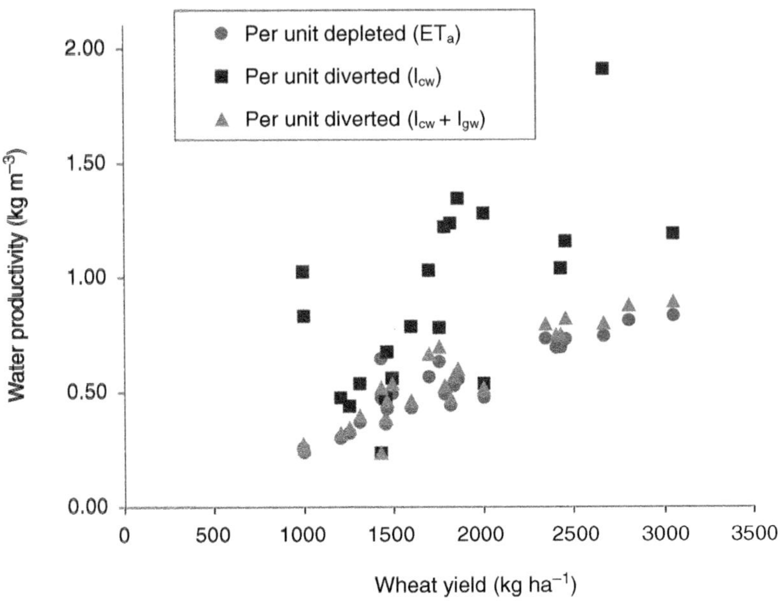

Fig. 18.6. Relationship between water and land productivity for a wheat crop across the Indus basin. Every point represents one canal command area.

WP_{Icw} is higher than WP_{ETa} because ET_a exceeds I_{cw} during the dry winter season. The spatial variability of WP_{Icw} is more than for WP_{ETa}, because ET_a and I_{cw} have a weak relationship due to unequal canal-water distribution throughout the basin (see Fig. 18.2). This brings us to the relationship between ET_a and I_{cw}, which is addressed in the classical irrigation-efficiency concept (e.g. Israelsen, 1950; Keller *et al.*, 1996; Seckler *et al.*, Chapter 3, this volume) as:

$$E_c = (ET_a - P_{net})/I_{cw} \times 100 \quad (18.10)$$

where the numerator represents the net irrigation requirements and the denominator the canal-water supply. When the efficiency, E_c, is lower than 100%, water that is not evaporated from moist soil or transpired by crops is considered as lost in the classical efficiency concept. This may be true if water that is drained from irrigation schemes flows out of the basin or is no longer available for further use by any other means. In most cases, however, drainage water rejoins the river downstream of an irrigation system.

In the absence of a surface-water drainage system, this 'lost' water stays in the system as groundwater. Irrigation systems such as in the Indus basin are underlain by a productive aquifer with high permeability, and here groundwater is transferred laterally and pumped up by shallow and deep tube wells. If this water is reused in an irrigation system, classical efficiency is not a suitable indicator of water productivity. Thus, canal water 'lost' from one irrigation command area may be reused in another.

Classical irrigation efficiencies were calculated for a set of canal commands for both rabi and kharif seasons to demonstrate how: (i) groundwater use has surprising effects on the irrigation efficiency; and (ii) ET_a relates to I_{cw} . Figure 18.7 shows that, in the wet summer kharif season, E_c ranges between 50 and 200%. The map shows that canal command areas with $E_c > 100\%$ can be found next to command areas with $E_c < 100\%$. This suggests that there is a net groundwater movement in the direction of the command area with the highest efficiency. Unfortunately, appropriate piezometric data were not available to verify these flow directions. But, if $ET_a >> (P + I_{cw})$, it is obvious that groundwater is an important source of irrigation.

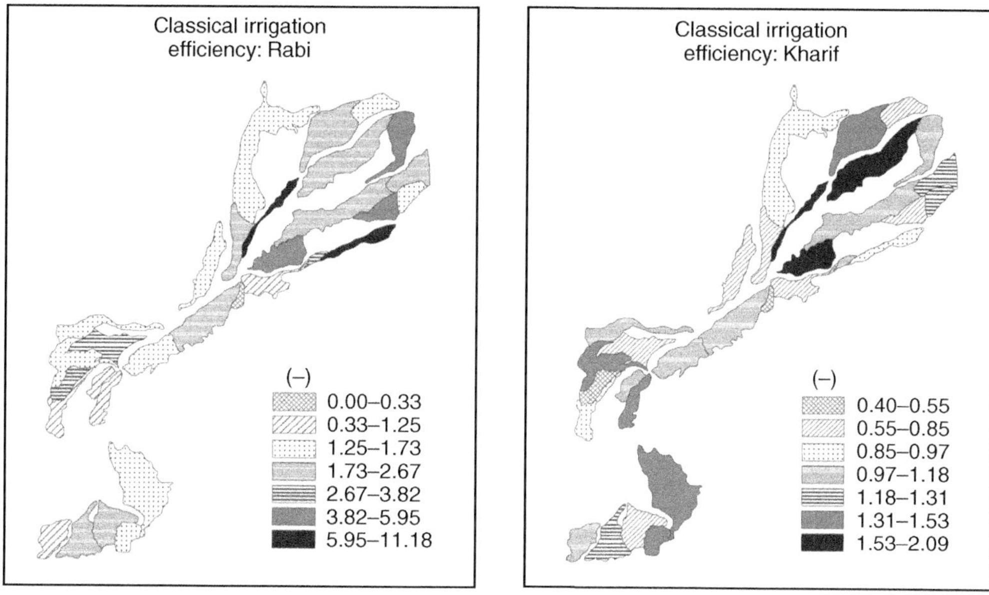

Fig. 18.7. Classical irrigation efficiencies for canal command areas in rabi (1993/94) and kharif (1994). The values are given as a fraction.

By comparing the situation in the two seasons, it appears that command areas having a low E_c in kharif – because precipitation P and canal water I_{cw} far exceed the actual crop evapotranspiration ET_a – respond with $E_c > 100\%$ during the rabi season. This is feasible if kharif water is carried over to the rabi season through soil moisture and groundwater storage mechanisms. Hence, Fig. 18.7 shows that recycling of water is a very important issue in the Indus basin, not only between adjoining command areas, but also between successive growing seasons. Thus, groundwater acts as a storage mechanism and a mediator for making canal operations more effective.

Spatial-scale Issues

How do the water productivity and efficiency change with scale? Our GIS database allowed us to aggregate various canal command areas. This has been done in the upstream to downstream direction, which allowed us to study the productivity at different spatial scales. The hydrological data were combined first and weighted according to the area, i.e. mixing-cell approach. Thereafter, the productivity was recalculated assuming that one is dealing with a unified and larger canal command area, instead of a mosaic of separated canal command areas. The smallest scale is 43,000 ha and the largest scale for a total of 32 combined canal command areas became 11.6 million ha. The total size of all canal command areas in the Indus basin is larger, but not all of them could be included, due to missing data. The wheat crop in the rabi season was chosen because it is the dominant winter crop and most canal water is used for the irrigation of wheat. Figure 18.8, which is a plot of the upscaled water productivity of wheat production during rabi 1993/94, shows two very important phenomena:

- WP_{Icw} has greater spatial scale variability than WP_{ETa}.
- WP_{Icw} and WP_{ETa} tend to have the same value at increasing scales.

The highest water productivity in the Indus basin occurs in smaller canal command areas, especially where the groundwater-resource ratio is high. But, as can be seen in Fig. 18.8, low water-productivity

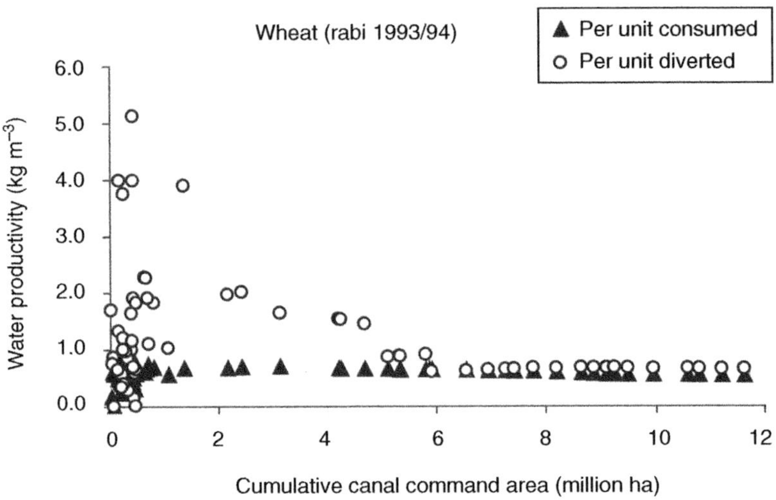

Fig. 18.8. Upscaling of water productivity over the canal command areas in the Indus basin system for the dry winter season, when wheat is predominantly cultivated.

values are also found at this smaller scale. Figure 18.8 shows that PW_{Icw} decreases from 2.0 kg m^{-3} at a scale of 2.0 million ha to 1.0 kg m^{-3} at the 6 million ha scale. These results arise from merging the more fertile soils with a high yield in Punjab in the upstream part of the basin, with the downstream areas receiving less canal water during rabi and being prone to salinity in Sindh.

Water-management interventions and water-saving techniques should therefore focus more on reducing the wide range in WP_{Icw} values at scales between, say, 10,000 to 1 million ha, and not aim to promote changes in one (small) area within the basin. An important conclusion from this work is that increasing WP_{Icw} in a poorly performing canal command area comes at the cost of highly productive systems elsewhere in the region. But it will result in less fluctuation of WP_{Icw} at a lower scale and reduces the scale below which variability becomes insignificant. One of the targets in water-resources management is to obtain the averages of $WP_{ETa} = 0.55$ and $WP_{Icw} = 0.66$ kg m^{-3} at the smallest possible scale, i.e. the scale above which no further changes in the values are likely to occur (correlation length in geostatistics).

Conclusions

Water productivity can be expressed per unit water diverted and per unit water depleted. Since actual crop yield and actual evapotranspiration both depend on plant physiological processes – stomata need to open for carbon inhalation and vapour exhalation – the productivity per unit depleted shows less variability than the productivity per unit diverted. The relationship between diversion and depletion is complex and not clear beforehand. It is demonstrated in this chapter that the ratio of crop yield to evapotranspiration is not conservative and there is some scope for improving productivity per unit depleted, by enhancing physical yield per unit land area.

The biggest challenge, though, is to increase the productivity per unit of water diverted. The results reveal that a significant variability exists due to variations in the classical irrigation efficiency. The variations average out if one moves to a larger scale. This can only be explained hydrologically if groundwater recycling occurs as a predominant process. Water budgets demonstrate that net groundwater use is a key component of the water balance. Productivity of water per unit consumed and per unit diverted become equal at a scale of 6 million ha, which proves that water in the Indus basin is not lost but is used by evaporative depletion elsewhere in the system. A significant transfer of water was detected from the wet summer season to the dry winter season. But groundwater may also flow to adjoining canal command areas. Piezometric information is required to verify this hypothesis.

The impact of small-scale interventions, such as alternate wet–dry phases in rice production, zero tillage, micro-water harvesting, etc., can help improve the local water productivity. There is, however, a possibility that they adversely affect water productivity elsewhere and may further enhance spatial differences in water productivity. Therefore, we recommend narrowing the amplitude of water productivity. That will ultimately lead to a smaller scale above which the average water productivity in the basin stabilizes. Interventions should start in the areas with the lowest water productivity.

Significant progress has been made in the development of frameworks for irrigation efficiency, performance ratios, etc. It is felt that carry-over groundwater from neighbouring irrigation schemes and from the preceding season needs to be more explicitly addressed in these analytical water-productivity frameworks.

This work has demonstrated how affordable images, such as NOAA-AVHRR or from alternative sensors, can help in providing a quick scan of parameters necessary for water-productivity assessment. More research needs to be done on the assessment of soil water-storage changes. Satellite data have been used to determine crop occurrence, actual evapotranspiration by crops, crop yield and, indirectly, net groundwater use. This helps in environments where data are not present or are difficult to access. Coarse images, such as NOAA, are suitable for getting an overall impression at scheme level. Smaller areas or specific crop types would require finer-resolution images, such as those available from Landsat and the Advanced Spaceborne Thermal Emission and Reflection Radiometer.

Acknowledgements

The authors wish to express their gratitude to the Punjab Irrigation and Drainage Authority (PIDA), Pakistan Meteorological Department (PMD), Water and Power Development Authority (WAPDA) and the Directorate of Economic and Marketing of the Provincial Agriculture Department for making available the required information.

References

Ahmad, M.D. and Bastiaanssen, W.G.M. (2003) Retrieving soil moisture storage in the unsaturated zone from satellite imagery and bi-annual groundwater table fluctuations. *Irrigation and Drainage Systems* (in press).

Ahmad, M.D., Bastiaanssen, W.G.M. and Feddes, R.A. (2002) Sustainable use of groundwater for irrigation: a numerical analysis of the subsoil water fluxes. *ICID Journal for Irrigation and Drainage* 51(3), 227–241.

Bastiaanssen, W.G.M. and Ali, S. (2003) A new crop yield forecasting model based on satellite measurements applied across the Indus basin, Pakistan. *Journal of Agriculture, Ecosystems and Environment* 94(3), 321–340.

Bastiaanssen, W.G.M., Ahmad, M.-ud-D. and Chemin, Y. (2002) Satellite surveillance of evaporative depletion across the Indus basin. *Water Resources Research* 38(12), 1273–1282.

Droogers, P. and Kite, G.F. (1999) Water productivity from integrated basin modeling. *Irrigation and Drainage Systems* 13, 275–290.

Habib, Z., Tahir, Z. and Khan, A.R. (1999) Across-the-basin-analysis of water and land utilization using spatio-temporal information techniques emphasize the need for integrated resources management. In: Musi, Fritsch and Pereira (eds) *Proceedings of the 2nd Inter-Regional ICID Conference on Water and Environment, Lausanne, September 1999.*

Hussain, I., Marikar, F. and Jehangir, W. (2000) *Productivity and Performance of Irrigated Wheat Farms Across*

Canal Commands in the Lower Indus Basin. IWMI Research Report 44, International Water Management Institute, Colombo, Sri Lanka.
Israelsen, O.W. (1950) *Irrigation Principles and Practices*, 2nd edn. John Wiley & Sons, New York.
Keller, A., Keller, J. and Seckler, D. (1996) *Integrated Water Resource Systems: Theory and Policy Implications.* Research Report 3. International Water Management Institute, Colombo, Sri Lanka, 15 pp.
Molden, D.J., Sakthivadivel, R., Perry, C.J., de Fraiture, C. and Kloezen, W.H. (1998) *Indicators for Comparing Performance of Irrigated Agricultural Systems.* Research Report 20, International Water Management Institute, Colombo, Sri Lanka, 26 pp.
Sarwar, A., Bastiaanssen, W.G.M., Boers, Th.M. and van Dam, J.C. (2000) Evaluating drainage design parameters for the Fourth Drainage Project, Pakistan by using SWAP model: Part 1: calibration. *Irrigation and Drainage Systems* 14, 257–280.
Tahir, Z. and Habib, Z. (2000) *Land and Water Productivity Trends Across Punjab Canal Commands.* International Water Management Institute, Working Paper 14. Pakistan Country Series No. 3, International Water Management Institute, Lahore, 35 pp.

Appendix 1: canal command wise water balances for rabi and kharif

A negative net groundwater use reveals recharge, a positive value relates to groundwater depletion. All data are expressed in gross canal command area.

	Rabi				Kharif			
Canal command area	Rain	Canal water	ET_a	Net groundwater use	Rain	Canal water	ET_a	Net groundwater use
1	26	55	346	+265	410	281	633	−58
3	36	212	336	+87	295	279	588	+15
4	25	59	320	+236	254	290	583	+39
5	25	64	347	+258	250	466	575	−141
9	25	256	367	+86	480	356	675	−161
10	25	140	359	+195	305	222	570	+43
11	44	165	300	+91	310	233	453	−90
13	25	234	350	+91	250	380	595	−35
17	25	30	302	+246	250	360	549	−61
19	25	612	222	−415	145	558	421	−282
20	25	225	277	+28	150	480	480	−150
22	25	146	363	+192	65	466	594	+64
23	25	217	344	+102	138	677	622	−193
24	25	138	335	+172	206	628	597	−237
25	25	29	347	+293	237	212	634	+184
31	25	97	391	+269	50	820	707	−163
32	25	203	372	+144	50	514	599	+35.3
33	25	254	404	+125	50	396	635	+189
34	25	120	422	+279	50	1425	720	−755
35	25	308	405	+71	50	583	597	−36
36	25	373	422	+24	50	577	697	+70
37	25	266	304	+13	50	322	500	+129
39	25	245	358	+88	127	367	619	+126
41	25	286	392	+80	150	415	756	+191
42	25	156	411	+230	120	820	790	−149
43	25	147	412	+240	58	563	776	+155

19 Improving Water Productivity through Deficit Irrigation: Examples from Syria, the North China Plain and Oregon, USA

Heping Zhang
CSIRO Plant Industry, Wembley, Australia

Abstract

Improving water productivity is urgently needed in water-scarce dry areas. This chapter discusses crop–water production functions, i.e. the relationships between yield and water supply and water productivity. Using data from Syria, the North China Plain and Oregon, USA, crop–water production functions are developed from which the productivity of the applied water can be derived. After an initial sharp increase, the productivity reaches its maximum at a given amount of supplied water to the plant and then decreases or remains at a relatively high level with further increasing water supply. This chapter demonstrates that deficit irrigation produces a higher overall grain yield with the same amount of water resources compared with full irrigation and, therefore, has a higher productivity. Deficit irrigation can be considered as a key strategy for increasing on-farm water productivity in water-scarce dry areas. The risk associated with deficit irrigation can be minimized through proper irrigation scheduling (when and how much to irrigate) and by avoiding water stress during the growth stages when the crop is especially sensitive to water stress.

Introduction

Water scarcity is a real threat to food production for millions of people in arid and semi-arid areas. As the world population continues to grow, the arable land area per capita will further decrease. The Food and Agriculture Organization (FAO, 1988) estimated that almost two-thirds of the increase in crop production needed in the next decades must come from higher yields per unit of land. Hence, rainfall and irrigation water must be used more efficiently and water productivity increased.

Theory of Crop–Water Production Function

The relationship between crop production and water received is called the crop–water production function. According to Vaux and Pruit (1983), research aimed at determining this function can be categorized into three groups, according to different considerations of what constitutes a desirable level of water use:

- Agronomists and other production-oriented scientists often aim for the level of

 Water Productivity in Agriculture: Limits and Opportunities for Improvement (eds J.W. Kijne, R. Barker and D. Molden)

water inputs necessary to achieve maximum yield per unit land area.

- Irrigation engineers, at least in theory, desire to maximize the efficiency of irrigation water use.
- Economists argue that water, to be used efficiently, should be applied up to the point where the price of the last unit of water applied is just equal to the revenue obtained as a result of its application.

A simple model of production can be used to demonstrate these three different goals, as presented in Fig. 19.1.

A production function in which crop yield (Y) is a function of the amount of water received by the crop in terms of rainfall (P) and irrigation (I) can be defined as follows:

$$Y = f(P, I) \qquad (19.1)$$

The average yield $\bar{Y}$, which is output divided by input, can be written as

$$\bar{Y} = Y/(P + I) \qquad (19.2)$$

The marginal yield ($\hat{Y}$) is defined as the change in production associated with the addition of one unit input. It can be written as

$$\hat{Y} = \partial Y/\partial(P + I) \qquad (19.3)$$

The maximum yield is achieved when the marginal yield is equal to zero. Maximum water-use efficiency requires that the derivative of the average yield is equal to zero,

$$(P + I)^{-1} [\partial Y/\partial(P + I) - (Y/(P + I)] = 0 \qquad (19.4)$$

Equation 19.4 shows that, as long as some quantity of water is applied, water-use efficiency is maximal where it is equal to the marginal production.

Case Studies of Crop–Water Production Functions

Crop–water production functions for wheat were derived from supplemental irrigation experiments conducted in Syria (Zhang and Oweis, 1999), the North China Plain (Zhang *et al.*, 1999) and Oregon state, USA (English and Nakamura, 1989) (Fig. 19.2a–d). The quadratic production function was used to describe the response of wheat yield to total applied water:

$$Y = b_0 + b_1(P + I) + b_2(P + I)^2 \qquad (19.5)$$

where Y is wheat yield (t ha^{-1}), I is the irrigation water (mm), P is precipitation (mm)

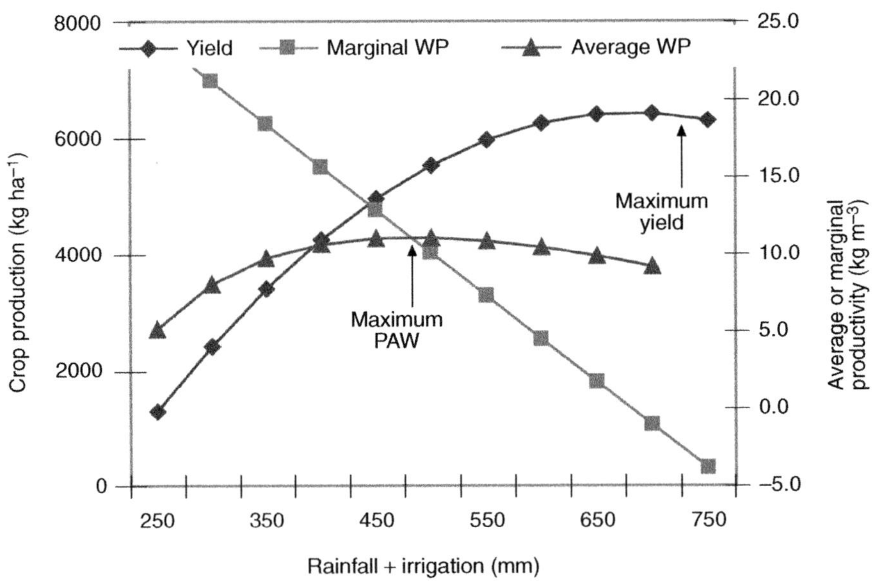

Fig. 19.1. Relation of crop production, productivity of applied water (PAW) and marginal productivity to the crop water supply. The arrows indicate that the maximum PAW value occurs at a lower value of applied water than maximum yield does.

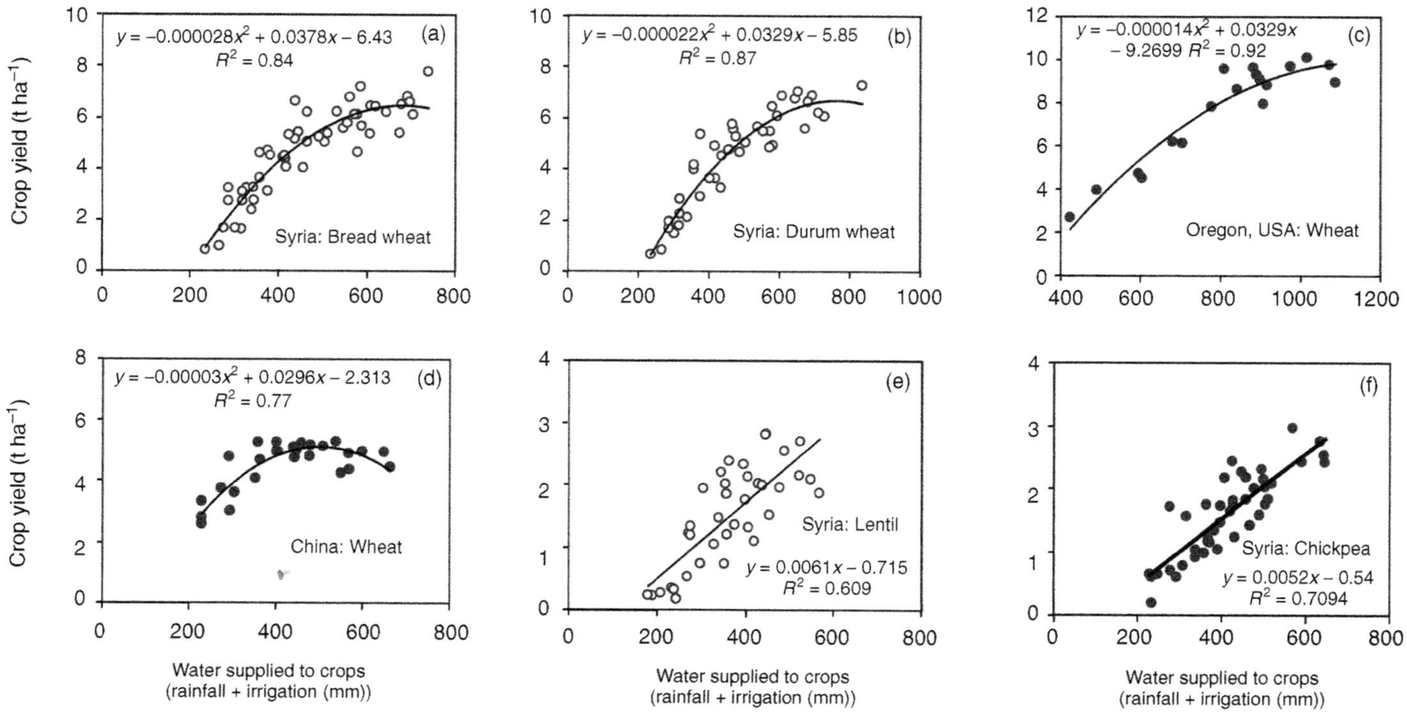

Fig. 19.2. Crop production functions for wheat in China, Oregon, USA, and Syria and for chickpea and lentil in Syria.

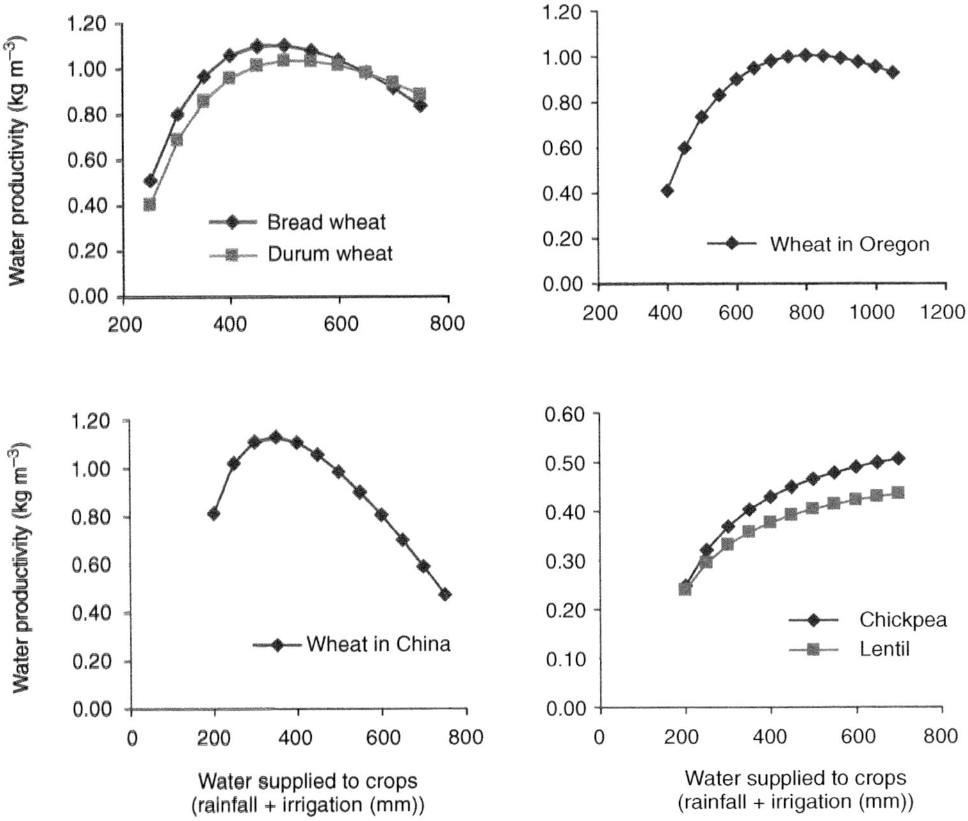

Fig. 19.3. Productivity of applied water for wheat and legumes from Fig. 19.2.

during the growing season, and b_0, b_1 and b_2 are the regression coefficients. The response of yield to total applied water showed very similar characteristics for wheat at all three locations. Initially, yield increased linearly with increasing water supply. As water supply increased further, yield reached a plateau and finally approached the maximum. Unlike wheat, the response of chickpea and lentil to the total amount of water received in northern Syria was linear (Fig. 19.2e–f). The difference in the response of yield to water supply might be related to the growth habit of the crops.

Water Productivity

The productivity of total applied water (PAW) is defined as crop yield per unit volume of water supply to the crops, following Molden (1997), and is estimated by dividing crop yield, estimated from crop production functions in Fig. 19.2, by total applied water (rainfall + irrigation). Figure 19.3 shows the relationship between PAW and the level of water application for wheat in northern Syria, the North China Plain and Oregon, USA, and for chickpea and lentil in northern Syria. The crop production functions in Fig. 19.2 were used to derive the productivity of the applied water. For wheat, PAW for these three locations, representing different climatic conditions, increases sharply at a low water-supply level and reaches a maximum at a certain level of water supply. After its maximum, PAW shows a decrease with increasing water supply, depending on the response of yield to water. The level of water application at the maximum PAW differs

Table 19.1. Comparison of water productivity (PAW) of irrigation levels for wheat and maize.

	Wheat, Texas, USA[a]		Wheat, Syria		Maize, Texas, USA[b]	
Irrigation level	Yield (t ha^{-1})	PAW (kg m^{-3})	Yield (t ha^{-1})	PAW (kg m^{-3})	Yield (t ha^{-1})	PAW (kg m^{-3})
Full	4.76	0.64	5.79	0.93	13.95	1.42
67% of full	4.74	0.76	5.24	1.19	11.36	1.53
33% of full	3.88	0.80	5.15	0.99	6.62	1.21
Rain-fed	2.19	0.61	3.27	0.93	1.36	0.43

[a]From Schneider and Howell (1996).
[b]From Howell *et al.* (1997).

considerably for the three locations. The most productive use of water was reached with about 440–500 mm of water supply (140–180 mm irrigation) in northern Syria, 400 mm (120–160 mm irrigation) in the North China Plain and 750–850 mm (350–450 mm irrigation) in Oregon, USA. For grain-legume crops in northern Syria, PAW gradually increases with increasing water supply and reaches a plateau at a maximum PAW. The maximum PAW is about 0.5 kg m^{-3} for chickpea and 0.4 kg m^{-3} for lentil.

Significant differences in the PAW have been observed between crops. In north Syria, wheat has a PAW (1 kg m^{-3}) twice as high as grain-legume crops (0.4–0.5 kg m^{-3}) (Zhang and Oweis, 1999; Zhang *et al.*, 2000). Although the three experiments represent very different climatic conditions, the maximum PAW for wheat is about 1–1.2 kg m^{-3}. Rice has a relatively low PAW of about 0.37–0.68 kg m^{-3} (Tuong and Bhuiyan, 1999). Maize has a relatively high PAW of about 1.2–1.5 kg m^{-3}. PAW values of 0.4 kg m^{-3} were reported for cotton (Droogers *et al.*, 2000).

Deficit Irrigation: an Efficient Way to Increase the Productivity of Applied Water

The relationships between crop yield (Fig. 19.2) and the productivity of applied water (Fig. 19.3) and water supply demonstrate that higher PAW is achieved at a water-supply level that is lower than that at maximum yield. Many irrigation experiments involving different irrigation levels have also shown that deficit irrigation usually has higher PAW than full irrigation. For example, two-thirds of full irrigation increased PAW by 19–28% for wheat and 8% for maize (Table 19.1). Using the principle developed by English and Raja (1996) and Zhang and Oweis (1999), we can derive different irrigation scenarios. Two of the most important scenarios are those for maximizing production and maximizing farmers' net profit under limited-water-resources conditions. The scenario for maximizing production is referred to as full irrigation (I_f) and the other scenario with water supply less than I_f is defined as deficit irrigation (I_d). The production, water application and water productivity for these two scenarios are presented in Table 19.2 for wheat in Syria and Oregon, USA, and maize in Zimbabwe. With the same amount of water available to the crops, I_d scenarios can improve the productivity of applied water by 12–20% for wheat. We conclude that deficit irrigation can increase the productivity of applied water by producing more yield with the same amount of water resources for crops.

The risk with deficit irrigation is low because the response curve of crop yield to water supply often has a wide plateau (Fig. 19.2); a considerable amount of water can be saved without a significant yield reduction compared with full irrigation. Zhang and Oweis (1999) reported that I_d strategy allows one to apply 40–70% less irrigation water for a grain-yield loss of only 13%. Similarly, English and Raja (1996) reported that deficit irrigation averaging 64% of full irrigation was found to be economically equivalent to full irrigation when water was the limiting factor, and deficit

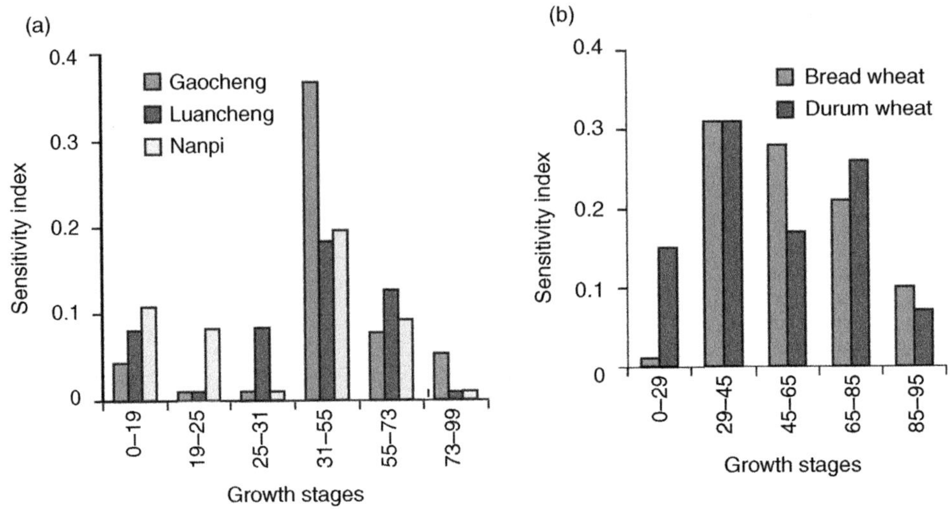

Fig. 19.4. Sensitivity indexes (λ value) of wheat to water stress during individual growing periods at three locations in (a) the North China Plain (winter wheat) and in (b) northern Syria (spring wheat). The growth stages are based on Zadoks *et al.* (1974).

irrigation in which only 30% of full irrigation was applied was found to be equivalent to full irrigation in land-limiting cases. If the saved water resources were allocated to other cropped areas, the total production and the productivity of the applied water would be increased, as indicated in Table 19.2. However, information is needed to guide farmers on when and how much to irrigate with deficit irrigation in order to reduce the unwanted effect of water stress on crop yield. Jensen (1968) developed a model to quantify the effect of water deficits during certain growth stages on grain yield, using the following equation:

$$\frac{Y}{Y_m} = \prod_{i=1}^{n}\left(\frac{ET_i}{ET_m}\right)^{\lambda_i} \qquad (19.6)$$

where Y is grain yield (t ha^{-1}), Y_m is the maximum yield from the plot without water stress during the growing season, ET_i is the actual evapotranspiration (mm) during the growing stage i, ET_m is the maximum evapotranspiration corresponding to Y_m, λ_i is the sensitivity index of the crop to water stress and i is the growth stage. Using Jensen's (1968) model, the sensitivity indexes (λ values) of crop to water stress at different crop growth stages were quantified for wheat in northern Syria (Zhang and Oweis, 1999) and in the North China Plain (Zhang *et al.*, 1999). These authors concluded that the most sensitive stages for water stress for wheat are from the stem-elongation to the grain-filling stage (Fig. 19.4). The variation of λ values indicates that crop grain yield depends not only on total water use during the growing season, but also on water use during different growth stages. For example, a 40% decrease in evapotranspiration (ET) during the period from heading to milking reduced grain yield by 15% for winter wheat in the North China Plain, while this deficit in ET during the period from winter freezing to reviving reduced grain yield by only 3%. Similarly, a 40% deficit in ET during the period of stem elongation to grain-filling reduced yield by 15–20% for spring wheat in northern Syria, while this deficit in ET at seedling stage and late grain-filling stage hardly affected the grain yield at all.

Water-stress sensitivity indexes have an important implication for irrigation scheduling, in particular for deficit irrigation. Since water stress during growth stages with high

Table 19.2. Scenario analysis of total production and productivity of applied water at full (I_f) and deficit (I_d) irrigation.

	Irrigation scenarios	Water applied (I) (mm)	Yield (Y) ($t\ ha^{-1}$)	Area irrigated (ha)	Area rain-fed (ha)	Total area (ha)	Yield from irrigated (t)	Yield from rain-fed (t)	Total yield[a] (t)	PAW[b] ($kg\ m^{-3}$)
Wheat, Syria	Full	330	6.4	1.00	1.06	2.06	6.4	3.4	9.8	0.94
	Deficit	160	5.6	2.06	0	2.06	11.8	0	11.8	1.12
Wheat, Oregon, USA	Full	740	9.9	1.00	0.43	1.43	9.9	1.9	11.8	0.88
	Deficit	520	9.2	1.43	0	1.43	13.3	0	13.3	0.98
Maize, Zimbabwe	Full	525	6.0	1.00	1.44	2.44	6.0	–	–	–
	Deficit	215	4.1	2.44	0	2.44	10.1	0	10.1	–

[a]For full-irrigation scenarios, total yield = $A_f \times Y_f + (A_d - A_f) \times Y_{rain\text{-}fed}$. For deficit-irrigation scenarios, total yield = $A_d \times Y_d$, where subscripts f and d represent full and deficit irrigation, respectively.

[b]For full-irrigation scenarios, PAW = total yield/$[I_f \times A_f + \text{rainfall} \times (A_d - A_f)]$. For deficit-irrigation scenarios, PAW = total yield/$[(I_d + \text{rainfall}) \times A_d]$.

Table 19.3. Amount (mm) and timing of deficit irrigation for high productivity of the applied water under different rainfall conditions.

	Rainfall (mm)	Deficit irrigation (mm)	Time of irrigation
Northern Syria	250	160–260	Stem elongation, booting, flowering and grain-filling
	300	110–210	Stem elongation, flowering and/or grain-filling
	350	60–160	Flowering and/or grain-filling
	400	0–110	Grain-filling
North China Plain	80	160–240	Stem elongation, booting, flowering and grain-filling
	120	120–180	Stem elongation, flowering and grain-filling
	160	100–160	Stem elongation and flowering

λ values has a much greater effect on final yield, to prevent stress during these stages irrigation would be advisable and consequently a higher PAW could be achieved, especially in the areas where water resources are limited. Based on the production functions and the water-productivity analysis, an optimal irrigation scheduling for wheat crops in the North China Plain and northern Syria is proposed in Table 19.3. Such a schedule can only be followed if farmers have full control over the timing and amount of irrigation water they apply. This is usually the case when irrigation water comes from shallow tube wells that are operated by the farmers. However, in countries where irrigation water is supplied through canals according to a strict rotational schedule (as is the case, for example, in much of the Indian subcontinent), there is no flexibility in the delivery of irrigation water. The implication is that under these circumstances deficit irrigation cannot be practised.

Conclusions

This chapter concludes that deficit irrigation leads to higher productivity of the water (rainfall and irrigation) than can be attained with full irrigation and can, therefore, be used for improving the productivity of water in semi-arid areas. Deficit irrigation requires more control over the amount and timing of water application than full irrigation practice. Information on when and how much to irrigate is needed in order to reduce unwanted effects of water stress on production. With reliable crop–water production functions and knowledge of the stages of the crop that are sensitive to water stress, optimal deficit irrigation can be scheduled with a minimum yield reduction compared with full irrigation and, therefore, limited water resources can be utilized more efficiently. In addition, when the crop production functions are known, it is possible to appropriately allocate limited water resources between crops where crops compete for scarce water in dry areas.

References

Droogers, P., Kite, G. and Murray-Rust, H. (2000) Use of simulation models to evaluate irrigation performances including water productivity, risk and system analyses. *Irrigation Science* 19, 139–145.

English, M.J. and Nakamura, B.C. (1989) Effects of deficit irrigation and irrigation frequency on wheat yields. *Journal of American Society of Civil Engineering* 115(IR2), 172–184.

English, M.J. and Raja, S.N. (1996) Perspectives on deficit irrigation. *Agricultural Water Management* 32, 1–14.

FAO (1988) *World Agriculture toward 2000: an FAO Study*. Belhaven Press, London.
Howell, T.A., Schneider, A.D. and Evett, S.R. (1997) Subsurface and surface microirrigation: corn – Southern High Plains. *Transaction of ASAE* 40, 635–641.
Jensen, M.E. (1968) Water consumption by agricultural plants. In: Kozlowski, T.T. (ed.) *Water Deficits and Plant Growth*, Vol. 2. Academic Press, New York, pp. 1–22.
Molden, D. (1997) *Accounting for Water Use and Productivity*. SWIM Paper 1, International Irrigation Management Institute, Colombo, Sri Lanka.
Schneider, A.D. and Howell, T.A. (1996) Methods, amounts, and timing of sprinkler irrigation for winter wheat. *Transaction of ASAE* 40, 137–142.
Tuong, T.P. and Bhuiyan, S.I. (1999) Increasing water-use efficiency in rice production: farm-level perspectives. *Agricultural Water Management* 40, 117–122.
Vaux, H.J. and Pruit, W.O. (1983) Crop–water production functions. In: Hillel, D. (ed.) *Advances in Irrigation*, Vol. 2. Academic Press, New York, pp. 61–97.
Zadoks, J.C., Chang, T.T. and Konzak, C.F. (1974) A decimal code for the growth stages of cereals. *Weed Research* 14, 415–421.
Zhang, H. and Oweis, T. (1999) Water-yield relations and optimal irrigation scheduling of wheat in the Mediterranean region. *Agricultural Water Management* 38, 195–211.
Zhang, H., Wang, X., You, M. and Liu, C. (1999) Water-yield relations and water-use efficiency of winter wheat in the North China Plain. *Irrigation Science* 19, 37–45.
Zhang, H., Pala, M., Oweis, T. and Harris, H. (2000) Water use and water-use efficiency of chickpea and lentil in a Mediterranean environment. *Australian Journal of Agricultural Research* 51, 295–304.

Appendix A
A Note on Transpiration

David Seckler[1]
International Water Management Institute, Colorado, USA

Crop losses to water shortage may exceed those from all other causes combined.
(Kramer, 1980)

Always the beautiful answer
Who asks a more beautiful question
(e.e. cummings)

Introduction

This note is the product of a dialogue in the International Water Management Institute (IWMI) over the simple question: 'Why do plants need so much water for transpiration?' 'Transpiration consists of the vaporization of liquid water contained in plant tissues and the vapor removal to the atmosphere' (Allen *et al.*, 1998).[2] Between 200 and 1000 kg of transpired water is lost to the atmosphere in the production of only 1 kg of plant biomass. And this is just for biomass; the amount of transpiration per unit

[1] It should be emphasized that this note is written by an economist, not a plant scientist or similar expert. I hope that what is sacrificed in terms of expertise may be partly compensated for by a somewhat different perspective on the issues and the need to write it in simple language more accessible to other interested laypersons. Fortunately, previous drafts of this note have been closely reviewed, corrected and contributed to by two real experts in the field: Richard G. Allen of Idaho State University and Terry Howell of the Agriculture Research Service of the US Department of Agriculture (USDA). I am enormously grateful for the time they devoted to this task and our interesting and enjoyable correspondence. I have also benefited from criticisms of a previous draft by Bruce Bugee of Utah State University and a constructive review of the final draft by Parviz Soltanpur of Colorado State University. Of course, all remaining errors, whether of commission or omission, remain my responsibility.

I have also benefited a great deal by discussions with Andrew Keller of Keller–Bliesner Engineering, Logan, Utah, concerning both this note and a companion paper to this one, which he has written: '*Note on crop yield to water relationships*'. We are grateful to the International Water Management Institute for supporting both of these notes, as part of a research programme in water productivity.

[2] This is the reference work on the subject and should be studied by anyone interested in the use of water in agriculture – which constitutes most of the developed water resources used in the world.

 Water Productivity in Agriculture: Limits and Opportunities for Improvement (eds J.W. Kijne, R. Barker and D. Molden)

of grain or other fruits is at least twice this amount. Clearly, this is an important question for an institution like IWMI, which is dedicated to more productive water use: 'More crop per drop'.[3]

One result of this study is that it is now easier to understand why the dialogue in IWMI never reached a satisfactory conclusion. While some of the central features of transpiration are clearly understood and command broad agreement among experts, other important aspects remain hazy and controversial. The problem, in other words, is that in some respects this 'beautiful question' has several 'beautiful answers', some of which are contradictory. This has made the present note exceptionally difficult and interesting to write. To accommodate this problem, extensive quotations and references are used in the text and footnotes, and I have used the personal pronoun to distinguish my own thoughts from those of authorities in the few cases where I thought they might be worth mentioning.

The Process of Transpiration

The *Columbia Electronic Encyclopedia* provides a lucid overview of the process of transpiration.[4]

> Transpiration, in botany, is the loss of water by evaporation in terrestrial plants. Some evaporation occurs directly through the exposed walls of surface cells, but the greatest amount takes place through the stomata, or intercellular spaces (on the leaves). Transpiration functions to effect the ascent of sap from the roots to the leaves (thus supplying the food-manufacturing cells with water needed for photosynthesis) and to provide the moisture necessary for the diffusion of carbon dioxide (CO_2) into and oxygen (O_2) out of these cells. The rate of transpiration is almost always far greater than the above functions would seem to warrant; in most plants 200–1000 lb. (or kg) of water are transpired for each pound (or kg) of solid material added to the plant. Various factors influence the transpiration rate. Photosynthesis, induced by light, has the effect of increasing the water pressure in the guard cells that border each stoma and that, in expanding, pull apart to widen the stomal aperture and thereby increase water loss. Low humidity promotes the diffusion of water vapour from the air passages inside the leaf into the outside air. A lack of water in the soil cuts down the water supply to the cells, thus lowering the water pressure and thereby limiting expansion of the guard cells. Therefore, the rate of transpiration is highest on a bright, dry day and lowest at night or in drought conditions. Morphological factors, such as reduced leaf surfaces, a heavy cuticle layer on the leaves, low numbers of stomata, and stomata recessed below the other epidermal cells, also lower the rate; plants, such as conifers and cacti, conserve water in these ways. Plants also lose some water by guttation, a process whereby water is exuded directly through pores called hydathodes. The reaction of a plant to excessive water loss is wilting and, eventually, death.

An important product of the process of photosynthesis is carbohydrate. The building blocks of carbohydrate are: (i) hydrogen, which plants acquire by breaking down water molecules; and (ii) carbon, which is acquired by breaking down molecules of CO_2. Water is obtained from the soil, through the plant roots. CO_2 is obtained from the atmosphere, through the stomata. The waste product of photosynthesis, O_2, is dispelled from the plant through the stomata into the air, where it is happily recycled by animals.

As noted above, the stomata are surrounded by guard cells, which may be pictured as doughnut-shaped bladders of water, with the pore of the stoma being the hole of the doughnut. Photosynthesis increases the water pressure in the guard cells (known as the turgor pressure), causing them to expand and thereby opening the stomata. This

[3] I would like to take this opportunity to set the record straight on this much-used slogan of IWMI's. It was invented by Chris Perry of IWMI around 1995. And of course it is subject to the usual constraints for environmental quality, equity, gender, etc.

[4] *The Columbia Electronic Encyclopedia*, 6th edn. (Encyclopedia. com), 'Transpiration'. This excellent source is freely available on the Internet. Here as elsewhere statements in brackets are mine.

enables the plant to absorb CO_2 and expel O_2. But, when the stomata are open, water vapour in the leaf is exposed to the atmospheric forces of evaporation. When soil water is plentiful, this does not normally present a problem for the plant: it simply absorbs sufficient amounts of water from the soil through the roots to compensate for the water losses of transpiration. But, if soil water is scarce, the plant could die of dehydration. The aptly named guard cells protect the plant against this fate, at least temporarily. With a relative shortage of soil water, the guard cells lose turgor pressure and the stomata close. While this protects the plant from dehydration, by reducing transpiration, it also reduces the intake of CO_2, the rate of photosynthesis and, hence, plant growth.

There is a high – indeed, almost perfect – correlation between the rate of transpiration and the rate of plant growth. However, 'correlation is not causation'. The question is whether this high correlation implies a causative role of transpiration in plant growth. The basic process of transpiration outlined above does not necessarily imply such a role. The exchange of gases with and the loss of water vapour to the atmosphere vary together because both processes are driven by atmospheric conditions and are conducted through and governed by the stomata. In this view, absorption of CO_2 is the central element of the process and transpiration is merely a consequence of the process. As Condon *et al.* (2002) put it, transpiration 'is the required unit of exchange for the acquisition of CO_2 by plants'.

There is no question that this view of transpiration is correct, so far as it goes; but is this all there is to it? Does transpiration provide beneficial functions for plant growth, at least to some degree and under some conditions? The answer to this question determines at least the theoretical potential for saving water in crop production by reducing transpiration (neglecting, for now, the question of whether this can be done in practical ways). This question of the beneficial functions of transpiration is the focus of the rest of this discussion.

Two Beneficial Functions of Transpiration[5]

Two possibly beneficial functions of transpiration are frequently mentioned in the literature.

First, transpiration helps to cool plants in extremely high temperatures. While this is true, most plants (unlike many animals) spend most of their time within their comparatively wide range of temperature tolerance. One authority says, 'leaves in the sun rarely are seriously overheated even when transpiration is reduced by wilting' (Kramer and Boyer, 1995). This is partly because there is almost always some moisture evaporating from the leaf. However, temperature can seriously affect plant growth beyond the limits of their tolerance. And the effects of temperature differ substantially among plants.[6]

Secondly, transpiration helps in the movement of sap, nutrients and moisture from the roots to the leaves (as stated in the above citation from the *Columbia Electronic Encyclopedia*). Since this second function seems to be controversial, it warrants a more extended discussion.

> Sap ascends (upward through the plant) at a rate of from 1 foot to 4 feet (30–122 cm) per hour; in the case of redwood it rises easily to a height of almost 400 ft. (120 m). The exact mechanisms behind this enormous lifting force are not certain, although several principles are thought to be involved. Chief among them is the pull of transpiration; as water evaporates from the leaf cells, they draw in liquid osmotically from the xylem tubes to replace it.

[5] Another excellent Internet source for this and the preceding section is *Kimball's Biology Pages*, based on a biology textbook by John W. Kimball. See 'Transpiration' and the associated hypertext.

[6] An excellent review of the possible effects of temperature and CO_2 changes on plant growth, associated with global warming, is found in H. Wayne Polley, Implications of atmospheric and climatic change for crop yield and water use efficiency, *Crop Science* 42, 131–140 (2002).

> Because of the great cohesiveness of water molecules, the resulting tension affects the entire continuous column of water down to the root tips, which in turn absorb more water from the soil.[7]

Also:

> The rise of water in plants depends chiefly on the attraction of water molecules for each other. In large masses of water this attraction is not obvious, but in long, slender tubes it is readily demonstrable, becoming stronger as the tube becomes slenderer. A column of water in a tube as slender as a plant vessel strongly resists being broken, and a pull at the top is transmitted throughout the column. Thus the water can be pulled upward, like a rope through a pipe ... This explanation of the rise of water in plants is called the *theory of cohesion*.[8]

Kimball (see note 5) reports experimental evidence on the astonishing pulling force of transpiration. A 150-foot-tall rattan plant was cut at the base, the stem placed in a sealed container of water. The plant continued to draw water from the container and 'the resulting vacuum becomes so great that the remaining water begins to boil spontaneously'. He also notes that coastal mangrove trees use this vacuum effect to desalt sea water through a membrane in their roots. The vacuum required for this task is around 500–800 lb. per square inch!

However, while transpiration-pull (as Kimball describes it) may be needed for very tall plants – like redwoods or rattan, where the pumping head is high – some scientists believe that root pressure and, possibly, other forces are able to perform this function for most (possibly shorter) plants. Kimball ('Root Pressure') directly addresses this theory and rejects it on the grounds that some plants have no root pressure, sometimes root pressure is negative and transpiration and root pressure are not well correlated.

In a previous draft of this note, I suggested a possible way to resolve this controversy through the following (mental or actual) experiment. Assume that the relative humidity (RH) of the ambient atmosphere of a plant is 100% and there are no other stresses on the plant. No transpiration would occur and yet the stomata would be open to receive CO_2 and expel O_2. How well would the plant do under these conditions? One of the reviewers reported that such an experiment was actually done in plant growth chambers, where all these factors are controlled: 'they grew cotton at various relative humidities and water and soil salinities. The 100% RH (relative humidity) treatments (little transpiration demand) did not bloom and set bolls well' (Terry Howell, personal communication).[9] Thus, assuming that there were no temperature or other stresses in this experiment, it appears that transpiration does perform some beneficial function in plant growth.

However, another reviewer of a draft of this note in which this experiment was reported said that 'only 1 to 2% of the transpiration is needed to move nutrients from the roots and for the water needed in photosynthesis'. And: 'Plants grow fine in 99% humidity assuming disease is prevented by other means' (Bruce Bugbee, personal communcation). It should be noted that the difference between the 100% RH in the experiment and the 99% RH in the reviewer's comments could provide the 1–2% of water needed for these purposes.[10] Even so, if this is the only beneficial function of transpiration, it does not represent enough water loss to worry about.

The discussion up to this point may be summarized by saying that, with the exception of the small amounts discussed above and the possible transport function in very tall plants, the consensus opinion of most

[7] The *Columbia Electronic Encyclopedia*, *op. cit*. 'Sap.'

[8] Daniel I. Arnon, *Encyclopedia Americana*, 'Plants.'

[9] As Richard G. Allen notes, this effect was probably caused by nutrient deficiency due to insufficient transpiration – as discussed in the section on 'The Third Function of Transpiration,' below.

[10] I am grateful to Andrew Keller for pointing this out.

experts is that transpiration is apparently just a 'necessary evil'. It is evil because of the tremendous loss of water to the atmosphere; it is necessary because of the need to keep stomata open to absorb CO_2. And that is all there is to it.

My Reservations

In my opinion, the above consensus creates a question that is even larger than that of transpiration itself. Given the crucial importance of water to plant survival and reproduction, how could plants evolve to be such wasteful users of such a scarce resource? One of the impressive features of the evolutionary record is how organisms adapt and evolve to make efficient use of resources in their specific environments. It is true that plants evolved in an aqueous environment, where water was not originally a constraint, and it is true that they have the formidable task of extracting very low densities of CO_2 out of the atmosphere, thus exposing themselves to evaporation losses.[11] However, while these are good reasons for this apparent waste of water, I do not find them compelling.[12] In fact, at one point, I thought that transpiration might represent a strong challenge to aspects of the theory of evolution itself.

This evolutionary question has driven a search for other possible functions of transpiration – a third function lurking in the background, so to speak, as in the film, *The Third Man*. As it turns out, there is indeed a third function of transpiration, which is well documented, is essential to plant growth and can account for most of the water used in transpiration. This is the function of transpiration in moving solutions of nutrients, not through the plant, but from the soil to the roots of the plant. That this important function is usually neglected in discussions of the functions of transpiration is perhaps accounted for by the fact that it occurs outside the plant itself, in the interface of the plant and its soil environment.

The Third Function of Transpiration[13]

Soil nutrients are dissolved in water and absorbed from this solution by the roots. Some of the nutrients are acquired by root interception from the soils immediately adjacent to the roots, but these soils typically do not have a sufficient amount of nutrients to meet plant needs. Nutrients are supplied to the roots from more distant soils by two different processes. One is by diffusion of nutrients (ions) through soil water to the roots, as

[11] Bugbee (B. Bugbee, personal communication) notes that because of the low concentration of CO_2, relative to O_2, in the atmosphere, it is '700 times' easier for animals to take oxygen in without losing water than for plants to take CO_2 in without losing water. Therefore, it is not correct to think of plants as inefficient water users. This is an excellent point; however, it would also be interesting to compare the different requirements of animals and plants for these two factors.

[12] For example, letting my imagination run, I wondered if plants could not use their remarkable ability to create semipermeable membranes to create one in the stomata that would let CO_2 in, let O_2 out and keep H_2O in. I asked a research chemist, Terry Krafft, about this and he said that it should not be difficult to create such a membrane. In fact, at a convention he saw a fish living happily in a sealed polymer bowl of this nature. But I forgot to ask him how they fed the fish! Given the ingenuity of plants and their enormous variety, I would not be astounded if someone actually found a dry-area plant with something like this membrane.

[13] The discussion of this subject is partly from Kramer and Boyer (1995, pp. 286–290). Strangely, even these authors appear to dismiss this bulk-flow function in a discussion of the alleged benefits of transpiration earlier in their book (p. 203). One gets the impression that the benefits section was written before the later one and they forgot to change it – which is perfectly understandable by anyone who has written a book. Much of the research summarized in this source is from Stanley A. Barber and his collaborators, referenced in connection with Table A.2, below.

determined by concentration gradients in the solutions. The other is by bulk flow of nutrient-carrying solutions to the roots. Transpiration plays a vital role in bulk flow by continually evacuating the water from spent solutions surrounding the roots, thereby generating convective flow of new solutions, carrying additional nutrients, to the roots. Since the solutions are fairly dilute, large amounts of water must be evacuated to generate these flows. Table A.1 provides a quantitative view of these effects.

Table A.1 shows that, when transpiration is at a moderate level of 500 times plant biomass, the soil solution is more than sufficient to supply calcium (Ca) and magnesium (Mg), but is considerably below the requirements for potassium (K) and phosphorus (P). In the latter case:

> If the inorganic ions are absorbed at a relatively greater rate than bulk flow can provide, as with phosphate and potassium, the concentration in the soil solution will decrease next to the root. In response, ions are released from the soil particles and tend to buffer the concentration. Nevertheless, there is a lowering at the root surface and ions will tend to move into the depletion zone by diffusion in addition to bulk flow.
>
> (Kramer and Boyer, 1995, p. 288)

Thus, transpiration, through bulk flow, is the first factor in providing nutrients to the roots, but, if this is insufficient, diffusion provides a localized and potentially limited supplementary supply of nutrients. These effects are partly governed by different diffusibilities among the nutrients. In addition, it was found that transpiration 'had little effect on uptake (of ions) by roots in low external solution concentrations, but had a significant effect when the external concentrations were high' (Kramer and Boyer, 1995, p. 289).[14] This seems to imply that the effect of transpiration is higher when soil moisture is lower. It was also found that the transpiration effect was strongest when plants were growing most rapidly.

Another authority stresses the importance of mass flow in the reproductive stage of plant growth:

> An important proportion of water-stress induced crop failure occurs when the stress coincides with flowering. In many instances, this is the result of reduced transport of critical nutrients to the developing reproductive structure. The most significant nutrients in this regard are boron, copper and calcium each of which is delivered to the grain solely in the transpiration stream. In the absence of these nutrients, reproductive growth is permanently impaired.[15]

He goes on to note that these problems can be alleviated by appropriate nutrient-management practices.

Table A.1. Supply of elements to maize roots by bulk flow caused by transpiration (from Kramer and Boyer, 1995).

Element	Plant dry matter (%)	Required concentration[a]	Actual concentration[b]
Ca	0.22	0.11	0.83
Mg	0.18	0.15	1.15
K	2.0	1.02	0.10
P	0.20	0.13	0.002

[a]Concentration of soil solution needed if transpiration is 500× (plant) dry matter (mM).
[b]Concentration of soil solution in 145 maize soils (mM).

[14] Richard G. Allen seriously questions this statement; however, I have left it in as a stimulus to future research.

[15] Patrick Brown of the University of California at Davis. Communication to an e-mail discussion on water productivity, May 2002.

Further evidence on the importance of solutions and concentrations is provided by the growth of plants in liquid solutions for research and practical purposes. Luttge and Higinbotham (1979, pp. 62–65) provide information on these solutions. For macronutrients, the concentrations vary from 24 to 224 p.p.m., and are less than 2 p.p.m. for micronutrients. Even in liquid solutions, without the resistance to water flows in soils, constant stirring of the solution is required to provide oxygen and avoid nutrient depletion in the solution adjacent to the roots.

Last, and perhaps most importantly, Barber (1995, p. 91, Table 4.4)[16] provides estimates of the amounts of major soil nutrients supplied to the roots by the three sources of root interception, mass flow (as he calls it) and diffusion, as shown in Table A.2.[17]

Conclusion

It appears that the solution to the dilemma of transpiration discussed in this note is, appropriately, a real, liquid solution. A major beneficial function of transpiration is to generate bulk flow of solutions containing soil nutrients to the plant roots. While this function is also performed by diffusion for some nutrients, Table A.2 indicates that it is indispensable for others – and the lack of any single nutrient has a large effect on plant growth. Most important from the present point of view is that the bulk-flow function can account for most of the water used by plants in transpiration.

While this conclusion is undoubtedly subject to qualification, depending on plant species, water conditions, soil fertility and fertilization practices, it appears to provide a strong answer to the original question of 'Why do plants use so much water in transpiration?' and an important counter to the 'necessary evil' argument. It is indeed perplexing that this function does not receive more attention in discussions of the functions of transpiration and that more research attention is not given to it.

Research on transpiration is, of course, particularly important in the case of rain-fed agriculture. Many ingenious methods of reducing transpiration in rain-fed agriculture have been proposed. One method is to capitalize on the different transpiration efficiencies of different kinds of plants (Terry

Table A.2. Relative significance of root interception, mass flow and diffusion in supplying maize with its nutrient requirements from a fertile Alfisol silt loam (kg ha^{-1}).

Nutrient	Nutrient needed[a]	Root interception	Mass flow	Diffusion
Nitrogen	95	2	150	38
Phosphorus	40	1	2	37
Potassium	195	4	35	156
Calcium	40	60	150	0
Magnesium	45	15	100	0
Sulphur	22	1	65	0

[a]Needed for 9500 kg of grain ha^{-1}.

[16] This is the most interesting and readable text on the subject I have found. See especially Chapter 4.

[17] Commenting on the zero values for diffusion of the last three nutrients in this table, Richard G. Allen wonders, how do these ions get into the water stream in the first place? He also wonders, how do pineapples, which have very low transpiration rates, get over the problem of potential nutrient deficiencies? All I can say is, as usual, very good questions!

Howell, personal communication).[18] Other methods range from the use of antitranspirants[19] to breeding plants for leaf curling and for resistance to water uptake (Richards *et al.* 2002).[20] All of these research efforts should be strongly supported. The role of transpiration in the bulk flow of plant nutrients should be considered as an integral part of this research programme. Hargreaves and Merkley (1998)[21] have indeed proposed that rates of fertilization of crops should be based on rates of transpiration, in order to avoid imbalances between these two factors. Bulk flow would appear to be an especially important factor to consider in rain-fed agriculture, where soil moisture and fertility tend to be low.

References

Allen, R.G., Pereira, L.S., Raes, D. and Smith, M. (1998) *Crop Evapotranspiration – Guidelines for Computing Crop Water Requirements*. Irrigation and Drainage Paper 56, FAO, Rome, pp. 1–15.

Barber, S.A. (1995) *Soil Nutrient Bioavailabilty*. John Wiley & Sons, New York.

Condon, A.G., Richards, R.A., Rebetzke, G.J. and Farquhar, G.D. (2002) Improving intrinsic water-use efficiency and crop yield. *Crop Science* 42, 122–131.

Hargreaves, G.H. and Merkley, G.P. (1998) *Irrigation Fundamentals*. Water Resource Publications, Westview Press, Boulder, Colorado, pp. 12–17.

Kramer, P.J. (1980) Drought, stress, and the origin of adaptations. In: Turner, N.C. and Kramer, P.J. (eds) *Adaptation of Plants to Water and High Temperature Stress*. John Wiley & Sons, New York, pp. 7–20.

Kramer, P.J. and Boyer, J.S. (1995) *Water Relations of Plants and Soils*. Academic Press, San Diego, p. 203.

Luttge, U. and Higinbotham, N. (1979) *Transport in Plants*. Springer-Verlag, New York.

Polley, H.W. (2002) Implications of atmospheric and climatic change for crop yield and water use efficiency. *Crop Science* 42,131–140.

Richards, R.A., Rebetzke, G.J., Condon, A.G. and van Herwaarden, A.F. (2002) Breeding opportunities for increasing the efficiency of water use and crop yield in temperate cereals. *Crop Science* 42, 111–121.

[18] For example, 'Plants are categorized as to their P [photosynthesis] mechanisms as C4 (sorghum, maize ...), C3 (cotton, wheat, rice, sugarbeet), or CAM (pineapple ...). The transpiration efficiency (P/T) is generally, CAM>C4>C3. C3 plants have a greater photosynthetic respiration. CAM species keep their stomates closed during the day.'

[19] Kramer and Boyer (1995, pp. 400–402) discuss the use of antitranspirants, which are chemicals that reduce transpiration rates. Unfortunately, they tend to reduce CO_2 absorption and plant growth even more than transpiration.

[20] This is an excellent survey of the problems and opportunities in crop breeding for this purpose.

[21] I am grateful to Andrew Keller for this reference.

Appendix B
Note on Agronomic Practices for Increasing Crop Water Productivity

Jacob W. Kijne
International Water Management Institute, Colombo, Sri Lanka

Of the many measures described in this book, the field-level agronomic or cultural measures may appear to be the easiest to implement and to be the least likely to cause undesirable externalities. The purpose of this note is to summarize what is reported about them dispersed throughout this book, and to assess whether indeed these measures are ready to be adopted on a wider scale. Are the agronomic practices perhaps location-specific or only applicable for a limited set of soil and crop conditions? Many of them are also known to have trade-offs in terms of higher labour and management demands, and further studies may have to be done before scientists feel confident to recommend them for wide-scale application.

The agronomic practices that are mentioned in the chapters of this book and that have the potential for increasing crop water productivity are summarized in Table B.1. The selection of the appropriate cultivar with its specific length of growing season, harvest index, stress tolerance and disease resistance, although directly affecting crop yield and water productivity, is not included in the table. Here we are concerned with measures which the farmer can take and which affect especially the partitioning of rainfall or irrigation between infiltration and runoff and the partitioning of evapotranspiration between evaporation and transpiration. These two effects are often interactive, as any measure that maximizes infiltration of water into the soil also minimizes water losses through surface evaporation and runoff and reduces soil erosion.

The irrigation method determines to what extent it is possible to reduce evaporation from the soil surface while maintaining adequate soil moisture levels in the root zone to avoid crop stress. Precision application with drip or subsurface irrigation is the method of choice; surface irrigation on precisely levelled fields is a distant second.

Quite a few of the measures require additional expense, skills, time and machinery. It is not only for commercial farmers that these measures are still suitable. But, with small and resource-poor farmers, it is even more important to provide adequate training and advice and also to ensure that maintenance and spare parts for the machinery are available whenever needed.

Obviously, not all measures are suitable under all circumstances. The list of measures is not to be taken as a menu from which to choose at liberty. For example, minimum tillage may not work on soils that tend to form a surface crust or a hardpan. In

Table B.1. A summary of the agronomic practices.

Category/item	Comments	Adoption
Water-related		
Alternate-row irrigation	Suitable for row crops, depends on irrigation system	Requires more labour and management than conventional furrow irrigation
Minimize preplanting irrigation	Requires control over irrigation supply	
Minimize time between preplanting irrigation and planting	Especially with rice; requires control over timing of irrigation	Constraints in water delivery, labour and farm machinery often interfere
Soil-related		
Reduced (or conservation) tillage	Requires farm machinery, such as shallow rotovator, inverted-T opener, combined with seed drill	Cost, availability and maintenance of machinery are constraints
Zero tillage	Placing seed on saturated soil without tillage	Unless the field is level, yields may be less
Raised beds	Requires tools to make the beds	Labour requirement if beds have to be remade each season
Broad beds and furrows	Requires tools to make the beds	As for raised beds
Row spacing and orientation	Affects interception of radiation and, if planting is on contours, reduces runoff	Requires flexible seed drill; may be more labour-intensive
Land levelling	Prevents ponding and unequal application of water	Requires skilled labour and machinery; needs to be repeated every 2–3 years
Mulching and residue management	Lowers evaporation from soil surface and reduces runoff	Gravel mulches, etc., are expensive
Application of organic matter (OM)	Increases water-holding capacity of soil; needs to be repeated often as OM in semi-arid tropics decomposes quickly	OM is scarce and often used for other purposes, e.g. as fuel
Plant-related		
Direct seeding of rice	Direct-seeded rice may have more diseases, insect and weed problems and, hence, give lower yields	Often done to reduce labour rather than for increasing WP; requires more weed control and pest control
Timely planting, etc.	Timely sowing, weed control, fertilizer application, nutrient management and best crop rotation raise yields	Requires good farming skills and extension services; labour-intensive

WP, water productivity.

general, a combination of measures may work best, such as combining laser levelling with minimum tillage and bed planting.

There are trade-offs between water conservation and yield, e.g. less frequent irrigation of grains, including rice, could lower yield, and also between water application and labour, e.g. excessive irrigation is often done to save the labour involved in levelling the fields. So much the more reason for being quite explicit with farmers about the consequences of adopting these water-productivity-enhancing measures. As mentioned by Barker *et al.* (Chapter 2, this volume), in promoting the adoption of new technologies, researchers and extension agents often focus on the higher yield potential, but ignore the opportunity cost of family labour and the increased management requirements. As argued by Wani *et al.* (Chapter 12, this volume), low adoption of improved agronomic practices is because insufficient attention was given to farmer participation, community action, etc. However, this attention can only be given if the impact of adoption is known – in other words, if the benefits and the costs are clearly articulated.

The inevitable conclusion, then, is that more work needs to be done before we can entreat farmers to adopt these measures. Perhaps it is not adoption that we should aim at but adaptation in a manner that is suitable for the specific set of conditions. The contribution that science should make in this process is to study the necessary conditions for success, which include the analysis of all the consequences of their introduction.

Index

Note: Page numbers in **bold** refer to figures in the text and those in *italics* to tables or boxes. The letter *n* following a page number indicates text in the footnotes.

abscisic acid (ABA) 108–109
access to water 9, 99, 100
Acrimonium implicatum 114
actual evapotranspiration
  defined 38–39
  remotely sensed data 291–292
adoption of technology *see* technology adoption
aerobic rice 61–62, 109
agricultural drought 112
agroforestry
  catchment experiments 225–226
  environmental functions 223
  manipulation of water use/root function 224–225
  potential benefits 219
  progress and challenges 225–227
  rainfall-use efficiency 219–221
  resource capture 223–224
  semi-arid environments 226–227
  tropical countries 226
  water productivity 221–223
agronomic practices
  summary 319–321
  see also resource-conserving technologies *and individual practices*
Alfisols 208, *209*
alkalinity
  chemical amelioration 76, *77*, 85–86
  hazards 71–72
allele pyramiding 118–121
allocation of water
  sectoral 30–31, 55, 280–283
  Thailand 280–283
almond trees 187
alternate row irrigation *320*
alternate wetting and drying (AWD) 59, **60**, 63
Andes, potato production 236–237
anthesis silking interval (ASI) 18
antitranspirants 318*n*
apoplastic invertase 110
ascorbate peroxidase 109
Asian Development Bank (ADB) 274
Australia 16, 61, 108, 223–224
available water *4*
available water supply (AWS) 45
AWD *see* alternate wetting and drying
Azadirachta indica 224

badia 186–187
banana 114
Bangkok metropolitan area 274
Bangladesh, floods 219
basin efficiency 44–47
  available water supply 45
  beneficial depletion 44–45
  economic 28–29
  global current and future projections 1995 166, *174*
  impact of improved on future water productivity 174–177
  types of river basin 46
basins 2–3
  beneficial water depletion 44–45
  classification 46
  closed *4*, 46, 276–277
  defined 5
  land-use decisions 16–17
  open 46
  options for productivity improvements 9, 10–12
  reallocation of water 15–16
  responses to water scarcity 28–29
  water accounting 4, 5–6
  see also named river basins
basmati rice 70, *77*, 78–79
BBF *see* broad-bed and furrow

bean, common *113*, 114
bed-planting systems 245–247, 249–250
benefit : cost analyses 26–28
berseem *77*
Bhakra irrigated area, India
climate and soils 257
water accounting **5**
watercourse characteristics 258, *259*
wheat yields 258–260
and canal water distribution 267–268, *269*, 270
factors affecting 260–266
Bharatiya Agro Industries Foundation (BAIF) 211–212
blue water 5, 147–148
boron 211–212
Brazil 62
bread wheat 193, **303–304**, **306**
broad-bed and furrow (BBF) 208, 210–211, *320*
bulk flow 315–317
Burkina Faso 145*n*, 152–155, 158

C_4 metabolism *57*, 62, 111, 222, 318*n*
calcium uptake 316
CAM species 318*n*
carbon allocation 109–110, 118
carbon dioxide, plant uptake 315
carbon isotope discrimination 109
cash crops 155, 159, 279–280
Centro Internacional de Agricultura Tropical (CIAT) *113*, 114
Centro Internacional de la Papa (CIP) *113*, 114, 229
cereals
global water consumption, current and future **172**
global water productivity, current and future 167–174
global yields, current and future **172**
imports to developing countries 29, **30**, 177
world prices *177*
see also individual cereals
CERES water balance models 203
CGIAR *see* Consultative Group on International Agricultural Research
Chaj Doab sub-basin, Pakistan 257
Chao Phraya basin 29, 64
crop choice 279–280
water allocation and efficiency 276–277, 283, 285
chickpea
agroclimate **129**
crop–water–production function **303**
drought-tolerance breeding 134, *137*, 138, 140
early sowing 192
integrated watershed management 208
productivity of applied water 304, 305
short-duration genotypes 132–133
simulated yields and yield gap 205–206
water management 188, *189*
chilli 208, 280
China
adoption of new technologies 63
cereal water productivity 1995 167–168, **169**
competition for water 55
Gansu Province 155
groundwater exploitation 54–55
Northern plain 302–305, 306, 308
potato production 230, *231*
rice production 54–55, 62, 63, 104, 242
water pricing 31–32
wheat production 302–305, 306, 308
Yellow River 32, 55
Zhanghe irrigation system 31–32, 55
see also Indo-Gangetic plains
CIAT *see* Centro Internacional de Agricultura Tropical
CIMMYT *see* International Maize and Wheat Improvement Center
CIP *see* Centro Internacional de la Papa
climate change 182
cold tolerance 108
committed water *4*
competition for water 16, 55, 163–164, 274
concepts 37
Conservation Agriculture *see* resource-conserving technologies
conservation tillage 155–156, 244–245, 319–320
Consultative Group on International Agricultural Research (CGIAR)
drought tolerance breeding 113–114
ecoregional programmes 241
germplasm screening 118–121
see also individual CGIAR centres
cost recovery *see* water pricing
cotton 89–90, 294, 295
cotton–wheat cropping *77*
cowpea 208, *209*
crassulacean acid metabolism (CAM) 318*n*
crop breeding *see* plant breeding
crop diversification 159, 279–280
crop simulation modelling 205–206, 235–237
crop yields
global 1995 and projected future **171–172**
rain-fed agriculture 130, 146, 150–152
supplemental irrigation 152–154
remotely sensed data 293–294
saline conditions 90–92
simulation modelling 205–206, 235–237
trade-off with water conservation 320–321
and water productivity 10, **12**
and water stress 115–116, 189
crop–water production functions
case studies 302–305, 308
theory 301–302
cropping systems
changes in 28–29
dry areas 190–191
saline conditions 70, 77
water productivity 104–106
crown pruning 225

dams, Thailand 274, 276
Decision Support System for Agricultural Technology (DSSAT) model 205–206
deep water capture 223–224
deficit irrigation 10, 188–189, 194–195, 305–308
deforestation 219, 226
delayed senescence 136, 138
depletion of water
beneficial 3–5, 44
non-beneficial 44–45
desalination 180
developed countries
cereal water productivity

estimated 1995 167–168
impact of changing water-use efficiency 174–177
projected 1995–2025 168–174
potato production 229, **230**
see also named countries
developing countries
cereal imports 29, **30**, 177
cereal water productivity
estimated 1995 167–168
impact of changing water-use efficiency 174–177
projected 1995–2025 168–174
potato production 229–230, *231*
see also named countries
digital terrain models (DTMs) 203, 205
direct seeding 60, 63, 106–107, *320*
directed mutagenesis 118–119
diversification, crop 159, 279–280
drainage
agroforestry 221
reuse in irrigation 92, 96–97
semi-arid tropics 150, **151**
drip kits 155
drought
definitions 112
plant responses 130, 131–132
semi-arid tropics 130
see also water stress
drought tolerance
complexity 114–116
crop screening 133–134
plant breeding
CGIAR programmes 113–114
germplasm screening 118–121
identifying genes for 118
impact of 139–140
inefficiencies 114–116
integrated strategies 140–141
plant traits 108–109, 114–116, 136–138
QTL mapping 116–118, 138–139
rice 58, 113
short-duration crops 132–133
drought-susceptibility index (DSI) 134
dry areas 179–180
additional water sources 180
crop improvement 113–114, 184, 190, 192–193
cropping patterns 190
deficit irrigation 305–308
effective water management 180–181
future directions and research issues 195–197
integrated approach to water management 183–184
small-scale irrigation systems 196
soil management 190–191
water scarcity 181–182
water-use efficient techniques 185–189, 193–195
DSI *see* drought-susceptibility index
DSSAT (Decision Support System for Agricultural Technology) model 205–206
durum wheat 193, 194, **303–304**, **306**

early growth vigour 138
economic efficiency (EE) 21–22
crop varietal improvement 25–26
irrigation systems 26–28
net private returns 23–24
net social returns 24
river basins 28–29
valuation of water 23, 46–47
economic water scarcity 2, 54
efficiency
classical concepts 38–40, 47–48
effective 41, 42
fractions 42–43
neoclassical concepts 40–43, 48
net 41
Egypt 93, 99, 187, 280
endophytic fungi 114
Energy Generation Authority of Thailand (EGAT) 276
environmental issues 24–25, 33, 63–64
estuaries, salinity 64
Eucalyptus camaldulensis 224
evapotranspiration
actual 38–39
net 39
remotely sensed data 291–292
semi-arid tropics 150, **151**
exchangeable sodium percentage (ESP) 72, 91
externalities 24, 48

faba bean 188, *189*
Faidherbia albida 222
farmers
adoption of new technology *see* technology adoption
impact of water reallocation 282
incentives for water productivity increases 13, 277
interactions with irrigation management 12–15
options for water productivity increases 8–9, 10
participatory irrigation management 32
participatory plant breeding **120**, 121
participatory technology development 194–195, 250–251
participatory watershed management 201, 208–209, 212, 214
resource-poor 251
risk perception 158
training 212
fertilizer use
dry areas 190–191
semi-arid tropics 153–154, 158
wheat production *262*, 263
fisheries 5–6, 279
flooding 219, 276
flooding tolerance 111–112
flowering, water stress 115, 316
food security 163, 182, 240
food-water model *see* IMPACT-WATER
forests 217, 218–219, 226
free markets 48–49
small-scale 82–84

Gal Oya Water Management Project 27–28
Gansu Province, China 155
Gediz basin, Turkey 28–29
gender 212, 214
genetic transformation 62, 107, 108, 110, 111
geographical information systems (GIS) 202–205, 289, **291**, **293**, 297

germplasm screening 118–121
Gini coefficient 261*n*
GIS (geographical information systems) 202–205, 289, **291**, **293**, 297
glyphosate resistance 107
grapes 12–13
green manures 107–108
Green Revolution 218
green water 5, 147–149, 157–158
greenhouse gases 63
Grevillea robusta 222
groundnut
  agroclimate 129
  drought-tolerance breeding 134, *137*, 138, 140
  short-duration genotypes 132–133
groundwater
  conjunctive use with surface water 32–33, 73, *74*, 82, **83**, 96–97, 99–100
  overexploitation 33, 54–55
  salinity 78, 82–84, 92–93
  transfers 82–84
  uptake by tree roots 93, 224
  wheat production 261, 265, 266
groundwater–resource ratio 293
guard cells 312–313
gypsum 76, 77, 85–86, 211

Hadejia-Jama'are, Nigeria 9
Han Dao rice 62, 109
harvest index 25–26, 109–110
  panicle 136
Haryana state, India
  climate 72–73
  saline/alkaline groundwater **71**
  salinity/alkalinity management 73–75, 76, 78–79, 80, 81, 82–85
head–tail equity ratio 261*n*
heat tolerance 109, 232, 234, 237
heat-shock proteins 109
hedgerows 220–221, 224, 225
herbicide resistance 107
herbicides 243, **244**, 245
Himalayas 219
hydraulic lift 221
hydrological drought 112

ICARDA *see* International Centre for Agricultural Research in the Dry Areas
ICRISAT *see* International Crop Research Institute for the Semi-Arid Tropics
IITA *see* International Institute for Tropical Agriculture
IMPACT-WATER model 164
  methodology, data and assumptions 164–166
  water productivity 1995–2025 167–174
    alternative scenarios 174–177
India
  adoption of new technology 63
  Bhakra irrigated area
    canal water distribution 267–268, *269*, 270
    climate and soils 257
    water accounting **5**
    watercourse characteristics 258, *259*
    wheat production factors 260–266
    wheat yields 258–260
  cereal productivity 1995 168, **169**
  flooding 219
  potato production 230, *231*
  Rajastan irrigation district 157
  saline groundwater **71**, 82–84
  salinity/alkalinity management 73–85
  salinization/desalinization cycles 72–73
  water budget simulation 202–205
  wheat production 256, 258–266
  see also Indo-Gangetic plains
Indo-Gangetic plains
  food demand 240
  potato production 230, 236
  Rice–Wheat Consortium 241
  rice–wheat cropping 239–240
    resource-conserving technologies 241–249
    technology adoption 249, 250–251
    water savings 249–250
  water demand 240
induced innovation theory 28
Indus basin, Pakistan 93
  crop yield data 293–294
  hydrological data 291–293
  water productivity 294–298
industry 9, 16, 24, 55
INERA *see* Institut National de l'Environment et de la Recherche Agricole
infiltration 91, 220
inflows *4*, 8*n*
INSTAT 145*n*
Institut de Recherche pour le Dévéloppment (IRD) 145*n*
Institut National de l'Environment et de la Recherche Agricole (INERA) 145*n*
institutional water management 29–33
integrated on-farm water management, dry areas 183–184
interception 220
intercropping systems 208, *209*
internal rate of return 26–28
International Centre for Agricultural Research in the Dry Areas (ICARDA) 179–180
  plant breeding programmes 113–114, 190, 192
  research areas 184
  supplemental irrigation research 185–186, 188–189, 194–195
International Crop Research Institute for the Semi-Arid Tropics (ICRISAT) 113–114
  integrated watershed management
    consortium approach 212
    farmer participation 201, 208–209
    microwatershed development 209–212
    monitoring and evaluation 212–213
    new science tools for evaluation and management 202–207
    on-farm trials 200, 201
    on-station research 207–208, *209*
    Vertisol technology package 200, 201, 207
  mandate crops 127
    characterization of drought-resistance traits 136–138
    drought tolerance screening 133–134
    improvement strategies 134–136
    integrated genetic improvement strategies 140–141
    molecular breeding techniques 138–139

short-duration genotypes 132–133
target environments 128–130
International Institute for Tropical Agriculture (IITA) *113*, 114
International Maize and Wheat Improvement Center (CIMMYT) 113–114
International Potato Center (CIP) *113*, 114, 229
potato research 232–235
SUBSTOR-Potato model 235–237
International Rice Research Institute (IRRI) 58, 107, 113–114
international trade 29, **30**, 177
Internet sources 312*n*, 313*n*
invertase gene 110
inverted-T openers 242–244
Iran 177
IRD *see* Institut de Recherche pour le Developpment
IRRI *see* International Rice Research Institute
irrigation
agronomic practices 58–59, **60**, 80–81, 319–320
benefit–cost analyses 26–28
canal-water distribution 99, 100, 261–263, 265–270
deficit 188–189, 194–195, 305–308
global water use 163
groundwater use 32–33, 261, 293, 296
management
interactions with farms 12–15
participatory 32
re-use of drainage 96–7, 180
saline conditions
economic efficiency 78–81
leaching requirement 75, 92, 94–96, 97–98
productivity increase 81–85
root-zone management 73–78
water-use practices 85–86
small-scale 155, 196
supplemental 181, 185–186, 188–189, 192–193, 194–195
water accounting 3–5
water pricing 31–32, 277–278, 285
water productivity 104–5
definitions 9, 295–296, 298
global current and future projections 168–170, **171–173**
options for improvement 9, 10–11
and spatial scale 57, *58*, 297–298
water savings 58–59, **60**, 64
irrigation efficiency (IE) 20–21
classical concept 20–21, 38–40, 47–48, 296–297
and economic efficiency 21–22
neoclassical concept 21, 40–43
saline conditions 70
isoproturon 245

Jordan 181, 187

karite 222
Kenya, Machakos 145*n*
agroforestry 220, 221, 224
rain-fed agriculture 152–155
rainfall patterns 150–151

land levelling *320*
rice–wheat cropping 248, 250
saline conditions 80–81
land-use changes 16–17
leaching, of salts 75, 92, 94–96, 97–98
leaf area 138
leaf cuticular membranes 108
lentil 188, *189*, **303**, **304**, 305
Leucaena leucocephala 224
lignin 109
line-source-sprinkler irrigation 134
Lower Jhelum canal system, Pakistan
canal water distribution 267–268, *269*, 270
climate and soils 257
watercourse characteristics 258, *259*
wheat productivity factors 260–266
wheat yields 258–260

Machakos, Kenya
agroforestry 220, 221, 224
rain-fed agriculture 150–155
rainfall patterns 150–151
Madagascar 26, *27*, 278*n*
Mae Klong basin, Thailand 277, 285
Maesopsis emini 224
magnesium uptake 316
maize
anthesis silking interval (ASI) 118
conservation tillage 156
global production *230*
improved watershed management 210
nutrient uptake *316*, 317
supplemental irrigation 153–154, *305*, *307*
water productivity *57*, 305
yield and water availability 150–151
Majjia valley, Niger 224
Malaysia, Muda irrigation scheme 59, 60
marginal/low-quality water use 180, 247
marker-assisted selection (MAS) 119, 139
Melia volkensii 226
methane emissions 63
Mexico 29, **30**, 245
microarray analysis 116, **117**
microcatchments 187
microclimate 108, 222
microirrigation 155
microwatershed 209–212
molecular biology 62, 103, 118–121, 138–139
monsoonal climate 72–73
montane forests 226
Muda irrigation scheme, Malaysia 59, 60
mulches 234–235, 242, 243, *320*
multiplier effect 46
mungbean 79, *80*
mustard *77*, 79, 80
mutagenesis, directed 118–119

national agricultural research systems (NARS) 192
National Oceanic and Atmospheric Administration-Advanced Very High Resolution Radiometer (NOAA-AVHRR) satellite 291–294
nere 222
Niger 224
Nigeria 9
Nile basin, Egypt 93, 99
nim trees 224

nitrogen fixation 108
nitrogen-use efficiency 108
nitrous oxide 63
NOAA-AVHRR satellite 291–294
North China Plain 302–304, 306, 308
nutrient uptake
  agroforestry 223–224
  bulk flow to roots 315–317
nutrient-budgeting 211–212, 213

Oregon state, USA 302–305, *307*
organic matter application *320*
Oryza glaberrima 107
Oryza rufipogon 111
outflows 3–6
  basin scale 6
  committed/uncommitted *4*
  field scale 3, **4**
  irrigation system 3, **5**
  rice production 56–60

Pakistan
  Indus basin 93
    crop yield data 293–294
    hydrological data 291–293
    water-productivity analyses 290, 294–298
  Lower Jhelum canal system 257
    canal-water distribution 267–268, *269*, 270
    climate and soils 257
    watercourse characteristics 258, *259*
    wheat productivity factors 260–266
    wheat yields 258–260, 268–270
  wheat production 256, 260–270
panicle harvest index 136
Parakramabahu, King of Sri Lanka 37
Parkia biglobosa 222
partial factor productivity 22–23
participation
  irrigation management 32
  plant breeding 121
  research and technology development 194–195, 208–209, 212, 250–251
pearl millet
  drought tolerance breeding 134, 135, *137*, 139–140
  ICRISAT target environment 128, **129**
  short-duration genotypes 132
  supplemental irrigation 208, *209*
Peru
  potato production 236–237
  potato research 232–234
Phalaris minor **244**, 245–246
phosphorus uptake 316
photosynthesis 11, *57*, 62, 111, 222, 318*n*
photosynthesis:transpiration ratio 109
pigeonpea
  drought-tolerance breeding 134, 136, *137*, 140
  ICRISAT target environment 129–130
  short-duration genotypes 132–133, 140
  sorghum/soybean intercrop 208, *209*, 210
  supplemental irrigation 208, *209*
plant breeding 25–26
  drought tolerance 113–116
    impact of 139–140
    integrated strategies 140–141
    molecular methods 118–121, 138–139
    plant traits 108–109, 114–116, 136–138
    screening and selection 118–121, 133–134
  dry areas 113–114, 184, 190, 192–193
  funding 26
  interactions with management 192–193
  molecular methods 103, 110, 112, 116–121, 138–139
  potato 232–234
  productivity increase 25–26, 103, 109–111
  rice 58, 62, 113
  short-duration 110–111, 132–133
plant traits
  drought tolerance 114–116, 136–138
  increasing production without increasing transpiration 109–111
  QTL mapping 110, 112, 116–118, 138–139
  reducing non-transpiration water use 106–107
  reducing transpiration 108–109
  salinity tolerance 58, 80, 85, 112
  waterlogging/flooding tolerance 111–112
pollution 49*n*
population, human 182, 240
potassium uptake 316
potato
  agronomic practices 234–235
  genotypic variation in water productivity 234, **235**
  global production 229, *230*
  heat-tolerance 232, 234
  production in developing countries 229–230, *231*
  productivity simulation modelling 235–237
  yield improvements 230–231
  yield and water supply 232–234
precision levelling 80–81
pricing *see* water pricing
productivity of applied water (PAW) 304–305
  and deficit irrigation 305–308
proteomics 116, **117**
puddling 59, 247–248, 250
Punjab Agricultural University 245–246

quantitative trait loci (QTL) 110
  drought tolerance 116–118, 135, 136, 138–139
  rice 110, 116
  salt tolerance 112

rainfall
  infiltration 91, 220
  interception 220
  partitioning 150, **151**
  productivity 152
  variability 149–151
rainfall-use efficiency 60, 219–221
rain-fed agriculture
  conservation tillage 155–156
  crop genetic enhancement 131–140
  dominance of 146
  hydroclimatic challenges 149–150
  rice 54, 106, 170–174
  risks 158, 159
  soil fertility management 153–154, 158
  supplemental irrigation 152–155, 181, 185–186, 192–193
    deficit 188–189, 194–195, 305–308
    microirrigation 155

water productivity
assessment 104, 152
global current and future projections 170, **172–173**, 174
yield improvement options 130, 150–152
yields 130, 146
see also watershed management
rain-out shelter 134
rainwater management
dry environments 186
evaluation 207–208, *209*
rationale for 147–149
saline environments 81
watershed scale 156–157
raised beds 60–61, 245–247, *320*
raised-bed-furrow 247
Rajastan district, India 157
Ranga Reddy district, Andhra Pradesh 208–209
RCTs *see* resource-conserving technologies
reallocation of water 9, 11–12, 15–17, 30–31, 55, 280–283
trade-offs between uses 9, 16–17
redwood 313–314
relative humidity 314
remotely sensed data 299
crop yields 293–294
hydrology 291–293
reservoir storage, global 166
resource-conserving technologies (RCTs) 241
adoption and participatory development 249, 250–251
bed-planting systems 245–247, 249–250
laser levelling 248, 250
low-quality water use 247
non-puddling for rice 247–248
reduced tillage 244–245
supplemental water use 248–249
surface seeding 242
water saving 249–250
zero tillage 242–244
reuse/recycling of water 38, 92, 96–97
Rhizobium 108
rice
aerobic 61–62, 109
drought tolerance 58, 115–116
genetic transformation 62, 107, 108, 110, 111
germplasm improvement 58, 62, 113
photosynthesis:transpiration ratio 109
semi-dwarf trait 110
short duration 110–111
submergence tolerance 111–112
water stress 115–116
rice production
adoption of new technologies 62–63
aerobic 61–62, 109
biotechnology 62
diversification to other crops 279–280
dry/direct seeding 60, 63, 106–107, *320*
global *230*
global prices, estimated and projected *177*
global water consumption, current and future 168–170, **171**
global water productivity, current and future projections 167–170, **171–173**
global yields, current and future projections 170, **171**
Indus basin 294, 295
land preparation 59, 63
puddling 59, 247–248, 250
rain-fed 54, 106, 170–174
risks 279–280
saline/alkaline conditions 75–77, 99
saturated soil culture 59, 60–61
sustainable management 63–64
water availability 54–55
water inputs 56–57
water outflows 56–60
water productivity 57–58
current and future projections 167–170, **171–173**
see also rice–wheat cropping
Rice–Wheat Consortium (RWC) 241–9
rice–wheat cropping 105
Indo-Gangetic plain 239–240
resource-conserving technologies (RCTs)
adoption 249, 250–251
bed-planting systems 245–247, 249–250
laser levelling 248, 250
low-quality water use 247
non-puddling of soils 247–248, 250
reduced tillage 244–245
supplemental water use 248–249
water use benefits 249–250
zero tillage 242–244
saline/alkaline conditions 75–77, 247
Rio Lerma-Chapala basin, Mexico 29
rippers 156
risk
rain-fed agriculture 158, 159
rice production 279–280
river basin *see* basins
root-zone salinity 73–78, 94–96, 98–99
roots
drought-tolerance traits 138
hydraulic lift 221
pressures 314
pruning 224–225, 226
soil nutrient uptake 315–317
trees 221, 223–224
water loss 109
row spacing/orientation *320*
runoff
agroforestry systems 220, *221*
harvesting 147, 157, 205
reuse in irrigation 92, 96–97
semi-arid tropics 150, **151**
RWC *see* Rice–Wheat Consortium

safflower 208
saline conditions
crop yield response functions 90–92
cropping systems 70, 77, 80
disposal of salts 84–85, 93, 100
groundwater 78, 82, 92–93
hazards 71–72, 91
Indus River basin 93
land levelling 80–81
productivity improvements 81–85
approaches to 69–71
economic 70, 78–81
rainwater conservation 81
rice production 75–77, 99, 247
root-zone management 73–78, 94–96, 98–99
seasonal cycles 72–73

saline conditions *continued*
soils 3, 72, 74–75, 91, 260–261
water application modes 33, 73–75, 82, **83**, 96–97, 99–100, 247
water diversion 99–100
salinity tolerance 80, 85
cotton 89–90
rice 58, 112
threshold values 90, 91–92
timing of water stress 79–80, 92
wheat 79, **90, 91**
SALLUS-TERRAE model 205
salt stress 79–80, 92
satellite data 299
crop yields 293–294
hydrology 291–293
saturated soil culture (SSC) 59, 60–61
scale issues 2–3, 8–9, 17, 23, 297–298
irrigation systems 57, *58*, 297–298
water accounting 3–6
water-use efficiency 275–276
SEBAL (surface energy-balance algorithm for land) 291–292
sectoral water allocation 9, 11–12, 15–16, 30–31, 55, 280–283
seed drill, inverted-T opener 242–244
semi-arid tropics (SAT)
climate 128, **129**, 149–150, 200
drought 128–130
rain-fed agriculture 156
conservation tillage 155–156
crop genetic enhancement 131–140
hydroclimatic challenges 149–150
rice 54, 106, 170–174
risk 158, 159
soil fertility management 153–154, 158
supplemental irrigation 152–155, 185–186, 188–189, 192–195, 305–308
water productivity 104, 152, 170, **172–173**, 174
yield improvement options 130, 150–152
yields 130, 146
water scarcity 200
watershed management 156–157
adoption of technology 200, 201, 208–209, 212, 214
farmer participation 201, 208–209
microwatershed development 209–212
model for 201–202
monitoring and evaluation 212–213
new science tools for evaluation and management 202–207
on-farm trials 200–201
on-station research 207–208, *209*
semi-dwarf trait 110
senescence, delayed 136, 138
sewage effluent, use of treated 180
shading, agroforestry 222
shared water 182
short-duration crops 110–111, 132–133, 140
shrimp farming 279
SIDA *see* Swedish International Development and Cooperation Agency
slopes, agroforestry 220–221, 225
socio-economic factors 24, 158–159, 182
soil
alternate wetting and drying (AWD) 59, **60**, 63
salinity/alkalinity 3, 72, 74–75, 91, 260–261
saturated culture (SSC) 59, 60–61
soil aeration 59, **60**, 63
soil conservation 155
soil evaporation 150, **151**, 220
soil fertility
boron and sulphur amendments 211–212
dry areas 190–191
semi-arid tropics 152–154, 153–154, 158
wheat production *262*, 263
soil infiltration 91, 220
soil leaching 75, 92, 94–96, 97–98
soil moisture, agroforestry 220–221
soil nutrients
boron 211–212
sulphur 211–212
uptake by roots 223–224, 315–317
watershed management 211–212, 213
soil puddling 59, 247–248, 250
soil redox potential 63
soil–water balance 290
simulation modelling 202–205
Solanum tuberosum see potato
sorghum
agroclimate 128–129
drought tolerance 136–138, 140
pigeonpea intercropping 208, *209*, 210
short-duration genotypes 132–133
stay-green genotypes 111, 136, 138
supplemental irrigation 153–154, 208, *209*
sorghum–wheat cropping *77*
South Africa, savannah 224
sowing
direct 60, 63, 106–107, *320*
surface 242
timing 189, 192, 243–244, *320*
soybean
pigeonpea intercrop 208
soil boron and sulphur treatments 211
yield simulations 205–206
SRI *see* System of Rice Intensification
Sri Lanka, Gal Oya Water Management Project 27–28
SSC *see* saturated soil culture
stay-green trait 111, 136, 138
steppe 186–187
stomatal function 105, 108–109, 312–313
storage tanks, subsurface 155
straw mulch 242, 243
sub-Saharan Africa
cereal water productivity 1995 167–168
cereal water productivity 1995–2025 **170–172**
dependence on rain-fed agriculture 146
rainfall partitioning 150, **151**
supplemental irrigation 152–154
suberin 109
submergence tolerance 111–112
subsoilers 156
SUBSTOR-Potato model 236–237
sucrose synthase 110
sugar cane 295
sulphur 211–212
supplemental irrigation 152–154, 181, 185–186
deficit 188–189, 194–195, 305–308

microirrigation 155
semi-arid tropics 208, *209*
socio-economic factors 158–159
and soil fertilty 158
wheat production 192–193
surface energy-balance algorithm for land (SEBAL) 291–292
surface runoff *see* runoff
surface seeding 242
sustainability 25, 28–29, 64, 218–219
Swedish International Development and Cooperation Agency (SIDA) 145*n*
Syria
deficit irrigation 194–195, 305–308
legume production 188, *189*, 192
soil management 190–191
water harvesting 187
wheat production 185–186, 188, 192–193, 194–195, 302–308
System of Rice Intensification (SRI) 26, *27*

tank irrigation systems 205, 206–207
Tanzania 149
technology adoption 26, *27*
constraints 26, 201
farmer incentives 12–13
Indo-Gangetic plains 249, 250–251
integrated watershed management 200, 201, 208–209, 212, 214
rice production 26, *27*, 62–64
social benefits 24
terracing 155
Thailand
Chao Phraya basin 29, 64, 276–277, 279–280, 283, 285
crop choice 279–280
Energy General Authority 276
Mae Klong basin 277, 285
types of river basin 285
water allocation 280–283
water laws 274, 283–284
water pricing 277–278, 285
water scarcity 274
water-use efficiency 275–277
Thompson, Sir G. 37
tillage
reduced (conservation) 155–156, 244–245, 319–320
rice production 59, 63
zero 155–156, 242–244, 249, *320*
timber production 226
tomato 208, *209*
total factor productivity (TFP) 22
trade, cereals 29, **30**, 177
transfer of water
large scale 180
non-agricultural uses 24
saline conditions 82–84, 99, 100
see also reallocation of water
transpiration 103–104
agroforestry system *221*
beneficial functions 313–315
and drought 131
physiological process 105–106, 312–313
plant traits minimizing 108–109
role in nutrient uptake 315–317
transpiration:photosynthesis ratio 109
trees
ecological functions 218–219, 226
hedgerow 220–221, 224, 225
interactions with crops 223–225
potential benefits in crop systems 217, 219–221
transpiration pull 313–314
Tropicultor, bullock-drawn 209, 210
tube wells 32–33, 85, 86, 276
Turkey 12–13, 28–29, 180

under-irrigation *see* deficit irrigation
USA
cereal productivity 1995 168, **169**
reallocation of water 283
water law and management 48, 281
wheat production 302–305, *307*

van der Heide, H. 273–274
varietal improvement 25–26
see also molecular biology; plant breeding
Vertic Inceptisols 208
Vertisols
ICRISAT technology package 200, 201, 207
productivity potential 207–208, *209*
Vitellaria paradoxa 222

WANA *see* West Asia and North Africa
WARDA *see* West Africa Rice Development Association
WATBAL model 202–204
water accounting 3–6, 23
basin scale 4, 5–6
definitions *4*
field scale 3, **4**
irrigation systems 3–5
water depletion 3–5
beneficial 3–5, 44–45
defined 3*n*, *4*
'water efficiency paradox' 37–38
water harvesting
downstream impacts 147, 157
dry areas 186–187
microirrigation 155
semi-arid tropics 52–53, 205, 206–207
socio-economic factors 158–159
water laws 281
implementation 284
Thailand 274, 283–284
water losses, irrigation 39
water markets 48–49
small-scale 82–84
water multiplier effect 46
water pricing 31–32, 48–49, 180–181
China 31–32
Thailand 277–278, 285
water productivity (WP)
concept and definitions 22–23, 182–183
economic 22–24, 46–47
partial factor productivity (PFP) 22–23
per kg water input
rice *57*
wheat *57*
per unit available water 7–8
per unit depletion *57*, 104, 105, 294–295
per unit diverted 6, *7*, *57*, 104, 105, 294–296
per unit rainfall 152

water productivity (WP) *continued*
  physical 22, 46–47
  scale of analysis 2–3, 8–9, 17, 23, 57, *58*, 297–298
water quality 33, 180
  wheat production 261, 263, 266
  see also alkalinity; salinity
water resource degradation 24–25, 33
water savings 13–14, 64
water scarcity 1–2, 37, 301
  dry areas 181–182
  economic 2, 54
  management-induced 2
  physical 54
  projections 147–148
  rice-growing areas 54
  semi-arid tropics 200
  socio-economic impacts 182
  Thailand 274
water stress
  complexity 127–128
  and flowering 115, 316
  indicators 189
  and plant growth cycle 114–115, 306, 308
  rice 115–116
  sensitivity index 306, 308
  and yield 115–116, 189
water table, shallow saline 78, 92–93
  see also groundwater
water taxes 274*n*
water valuation 23, 46–47
water wholesaling 277–278, 285
water withdrawals
  blue water 147–149
  global estimated 1995 166, *174*
  global projections 147–149, 166, 174
  green water 148–149
  impacts of future global reductions 174–177
water-use efficiency
  scale of analysis 275–276, 297–298
  semi-arid tropics 152
  Thailand 275–277
waterlogging tolerance 111–112
watershed
  functions of trees 218–219, 225–226
  water budget simulation 202–205
watershed management 156–157
  adoption of new technology 200, 201, 208–209, 212, 214
  farmer participation 201, 208–209
  microwatershed development 209–212
  model for 201–202
  monitoring and evaluation 212–213
  new science tools for evaluation and management 202–207
  on-farm trials 200–201
  on-station research 207–208, *209*
weed management
  rice 63, 64, 107
  wheat 245–246
  zero tillage 243, **244**
wells 32–33, 85, 86, 276
West Africa, agroforestry 224
West Africa Rice Development Association (WARDA) 107
West Asia and North Africa (WANA)
  additional water sources 180
  crop germplasm improvement 192–193
  cropping systems 190
  effective water management 180–181
  future directions and research issues 195–197
  integrated approach to water management 183–184
  simulated cereal water productivity 1995–2025 **170–171**
  soil management 190–191
  water scarcity 179, 181–182
  water-use efficient techniques 185–189, 193–195
Western Australia 16
wheat
  crop–water production functions 302–304
  lodging 246
  salinity tolerance 79, **90**, **91**
  water stress 186, 306, 308
  weed management 243, **244**, 245–246
wheat production
  bed-planting 245–247
  dry areas 185–186
  fertilizer use *262*, 263
  global *230*
  India and Pakistan 256
  Indus basin 294, 295–298
  research literature 256–257
  saline/alkaline conditions 75–77, 78, 82
  seeds 263
  supplemental irrigation 155, 188, 192–193, 304–308
  surface seeding 242
  water productivity *57*, 193–195
  water quality 261, 263, 266
  yields 256
    factors affecting 260–264
    and irrigation water management 261–263, 267–268, *269*, 270
    variability 258–260, 268
    yield function analysis 264–266
  see also rice–wheat cropping
women 212
World Bank 31

Yellow River, China 32, 55
yield-function analysis 264–266
yield-gap analysis 205–206
youth 212

zero tillage 155–156, 242–244, 249, *320*
Zhanghe irrigation system, China 31–32, 55
Zimbabwe 305, *307*

Printed and bound by CPI Group (UK) Ltd, Croydon, CR0 4YY
16/09/2026
14963435-0001